T0315144

# THE SEA

**Ideas and Observations on Progress in the Study
of the Seas**

# THE SEA

## Ideas and Observations on Progress in the Study of the Seas

## Editorial Board:

M. N. HILL
*Department of Geodesy and Geophysics*
*Madingley Rise, Cambridge, England*

EDWARD D. GOLDBERG
*University of California*
*La Jolla, California*

C. O'D. ISELIN
*Woods Hole Oceanographic Institution*
*Woods Hole, Massachusetts*

W. H. MUNK
*Institute of Geophysics and Planetary Physics*
*University of California*
*La Jolla, California*

# THE GLOBAL COASTAL OCEAN

## MULTISCALE INTERDISCIPLINARY PROCESSES

*General Editor*

**M.N. Hill**

## The Sea

**Ideas and Observations of Progress in the Study
of the Seas**

**Volume 2
The Composition of Sea-Water
Comparative and Descriptive Oceanography**

Harvard University Press
Cambridge, Massachusetts and London

First published 1963 by John Wiley & Sons, Inc.

Copyright assigned 2003 to the President and Fellows of Harvard College.

First Harvard University Press publication 2005.

Printed in the United States of America.

All Rights Reserved.

ISBN: 0-674-01728-5

Library of Congress catalog card number: 62018366

# CONTRIBUTORS TO VOLUME 2

W. L. BELSER, Division of Life Sciences, University of California, Riverside, California

R. J. H. BEVERTON, Ministry of Agriculture, Fisheries and Food, Fisheries Laboratory, Lowestoft, Suffolk, England

E. BRINTON, Scripps Institution of Oceanography, University of California, La Jolla, California

W. S. BROECKER, Department of Geology, Lamont Geological Laboratory, Columbia University, Palisades, New York

W. M. CAMERON, Marine Sciences Branch, Department of Mines and Technical Surveys, Ottawa, Ontario

D. E. CARRITT, Massachusetts Institute of Technology, Cambridge, Massachusetts, and Woods Hole Oceanographic Institution, Woods Hole, Massachusetts

G. E. R. DEACON, National Institute of Oceanography, Wormley, Godalming, Surrey, England

R. S. DIETZ, U.S. Navy Electronics Laboratory, San Diego 52, California

E. W. FAGER, Scripps Institution of Oceanography, University of California, La Jolla, California

EDWARD D. GOLDBERG, University of California, La Jolla, California

H. F. P. HERDMAN, National Institute of Oceanography, Wormley, Godalming, Surrey, England

JOHN D. ISAACS, Scripps Institution of Oceanography, University of California, La Jolla, California

M. W. JOHNSON, Scripps Institution of Oceanography, University of California, La Jolla, California

B. H. KETCHUM, Woods Hole Oceanographic Institution, Woods Hole, Massachusetts

J. A. KNAUSS, Graduate School of Oceanography, Narragansett Marine Laboratory, University of Rhode Island, Kingston, R.I.

Y. MIYAKE, Meteorological Research Institute, Mabashi, Suginami-Ku, Tokyo

E. Steemann Nielsen, Danmarks Farmaceutiske Højskole, Botanisk Laboratorium, Universitetsparken, Copenhagen

June Pattullo, Department of Oceanography, Oregon State University, Corvallis, Oregon

L. Provasoli, Haskins Laboratories, 305 East-43rd Street, New York 17, N.Y.

D. W. Pritchard, Chesapeake Bay Institute, The Johns Hopkins University, 121 Maryland Hall, Baltimore 18, Maryland

A. C. Redfield, Woods Hole Oceanographic Institution, Woods Hole, Massachusetts

J. L. Reid, Scripps Institution of Oceanography, University of California, La Jolla, California

F. A. Richards, Department of Oceanography, University of Washington, Seattle 5, Washington

G. A. Riley, Bingham Oceanographic Laboratory, Yale University, Yale Station, New Haven, Connecticut

J. H. Ryther, Woods Hole Oceanographic Institution, Woods Hole, Massachusetts

M. B. Schaefer, Institute of Marine Resources, Scripps Institution of Oceanography, University of California, La Jolla, California

G. H. Volkmann, Woods Hole Oceanographic Institution, Woods Hole, Massachusetts

W. S. von Arx, Woods Hole Oceanographic Institution, Woods Hole, Massachusetts

W. S. Wooster, Scripps Institution of Oceanography, University of California, La Jolla, California

# PREFACE

Oceanography has surged forward as a subject for research during the past twenty years, and great progress has been made in our understanding of the structure of the water-masses, of the crust of the earth beneath the oceans, and of the processes which are involved in creating these structures. This progress has been particularly rapid in the recent past for two prime reasons. Firstly, techniques of investigation have become available which have made many hitherto intractable problems capable of solution. For example, it has only recently become possible to measure the value of gravity in a surface ship, to determine the deep currents of the oceans by direct methods, or to analyze wave spectra with the detail which modern electronic computers allow. The second reason for the present rapidity of progress lies in the increasing availability of the resources for both theoretical and practical marine investigations; the international character of oceanography and the growing world-wide interest in fundamental research has resulted in the provision of opportunities hitherto impossible.

Some years ago Dr. Roger Revelle of the Scripps Institution of Oceanography suggested to us that perhaps the production of treatises on the new developments in oceanography were not keeping pace with the progress in the subject. It was true that papers were appearing in healthy numbers throughout the journals of the world, but there were no recent comprehensive works such as *The Oceans*, written by Sverdrup, Johnson and Fleming and published in 1942. This work had been, and still is, of value to many, if not all, oceanographers; Revelle suggested that we produce another such volume containing ideas and observations concerning the work accomplished during the twenty years since this masterpiece. It was suggested that this new work should not attempt to be a textbook but a balanced account of how oceanography, and the thoughts of oceanographers, were moving.

It was early apparent that a work of this nature would have to be from the pens of many authors; the subject had become too broad, the oceanographers perhaps too specialized, to allow a small number of contributors to cover a vast field of study. It was also apparent that we could do no more than include biology in so far as it was directly related to physical, chemical and geological processes in the ocean and on the ocean floor. Marine biology could alone fill that space which we thought was the maximum we should allow for these volumes. Biology is, therefore, somewhat scattered through this work; so it has to be, since the contributions of biologists extend throughout the disciplines embraced by the contents of these volumes.

It has been said in the past that oceanography as a subject did not, or even should not, include the study of the mode of formation and the structure of the muds and hard rocks forming the floor of the oceans and the seas. This exclusion is no longer possible, and we have included in these volumes geological and geophysical ideas and observations concerning the earth beneath the seas. In the great oceanographical (or oceanological) establishments of the world the earth sciences relating to the oceanic environment are pursued alongside one another without discrimination as to whether the air over the sea, the water, or the solids beneath the sea could claim prior importance. We have adopted the same breadth of outlook in these volumes.

With a composite work such as we have tried to produce, it has proved difficult to ensure a precise balance in the emphasis given to our wide variety of topics. Sometimes, maybe, there will be too much written about a narrow field; sometimes we shall perhaps be criticized for the converse, too little written about a broad field. At least, however, we hope we have covered most of the topics suitable for a work on the new developments in our subject.

When we drew up the first list of contents of this work we had supposed that we should be producing one volume. We soon found, however, that we had omitted a number of topics, and the contents list consequently increased. We also found that our contributors could not possibly have provided, within the limits we originally proposed, adequate discourses on their various topics. It was apparent that we should have to subdivide *The Sea* into three volumes. The first consists of new thoughts and ideas on Physical Oceanography; the second on the Composition of Sea-Water and Comparative and Descriptive Oceanography; and the third on the Earth Beneath the Sea and History. The three volumes, therefore, are divided from one another by major divisions of oceanography, and we believe for this reason that our readers will not be inconvenienced by constant cross-referencing from one volume to another. This clear division has also allowed us to index the volumes separately; it should be possible to select the volume containing information on any particular topic without difficulty. We have no overall index, since for a work of this magnitude it would be cumbersome.

We considered that for ease of production we should start a new numbering sequence at the beginning of each chapter for figures, equations and tables. In order to avoid confusion in referring back, we have printed the chapter number and section number on alternate pages.

In collecting the material for these volumes we have had most helpful cooperation from our contributors. In some topics, however, we have failed to find authors, or potential authors have withdrawn. We must admit, therefore, to omissions, some of which are conspicuous and important. By further effort we could doubtlessly have repaired these gaps in subject matter, but this would have delayed publication, and we believe the contributions we already have make a valuable collection. It is possible that in the future a fourth volume might be produced to deal with topics omitted or new since we first started collecting material for *The Sea*.

We are greatly indebted to the secretarial assistance provided by Mrs. Joyce

Nightingale and Mrs. Joyce Day. Mr. Cameron D. Ovey has provided editorial assistance of outstanding value, and Dr. J. E. Holmstrom has produced an index of great merit. To them and to the publishers and printers we extend our thanks.

*May, 1962*                                                                                    M.N.H.

# CONTENTS

## THE COMPOSITION OF SEA-WATER

## COMPARATIVE AND DESCRIPTIVE OCEANOGRAPHY

# The Composition of
# Sea-Water

# I. CHEMISTRY

## 1. THE OCEANS AS A CHEMICAL SYSTEM

E. D. Goldberg

One hundred years ago Forchhammer (1865) elegantly posed the theme of marine chemistry: "Thus the quantity of the different elements in sea water is not proportional to the quantity of elements which river water pours into the sea, but inversely proportional to the facility with which the elements in sea water are made insoluble by general chemical or organo-chemical actions in the sea." Since this time the many complex chemistries involved in giving sea-water its rather curious composition have been investigated by workers from the varied branches of marine science. Inroads have been made through a number of complementary attacks which may be systematized into four groups: (1) the speciation and isotopic compositions of elements in sea-water; (2) the relative reactivities of the elements; (3) the chemical reactions in the oceans and the compositions of the end-products; and (4) the spatial and temporal distributions of the reactants and products. It is the purpose of this presentation to elaborate upon the hypothesis of Forchhammer by consideration of examples and problems from these four types of studies.

### 1. The Composition of Sea-Water

Table I purports to give the average composition of ocean water, although such a tabulation suffers from a number of deficiencies (Goldberg, 1957). First of all, many of the values are from a single set of analyses of surface waters, samples possibly quite uncharacteristic of the marine hydrosphere as a whole. The upper 100 m or so, containing only a few per cent of the volume of the oceans, form an environment for intense biological activity which can cause large-scale fractionations of certain materials (see Chapter 2 by Redfield, Ketchum and Richards). Elements involved in the biochemical cycles, such as iron and silicon, can differ from one water-mass to another by a factor of $10^3$ or even higher. Finally, some of the analyses failed to assay all of the various forms, dissolved and/or particulate, of the element in question. Since chemical reactions in the oceans take place at phase discontinuities, namely the atmosphere–hydrosphere, biosphere–hydrosphere or sediment–hydrosphere, a knowledge of the chemical make-up of the water at the reaction site is critical for significant studies.

In spite of these limitations such a listing is important for initial considerations of the speciation of elements in sea-water. To consider equilibrium reactions in the oceans, as well as to gain an insight into the stability of various dissolved species, a knowledge of the forms in which the reacting elements

[*MS received October, 1961*]

## TABLE I

### Geochemical Parameters of Sea-Water

| Element | Abundance, mg/l. | Principal species | Residence time, years |
|---|---|---|---|
| H | 108,000 | $H_2O$ | |
| He | 0.000005 | He (g) | |
| Li | 0.17 | $Li^+$ | $2.0 \times 10^7$ |
| Be | 0.0000006 | | $1.5 \times 10^2$ |
| B | 4.6 | $B(OH)_3$; $B(OH)_2O^-$ | |
| C | 28 | $HCO_3^-$; $H_2CO_3$; $CO_3^{2-}$; organic compounds | |
| N | 0.5 | $NO_3^-$; $NO_2^-$; $NH_4^+$; $N_2$ (g); organic compounds | |
| O | 857,000 | $H_2O$; $O_2$ (g); $SO_4^{2-}$ and other anions | |
| F | 1.3 | $F^-$ | |
| Ne | 0.0001 | Ne (g) | |
| Na | 10,500 | $Na^+$ | $2.6 \times 10^8$ |
| Mg | 1350 | $Mg^{2+}$; $MgSO_4$ | $4.5 \times 10^7$ |
| Al | 0.01 | | $1.0 \times 10^2$ |
| Si | 3 | $Si(OH)_4$; $Si(OH)_3O^-$ | $8.0 \times 10^3$ |
| P | 0.07 | $HPO_4^{2-}$; $H_2PO_4^-$; $PO_4^{3-}$; $H_3PO_4$ | |
| S | 885 | $SO_4^{2-}$ | |
| Cl | 19,000 | $Cl^-$ | |
| A | 0.6 | A (g) | |
| K | 380 | $K^+$ | $1.1 \times 10^7$ |
| Ca | 400 | $Ca^{2+}$; $CaSO_4$ | $8.0 \times 10^6$ |
| Sc | 0.00004 | | $5.6 \times 10^3$ |
| Ti | 0.001 | | $1.6 \times 10^2$ |
| V | 0.002 | $VO_2(OH)_3^{2-}$ | $1.0 \times 10^4$ |
| Cr | 0.00005 | | $3.5 \times 10^2$ |
| Mn | 0.002 | $Mn^{2+}$; $MnSO_4$ | $1.4 \times 10^3$ |
| Fe | 0.01 | $Fe(OH)_3$ (s) | $1.4 \times 10^2$ |
| Co | 0.0005 | $Co^{2+}$; $CoSO_4$ | $1.8 \times 10^4$ |
| Ni | 0.002 | $Ni^{2+}$; $NiSO_4$ | $1.8 \times 10^4$ |
| Cu | 0.003 | $Cu^{2+}$; $CuSO_4$ | $5.0 \times 10^4$ |
| Zn | 0.01 | $Zn^{2+}$; $ZnSO_4$ | $1.8 \times 10^5$ |
| Ga | 0.00003 | | $1.4 \times 10^3$ |
| Ge | 0.00007 | $Ge(OH)_4$; $Ge(OH)_3O^-$ | $7.0 \times 10^3$ |
| As | 0.003 | $HAsO_4^{2-}$; $H_2AsO_4^-$; $H_3AsO_4$; $H_3AsO_3$ | |
| Se | 0.004 | $SeO_4^{2-}$ | |
| Br | 65 | $Br^-$ | |
| Kr | 0.0003 | Kr (g) | |
| Rb | 0.12 | $Rb^+$ | $2.7 \times 10^5$ |
| Sr | 8 | $Sr^{2+}$; $SrSO_4$ | $1.9 \times 10^7$ |
| Y | 0.0003 | | $7.5 \times 10^3$ |
| Zr | | | |
| Nb | 0.00001 | | $3.0 \times 10^2$ |
| Mo | 0.01 | $MoO_4^{2-}$ | $5.0 \times 10^5$ |
| Tc | | | |
| Ru | | | |

## TABLE I (*continued*)

| Element | Abundance, mg/l. | Principal species | Residence time, years |
|---|---|---|---|
| Rh | | | |
| Pd | | | |
| Ag | 0.0003 | $AgCl_2^-$; $AgCl_3^{2-}$ | $2.1 \times 10^6$ |
| Cd | 0.00011 | $Cd^{2+}$; $CdSO_4$ | $5.0 \times 10^5$ |
| In | < 0.02 | | |
| Sn | 0.003 | | $5.0 \times 10^5$ |
| Sb | 0.0005 | | $3.5 \times 10^5$ |
| Te | | | |
| I | 0.06 | $IO_3^-$; $I^-$ | |
| Xe | 0.0001 | Xe (g) | |
| Cs | 0.0005 | $Cs^+$ | $4.0 \times 10^4$ |
| Ba | 0.03 | $Ba^{2+}$; $BaSO_4$ | $8.4 \times 10^4$ |
| La | 0.0003 | | $1.1 \times 10^4$ |
| Ce | 0.0004 | | $6.1 \times 10^3$ |
| Pr | | | |
| Nd | | | |
| Pm | | | |
| Sm | | | |
| Eu | | | |
| Gd | | | |
| Tb | | | |
| Dy | | | |
| Ho | | | |
| Er | | | |
| Tm | | | |
| Yb | | | |
| Lu | | | |
| Hf | | | |
| Ta | | | |
| W | 0.0001 | $WO_4^{2-}$ | $1.0 \times 10^3$ |
| Re | | | |
| Os | | | |
| Ir | | | |
| Pt | | | |
| Au | 0.000004 | $AuCl_4^-$ | $5.6 \times 10^5$ |
| Hg | 0.00003 | $HgCl_3^-$; $HgCl_4^{2-}$ | $4.2 \times 10^4$ |
| Tl | < 0.00001 | $Tl^+$ | |
| Pb | 0.00003 | $Pb^{2+}$; $PbSO_4$ | $2.0 \times 10^3$ |
| Bi | 0.00002 | | $4.5 \times 10^5$ |
| Po | | | |
| At | | | |
| Rn | $0.6 \times 10^{-15}$ | Rn (g) | |
| Fr | | | |
| Ra | $1.0 \times 10^{-10}$ | $Ra^{2+}$; $RaSO_4$ | |
| Ac | | | |
| Th | 0.00005 | | $3.5 \times 10^2$ |
| Pa | $2.0 \times 10^{-9}$ | | |
| U | 0.003 | $UO_2(CO_3)_3^{4-}$ | $5.0 \times 10^5$ |

exist is essential. Further, coherence in geochemical behavior can often be postulated for dissolved species of the same form and charge.

Sillén (1961) has attempted to define the thermodynamically stable dissolved species in sea-water, where the appropriate physico-chemical data exist. His computations rely upon an oceanic model in which the dissolved ions and molecules are in equilibrium with known or assumed solid phases. He used average values for the acid–base and redox equilibria as

$$pH = 8.1 \pm 0.2 \quad \text{and} \quad pE = 12.5 \pm 0.2.$$

Table I lists the most probable major forms of the elements, based in part upon this work of Sillén.

He points out that hydroxide complexes are significant for all ions of oxidation number greater than 2. The more abundant chloride ion can compete with hydroxide in forming complexes with metals under the following conditions. Let the stability constants of the chloride and hydroxide be given by $K_{Cl} = [MCl]/[M][Cl^-]$ and $K_{OH} = [MOH]/[M][OH^-]$ respectively, where M is a given metal ion. For MCl to be more abundant than MOH, it is clear that

$$\log K_{Cl} - \log K_{OH} > \log[OH^-] - \log[Cl^-] = -5.4.$$

Sillén noted that this equation can be satisfied by only a few metal ions, grouped in the middle of the periodic table: $Ag^+$, $Hg^{2+}$ and $Au^+$ and perhaps $Cd^{2+}$ and $Pb^{2+}$. $Mg^{2+}$ is possibly the only major ion involved in fluoride complexing.

Ion-pair formation between sulfate and divalent metal ions can account for around 10% of the concentrations of the alkaline earth metals (excluding beryllium) as well as of nickel, copper, cobalt and zinc (Goldberg and Arrhenius, 1958). Of particular interest in a later section is the very strong carbonate complex of uranium, $UO_2(CO_3)_3^{4-}$, which makes this element rather unreactive in the marine environment.

The lack of available data makes prediction upon bromide, phosphate, carbonate, etc. complexes impossible. Further, the existence of dissolved fatty acids in microgram concentrations (Williams, 1960), as well as of other organic compounds, can give rise to strong complexes with metal ions. For example, Laevestu and Thompson (1958) postulate that the markedly high iron abundances in coastal waters are due to the formation of ferric-organo complexes. However, firm conclusions as to the importance of such organic complexes in open ocean waters cannot be reached inasmuch as the character of the bulk of the organic compounds, whose concentrations attain values of 2–3 mg of carbon per liter of sea-water, has as yet not been defined.

Although most of the dissolved constituents of sea-water have had many thousands of years to reach equilibrium, nonetheless, thermodynamically unstable species, such as manganous, iodide and arsenite ions, persist. Their occurrence can be attributed to a lack of reaction sites where equilibrium might be obtained, i.e. the water-mass containing ions of this type does not encounter a surface of the lithosphere, biosphere or atmosphere at which the energetically possible reactions can proceed.

Manganous ion is readily seen to be thermodynamically unstable in sea-water from the reaction

$$2OH^- + Mn^{2+} + \tfrac{1}{2}O_2 = MnO_2 + H_2O,$$

which has a free energy of $-9$ kcal at $25°C$ at a pH of 8, a manganous ion concentration of $10^{-9}$ molar and a partial pressure of oxygen of 0.25 atm. These concentrations are similar to those in near-bottom sea-water. Although the bottom temperatures are near $0°C$, this difference does not seriously affect the free energy value cited above. Manganese occurs in crustal rocks primarily in the divalent state and most probably enters the marine environment as a dissolved species at this oxidation level. In order for manganese to reach its thermodynamically stable form, a reaction surface is undoubtedly necessary. The widespread occurrence of tetravalent manganese as $\gamma$-manganese oxides (Hans Wedepohl, *in litt.*) in the ferromanganese minerals of pelagic sediments, which form on the sea floor in areas characterized by having oxygenated water layers above them and by low rates of accumulation of other components in the deposits, suggests that the associated iron oxides provide the necessary catalytic sites. Iron occurs in sea-water principally as solid oxide phases which accumulate on the sea floor. The subsequent formation of the mineral surface provides then further surfaces for the oxidation of the manganese ions.

In the case of iodine we have

$$IO_3^- + 6H^+ + 6e^- = I^- + 3H_2O$$

with a log $K$ of 110.1. By inserting the previously cited values of pH and pE, it follows that

$$IO_3^-/I^- = 10^{13.5}$$

and at equilibrium iodate would be the most abundant species of this element. However, iodide is ubiquitously present in sea-waters, an observation initially made by Winkler (1916) and more recently confirmed by Barkley and Thompson (1960). It will be quite rewarding to seek out those factors that give rise to the existence and maintainance of the reduced form of iodine. The marine biosphere must be intimately involved in the iodine cycle. Plants and animals of the sea contain reduced iodine in the forms of iodo tyrosines and thyroxine. The forms of iodine assimilated from sea-water, as well as those regenerated upon the destruction of the organic phases, provide the bases for future significant investigations.

These simple thermodynamic considerations are of further import in the understanding of inorganic precipitation processes in sea-water. For example, although divalent lead in the forms of $Pb^{2+}$, $PbOH^+$ and $PbSO_4$ are the main species in solution (Sillén, *op. cit.*), $PbO_2$ would be the stable solid phase in equilibrium in oxidizing environments, especially if it forms solid solutions with some other phase or phases. Such a concept is suggested by the marked concentrations of lead in the ferromanganese minerals, where it attains values of several tenths of a per cent (as opposed to a sea-water concentration of 0.03 μg/l.) and may form solid solutions with the manganese oxides.

The fractionation of isotopes by geochemical and biological processes has proven especially useful for determining the nature of the mechanisms involved. The isotopic compositions of sea-waters, for example, are strongly influenced by their previous involvement in condensation, evaporation, freezing and melting processes. Both hydrogen and oxygen have a pair of stable isotopes, $^1H$ (protium) and $^2H$ (deuterium), and $^{16}O$ and $^{18}O$, which are significantly fractionated in a similar way in oceanic waters (Epstein and Mayeda, 1953; Friedman, 1953). Craig (1961) has pointed out that the relationship

$$X_D = 0.6359 X_O - 1105.6$$

is approximately obeyed by sea-waters, where $X_D$ and $X_O$ are the atom fractions of deuterium and $^{18}O$ in atomic parts per million.

The water vapors over the surfaces of the ocean are depleted in the heavier isotopes relative to the condensed waters. With subsequent precipitation from the vapor phase, the first rain preferentially accumulates the waters containing deuterium or $^{18}O$, causing the vapor to be further depleted in these isotopes. Thus surface waters of the oceans are enriched in the heavier isotopes, a process which resembles a multiple-stage fractionation (Epstein and Mayeda, op. cit.)

Friedman found that in general equatorial surface waters are enriched in deuterium relative to higher latitude waters and pointed out that the evaporated lighter waters were deposited subsequently as rain nearer the polar regions. Also, surface waters of the ocean tend to be enriched in the heavier isotopes relative to deeper waters and such results are attributed to the polar origins of the deeper waters.

The isotopic composition of the dissolved oxygen gas has been determined by Rakestraw et al. (1951), who found that the $^{18}O/^{16}O$ ratio mirrored the total dissolved oxygen with a maximum in the ratio as a function of depth closely coinciding with the oxygen minimum. They attribute these results to a preferential metabolism of $^{16}O$ over $^{18}O$ by marine vegetation, plankton, bacteria and other sea life.

The isotopes of nitrogen and sulfur also show promise of extending our knowledge of oceanic chemical reactions. Richards and Benson (1961), on the basis of $N_2/A$ and $^{29,\ 28}N_2/^{28,\ 28}N_2$ ratios, have postulated the existence of nitrogen gas produced by the reduction of nitrate ions in the anaerobic environments, the Cariaco Trench, the Caribbean Sea and Dramsfjord, Norway. This biogenic nitrogen attains values of 0.5 ml/l. in the more reducing zones of these waters. Thode, MacNamara and Fleming (1953) have observed a fractionation of the sulfur isotopes, $^{32}S$ and $^{34}S$, in the precipitation of sulfides and sulfates in the marine environment in which the sulfides have been depleted and the sulfates enriched in the heavier isotope.

## 2. Reactivities of the Elements

If we consider the oceans as a dynamic chemical system, the reactions of primary interest are those that influence or control the concentrations of elements. The life processes have long been known to be responsible for the most dramatic compositional changes and for the regulation of the abundances of such elements as oxygen, phosphorus, silicon, nitrogen and carbon. An insight into the important inorganic reactions has been gained through considerations of the relative reactivities of the elements based on the average times spent before removal to the sea floor or on the degree of saturation attained by certain elements in sea-water.

### A. Residence Times

Barth (1952) first proposed the concept of residence (passage) time of an element in the marine environment by assuming an ocean in steady state in which the amount of a given element introduced per unit time is compensated by an equal amount deposited in the sediments. The residence time of an element, $\tau$, can then be defined as the average time it remains in sea-water before removal by some precipitation process. Thus,

$$\tau = \frac{A}{(dA/dt)},$$

where $A$ is the total amount of the element in the dissolved and particulate states in the oceans and $dA/dt$ is the amount of the element introduced or precipitating per unit time. It is further assumed that there is a complete mixing of the element in question within the world ocean in times that are short with respect to the residence times.

Barth made his computations on the basis of river influx of *dissolved* substances, neglecting the introduction of the suspended load as well as any solid materials introduced into the oceans via the atmosphere. He used Clarke's (1924) figure for the total amount of dissolved substances supplied by the rivers, i.e. $2.73 \times 10^{15}$ g/year. The absolute concentrations of elements in rivers can not be used directly to calculate $dA/dt$ inasmuch as significant amounts of materials are cycled from the sea via the atmosphere and rains to the rivers and subsequently returned to the oceans. Conway (1943) has pointed out that nearly all of the chlorine and about 53% of the sodium in rivers is of cyclic origin. Barth took cognizance of Conway's data to obtain the net transfer of material between the oceans and continents, and was able to derive the residence times of five elements (Table II).

Goldberg and Arrhenius (1958) independently calculated the residence times on the total rate of sedimentation in the marine environment, using a derived figure of $2.5 \times 10^{15}$ g of weathered products and pyroclastics entering the ocean per year. Assuming the composition of such materials to be the same as crustal rocks, residence times for many of the elements whose sea-water concentrations are available have been calculated (Table I).

## TABLE II

#### The Residence Times of the Elements in Sea-Water as Calculated by River Input and Sedimentation

| Element | Supply to oceans × $10^{14}$ g/year | | Oceanic content × $10^{20}$ g | Residence time × $10^6$ years | |
|---------|-------|----------------------|-------|-------------|---------------|
|         | Total | Cyclic salt corrected |       | River input | Sedimentation |
| Na | 1.58 | 0.71 | 147.8 | 210 | 260 |
| Mg | 0.93 | 0.82 | 17.8 | 22 | 45 |
| K  | 0.58 | 0.54 | 5.3 | 10 | 11 |
| Ca | 5.58 | 5.53 | 5.6 | 1 | 8 |
| Si | 1.48 | 1.48 | 0.052 | 0.035[a] | 0.008 |

[a] This value was obtained from the influx value given by Barth (1952) and the amount of silicon in the oceans from Table I. Barth gives only an upper limit for this latter number.

In spite of this drastically oversimplified model of the oceans there is a remarkable agreement between these two sets of data (Table II). The figures span a time range of six orders of magnitude—sodium with $2.6 \times 10^8$ years to aluminum with 100 years. At this point it should be noted that one further consideration must be met in the utilization of these results. Both $A$ and $dA/dt$ can not have changed in times of the order of $3\tau$. Even in the case of sodium this assumption appears to be valid, for about $10^9$ years as the age of the oceans agrees with present-day geological evidence.

The elements with long residence times, the alkali metals and alkaline earths (excluding beryllium) are characterized by a lack of reactivity in the marine environment. The decrease in residence times of the alkali metals with increasing atomic number, going from sodium to cesium, reflects most probably their involvement in ion-exchange equilibria with the clay minerals on the sea floor, reactions proposed by Grim (1953) to regulate primarily their oceanic concentrations. For positively charged monovalent ions, retention by the clay phases increases with increasing radius (i.e. atomic number). Thus the sequential decrease in residence times of these elements is in accord with their known aqueous chemical behavior.

On the other hand, elements such as Be, Al, Ti, Cr, Fe, Nb and Th have residence times under 1000 years, periods of the order of or less than the mixing times for oceanic water-masses. These elements in part enter the oceans as particulate phases from the continents or from volcanic activity (i.e. the clay minerals, feldspars, augite, volcanic glasses, magnetite, etc. which rapidly settle to the sedimentary deposits). Further, some of these elements, Al, Ti and Fe, are reactants in the formation of such authigenic substances as the ferromanganese minerals, zeolites, glauconites, etc. Thus, their entry into the oceans as solids and/or their high chemical reactivity matches their short residence times.

For such elements, one can expect varying degrees of accumulation in solution going from ocean to ocean or water-mass to water-mass. Goldberg and Koide (1962) have pointed out that the thorium contents of authigenic minerals, presumably reflecting the thorium input to the overlying waters, vary as a function of geographic location. The thorium concentrations were normalized to ionium, $^{230}$Th, whose rate of production in sea-water is essentially constant owing to the uniform distribution of its radioactive predecessor, $^{238}$U. Hence, variations in the $^{230}$Th/$^{232}$Th ratio can be attributed to differences in the inputs of $^{232}$Th to different water-masses. A measure of the thorium introduced into oceanic areas may be found in considerations of the relative amounts of continental run-off waters. Lyman (1959), using the data of L'Vovich, gives the areas of the oceans and the respective values of the areas draining into them (Table III). The Pacific Ocean receives but one-sixth of the drainage accepted by the Atlantic on an areal basis. Thus, the amount of thorium entering the Atlantic per unit area per unit time can be assumed to be greater than that going into the Pacific.

TABLE III

Surface Values of the Ionium/Thorium Ratio and Thorium Concentrations in Marine Sediments Compared with Oceanic Areas and the Land Areas Draining into Them (Lyman, 1959)

| Ocean | Area, $km^2 \times 10^3$ (1) | Land area drained, $km^2 \times 10^3$ (2) | $\dfrac{(2)}{(1)}$ | $\dfrac{Io}{Th}$ |
|---|---|---|---|---|
| Atlantic | 98,000 | 67,000 | 0.68 | |
| North | | | | 1.5–6 |
| South | | | | 9.4–19 |
| Pacific | 165,000 | 18,000 | 0.11 | |
| South | | | | 143–158 |
| East Equatorial | | | | 40–58 |
| Northeast Coastal | | | | 16–25 |
| Northeast | | | | 48–57 |
| Mid-Central | | | | 30 |
| Indian | 65,500 | 17,000 | 0.26 | 20–27 |
| Antarctic | 32,000 | 14,000 | 0.44 | |

It is quite clear from the data of Goldberg and Koide in Table III that the relative rate of thorium introduction (normalized to ionium production) into the Atlantic exceeds that of the Pacific. $^{230}$Th/$^{232}$Th ratios for surface authigenic minerals vary in the Pacific between 16 and 158 while the Atlantic values span a range of 1.5–19 (in units of disintegrations per unit time of $^{230}$Th/ disintegrations per unit time of $^{232}$Th). The well known observation that more and larger rivers drain into Northern Hemispheric areas is reflected in the higher ionium/thorium ratios in southern Pacific and southern Atlantic surface deposits as compared to their northern counterparts.

A question pertinent to the marine geochemistry of thorium involves the path of this element from run-off waters to the sea-floor precipitates. Previously, it was proposed that the variations in the ionium/thorium ratio, as well as in the lead isotopic ratios, in surface sediments were related to the direct input of thorium and lead to the bottom waters from the continents (Goldberg, Patterson and Chow, 1958). However, it was difficult to reconcile the geographic distributions of the ratios with the prevalent ideas of bottom-water circulation, especially with respect to the rather well-defined differences in lead and thorium isotopic ratios between northern and southern oceanic waters with the equator acting as a boundary.

Perhaps, as in the case of barium (see page 20) and $^{210}Pb$ (Rama, Koide and Goldberg, 1961), $^{232}Th$ is conveyed from shallow to deeper waters as a result of biochemical and inorganic processes and is initially introduced to the open ocean in surface waters. Such a path for thorium would not conflict with the known travels of surface waters where the equatorial regions form a barrier to the transfer of such waters from the Northern to the Southern Hemisphere. On the other hand, near-bottom waters, which give the authigenic minerals their lead and thorium isotopes, apparently travel from south to north (at least in the Pacific) crossing the equator along the way. If river run-off were introduced directly into deep-sea waters, it would be very difficult to account for the observed differences in the isotopic ratios in the sediments.

## B. Degrees of Undersaturation

A second, but less satisfying measurement of the reactivities of elements in sea-water can be gained on the basis of the degree of undersaturation, especially in the case of metallic ions and dissolved gases. The upper limit of concentration that a metal ion could attain would conceivably be regulated by the solubility of its least soluble compound, as determined by the anionic species present, if no other chemical reaction reduced its abundance. As a first approximation it might be deduced that those elements with the highest degrees of undersaturation would be the most reactive, while those near or at saturation would be essentially inert in the marine environment.

Krauskopf (1956) calculated the theoretical maximum concentrations that a group of metal ions might reach on the basis of the compounds formed with the major anions from sea-water which, from existing data in the literature, would result in the lowest amount of metal ion in solution. His data, which took into account the activity coefficients of the ions, are given in Table IV.

He further initiated an experimental attack upon this problem. To sea-water was added a solution of a given metal ion until a precipitate formed; adequate precautions were taken to maintain the pH of the sea-water relatively constant (between 7.8 and 8.2, values characteristic of open ocean water). The amount of the metal ion in solution was then determined, presumably an equilibrium value, but in any case an upper limit. Unfortunately, the characters of the solid phases were not established. An approach to equilibrium from the opposite direction was then made by adding some of the precipitate to a fresh sea-water aliquot and measuring the metal ion in solution until it reached a constant value. Such concentrations should at least be minimal values. Where the two values did not coincide, it was presumed that the equilibrium figure must lie somewhere in between. His data are collated in Table IV.

TABLE IV

Comparison of Calculated and Observed Concentrations of Metals at Saturation with the Observed Concentrations in Sea-Water. (Adapted from Krauskopf, 1956.)

| Metal ion | Insoluble compound | Solubility product | Concentration in saturated solution | | Sea-water concentration, mg/l. | Ratio of measured concentration to sea-water concentration | Residence time, years |
|---|---|---|---|---|---|---|---|
| | | | Calcd., mg/l. | Measured, mg/l. | | | |
| Pb | PbCO$_3$ | $1.5 \times 10^{-13}$ | 0.01 | 0.3–0.7 | 0.00003 | 4000–10,000 | $2.0 \times 10^3$ |
| Ni | Ni(OH)$_2$ | $1.6 \times 10^{-16}$ | 150 | 20–450 | 0.002 | 10,000–225,000 | $1.8 \times 10^4$ |
| Co | CoCO$_3$ | $8 \times 10^{-13}$ | 0.02 | 25–200 | 0.0005 | 50,000–400,000 | $1.8 \times 10^4$ |
| Cu | CuCO$_3$ | $2.5 \times 10^{-10}$ | 5.7 | 0.4–0.8 | 0.003 | 133–266 | $5.0 \times 10^4$ |
| Ba | BaSO$_4$ | $1 \times 10^{-10}$ | 0.03 | 0.11 | 0.03 | 3.7 | $8.4 \times 10^4$ |
| Zn | ZnCO$_3$ | $2 \times 10^{-10}$ | 4.6 | 1.2–2.5 | 0.01 | 120–250 | $1.8 \times 10^5$ |
| Cd | Cd(OH)Cl | $3.2 \times 10^{-11}$ | 105 | 4–1000 | 0.0001 | 40,000–10,000,000 | $5.0 \times 10^5$ |
| Ca | CaCO$_3$ | $5 \times 10^{-9}$ | 70 | 100–480 | 400 | 0.25–1.2 | $8.0 \times 10^6$ |
| Sr | SrCO$_3$ | $3$–$16 \times 10^{-10}$ | 9–44 | 22 | 8 | 2.75 | $1.9 \times 10^7$ |
| Mg | MgCO$_3$·H$_2$O | $1 \times 10^{-5}$ | 84,000 | 36,000 | 1350 | 27 | $4.5 \times 10^7$ |

There are clouds of uncertainty about both the theoretical and measured figures. In the former case, the activity coefficients are at best rough figures. The effects of complex ion formation have been disregarded. Perhaps the presumed solid phases are in error. For example, silicate or phosphate associations may yield with the metal ions a precipitate giving a lower value of the dissolved cation. The solid phases may not be the simple compounds tabulated, but possibly in some cases basic carbonates. In neither the experimental nor theoretical cases were the effects of temperature taken into account. Both dealt with temperatures 18°–25°C higher than the average oceanic value.

Overriding these limitations, certain uniformities in the results affirm their utility. In spite of the onerous burden of assumptions in obtaining both the measured and calculated values of maximum concentrations, there is a close agreement between them in most cases. The measured values are as a rule higher than the calculated ones; a result not unexpected if these metal ions engage in complex formation with such anionic species as chloride and sulfate. A most important deduction is that there can be little doubt that the concentrations of the tabulated metal ions, excluding calcium, strontium and possibly barium in deep waters, cannot be controlled by solubility equilibria. Finally, there is a respectable correlation between the residence times and the ratio of measured maximum concentration to the observed concentration in sea-water, an indicator of the degree of undersaturation.

These reactivity considerations have been concerned with metallic ions which have been derived from rock-weathering processes. These concepts may also be applied to the dissolved gases that enter the oceans via the atmosphere. The chemical passivity of the dissolved rare gases and nitrogen is reflected in the fact that they subsequently undergo little, if any, concentration changes in their water-masses. However, dissolved gaseous oxygen is found often in highly undersaturated states, and is even absent in some water, owing to its highly reactive nature in the biochemical cycles of the sea.

With the background of relative reactivities as ascertained by residence times and degrees of undersaturation, we can now attempt to seek out some of the inorganic and biochemical processes that regulate the intriguing chemical make-up of sea-water.

### 3. Chemical Reactions in the Oceans and the Composition of the End-Products

Although many chemical reactions proceed in the marine environment with rates at which discernible amounts of material accumulate on a square centimeter of sea floor in times of the order of hundreds or thousands of years, nonetheless, the very nature of the end-products or the chemical species within them can be decisive in reconstructing the chemical system in which they formed.

Krumbein and Garrels (1952), for example, have pointed out that marine phosphorites probably form in restricted basins in which the pH is relatively low, i.e. environments near or at anaerobism with pH values below 8, slightly less than that of normal sea-water. These minerals have the general formula $Ca_{10}(PO_4, CO_3)_6F_{2-3}$ in which the excess positive charges resulting from the substitution of carbonate for phosphate are compensated by excess fluorine or hydroxyl groups. Extensive deposits occur on the sea-floor of southern California where an area of about 6000 square miles contains phosphorite in the forms of nodules, slags and oolites (Emery, 1960). Emery also notes that about 98% of the material is in water depths from 100 to 1000 ft and that foraminiferal sands are often associated with the deposits.

Several recent investigations tend to confirm the hypothesis of Krumbein and Garrels. Altschuler, Clarke and Young (1958) find that the uranium in southern Californian phosphorites exists in the tetravalent state to the extent of 55 to 74% of the total uranium. The reduced uranium occurs in the apatites substituting for calcium in the lattice while the uranyl ions are apparently taken up by adsorption on the surfaces of the apatite crystallites (Sheldon, 1959). Further evidence for a reducing environment comes from the finding of a wooden log, mineralized by phosphorite intrusions, dredged from a 410 m terrace in the Gulf of Tehuantepec (Goldberg and Parker, 1960). The sea-water in contact with the mineralizing log was depleted in oxygen and contained maximal values of phosphate ions in the sea-water column. This area is quite active biologically and the deficiency in oxygen in the waters results from the combustion of the rather large amounts of organic matter falling from the euphotic zone.

It is important to ascertain precisely the site of the phosphorite formation—the sediment–water interface or an environment within the sediments themselves. Do these minerals receive their components directly from the overlying marine waters or from the substances in the deposits? Recent work on the state of uranium in Black Sea waters bears to these points. Kolyadin *et al.* (1960) sought the mode of occurrence of uranium in the deep reducing waters where the hydrogen sulfide contents approach 8 ml/l. and the oxygen concentrations are below the limits of analytical sensitivity. They found the uranium to be ionically dispersed and in the hexavalent state. Such an observation corresponds with physico-chemical calculations, as the authors point out that the reaction

$$U(OH)_4(s) + 3CO_3{}^{2-} = UO_2(CO_3)_3{}^{4-} + 2H_2O + 2e$$

requires a redox potential of 0.4–0.5 V for the uranium and carbonate concentrations of sea-waters. However, the redox potential in the Black Sea does not exceed 0.2 V, a value higher than that met with in the superficial waters above the phosphorite deposits.

Thus, we are led to the hypothesis that the reduction of uranium must take place within the reducing atmosphere of the sedimentary environment where high redox potentials can be attained and/or high concentrations of hexavalent uranium can be amassed. In either case our concern is with reactions that take place within the solid phases. The high uranium concentrations found by Manheim (1961) in the zones adjacent to the most stagnant basins of the Baltic Sea, as well as the unusually high uranium concentrations in phosphorites, which are usually of the order of several hundred parts per million, emphasize the above duality.

These observations coincide with the geological evidence on the deposition of phosphorite (McKelvey *et al.*, 1953). Marine apatites form in coastal deposits on the east sides of continents where upwelled waters give rise to the abundant production of organic matter. The rapid accumulation in the sediments of the organic phases results in a depletion of oxygen, as well as furnishing a source of phosphate. The sedimentary environments in which the decomposition of the organic phases takes place must also have the necessary redox potential for the reduction of hexavalent uranium as well as those solid phases which can pick up this uranium from sea-water.

The occurrence of tetravalent uranium in certain environments of the oceans is probably more widespread than has been heretofore realized. For example, the author has found within the skeletal structure of corals, from modern to thousands of years in age, not immodest amounts of this species. The implication that highly reduced conditions may prevail within calcium carbonate lattices, resulting most probably from the decomposition of organic matter, invites further investigation.

The authigenic clay mineral glauconite often occurs in deposits with the marine phosphorites and the conditions that lead to its formation may be inferred in part from such an association. Bank tops, ridge crests, hills that rise above shelves and some slope areas are hosts to this mineral at depths from 50 to more than 2000 m (Emery, 1960). Glauconite is a monoclinic mica, often disordered and interlayered, whose composition is characterized by 7–8% $K_2O$ and an iron content of 20–25% with the $Fe^{3+}/Fe^{2+}$ ratio varying between 3 and 9.

Burst (1958) suggests four pathways for its formation: (1) the transformation of fecal pellets or coprolites; (2) the conversion of materials filling foraminiferal tests; (3) the conversion of biotite booklets which yields rounded pellets retaining some of the book-like laminations of the parent material; and (4) the agglomeration of shale pellets or bottom clays with a subsequent transformation to glauconite.

If a reducing environment *within* the sedimentary phases is postulated for glauconite, as in the case of the phosphorites, a number of rather anomolous observations on this clay mineral can be brought into accord. For example, Emery (*op. cit.*) points out that the areas in which glauconite is most abundant are those in which the sea-water is oxygenated. In such environments the accumulation of large amounts of organic matter on the sea-floor can serve as a buffer against the intrusion of oxygen into the deposit site and can thus allow reducing atmospheres to be built up.

The existence of ferrous iron in the structure, where normally present in aerated waters are the oxides of ferric iron, directs one to a reducing environment for glauconitization—the interior of a foraminiferal shell, a fecal pellet, a clay agglomerate containing organic phases. Variations in the $Fe^{3+}/Fe^{2+}$ ratio probably stem not only from the existing redox potential of a given environment, but also from the amounts and types of organic matter that can form complexes with the various species of iron and hence can determine the relative availability of both the oxidized and reduced forms. Further, the character of the argillaceous progenitors of glauconite will strongly influence the chemical composition of the resultant mineral, as described by Burst in his rather extensive studies.

Finally, it is inviting to seek out further evidence to confirm or deny this concept of a reducing environment in the sediment for glauconitization. Investigations on the presence or absence of tetravalent uranium in the mineral would be most revealing in this regard.

Of equal interest to the reactions in anaerobic environments are those reactions occurring under oxidizing conditions at the sea-water–sediment interfaces in the pelagic and certain in-shore areas. Here, the ubiquitous ferro-manganese minerals accumulate in sites of limited sedimentation. Menard and Shipek (1958) estimate that between 20 and 50% of the deep-sea basement in the southwestern Pacific is covered with these minerals, on the basis of bottom photography and their occurrence in cores. They are found as nodular con-cretions which exist in sizes from millimeters to meters, as coatings about rocks and shells and as fine-grained dispersions in unconsolidated sediments. The nodules often form about a nucleus of phillipsite, pumice, shell fragments, lithified sediment or the refractory biogenous remains, ear-bones of whales or sharks' teeth. Growth is usually radial about the nucleus with easily visible concentric layers.

The rates of build-up of the ferromanganese minerals are extremely low—of the order of hundredths of millimeters to millimeters per thousand years. For example ionium/thorium geochronological measurements on a nodule from the Blake Plateau (29° 18′N, 52° 20′W; depth 5400 m, Lamont Theta Trawl No. 4) indicated a rate of accumulation of 0.1 mm/$10^3$ years. Perhaps such chemical growth, representing one of the lowest chemical reaction rates as-certained in nature, can better be expressed in terms of atomic layers per day in which the results range between 1 and 100.

The two principal metals in the minerals are iron and manganese, occurring as oxides or hydroxides, in similar amounts, although, less generally, either may be dominant. These phases act as hosts for a suite of metals including copper, nickel, cobalt, lead and the rare earths. These guest metals are mod-erately to highly reactive in sea-water and their assimilation by these wide-spread minerals may well account for their states of undersaturation in sea-water and their rather modest residence times.

Their rather unusual chemical composition (Table V) unlike that of any

TABLE V

Average Composition of Ferromanganese Minerals from the
Pacific Ocean

| Element | Weight % |
|---------|----------|
| Fe | 14 |
| Mn | 19 |
| Ni | 0.4 |
| Co | 0.3 |
| Cu | 0.5 |
| Ti | 0.8 |
| Zn | 0.04 |
| Pb | 0.1 |
| P | 0.5 |
| Al | 0.7 |
| Zr | 0.006 |

terrestrial minerals, coupled with their widespread distribution on the sea-floor, affirms their authigenic character. Further support comes from the previously cited lead-isotopic analyses of Chow and Patterson (1962), which indicate a geographic dependence of the relative amounts of radiogenic lead isotopes in the minerals. This isotopic composition of the lead in a given body of water, which furnishes these atoms to the iron–manganese accretions, is related to the input of the element from a specific land-mass; different continental areas yield different isotopic assays of lead based upon their rock composition.

The overall chemistry for the oxidation of manganese in the marine environment has been discussed in a previous section. Of significance is the hypothesis that such a reaction requires a surface, i.e. the ferromanganese mineral formation occurs at the liquid–sediment interface rather than in the water itself or in a micro-environment. Whether the oxidation of manganous ions taken up originally by the weathering waters from crustal rocks occurs in shallow-water deposits (Manheim, 1961), pelagic sediments or even on desert rocks[1], the necessity of an available surface free of even moderate accumulations of other sedimentary phases is evident.

Certain chemical characteristics of this deposition may well lend themselves as sensitive indicators of the redox potential of the environment. Cobalt and nickel, like manganese, occur in crustal rocks in the divalent state and are probably transported to the oceans in this form. Cobalt exists in sea-water at a concentration of about one-seventh that of nickel (Taivo Laevestu, *in litt.*) on the basis of recent refined analyses. Although cobalt and nickel show a strong geochemical coherence in behavior during the major sedimentary cycle, the ferromanganese phases contain nearly as much cobalt as nickel on the average (Table V). Further, cobalt shows a wider spectrum of concentrations in nodules than does nickel.

An explanation for these abundances may well be found in the greater ease of oxidation of cobalt from the divalent state to the insoluble trivalent form CoOOH and a subsequent accommodation of the CoOOH as a solid solution in the FeOOH of the ferromanganese minerals. Goldberg (1954) has pointed out a co-variance of cobalt with iron in manganese nodules, whereas nickel apparently follows the manganese concentrations.

This concept may be seen more advantageously from the equation of Sillén (1961) on the reduction of cobaltous ion in sea-water at 25°C:

$$CoOOH(s) + 3H^+ + e^- = Co^{2+} + 2H_2O \qquad \log K = 29.3.$$

Utilizing $\log [Co^{2+}] = -8$ and an activity coefficient for the cobaltous ion such that $\log \gamma_{Co} = -0.8$, then $\log [Co^{2+}] + \log \gamma_{Co} - 3 \log [H^+] - \log e^- = 28$ and the reaction has a tendency to proceed to the left. However, assuming a relatively constant cobalt and hydrogen ion concentration in the oceans, the redox potential could conceivably determine the amount of cobalt amassed by a

---

[1] The so-called desert varnish, described by Engel and Sharp (1958), has a markedly similar composition to that of the ferromanganese minerals and probably forms by a similar chemistry.

given ferromanganese mineral. Where the redox potential of the environment is relatively low, the cobalt/nickel ratios of sea-water would be expected in the minerals, as only the divalent ions are involved. High values for the ratio would be indicative of more strongly oxidizing conditions.

If such an approach is valid, the sensitivity of delimiting the physico-chemical conditions of the depositional environment may be enhanced by investigating other metals which are derived from crustal rocks in a reduced state and are amenable to oxidation. Lead and cerium, which enter the oceans most probably as plumbous and cerous ions, have the higher oxidation state solids $PbO_2$ and $CeO_2$. Where previous considerations have indicated $PbO_2$ is the stable particulate phase to be expected in equilibrium with plumbous ions in the oceans, the oxidation of cerium is not evident from thermodynamic considerations. However, the reaction may proceed by a combination of the cerous and manganous oxidations to form a solid solution of the ceric oxides with the manganese dioxides giving a thermodynamically possible reaction.

The cerium concentrations in these minerals are especially valuable as they can be normalized to those of its periodic-table neighbor lanthanum, whose trivalent state chemistry would follow that of cerous ion in the major sedimentary cycle. Present evidence suggests that an oxidation of cerium does occur. In ferromanganese minerals, recovered from the sea-floor, the cerium/lanthanum ratio has a value around six, whereas in crustal rocks of the earth's surface, the ratio is nearer two. An extreme case has been noted in a Triassic ferromanganese nodule from Timor where the ratio was an order of magnitude higher than in the more recent accretions.

Hence, one would expect high cobalt/nickel ratios to be accompanied by high lead concentrations and for especially strong oxidizing conditions at the reaction site high cerium/lanthanum ratios.

As a final example in this section, the inorganic precipitation of calcium carbonate appears to characterize an oceanic chemical system, limited to coastal waters in tropical or semi-tropical environments, in which the solubility product of this substance is exceeded. The most studied case in point involves the oolitic aragonitic sands of the Great Bahama bank, which are widely distributed over a 100,000-square-mile area of the continental shelf between Florida and Hispanolia. These oolites are readily distinguishable as resulting from an inorganic precipitation process by their distinctive chemical and isotopic compositions. Their strontium and uranium concentrations are markedly higher than most biogenous carbonates ($2.4\%$ $SrCO_3$ and 3.0 ppm of uranium; Tatsumoto and Goldberg, 1959). The isotopic ratios $^{13}C/^{12}C$ and $^{18}O/^{16}O$ in the crystals are unique and extend over a more limited range than biologically precipitated materials (Lowenstam and Epstein, 1957).

The oolites precipitate from fairly shallow depths—most of the area of the Great Bahama Bank is less than 5 m deep. Newell, Purdie and Imbrie (1960) point out that oolite formation is most extensive in and just below the intertidal zone. The factors encouraging the precipitation include not only the topography and the warm temperatures of the waters but also the intense biological activity in this environment. The cool, $CO_2$-rich waters that enter the shoals from the open sea are heated by sunlight. The resultant temperature increases cause the solubility product of calcium carbonate to fall. Further, the carbon dioxide content decreases and is accompanied by the consequential increase in carbonate ion concentration. The extensive photosynthesis in the area also causes a decrease in the carbon dioxide concentration. The net effects then are the reduction in the solubility of calcium carbonate and an increase in the concentration of the reactants, carbonate and calcium ions.

## 4. Space and Time Distributions of the Elements

Depth distributions of such chemical species as phosphate, nitrate and dissolved oxygen gas have proven especially valuable in understanding the major biochemical cycle in the sea, the primary production of organic matter by photosynthesis in surface waters and its subsequent combustion by oxygen at greater depths. Similarly, other vertical abundance profiles of reactive inorganic elements have been useful for elucidating their chemical reactions during residence in the marine biosphere.

For example, the geochemical behavior of barium is strongly shown in its concentration increase with depth in the oceans. Chow and Goldberg (1960) found surface concentrations were of the order of 10 μg of barium per liter whereas at depths of 4000–5000 m values three to six times higher were encountered (Table VI). The maximal values barium can attain, if its concentrations are determined by the precipitation of barium sulfate, go through a minimum as a function of depth. The solubility product of barium sulfate, at one atmosphere, decreases going from 25° to 0°C by a factor of 2 (Bjerrum *et al.*, 1958), accounting for an initial decrease in barium with depth. Using the data of Owen and Brinkley (1941), Chow and Goldberg calculated that the

TABLE VI

The Barium Concentration in the Pacific Ocean Waters

| Station<br>Latitude<br>Longitude | Downwind 6<br>11° 00′N<br>128° 30′W | | Downwind 7<br>07° 08′N<br>129° 16′W | | Downwind 8<br>03° 14′N<br>130° 31′W | | Downwind 17<br>34° 50′S<br>135° 53′W | |
|---|---|---|---|---|---|---|---|---|
| | Depth,<br>m | Ba,<br>μg/l. | Depth,<br>m | Ba,<br>μg/l. | Depth,<br>m | Ba,<br>μg/l. | Depth,<br>m | Ba,<br>μg/l. |
| | 0 | 12 | 705 | 32 | 0 | 12 | 4 | 10 |
| | 2698 | 32 | 3107 | 43 | 517 | 16 | 518 | 12 |
| | 4752 | 46 | 4232 | 51 | 1051 | 31 | 865 | 12 |
| | | | | | 2167 | 29 | 1937 | 27 |
| | | | | | 3002 | 43 | 3470 | 28 |
| | | | | | 4392 | 63 | 4185 | 30 |

pressure effect upon the saturation value of barium, assuming a constant sulfate concentration, results in a value 2.5-fold greater at depths around 5000 m than at the surface. Thus, if one uses a sulfate ion concentration in sea-water of 28 millimolar and activity coefficients of divalent ions of 0.1, a theoretical value of 70 μg of barium per liter is computed for oceanic depths of 4000 to 5000 m. Although saturation values of barium are not approached in surface ocean waters, they may be reached at greater depths.

What types of reaction then govern the depth distribution of barium? The gross similarity of the concentration profile of this element with those of the nutrient species suggests a relationship of barium to the biochemical cycles in the sea. The release of high concentrations of sulfate ions during the oxidation of organic sulfur in biological materials can result in a subsequent formation of barium sulfate within the decomposing biophase. As simultaneous sinking and combustion occur continuously in this organic micro-environment, part of the incorporated barium may be returned to sea-water, where this element is in an undersaturated state. The net result of such a process would be the conveyance of barium from shallow to deep waters. Part of the barium may end up in the sediments. In fact, concentrations of this element are markedly higher in pelagic sediments below biologically productive oceanic areas than in bottom samples below the more barren seas (Goldberg and Arrhenius, 1958; Goldberg, 1958). The high barium contents are associated with both siliceous and calcareous deposits, although the concentration of this element is not markedly high in the siliceous or calcareous hard parts of organisms. The barium is probably accumulated in the sediments through these chemical reactions involving the organic debris.

A second example of a vertical distribution reflecting chemical processes in the ocean may be found in the depth profiles of $^{226}$Ra, which show a similar distribution to those of barium, i.e. an increase in the abundance of this isotope going from the surface to less shoal waters. This radionuclide is a member of the $^{238}$U series and its immediate parent is $^{230}$Th (ionium):

$$^{238}\text{U} \xrightarrow[\alpha]{4.5 \times 10^9 \text{ y}} {}^{234}\text{Th} \xrightarrow[\beta]{24 \text{ d}} {}^{234}\text{Pa} \xrightarrow[\beta]{6.7 \text{ h}} {}^{234}\text{U} \xrightarrow[\alpha]{250{,}000 \text{ y}} {}^{230}\text{Th} \xrightarrow[\alpha]{80{,}000 \text{ y}} {}^{226}\text{Ra}$$

$$\Big\downarrow {\scriptstyle 1600 \text{ y}} \; \alpha$$

$$^{210}\text{Pb} \xleftarrow[\alpha]{1.64 \times 10^{-4} \text{ s}} {}^{214}\text{Po} \xleftarrow[\beta]{26.8 \text{ m}} {}^{214}\text{Bi} \xleftarrow[\beta]{19 \text{ m}} {}^{214}\text{Pb} \xleftarrow[\alpha]{3.1 \text{ m}} {}^{218}\text{Po} \xleftarrow[\alpha]{3.8 \text{ d}} {}^{222}\text{Rn}$$

Koczy *et al.* (1957) noted that the $^{226}$Ra is in excess by about six-fold over the amount which could be supported by $^{230}$Th. Ionium is removed quite rapidly from sea-water after its formation (the residence time of thorium is about 350 years) to the solid phases on the sea-floor.

To account for these increases, Koczy and co-workers suggested that the radium diffuses into the sea-water subsequent to its birth from ionium in the sediments and was subject to a vertical transport from the bottom due to eddy diffusivity. The total amount of radium given off by the sediments was calculated to vary between 1.2 and $1.8 \times 10^{-10}$ g of Ra/m$^2$/year. Further, many of the profiles show a clear surface depletion which is explained by an incorporation of radium into organisms and a subsequent release at depth during the oxidative decomposition of the biomass. Koczy and Titze (1958) found the highest radium contents ($0.275 \times 10^{-12}$ g of Ra/g of sample) in fresh plankton composed primarily of diatoms. This number is several orders of magnitude higher than normal sea-water concentrations ($10^{-16}$ g of Ra/ml).

The diffusion of radium from the sediments is supported by a number of examples indirectly indicating the migration of radium within the sediments (Arrhenius and Goldberg, 1954; Kroll, 1953). But more definitive evidence arises from studies of radioactive equilibrium in the $^{238}$U series in deep-sea cores. Table VII gives the $^{230}$Th and $^{210}$Pb activities as a function of depth in Core Monsoon 49G (14° 27′S, 78° 03′E; 5214 m depth). It is evident that $^{210}$Pb is not in radioactive equilibrium with ionium down to depths of 15 cm. The only nuclide between these two series members whose half-life is long enough to allow diffusion is radium.

## Table VII

Ionium ($^{230}$Th) and $^{210}$Pb Activities in the Leachates from Core Monsoon 49G
(14° 27′S, 78° 03′E; 5214 m depth)

| Depth in core, cm | $^{230}$Th, dpm/g [a] | $^{210}$Pb, dpm/g |
|:---:|:---:|:---:|
| 0–3 | 77 | 38 |
| 3–6 | 79 | 48.5 |
| 6–9 | 68 | 43 |
| 9–12 | 55 | 43 |
| 12–15 | 51 | 51 |
| 15–18 | 47 | 44 |
| 18–21 | 38 | 38 |
| 50–53 | 16 | 18 |
| 80–83 | 6 | 6 |
| 100–103 | 4 | 3 |

[a] Disintegrations per minute per gram.

Attempts to calculate the rate of diffusion of radium in sediments were made by Koczy and Bourret (1958) on the assumption of constant rates of sediment precipitation, radium accumulation and radium diffusion. In a single core they obtained a value of radium diffusion of $10^{-10}$ g of Ra/m$^2$/year, a value nearly identical with the results of the eddy diffusivity calculations.

Although a definitive knowledge of the composition of the oceans in past times is lacking, all evidence derived from chemical, geological and biological reasoning suggests that at best only modest deviations from the make-up of to-day's water have taken place (Rubey, 1951). Recent experimental approaches to delimiting such changes have centered about two bases: the variations in the temperature of the oceans with time and the composition of fossil skeletal remains of marine organisms. The utilization of the chemical composition of authigenic minerals to enter such investigations has not as yet been realized but clearly is a most promising possibility.

Determinations of the $^{18}O/^{16}O$ ratio in the calcium carbonate tests of Foraminifera have enabled Emiliani (1955) to establish fluctuations in oceanic temperatures during the past. Surface sea-water values have shown variations of about 6°C through the Pleistocene as evidenced from the Caribbean Sea and Atlantic samples. Further, this investigator (1954) has also indicated that the bottom-water temperatures have continuously decreased from the Oligocene through the Miocene into the late Pliocene by about 8°C. Also, it appears that the temperatures were more uniform during the Tertiary.

Such temperature variations would be accompanied by minor changes in the composition of sea-water. The dissolved gases, whose marine concentrations are in part determined by the temperature at which equilibrium took place in the water–atmosphere system, would show decreasing concentrations with increasing temperatures. Further, the calcium contents, whose levels are regulated by the precipitation of calcium carbonate, should have decreased during the warmer period. Bramlette (1961) has pointed out that there would have resulted a more uniform areal distribution of the supply of calcium carbonate from surface waters and also the bottom warm waters would greatly reduce in the Tertiary the solution of calcium carbonate in the deeper deposits.

The composition of the ocean during the past with respect to certain minor elements which substitute in skeletal materials for the major substances can, in principle, be determined by the analyses upon fossils. For example, Lowenstam (1959), on the basis of $^{18}O/^{16}O$ ratios, the Sr/Ca ratios and the $MgCO_3$ contents in fossil brachiopods from times as far back as the Mississippian, asserts that strontium and magnesium abundances have remained essentially constant during the last $2.5 \times 10^8$ years.

## References

Altschuler, Z. S., R. S. Clarke and E. J. Young, 1958. Geochemistry of uranium in apatite and phosphorite. *U.S. Geol. Surv. Prof. Paper* 314 D, 90 pp.

Arrhenius, G. and E. D. Goldberg, 1954. Distribution of radioactivity in pelagic clays. *Tellus*, **7**, 226–231.

Barth, T. F. W., 1952. *Theoretical Petrology.* John Wiley and Sons, New York.

Barkley, R. A. and T. G. Thompson, 1960. The total iodine and iodate-iodine content of sea-water. *Deep-Sea Res.*, **7**, 24–34.

Bjerrum, J. C., G. Schwarzenbach and L. G. Sillén, 1958. Stability Constants, Part II (Inorganic ligands). *Chem. Soc. Spec. Pub. 7.*

Bramlette, M. W., 1961. "Pelagic sediments". In "Oceanography", *Amer. Assoc. Adv. Sci.*, 345–366.

Burst, J. F., 1958. Glauconite pellets: their mineral nature and applications to stratigraphic interpretations. *Bull. Amer. Assoc. Petrol. Geol.*, **42**, 310–327.

Chow, T. J. and E. D. Goldberg, 1960. On the marine geochemistry of barium. *Geochim, et Cosmochim. Acta*, **20**, 192–198.

Chow, T. J. and C. C. Patterson, 1962. The occurrence and significance of lead isotopes in pelagic sediments. *Geochim. et Cosmochim. Acta*, **26**, 263–308.

Clarke, F. W., 1924. Data of Geochemistry. *U.S. Geol. Surv. Bull.*, 770.

Conway, E. J., 1943. The chemical evolution of the ocean. *Proc. Roy. Irish Acad.*, **48B**, 161–212.

Craig, H., 1961. Standard for reporting concentrations of deuterium and oxygen-18 in natural waters. *Science*, **133**, 1833–1834.

Emery, K. O., 1960. *The Sea off Southern California*. John Wiley and Sons, New York.

Emiliani, C., 1954. Temperature of Pacific bottom waters during the Tertiary. *Science*, **119**, 853–855.

Emiliani, C., 1955. Pleistocene temperatures. *J. Geol.*, **63**, 538–578.

Engel, C. G. and R. P. Sharp, 1958. Chemical data on desert varnish. *Bull. Geol. Soc. Amer.*, **69**, 487–518.

Epstein, S. and T. Mayeda, 1953. Variation of $O^{18}$ content of waters from natural sources. *Geochim. et Cosmochim. Acta*, **4**, 213–224.

Forchhammer, G., 1865. On the composition of sea water in the different parts of the ocean. *Phil. Trans. Roy. Soc. London*, **155**, 203–262.

Friedman, I., 1953. Deuterium content of natural waters and other substances. *Geochim. et Cosmochim. Acta*, **4**, 89–103.

Goldberg, E. D., 1954. Marine Geochemistry. *J. Geol.*, **62**, 249–265.

Goldberg, E. D., 1957. Biogeochemistry of trace elements. *Geol. Soc. Amer. Mem.*, **67**, 345–358.

Goldberg, E. D., 1958. Determination of opal in marine sediments. *J. Mar. Res.*, **17**, 178–182.

Goldberg, E. D. and G. O. S. Arrhenius, 1958. Chemistry of Pacific pelagic sediments. *Geochim. et Cosmochim. Acta*, **13**, 153–212.

Goldberg, E. D. and M. Koide, 1962. Geochronological studies of deep-sea sediments by the Io/Th method. *Geochim. et Cosmochim. Acta*, **26**, 417–450.

Goldberg, E. D. and R. H. Parker, 1960. Phosphatized wood from the Pacific sea floor. *Bull. Geol. Soc. Amer.*, **71**, 631–632.

Goldberg, E. D., C. C. Patterson and T. J. Chow, 1958. Ionium–thorium and lead isotope ratios as indicators of oceanic water masses. *Second U.N. Intern. Conf. on the Peaceful Uses of Atomic Energy*, Geneva.

Grim, R. E., 1953. *Clay Mineralogy*. McGraw-Hill, New York.

Koczy, F. and R. Bourret, 1958. Radioactive nuclides in ocean water and sediments. Progress Report, The Marine Laboratory, Miami, Florida (1958).

Koczy, F., E. Picciotto, G. Poulaert and S. Wilgain, 1957. Mesure des isotopes du thorium dans l'eau de mer. *Geochim. et Cosmochim. Acta*, **11**, 103–129.

Koczy, F. and H. Titze, 1958. Radium content of marine shells. *J. Mar. Res.*, **17**, 302–311.

Kolyadin, L. B., D. S. Nikolayev, S. M. Grashchencko, Y. V. Kuznetsov and K. Lazarev, 1960. Modes of occurrence of uranium in the water of the Black Sea. *Doklady Akad. Nauk S.S.S.R.*, **132**, 16–18.

Krauskopf, K. B., 1956. Factors controlling the concentrations of thirteen rare metals in sea water. *Geochim. et Cosmochim. Acta*, **9**, 1–33.

Kroll, V. S., 1953. On the age determination of deep sea sediments by radium measurements. *Deep-Sea Res.*, **1**, 211–215.

Krumbein, W. C. and R. M. Garrels, 1952. Origin and classification of chemical sediments in terms of pH and oxidation reduction potentials. *J. Geol.*, **60**, 1–33.

Laevestu, T. and T. G. Thompson, 1958. Soluble iron in coastal waters. *J. Mar. Res.*, **16**, 192–198.

Lowenstam, H., 1959. $O^{18}/O^{16}$ ratios and Sr and Mg contents of calcareous skeletons of recent and fossil brachiopods and their bearing on the history of the oceans. *Preprints Intern. Oceanog. Cong. New York, 1959*, 71–72.

Lowenstam, H. and S. Epstein, 1957. On the origin of sedimentary aragonite needles of the Great Bahama Bank. *J. Geol.*, **65**, 364–375.

Lyman, J., 1959. Chemical Considerations. In "Physical and Chemical Properties of Sea Water". *Nat. Acad. Sci.-Nat. Res. Council. Pub. No. 600.*

Manheim, F. T., 1961. A geochemical profile in the Baltic Sea. *Geochim. et Cosmochim. Acta*, **25**, 52–70.

Menard, H. W. and C. J. Shipek, 1958. Surface concentrations of manganese nodules. *Nature*, **182**, 1156–1158.

McKelvey, V. E., R. W. Swanson and R. P. Sheldon, 1953. The Permian phosphorite deposits of the Western United States. *19th Intern. Geol. Cong. (Algiers). C. R. Section XI*, 45–64.

Newell, N. D., E. Purdie and J. Imbrie, 1960. Bahamian oolitic sand. *J. Geol.*, **68**, 481–497.

Owen, B. B. and S. R. Brinkley, 1941. Calculation of the effect of pressure upon ionic equilibria in pure water and salt solutions. *Chem. Rev.*, **29**, 461–474.

Rakestraw, N. W., D. P. Rudd and M. Dole, 1951. Isotopic composition of oxygen in air dissolved in Pacific Ocean water as a function of depth. *J. Amer. Chem. Soc.*, **73**, 2976.

Rama, Koide M. and E. D. Goldberg, 1961. Lead-210 in natural waters. *Science*, **134**, 98–99.

Richards, F. A. and B. B. Benson, 1961. Nitrogen/argon and nitrogen isotope ratios in two anaerobic environments, the Cariaco Trench in the Caribbean Sea and Dramsfjord, Norway. *Deep-Sea Res.*, **7**, 254–264.

Rubey, W. W., 1951. Geologic history of sea water. *Bull. Geol. Soc. Amer.*, **62**, 1111–1147.

Sheldon, R. P., 1959. Geochemistry of uranium in phosphorites and black shales of the Phosphoria formation. *U.S. Geol. Surv. Bull.* 1084b., 83–115.

Sillén, L. G., 1961. The physical chemistry of sea water. In "Oceanography". *Amer. Assoc. Adv. Sci.*, 549–582.

Tatsumoto, M. and E. D. Goldberg, 1959. Some aspects of the marine geochemistry of uranium. *Geochim. et Cosmochim. Acta*, **17**, 201–208.

Thode, H. G., J. MacNamara and W. H. Fleming, 1953. Sulphur isotope fractionation in nature and geological and biological time scales. *Geochim. et Cosmochim. Acta*, **3**, 235–243.

Williams, P. W., 1960. Organic acids found in Pacific Ocean waters. Thesis. University of California, La Jolla, California.

Winkler, L. W., 1916. Der Jodid-und Jodat Iogehalt des Meerwassers. *Z. angew. Chem.*, **29**, 205–207.

## 2. THE INFLUENCE OF ORGANISMS ON THE COMPOSITION OF SEA-WATER

A. C. REDFIELD, B. H. KETCHUM and F. A. RICHARDS

### 1. Introduction

A number of components of sea-water enter into biochemical processes to such a degree that their concentrations are highly variable when compared to the total salinity. Whereas the distribution of the major constituents of sea-water are to be accounted for on physical and geochemical principles, additional considerations of a biological nature need to be taken into account where the biologically active components are concerned. Since the changes in such components may take place rapidly, relative to the life of a water mass, their study can illuminate the physical description of the sea.

The object of this chapter is to discuss the special considerations which are required for this purpose. They are of two sorts; the stoichiometric relations which arise from the specific composition of marine organisms, and the dynamic equilibria between biological and physical processes which determine the concentration of elements present at any point in the sea.

The influence of organisms on the composition of sea-water is determined by physiological influences and consequently exhibits the regularity inherent in organic processes. Elements are withdrawn from sea-water by the growth of marine plants in the proportions required to produce protoplasm of specific composition and are returned to it as excretions and decomposition products of an equally specific nature. While significant differences may occur in the requirements of different individuals and species, the statistical effects produced by the entire population present in any body of water have some regularity. Furthermore, ecological principles indicate that certain similarities will occur in populations of any sort when one considers the proportionate activity of the primary producers and of the predators, which occupy later positions in the food chain.

The dissolved components of sea-water are transported from place to place by advection and move from one parcel of water to another by eddy diffusion. These agencies act equally on all dissolved constituents of the water and their effects may be traced by measuring the concentrations of any one of the constituents, such as the chlorides. The biologically active constituents may move from one water layer to another in additional ways: namely, by the sinking of organized matter under the force of gravity and by the active vertical migration of organisms. Such movements of organic matter from one water layer to another provide a fractionating mechanism by means of which the difference in the distribution of conservative and nonconservative properties of sea-water may be explained.

Because of fractionation the biologically active elements circulate in a different pattern than does the water itself or its inactive solutes. We may speak of a biochemical circulation as distinct from, though dependent on, the physical circulation of the water. Since the elements required for the construction of protoplasm are drawn from the water in proportions which have some

[MS received August, 1960]

26

uniformity, they are distributed in somewhat similar patterns by the bio-chemical circulation.

## 2. The Biochemical Cycle

The exchange of chemical elements between sea-water and the biomass is a cyclic process. The cycle may be broken down into two phases, synthesis and regeneration. Elements are withdrawn from the water during the synthetic phase by photosynthesis in the proportions required for the growth of the primary producers, which are predominantly the phytoplankton. These elements are ultimately returned to the water in the regenerative phase as the decomposition products and excretions of both the primary producers and the subsequent members of the food chain which prey upon them, including the microorganisms which complete the decomposition of organic debris.

Oxygen occupies a unique position in the biochemical cycle because this element is set free in the course of photosynthesis. The concentration of oxygen consequently increases in the water when synthesis takes place and this oxygen becomes quantitatively available for the subsequent oxidation of the products of this synthesis.

### A. The Elementary Composition of Plankton

The proportions in which the elements of sea-water enter into the bio-chemical cycle is determined by the elementary composition of the biomass. Since the plankton comprises the bulk of the biomass these proportions are indicated by the analysis of plankton samples. Earlier analyses of net plankton by Redfield (1934) were substantiated by more extensive data assembled by Fleming (1940) who obtained the atomic ratios for the principal elements present in the organic matter given in Table I. The elementary composition of the zoo-plankton and phytoplankton is very similar and the average ratios may be taken as representative of the biomass as a whole.

TABLE I

Atomic Ratios of the Principal Elements Present in Plankton

|              | C   | N    | P |
| ------------ | --- | ---- | - |
| Zooplankton  | 103 | 16.5 | 1 |
| Phytoplankton | 108 | 15.5 | 1 |
| Average      | 106 | 16   | 1 |

The oxygen set free in the synthesis or consumed in the decomposition of the biomass would be 212 atoms for each atom of phosphorus if the oxidation of carbon alone were considered. If, in addition, four atoms of oxygen are con-sumed in oxidizing each atom of nitrogen, the oxidative ratio for phosphorus,

$\Delta O/\Delta P$, is equal to $-276$. This appears to be the most appropriate estimate for general use.

These ratios provide a stoichiometric basis for evaluating the general proportions in which the major nutrients present in sea-water may be expected to change as the result of biological activity. The observed changes in the composition of sea-water in the ocean support this generalization. Their use depends on the assumption that the composition of the plankton is statistically constant. There are, however, a number of considerations which indicate that substantial departures may be found under special circumstances. Riley (1951; 1956a) has discussed the effects of the varying composition of the plankton with respect to lipoid content, skeletal material, etc. He considers that the oxidation of organic matter may lead to $\Delta O/\Delta P$ ratios which vary between $-250$ and $-300$ and that more extreme variations may be noted.

The species composition of the biomass is observably variable both in time and place, and each species may be expected to have a composition which differs somewhat from others (Vinogradov, 1953). The chemical composition of a given species may be expected to vary also with the general nutritional conditions under which it grows. For these reasons statistical uniformity in composition is probably approached only in large masses of water, where deviations of this sort are averaged out.

Analysis of plankton from Long Island Sound by Harris and Riley (1956) indicates that on the average phytoplankton contains nitrogen and phosphorus in the ratio of 16:1 atoms, in agreement with Fleming's value. The zooplankton in contrast yields an average N/P ratio of 24:1. The individual collections made at different times of year differ from the average by about 25%. In the case of the zooplankton the ratios have a seasonal trend, the higher values being obtained in winter and spring.

## a. Effect of nutrient deficiencies on composition

It has been demonstrated repeatedly in culture experiments that the elementary composition of unicellular algae can be varied by changing the composition of the medium in which they grow. If one element is markedly deficient in the medium, relative to its need by the organism, cell growth and cell division can proceed for a limited period of time. The cells produced under these conditions contain less of the deficient element than do normal cells. When an element is provided in excess in the medium, luxury consumption can increase its content in the cells. It has been shown with radiophosphorus that the excess is readily exchangeable with the medium (Goldberg et al., 1951; Rice, 1953).

The data in Table II illustrate the extremes in phosphorus : carbon : nitrogen ratios which can be produced when deficiencies in phosphorus and nitrogen develop in culture experiments. Perhaps because of the composition of the culture medium, the normal cells contained more than twice as much phosphorus as is observed in phytoplankton populations growing in the marine environment. Deficiency of phosphorus in the medium reduced the ratio of

TABLE II

Experimental Variation of the C : N : P Ratios (by Atoms) in Cultures of the
Freshwater Alga, *Chlorella pyrenoidosa* (after data of
Ketchum and Redfield, 1949)

| Conditions | C | N | P |
|---|---|---|---|
| Normal cells | 47 | 5.6 | 1 |
| Phosphorus deficient cells | 231 | 30.9 | 1 |
| Nitrogen deficient cells | 75 | 2.9 | 1 |

phosphorus to carbon in the cells to about one-fifth and the deficiency of
nitrogen in the medium reduced the ratio of nitrogen to carbon to about one-
fourth of that of the normal cells. Relative to the phosphorus content the nitrogen content of the cells could be varied experimentally thirteen-fold.

In natural waters the concentrations of available nitrogen and phosphorus
are greatly reduced during periods of active growth. One or other element
may be almost absent while an excess of the other may remain in the water.
In Long Island Sound, nitrogen appears to be the element available in minimum
proportions relative to the needs of the phytoplankton and its ratio to phosphorus varies from nearly zero to about 8 : 1 (Riley and Conover, 1956). The
ratio of nitrogen to phosphorus in the plankton was found by Harris and Riley
(1956) to average 16.7 : 1 with relatively small variations. Similarly, in the
coastal waters south of Long Island the ratio of nitrogen to phosphorus is
lower than that of the requirement of normal plankton. In spite of the ano-
malously low ratio of nitrogen to phosphorus in the water, the plants apparently
assimilate these elements in the normal ratio of 15 atoms to 1 until very low
concentrations are reached (see Fig. 1). However, when the nitrogen is nearly
depleted from the water the phytoplankton cells can apparently continue to

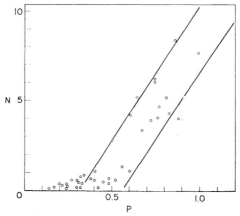

Fig. 1. Concentration of phosphate phosphorus and nitrate nitrogen in coastal waters
south of Long Island. Units: mg atoms/m$^3$. Slope of envelopes, $\Delta N/\Delta P = 15$. (After
Ketchum *et al.*, 1958.)

assimilate phosphorus, which is then present in excessive proportions in the water.

The ability of phytoplankton to continue to form cells of normal elementary composition, when growing in the sub-surface layers of the sea from which one essential nutrient is practically exhausted, may be explained by the consideration that mixing processes supply nutrients from deeper layers. Eddy diffusion may deliver nitrogen and phosphorus to the euphotic zone in a higher ratio than actually occurs in this layer, as Ketchum, Vaccaro and Corwin (1958) have pointed out. The decomposition of cells *in situ* will also regenerate nitrogen and phosphorus in the higher ratio characteristic of their composition. Under steady-state conditions the net growth of the population is frequently limited by the rate of diffusion of nutrients into the zone of active growth where the phytoplankton assimilates the elements as fast as they are supplied. The nutrients present in the water represent merely the residue of elements supplied which are not required to form cells of normal composition.

Thus, while culture experiments readily show the effects of nutrient deficiencies in the medium on the composition of the organisms, it is by no means certain that such deficiencies are developed by natural populations.

### B. The Changes in the Nutrient Content of Sea-Water Produced by Biological Activity

The principal sources of the major elementary components of marine organisms are the carbonate, phosphate and nitrate ions of sea-water. The synthetic process by which these nutrients are removed from the water is limited to the sub-surface layers of the sea into which adequate light may penetrate. The regenerative processes being independent of light may take place at any depth. Owing to the sinking of organized matter under the force of gravity and the vertical migrations of organisms, significant quantities of organic matter are carried downward to decompose at depth. Consequently the cycle does not run to completion in the euphotic zone, where synthesis exceeds regeneration sufficiently to produce the organic matter which decomposes at depth.

The result is a distribution of the biologically active elements in sea-water which is distinct from that of the major salts which results from purely physical motions. The nutrient elements are not only withdrawn from the euphotic zone in quantities sufficient to form the biomass locally present, but are also transported to depth where the water is enriched by the regenerative process. The distribution of oxygen is affected in a converse way. The excess synthesis in the sub-surface layer produces oxygen which causes this layer to be supersaturated, while the regeneration at depth reduces the oxygen content of the water in proportion. The transition level where synthesis balances regeneration is marked by the compensation point where the sea-water is exactly saturated with oxygen.

If the differences in the concentration of nutrient elements which are found at different depths in the ocean are due to the decomposition of organic matter synthesized near the surface, these differences should not only reflect the quantities of organic matter which have decomposed but also the proportions of the elements present in the plankton from which it has been derived. The latter expectation is confirmed by the data presented in Figs. 2 and 3, in which

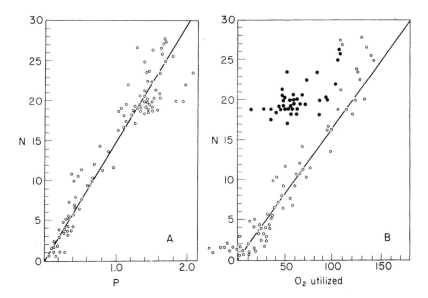

Fig. 2. A. Correlation between nitrate nitrogen and phosphate phosphorus in waters of western Atlantic. B. Correlation between nitrate nitrogen and apparent oxygen utilization in same samples. Open circles represent samples from above 1000 m.
     Concentrations in mg atoms/m³. Phosphorus corrected for salt error. (After Red-field, 1934.)

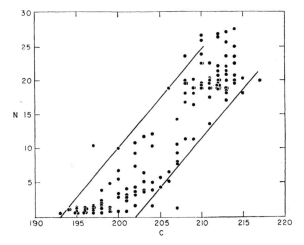

Fig. 3. Correlation between nitrate nitrogen and carbonate carbon in waters of western Atlantic. Concentrations in mg atoms/m³. (After Redfield, 1934.)

each point represents the concentrations of a pair of elements determined from a single sample of sea-water collected at varying depth in the western North Atlantic. In general the low values represent samples from small depths and the high values from greater depths. The slope of the line drawn through the points gives the ratio of the difference in concentration of the elements in question. These differences are assumed to be due to differences in the quantities of organic matter of uniform composition which have decomposed in the different samples. The ratios may be combined to give the relative proportions of the several elements in question in the decomposed material.

Table III gives these ratios as determined from the analysis of sea-water and from analyses of plankton collections. The agreement so far as phosphorus, nitrogen and carbon are concerned leaves little doubt that the changes in concentration at depth arise mainly from the decomposition of plankton and that the ratios are sufficiently precise to be useful in the analysis of oceanographic problems. Similar treatment of data from the Pacific, Indian and South Atlantic Oceans and the Barents Sea shows that the ratio $\Delta N/\Delta P = 15$ applies in sea-water on a world-wide scale.

TABLE III

Ratios of the Elements Involved in the Oxidation of Organic Matter in Sea-Water at Depth and Those Present in Plankton of Average Composition, by Atoms (after Richards and Vaccaro, 1956)

| Sea-water analyses | $\Delta O$ | $\Delta C$ | $\Delta N$ | $\Delta P$ | Ref. |
|---|---|---|---|---|---|
| Northwest Atlantic | $-180$ | 105 | 15 | 1[a] | Redfield (1934) |
| Cariaco Trench, upper layers | $-235$ | — | 15 | 1 | Richards & Vaccaro (1956) |
| Plankton analyses | $-276$[b] | 106 | 16 | 1 | Fleming (1940) |

[a] Corrected for salt error after Cooper (1938).
[b] Estimated assuming 2 atoms $O \backsimeq 1$ atom C and 4 atoms $O \backsimeq 1$ atom N.

It should be emphasized that these ratios do not represent the proportions in which the elements are available in sea-water, but rather the ratios of change in their concentration which result from biological activity.

### a. The oxidative ratio

The relation of the change in concentration of oxygen in sea-water, shown in Table III, to that of the nutrient elements does not agree well with the expectation raised by the statistical composition of plankton. This discrepancy requires examination.

Oxygen and carbon dioxide are exchanged with the atmosphere when sea-water is at the surface and their concentrations in water sinking to depth is controlled by this exchange. In order to determine the quantity of oxygen which has disappeared from a sample of water collected at depth, it is necessary to estimate its oxygen content at the time it sank from the surface. The difference between the oxygen concentration of water in equilibrium with a normal atmosphere at the temperature and salinity observed *in situ* and the oxygen concentration as measured is referred to as the *apparent oxygen utilization*, or A.O.U. The conditions specified are met only approximately in nature.

Surface sea-water may be substantially supersaturated with oxygen under conditions favorable for growth, or may be undersaturated under the turbulent conditions obtaining in winter (Redfield, 1948). Some undersaturation is to be expected in high latitudes at the time when deep water is formed. Seasonal variations in atmospheric pressure may also influence the result (Carritt, 1954). Finally, because the saturation values for oxygen are not linear functions of temperature and salinity, the oxygen content of a mixture of sea-waters equilibrated under different conditions would be different from that calculated from the temperature and salinity of the mixture. These considerations introduce errors into the estimation of the utilization of oxygen and its ratio to the regeneration of nitrogen and phosphorus. Probably these errors are not greater than the analytical errors in the determination of phosphate phosphorus or nitrate nitrogen with which they are used, and are not large enough to account for the discrepancy in the ratio as estimated from analyses of plankton and of sea-water.

A more significant source of error in the estimation of oxidative ratios from sea-water analysis is the presence of nutrients in the water which have not been derived from the oxidation of organic matter. The deep water of the oceans is formed in high latitudes in winter where low light intensity restricts the growth of plants. Consequently on sinking it may contain substantial quantities of nutrients which are not in organic form. The presence of such preformed nitrate in the deep water is indicated in Fig. 2B, in which samples from great depth, indicated by solid circles, are seen to contain nitrate in much greater quantity relative to the utilization of oxygen than obtains in the upper 1000 m of water. Preformed nutrients are probably present in smaller quantities in these upper layers. Their presence causes estimates of the oxidative ratio to be too low and to be unreliable. Ratios obtained from the composition of plankton are not subject to this error and consequently are to be preferred.

The $\Delta O/\Delta P$ ratio of $-276$ will consequently be employed in the various estimates to be discussed. The use of ratios derived from sea-water in certain earlier studies (Redfield, 1942, 1948) led to estimates which require revision.

## b. Preformed nutrients and nutrients of oxidative origin

The nutrients present in a sample of sea-water may be separated into two fractions: (1) nutrients of oxidative origin which have been regenerated from organic matter, and (2) preformed nutrients which were present as such in the water at the time it sank from the surface. The quantity of a nutrient of oxidative origin may be estimated from the apparent oxygen utilization by applying the $\Delta O/\Delta P$ ratio. The difference between this quantity and the total quantity of nutrient present in inorganic form gives the quantity of preformed nutrient present. In estimating the fraction of oxidative origin, it is convenient to note that a $\Delta O/\Delta P$ ratio of $-276$ corresponds closely to the production of 1 mg atom/m$^3$ of phosphorus in the consumption of 3 ml O$_2$/l.

The distribution of these fractions of inorganic phosphorus at two stations in the North Atlantic is shown in Fig. 4. At the station in the subtropics (a)

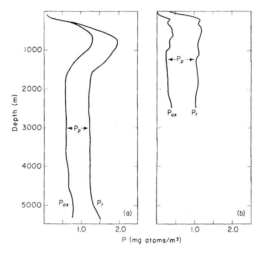

Fig. 4. Fractions of total phosphorus, P$_t$, present as phosphorus of oxidative origin, P$_{ox}$, and as preformed phosphorus, P$_p$, at stations in North Atlantic.

(a) *Crawford* Station 364; 16° 16′N, 54° 44′W.
(b) *Discovery* Station 3849; 55° 25′N, 33° 12′W.

more than half the phosphorus in the deep water is preformed and this fraction is present in practically the same concentration at all depths below 700 m. Above this depth the preformed phosphorus diminishes and is absent in the layers immediately below the surface. At the station in high latitude (b) about three-quarters of the total phosphorus is present in the preformed fraction and it is only in the surface layer that its concentration is reduced by assimilation by the phytoplankton.

As defined, the preformed nutrient concentration is a conservative property, being uninfluenced by the oxidation of organic matter which was present in the water mass at the time of its sinking or which has accumulated in the water at depth. Its value depends on the degree to which conditions at the sea surface influence the synthetic processes at the time the water masses are formed. In contrast, the apparent oxygen utilization, or its equivalent in nutrient, provides a measure of the effect of biological activity on the composition of the water since it sank below the influence of the atmosphere.

The validity of the concept of preformed nutrients and the method of their estimation may be tested by examining the values in water masses which may be recognized by their composition (see Table IV). In the deep water masses of the Atlantic, characteristic mean values for preformed phosphorus may be assigned which differentiate the North Atlantic deep water, the circumpolar water, and the Antarctic deep water. The dispersion of measurements for samples from each of these water masses is not greater than that of the measurement of inorganic phosphorus.

TABLE IV

Preformed Phosphorus Content ($P_p$) of Deep Water Masses of Atlantic Ocean (mg atoms/m³)

| Water mass | Criteria | No. samples | $P_p$ mean | Dispersion ($\sigma$) | |
|---|---|---|---|---|---|
| | | | | Individual estimate | Analytical method |
| North Atlantic deep water 59°N–34°S | S 34.85–35.0‰ T 2–4°C | 102 | 0.74 | ± 0.13 | ± 0.12 |
| Circumpolar water 24°S–69°S | S < 34.75‰ T 0–1.99°C | 41 | 1.50 | ± 0.19 | ± 0.23 |
| Antarctic deep water 49°S–69°S | S < 34.75‰ T < 0°C | 26 | 1.67 | ± 0.13 | ± 0.23 |

## c. Regeneration of nutrients

It has been implicit in the foregoing discussion that the oxidation of organic matter has gone to completion in the deep water. This is not absolutely correct since the deep sea is inhabited by organisms and various organic residues may be detected in its waters. Intermediate products of decomposition may occur in significant quantities in places where organic matter is decaying in quantity, and their distribution in the sea provides useful information on the regenerative part of the biochemical cycle. Although a variety of organic materials may be detected in sea-water (Duursma, 1960), discussion will be limited to the few which have been sufficiently studied to provide insight into the regeneration of phosphate and the nitrogenous nutrients.

The phosphorus of organisms is present for the most part in organic combination, forming molecules many of which are known to be of great physiological importance. On decomposition such molecules are liberated into the water as dissolved substances. On the oxidation of these materials, presumably by bacterial action, the phosphorus is released as phosphate in the ionic form. The method traditionally employed on shipboard measures only the phosphorus which has been completely regenerated, i.e. that present as inorganic phosphate ions. Treatment of a sample with strong oxidizing agents decomposes the organic matter completely, enabling the total phosphorus to be measured in ionic form. The difference yields the phosphorus present in organic combination. This fraction may be divided further by filtration into a portion present as particulate matter, including living microorganisms and debris, and a portion representing the phosphorus present in combination with dissolved organic molecules.

The total phosphorus content of a sample of sea-water may thus be separated into several fractions according to the following scheme:

| | |
|---|---|
| Total phosphorus ($P_t$) | Analysis after oxidation |
|   Inorganic phosphorus ($PO_4P$) | Direct analysis |
|     Phosphorus of oxidative origin | $P_{ox} = \text{A.O.U.}/276$ |
|     Preformed phosphorus | $P_p = PO_4P - P_{ox}$ |
|   Organic phosphorus | $P_t - PO_4P$ |
|     Dissolved organic phosphorus | } Separated by filtration |
|     Particulate organic phosphorus | |

The phosphorus of oxidative origin is merely the apparent oxygen utilization expressed as its equivalent in regenerated phosphorus. This notation is advantageous in reducing all aspects of the phosphorus cycle to common terms. The methods for the determination of phosphate concentration in sea-water are not precise, being subject to errors as large as 10%. Consequently, estimates of the fractions are only approximate and frequently require statistical treatment.

In the process of regeneration, dissolved organic phosphorus represents an intermediate product in the liberation of phosphorus of oxidative origin from the particulate phosphorus of living cells and their debris. In Fig. 5 the distribution of phosphorus among these fractions at various depths and seasons at a

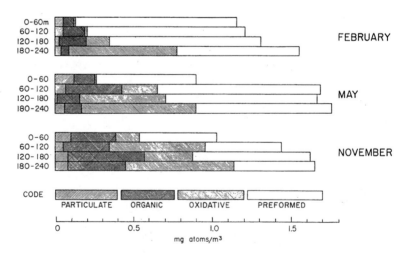

Fig. 5. Seasonal distribution of phosphorus fractions in Gulf of Maine. (From data of Redfield, Smith and Ketchum, 1937.)

station in the Gulf of Maine is shown. In spring, when active growth commences, most of the phosphorus is present in inorganic form; particulate phosphorus occurs chiefly in the photosynthetic zone (0–60 m) and very little dissolved organic phosphorus is present. During the summer and autumn the dissolved organic fraction becomes much greater and is distributed equally at all depths, providing evidence that phosphorus is carried downward by the sinking or migration of organisms. In winter the organic phosphorus decreases as the regeneration of inorganic phosphorus proceeds and in this form phosphorus is returned to the surface layer by the mixing of the water.

Measurements of the rate of phosphate regeneration in the laboratory indicate that up to one-half the total content of decomposing plankton appears in soluble form in the first day (Cooper, 1935; Seiwell and Seiwell, 1938). Vinogradov (1953) estimates that about half the nitrogen and phosphorus in algae is water-soluble. The remainder appears to exist in stable compounds that are decomposed after longer periods. These compounds appear in the water as the soluble organic fractions of phosphorus and nitrogen.

In the upper layers of the ocean, where organisms are decomposing in sufficient quantity, a substantial fraction of the phosphorus may be present in organic combination. Ketchum, Corwin and Keen (1955) found that nearly 50% of surface samples from the North Atlantic contained more than 0.25 mg atom/m³ of dissolved organic phosphorus. As depth increases, however, the proportions of samples in which statistically significant quantities of organic phosphorus were present diminished and in waters deeper than 1000 m none could be detected with certainty.

Rakestraw (1947) found that when samples of water from the oxygen minimum layer and from greater depths were left in the dark at the *in situ* temperature for nearly two years the oxygen content did not change after the first 50 days of storage. During the initial period oxygen was consumed, but not in excess of a few per cent of the total content, so that evidently the fraction of oxidizable organic matter in the deep water is very small.

These two lines of evidence justify the assumption that, in the deep ocean, the oxidation of organic matter has gone so nearly to completion that the unoxidized residues may be ignored for many practical purposes.

The regeneration of nitrate in sea-water is somewhat more complicated than that of phosphate. Dissolved nitrogen compounds are present in sea-water, Robinson and Wirth (1934) having found some 23 mg atoms/m³ of organic nitrogen in Puget Sound and 7 mg atoms/m³ in the off-lying Pacific Ocean. Nitrogen is released from organic combination as ammonia and is subsequently oxidized to nitrite, and then to nitrate. The steps in this process have been followed in laboratory experiments by von Brand and Rakestraw (1941). These experiments showed the successive appearance and disappearance of the fractions as decomposition proceeded. It required three or four months for the regeneration of nitrate to be completed.

In the sea, ammonia and nitrite appear as intermediate products at times and places where organic matter is decomposing in quantity. Ammonia is the first inorganic product in the regeneration of nitrate. In shallow coastal waters in temperate regions it is present in very small quantities at the end of winter, but increases as the season advances, and may become the principal form in which nitrogen is available in the water (Cooper, 1933). Under such conditions it may be absorbed directly by the phytoplankton without being oxidized to nitrite and nitrate (Harris, 1959). During the autumn and winter the ammonia decreases and nitrate becomes the major fraction of inorganic nitrogen in the water. In the deeper water of the Gulf of Maine ammonia appears in spring immediately below the photosynthetic zone and increases during the summer to occupy the entire water column (Redfield and Keys, 1938). In the deep sea ammonia does not ordinarily occur in significant quantities below the photosynthetic zone, except in anoxic basins.

Nitrite appears to be more evanescent than ammonia. Both fractions tend to accumulate below the euphotic zone where regeneration may be assumed to be intense. Under these conditions its concentration is much smaller than that of ammonia. In oceanic waters a second zone of increased nitrite concentration may be found at greater depths if oxygen is greatly reduced. This accumulation is attributed by Brandhorst (1959) to the reduction of nitrate by denitrifying bacteria.

Ammonia and nitrite are intermediate products of the nitrogen cycle which are present where organic matter is decomposing in quantity. In the deep sea where the increment of organic matter is small, relative to the rate of renewal of the water, the regeneration process runs nearly to completion and inorganic nitrogen is present almost entirely in the form of nitrate. Consequently, these intermediate products may frequently be neglected in examining the influence of organic activity on the chemistry of the deeper ocean waters.

There is some evidence that the regeneration of nitrate occurs more slowly than that of phosphate in shallow waters. This has been explained by the time required for bacteria to complete the several stages leading to the formation of nitrate (Atkins, 1926). It has also been observed that the maximum concentration of nitrate in the nutrient-rich layers of the ocean occurs at a greater depth than does that of phosphorus. This fact has been attributed to the more rapid release of phosphorus from the plankton as it sinks. The matter is of interest in the present connection since it provides a mechanism by which nitrogen and phosphorus may be fractionated and thus may explain the variations in the ratios of these elements in sea-water.

Frequently the minimum concentration of oxygen at stations in the deep ocean does not coincide in depth with that of the maximum concentration of phosphorus or nitrogen. Such deviations may arise because of the presence of preformed nutrients in the water or because the initial oxygen content of the water at different depths has been different owing to its temperature at the time when it sank below the surface. It is only when such influences have been eliminated that departures may be interpreted as indicating differences in the composition of the organic materials being oxidized at different depths. Using procedures to which this stricture does not apply, Riley (1951) has estimated the oxidative ratios for ocean water and has found them to vary substantially at different depths (see page 68). His investigations throw doubt on the strict application of the statistical composition of the plankton to estimations of the oxidative ratios of nitrogen and phosphorus at different depths in the sea.

### C. The Availability of Nutrient Elements in Sea-Water

In considering the chemical changes in sea-water produced by organic activity it is convenient to establish a norm from which variation in concentration can be measured. Table V has been prepared for this purpose. The norm for phosphorus is based on conditions observed in the deeper water of the Southern Ocean. Nitrogen is derived from this value employing the ratio of 15 : 1 proposed by Cooper (1937, 1938a) as a normal ratio from which anomalies may be measured. The carbon concentration is the value given by Sverdrup, Johnson and Fleming (1942) for water of 19‰ chlorinity and the oxygen saturation value corresponds to water of this chlorinity in equilibrium with the atmosphere at 2°C. Table V shows the ratios in which these elements are available in "average" sea-water and in which they are utilized by plankton in forming organic matter, and also the ratio of availability to utilization. A number of interesting generalizations may be developed from these estimates.

### a. Limiting factors

Liebig's law of the minimum implies that in the growth of a crop of plants, when other factors such as light and temperature are favorable, the nutrient available in smallest quantity relative to the requirement of the plant will limit the crop. The ratios presented in Table V define concretely the relation of availability of the several elements present in sea-water to their utilization. Nitrogen and phosphorus appear to occur in sea-water in just the proportions in which they are utilized by the plankton, a fact first noted by Harvey (1926)

TABLE V

Availability of Nutrient Elements in "Average" Sea-Water ($S = 34.7‰$, $T = 2°C$) and the Ratios of Their Availability and Utilization by Plankton

| | Availability in "average" sea-water | | Utilization by plankton | Ratio of availability to utilization |
|---|---|---|---|---|
| | mg atoms/m$^3$ | ratio | ratio | |
| Phosphorus | 2.3 | 1 | 1 | 1 |
| Nitrogen | 34.5 | 15 | 16 | 0.94 |
| Carbon | 2340 | 1017 | 106 | 9.6 |
| Oxygen saturation value | 735 | 320 | 276 | 1.16 |

who observed the simultaneous exhaustion of nitrate and phosphate in the waters of the English Channel during the growing season. Harvey's observations may be extended to the open ocean as shown in Fig. 2 and by other data reviewed by Redfield (1934).

In contrast, carbon is present in average sea-water in about ten times the quantity which can be utilized if growth is limited by the phosphorus or nitrogen available. This is illustrated in Fig. 3, which shows that the total carbonate in ocean water is reduced by only 10% with the exhaustion of the nitrate nitrogen. Clearly carbon does not become a limiting factor in the growth of marine plants in the sea. In a similar way small residues of phosphate or nitrate are frequently observed in surface waters from which the other nutrient has become exhausted by phytoplankton growth. This may be explained by small variations in the ratios in which these elements are available and are utilized by the population locally present. Thus, under the conditions specified in Table V, a residue of phosphorus amounting to 0.15 mg atom/m³ would remain in the water after the complete exhaustion of nitrogen.

A number of elements known to be of biological importance, such as iron, manganese, copper, zinc, cobalt and molybdenum, are available in sea-water in very small concentrations and might act as limiting factors in the growth of phytoplankton. By adding these elements to samples of surface water from the Sargasso Sea, Menzel and Ryther (1961) have shown that the addition of iron will stimulate the growth of the native plankton when it has come to an end after the exhaustion of the nutrients naturally present. Additions of other trace elements did not produce this effect. In similar experiments with water from the English Channel, Harvey (1947) found that manganese might act as a limiting factor in the growth of *Chlamydomonas*. Similar effects due to the shortage of vitamins and other organic constituents are discussed in Chapters 8 and 9. While it is clear that iron and manganese, and perhaps other essential elements and compounds, are present in sea-water in critically small quantities and may, under local conditions, become limiting, the fact that nitrogen and phosphorus become exhausted so completely over wide areas of the surface waters of the ocean is evidence that their concentrations are in general the factors which control the quantity of phytoplankton produced.

## b. Potential fertility

Harvey (1947a) has proposed that the total phosphorus or total combined nitrogen in natural waters may be used to distinguish their potential fertility. The potential fertility may be defined as the quantity of organic matter which could be produced by photosynthesis from a unit volume of sea-water if it were brought from depth to the surface and illuminated there until the limiting nutrients were exhausted. Assuming the carbon of the organic matter to be 50% of its dry weight it would follow from the information in Table V that the nitrogen would be exhausted when there had been produced organic matter of dry weight equal to 5.48 g/m³ and containing 2.74 g C/m³. Assuming a dry-weight/wet-weight ratio of 0.2, the wet weight of plankton would be 28 g in 1 million g of sea-water or about 1 part in 36,000.

This estimate serves to indicate the maximum standing crop that could occur on the average in the photic zone under ideally efficient conditions. Actually, the water becomes impoverished while it is at the surface by the sinking of organic matter so that for this and other reasons such crops are not to be expected. Ryther (1960) has estimated that the plankton of oceans as a whole contains 3 g/m² carbon. Assuming this to be concentrated in the upper 100 m, the wet weight of plankton in the water would be equivalent to about 1 part in 3 million. These estimates serve to emphasize the low concentrations at which organisms may be expected to occur in sea-water and the inefficiency with which the potential fertility of the oceans is realized.

### c. Available oxygen

The concentration of free oxygen in the water at the sea surface is determined by the solubility of the gas when in equilibrium with the atmosphere. The solubility is influenced greatly by temperature and to a less degree by salinity. In water which has sunk below the euphotic zone, the dissolved oxygen is removed by the oxidation of organic matter. The oxygen content of the deep water of the oceans is consequently reduced in proportion to the quantity of nutrients which has been regenerated within it.

The composition of average sea-water, shown in Table V, indicates that such water when equilibrated with the atmosphere at a temperature of 2°C, which approximates that of the deep water of the oceans, will contain 735 mg atoms/m³ of oxygen. This is only 16% more than the quantity required to oxidize all of the organic matter that could be formed from the limiting amounts of phosphorus and nitrogen present.

The concentrations of phosphorus found in the waters of the Pacific and Indian Oceans are more than twice those of the North Atlantic. The effect of this fact on the demand which the oxidation of organic matter may make on the available oxygen content of the water is illustrated in Table VI. The upper part of this table shows the excess oxygen remaining after the oxidation of all the organic matter which could be formed from the limiting quantity of phosphorus present. It may be seen that, in the Atlantic, oxygen is available in large excess and the water would be 53% saturated after oxidation had gone to completion. In the North Pacific, in contrast, more oxygen would be required than is available and the deep water would become anoxic under the assumed conditions.

As a matter of fact, the oxygen of the deep oceanic water is not as completely utilized as these estimates indicate. The reason is that only a part of the nutrients available in the surface waters of high latitudes, where these water masses are formed, is used to form organic matter at the time when the water sinks. The residue of preformed nutrients does not contribute to the utilization of oxygen. In the lower part of Table VI allowance has been made for the preformed phosphorus present in estimating the equivalent oxygen utilization. It

TABLE VI

Comparison of the Excess Oxygen Available in Deep Water of the Oceans
(mg atoms/m³)

|  |  | North Atlantic | North Pacific |
|---|---|---|---|
| Phosphorus | $PO_4P$ | 1.25 | 3.00 |
| Equivalent $O_2$ utilization | $276 \times PO_4P$ | 345 | 828 |
| $O_2$ saturation value | $100\% \ O_2$ | 735 | 735 |
| Excess $O_2$ | $100\% \ O_2 - 276 \times PO_4P$ | 390 | −93 |
| $O_2$ saturation |  | 53% | −13% |
|  |  |  |  |
| Preformed phosphorus | $P_p$ | 0.75 | 1.50 |
| P of oxidative origin | $P_{ox} = PO_4P - P_p$ | 0.50 | 1.50 |
| Equivalent $O_2$ utilization | $276 \times P_{ox}$ | 138 | 414 |
| Excess $O_2$ | $100\% \ O_2 - 276 \times P_{ox}$ | 597 | 321 |
| $O_2$ saturation |  | 81% | 44% |

may be seen that with this allowance the deep water of the North Atlantic should contain 81% of its original oxygen content and the Pacific about 44%. These estimates approximate the magnitudes actually observed.

These estimates only take account of the oxidation of organic matter which is carried to depth with the movement of the water. Organic matter which sinks from the euphotic zone to depth before decomposing creates an additional demand on the oxygen supply. This effect is important at intermediate depths and contributes to the formation of the oxygen minimum layer. In the Pacific large areas exist at such depths where the oxygen is very nearly exhausted.

The margin of safety against the development of anoxic conditions in the deep ocean is not large. It depends on the limited quantities of nutrients present in ocean water and the incomplete absorption of these elements from the water during its circulation through the euphotic zone.

### D. Anoxic Conditions

Where the accumulation of organic matter is great, the oxygen dissolved in the water becomes completely exhausted, leading to a condition variously designated as anaerobic, anaeric or anoxic. Under this condition, the oxidation of organic matter continues by means of anaerobic bacterial processes in which sulfate, nitrate, nitrite and carbon dioxide serve as hydrogen acceptors. The reduced products of these substances accumulate in the water in addition to the products of the oxidation of the organic matter. The result is to modify the proportions of the components of the water in ways which differ from those characteristic of regeneration in the presence of free oxygen.

The principal reactions taking place in the presence and in the absence of oxygen may be formally represented by the following equations, in which

$CH_2O$ represents the carbon compounds of the organic matter and $NH_3$ the nitrogen liberated by its oxidation.

Oxidation by oxygen:

$$CH_2O + O_2 \longrightarrow CO_2 + H_2O$$
$$NH_3 + 2O_2 \longrightarrow HNO_3 + H_2O$$

Denitrification:

$$5CH_2O + 4HNO_3 \longrightarrow 5CO_2 + 2N_2 + 7H_2O$$
$$5NH_3 + 3HNO_3 \longrightarrow 4N_2 + 9H_2O$$

Sulfate reduction:

$$2CH_2O + H_2SO_4 \longrightarrow 2CO_2 + H_2S + 2H_2O$$
No oxidation of $NH_3$

The order and extent to which these steps proceed depends on the free energy available from the respective reactions and on the concentration of the reactants. The free energy decreases in the order oxygen > nitrate > sulfate > carbonate when these serve as hydrogen acceptors (McKinney and Conway, 1957). Consequently, oxygen should be utilized first if it is available. With the exhaustion of oxygen, the oxidation of organic matter should continue by the reduction of the nitrate produced while oxygen was available. Following the exhaustion of the nitrate, oxidation continues by the reduction of sulfate. The free energy available from sulfate reduction is apparently insufficient to oxidize $NH_3$ and, consequently, this product should accumulate during sulfate reduction. Thus the oxidation of organic matter may be expected to proceed in three stages—oxidation by free oxygen, denitrification and sulfate reduction—separated by the exhaustion of free oxygen and of nitrate. In nature, these stages may overlap to some extent. Nitrite is produced and should be present as an intermediate product both of the formation of $NO_3$ and of its reduction.

The limiting concentrations of oxygen below which denitrification occurs in sea-water are not known. In fresh water nitrate reduction takes place only when the oxygen concentration falls below a few tenths of a milligram per liter, depending on the nitrate concentration (Langley, 1958). Relatively large concentrations of nitrite (up to 2.5 mg atoms $N/m^3$) occur below the thermocline in parts of the Pacific where the oxygen is depleted, but only if the oxygen concentration is less than 1.0 ml/l. The presence of nitrite in these places is attributed to the reduction of nitrates as the initial step in denitrification (Brandhorst, 1959).

Free nitrogen appears to be the final product of denitrification. In the Black Sea, the concentrations of gaseous nitrogen increase with depth and supersaturation occurs in the deeper water (Kriss, 1949). In the anoxic water of the Cariaco trench and of the Dramsfjord, Richards and Benson (1961) have examined the accumulation of free nitrogen in excess of that expected from its solubility by comparing the observed nitrogen/argon ratios with those in sea-water equilibrated with air. The excess quantities of nitrogen found correspond approximately with the amounts estimated to be produced by the decomposition of organic matter as indicated by changes in the oxygen, sulfide and phosphorus content of the water. They found that nitrate and nitrite were present in mere traces in, or were absent from, these anoxic waters. Ammonia, on the other hand, was present and varied in concentration in proportion to the accumulation of sulfides. This observation indicates that sulfate does not act as a hydrogen acceptor in oxidizing the ammonia liberated from organic matter during sulfate reduction.

The most conspicuous change in the composition of sea-water produced under anoxic conditions is the appearance of hydrogen sulfide. The principal source of sulfide sulfur in marine waters is the reduction of sulfates rather than the sulfur of decomposing organic matter. In the Cariaco Trench, the presence of 24 mg atoms/m³ of sulfide sulfur is accompanied by an accumulation of only one-tenth this amount of phosphate phosphorus. If both elements had been derived from the decomposition of plankton, its organic matter must have contained ten times as much sulfur as phosphorus. Data on the sulfur content of plankton are lacking, but in the flesh of fish the sulfur present is somewhat less than the phosphorus (Vinogradov, 1953). Consequently, it is unlikely that much of the sulfide has been derived from the sulfur content of decomposing organisms.

Data from the Black Sea demonstrate in a more positive way that sulfate is the source of the sulfide present in this anoxic basin. Skopintsev (1957) and Skopintsev $et$ $al.$ (1958) have found that as the sulfide concentration increases with depth, the ratio of sulfate to chlorinity decreases, indicating that sulfate

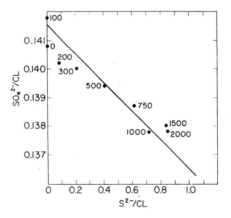

Fig. 6. Relation of sulfide/chloride and sulfate/chloride ratios in waters of the Black Sea. Numbers indicate depths of samples. Slope of line corresponds to $\Delta S^{2-}/\Delta SO_4^{2-} = -1.0$. (From data of Skopintsev $et$ $al.$, 1958.)

has disappeared from the water. Fig. 6 shows that these changes occur in approximately the ratio required if one sulfide ion is formed from each sulfate ion which disappears. By taking the accumulation of phosphorus as an index of the quantity of organic matter which has been oxidized in the water of the

Cariaco Trench and Black Sea, Richards and Vaccaro (1956) have shown that organic decomposition proceeds in proportion to the apparent oxygen utilization until the supply of oxygen is exhausted. After this, the phosphorus continues to increase at a rate which indicates that in its regeneration each sulfide ion is equivalent to approximately four oxygen atoms in accordance with the stoichiometry of the equation for sulfate reduction.

The reduction of carbon dioxide to methane is well known to occur under anoxic conditions in freshwater systems, where it gives rise to the production of marsh gas. We know of no direct evidence that this takes place in sea-water. Perhaps this is because sulfate is present in sea-water in excess of the amounts required for the oxidation of the organic matter.

Carbon dioxide is produced by the oxidation of organic matter by any of the hydrogen acceptors. Because the accumulation of organic matter is greater in anoxic basins, its decomposition produces larger changes in the total concentration than is usual in ocean waters. In the Black Sea, the total carbon present as $HCO_3^-$ and $CO_3^{2-}$ increases from 3275 mg atoms/m³ in the surface water to 4263 mg atoms/m³ at a depth of 2000 m (see Fig. 7). This is an increase of 30%.

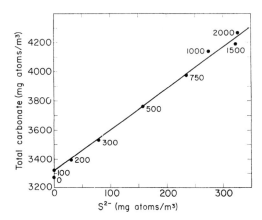

Fig. 7. Relation of sulfide sulfur and total carbonate carbon in waters of the Black Sea. Numbers indicate depth of samples. Slope of line corresponds to $\Delta S^{2-}/\Delta C = 0.36$. (From data of Skopintsev et al., 1958.)

In contrast, the oxidation of organic matter in the depths of oceanic waters increases the total carbonate by only 10%. It follows that the quantity of organic matter which has decomposed in the deep water of the Black Sea is about three times that characteristic of the open ocean basins.

The change in total carbonate carbon with increasing depth is proportional to the change in sulfide content and proceeds in the ratio $\Delta S/\Delta C = 0.36$ (see Fig. 7). This is somewhat less than the ratio of 0.5 given by the equation for sulfate reduction.

When separate components are considered, the changes in chemical composition of sea-water in anoxic basins thus appear to proceed in about the proportions required by the equations shown on page 43. If it is assumed that decomposed organic matter liberates phosphorus and carbon in the ratio of 1 to 106, the expected ratios of the simultaneous changes in the several components during each step may be estimated from these equations to be as shown in Table VII.

TABLE VII

Ratio of Change in the Atomic Concentration of Products of Decomposition of Organic Matter in the Presence of Oxygen and Under Anoxic Conditions

| | $\Delta O$ | $\Delta CO_2\text{-}C$ | $\Delta NO_3\text{-}N$ | $\Delta N_2\text{-}N$ | $\Delta NH_3\text{-}N$ | $\Delta S^{3-}$ | $\Delta P$ |
|---|---|---|---|---|---|---|---|
| Oxygen present | $-276$ | 106 | 16 | 0 | 0 | 0 | 1 |
| Oxygen absent | | | | | | | |
|    Denitrification | 0 | 106 | $-70.7$ | 86.7 | 0 | 0 | 1 |
|    Sulfate reduction | 0 | 106 | 0 | 0 | 16 | 53 | 1 |

By employing these ratios, the changes in concentration of the products of decomposition during the successive stages may be computed and the results shown as a function of the increasing phosphate phosphorus concentrations, as in Fig. 8. It is necessary to know only the initial concentration of oxygen in the water to obtain quantitative values applicable to any natural situation. Such diagrams are not very exact in respect to the products of denitrification because the separate stages probably overlap, and the completeness of the reactions involving these fractions is not understood. The presence of nitrite as

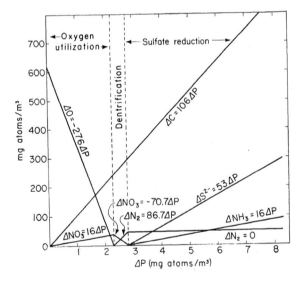

Fig. 8. Simultaneous changes in concentration of products of decomposition of organic matter under oxic and anoxic conditions. Calculated assuming initial oxygen content of 615 mg atoms/m³, appropriate to surface water of Black Sea.

an intermediate product and the possibility that preformed nitrate undergoes denitrification are also neglected.

In Table VIII, the concentrations of various products of decomposition observed in the deep water of the Black Sea and of the Cariaco Trench are compared with the concentrations calculated in this way. The agreement is sufficiently good to indicate that the conditions observed in anoxic basins have resulted in the main from the processes defined by the equations and that the quantitative relations depend on the elementary composition of the plankton.

TABLE VIII

Comparison of Observed and Calculated Values for Concentration of Products of Decomposition in Anoxic Basins, mg atoms/m³

| Depth<br>Oxygen at surface | Black Sea<br>750 m<br>615 | | Cariaco Trench<br>915 m<br>471 | |
| --- | --- | --- | --- | --- |
| Concentrations at depth | Observed | Calcd. | Observed | Calcd. |
| $PO_4$-P | 7.08 | — | 2.42 | — |
| $CO_3$-C | 696 [a] | 750 | — | 257 |
| $NO_3$-N | — | 0 | — | 0 |
| $N_2$-N | 44 | 43.8 | 18.6 [a] | 33.6 |
| $NH_3$-N | 52.8 | 69.4 | 10.6 | 5.1 |
| $S^{2-}$-S | 235 | 230 | 20.4 | 17.0 |

[a] Corrected for preformed carbonate and nitrogen of surface water.

Data from Danil'chenco and Chigirin (1929, 1929a), Skopintsev et al. (1958), Richards and Vaccaro (1956) and Richards and Benson (1961).

Sulfate is one of the major constituents of sea-water and is present in quantities sufficient to oxidize more than one hundred times the amount of organic matter than can be formed from the phosphorus and nitrogen available. The sulfate reduced under anoxic conditions in the Black Sea is less than 5% of the content of the surface water (see Fig. 6). Under natural conditions in marine waters, sulfate reduction is probably limited only by the amount of organic matter which accumulates. This need not be true of fresh waters in which the available sulfate is more readily exhausted.

The oxygen content of the atmosphere is thought to have been produced in large part by the photosynthetic reduction of carbon dioxide. It has recently been suggested that the reduction of sulfates in anoxic basins or in marine sediments may have contributed substantially in making oxygen available to the atmosphere (Redfield, 1958). The supposition is that sulfides produced in the process of sulfate reduction have been entrapped in the sediments and that the carbon dioxide released at the same time remained in solution and, on ultimate reduction by photosynthesis, supplied oxygen to the atmosphere.

The composition of sedimentary rocks indicates that quantities of oxygen equivalent to those now present in the atmosphere may have been liberated in this way. Since this process would operate only to the extent that the supply of oxygen was inadequate, it may have served to regulate the composition of the atmosphere.

### E. Silicon in the Biochemical Cycle

As a nutrient silicate silicon differs from the inorganic compounds of nitrogen and phosphorus in not being a universal requirement of living matter. It is present, however, in large quantities in the tests of diatoms, which dominate the phytoplankton in the cooler ocean waters. Analyses of diatoms show Si/P ratios varying between 16 and 50, depending on the species (Vinogradov, 1953). Silicon is transported from the surface to the depths by the sinking of these organisms and is liberated there in the course of regeneration.

Studies on the seasonal changes in nutrient concentrations in the water of the English Channel show that the silicate cycle corresponds in many ways with the cycle of phosphorus; silicate and phosphate being depleted or regenerated simultaneously (Atkins, 1930).

The relationship of the distribution of silicate to other elements involved in the biochemical cycle is less obvious in the open oceans than in locally restricted waters. However, Richards (1958) has demonstrated a linear relationship in the simultaneous changes in the concentration of silicate silicon, inorganic phosphate and nitrate nitrogen in much of the water column at stations in the Western Atlantic. The ratios of change were approximately $\Delta O : \Delta Si : \Delta N : \Delta P = -272 : 15 : 16 : 1$. In these waters, silicon appears to enter the biochemical cycle in about the same proportions as nitrogen.

In the oceans, in general, the concentrations of silicate vary greatly in their proportion to the phosphate and nitrate present. Such variation arises from the fact that in different parts of the oceans the proportions of diatoms to other phytoplankton, which does not require silicon, differ greatly. Consequently, the statistical composition of the plankton is variable in respect to silicon. Moreover, the solution of silicate from the diatom tests may be expected to follow a different course from the regeneration of nitrogen and phosphorus, which is related more directly to the oxidation of the organic matter. Consequently, the silicate may be set free at different depths than are nitrate and phosphate. Finally, animals which feed on phytoplankton have no use for silicon. The tests of diatoms will be rejected and will tend to sink to dissolve at depth while nitrogen and phosphorus will be retained by the animal and be regenerated from its excretions and decomposition products in the upper layers of water.

For these reasons, the fractionation of silicate by biological action probably follows a different course from that of the other nutrients. A better analysis of the factors responsible for the distribution of silicate in the oceans might add greatly to our understanding of the biochemical circulation.

## 3. The Biochemical Circulation

The selective absorption of certain elements by organisms, combined with the movement of organized matter from one layer of water to another by sinking or active migration, provides a mechanism for the fractionation of the components of sea-water and the redistribution of these elements in characteristic patterns. The circulation of nonconservative elements and the resulting pattern of their distribution depends on biological as well as hydrographic processes and they thus differ in detail from that of water and its inert solutes.

The concentration of a conservative element at any point in the sea is determined by a dynamic equilibrium between the effects of eddy diffusion and advection. In the case of nonconservative elements the effects of biological processes on the equilibrium must also be taken into account. The theory of this interaction, as developed by Sverdrup (1938), is essentially as follows.

Consider the water in a unit cube of space within the sea in which the concentration of some nonconservative property is denoted by $N$. The value of $N$ is subject to change by biological processes, such as the assimilation or regeneration of nutrients, the consumption of oxygen, etc., which change $N$ at a rate denoted by $R$. In addition, $N$ will be altered by interchange of water with adjacent cubes if they contain water with different concentrations of $N$. These interchanges are of two sorts, eddy diffusion and advection.

If the interchanges along the $x$-axis are considered, the effect of eddy diffusion on the values of $N$ is given by $(A_x/\rho)\ \partial^2 N/\partial x^2$ in which $A_x$ is the coefficient of diffusion in the $x$-axis and $\rho$ is the specific gravity of sea-water. The effect of advection is given by $V_x \cdot \partial N/\partial x$, in which $V_x$ is the component of current velocity in the direction $x$. Similar expressions define the effect of water movement in the $y$- and $z$-axes, and the complete expression is

$$\frac{\partial N}{\partial t} = R + \frac{A_x}{\rho} \cdot \frac{\partial^2 N}{\partial x^2} + \frac{A_y}{\rho} \cdot \frac{\partial^2 N}{\partial y^2} + \frac{A_z}{\rho} \cdot \frac{\partial^2 N}{\partial z^2} - V_x \cdot \frac{\partial N}{\partial x} - V_y \cdot \frac{\partial N}{\partial y} - V_z \cdot \frac{\partial N}{\partial z}. \quad (1)$$

In its complete form the equation is too complicated to be readily usable. With various assumptions it may be simplified and applied to the analysis of oceanographic observations. Since oceanographic data commonly consist of measurements at discrete points, practical application is most easily made by expressing the relations in terms of Eulerian equations in finite difference form. The papers of Riley (1951, 1956, 1956a) may be consulted for a discussion of its application.

The utility of equations for the dynamic equilibrium affecting the concentration of nonconservative properties is of three sorts. When the physical terms can be evaluated the expression may be used to determine the rates of the biological processes expressed by $R$. Conversely, if the value of $R$ can be evaluated from biological data, information on the physical terms may be obtained. When, as is often the case, data are insufficiently numerous or precise to enable quantitative estimates to be obtained, the equations provide a guide to the intuitive reasoning with which observed conditions are interpreted.

The dynamic equilibria obtaining in a series of situations of increasing complexity will be discussed and illustrated in the following pages. These will be (1) cases in one dimension in which the distribution of nonconservative elements results from a balance between the effects of biological activity and vertical diffusion; (2) cases in two dimensions in which advective motion in one horizontal direction must also be considered; and (3) cases in which motion in all three dimensions is taken into account.

### A. Vertical Exchange under Steady-State Conditions

In many situations, the effect of horizontal motions on the composition of sea-water is negligible. This is obviously true where the boundaries of a basin prevent substantial advection, as in an enclosed basin. It happens frequently that conditions are uniform over great horizontal distances, or that the horizontal velocities are very small, in either of which case the influence of advection is small and may be disregarded. If, in addition, conditions do not change with the season so that a steady state exists, in which $\partial N/\partial t = 0$, the dynamic balance on which the distribution of nonconservative elements depends is reduced to the terms defining the rate of change in concentration due to the sinking and regeneration of organic matter, and that due to the vertical diffusion of the regenerated elements toward the surface.

### a. Stagnant basins

Deep basins in which shallow sills limit the horizontal circulation to a layer near the surface provide the simplest and the most extreme cases of the accumulation of nonconservative elements under the influence of the vertical circulation.

The Cariaco Trench, described by Richards and Vaccaro (1956), is an example of such a basin. It is a depression about 1400 m deep in the continental shelf off Venezuela. The water of the Caribbean flows freely across this depression down to the sill depth at 150 m and is not obviously influenced by the conditions within the depression. Below 250 m the temperature and salinity are nearly uniform and show that the water is isolated from the Caribbean water of comparable depths and can exchange with its surroundings only by eddy diffusion in the vertical across the transition zone separating it from the water above sill depth. In the depths of the Cariaco Trench the total phosphorus concentration is 2.6 mg atoms/m³; at sill depth it is 1.4 mg atoms/m³. Consequently 1.2 mg atoms/m³ have accumulated in the deep water. Water at sill depth, if carried downward by eddy diffusion, would contain only enough oxygen to produce 0.4 mg atom/m³ phosphorus of oxidative origin. Consequently the oxygen has been completely exhausted by oxidation of the excess of organic matter settling from the surface layer. The remaining organic matter has been oxidized by oxygen derived from the reduction of nitrate and sulfate as discussed on page 43.

The accumulation of phosphorus in the depths of the basin may be considered to result from a balance between the gains from the regeneration of phosphate from sinking organic matter and the losses by eddy diffusion. The steady-state equilibrium may be expressed by

$$\frac{\partial Q_R}{\partial t} - A \cdot \frac{\partial N}{\partial z} = 0 \tag{2}$$

in which $\partial Q_R/\partial t$ is the rate at which the quantity of phosphorus in the water below unit area of the transition zone is increased by regeneration, $A$ is the coefficient of diffusion and $\partial N/\partial z$ is the gradient of phosphorus concentration in the transition zone.

Measurements of the assimilation of carbon in the euphotic zone over the Cariaco Trench indicate that 0.48 g C/m² is absorbed per day. This is equivalent to the production of organic matter containing $4.4 \times 10^{-10}$ mg atom/cm² of phosphorus each second. How much of this organic matter sinks below sill depth is unknown, but Riley (1951) estimates that in the sea only about one-tenth of photosynthetic production sinks to decompose at depths greater than 200 m. Consequently we may assume for purposes of illustration that $\partial Q_R/\partial t = 0.44 \times 10^{-10}$ mg atom/cm² sec. The hydrographic data obtained by Richards and Vaccaro show that at sill depth the gradient in total phosphorus concentration, $\partial N/\partial z$, is $0.75 \times 10^{-10}$ mg atom/cm³ cm. Applying these values to equation (2) the coefficient of eddy diffusion $A$ is found to be approximately 0.6 g/cm/sec. This estimate agrees well with values for intermediate depths in the Atlantic Ocean obtained by Riley (1956) by procedures to be discussed in the following section (Table XI).

This example illustrates how a physical characteristic of the motion of water may be evaluated by the use of biochemical information.

## b. The half-life period

Estimates of the average time a constituent of the water remains below sill depth in a stagnant basin may be made by dividing the content of the basin, $x$, by the rate at which the content is changed by some measured process, $dx/dt$. The time is appropriately expressed as the half-life period, $\tau$, which equals $0.693x/(dx/dt)$. The content of the basin is given by $N \cdot h$, where $N$ is the *mean* concentration of $N$ below sill depth and $h$ is the depth of water below this level. The content is changed at a rate given by $dQ_R/dt$ or by $A \cdot dN/dz$, which are equal according to equation (2). Consequently,

$$\tau_N = 0.693 \frac{N \cdot h}{dQ_R/dt} \quad \text{or} \quad 0.693 \frac{N \cdot h}{A \cdot dN/dt}.$$

In the case of the Cariaco Trench, $N = 2.4 \times 10^{-6}$ mg atom P/cm², $h = 1.2 \times 10^5$ cm and $dQ_R/dt = 0.44 \times 10^{-10}$ mg atom P/cm² sec. Consequently the half-life period of phosphorus below sill depth, $\tau_N$, is $0.45 \times 10^{10}$ sec or 142 years. This estimate is subject to error in the assumption that one-tenth of the organic matter produced in the euphotic zone sinks and decomposes below sill depth. As an expression of the average for all depths in the water column, it undoubtedly overestimates greatly the rate of exchange of materials in the deeper layers of the basin. It serves to indicate, however, how very slowly exchange takes place in a stagnant basin of this sort.

It is evident that accumulation of a constituent of sea-water can occur locally only if the constituent circulates through the region more slowly than does the water. If the constituent circulated at the same rate as the water, there would be no change in its concentration. This consideration is overlooked when estimates of the exchange of water are based on the accumulation of nonconservative elements.

The principle involved may be illustrated by a simplified model of a stagnant basin in which a sharp transition zone at sill depth separates an upper layer of water containing a nonconservative element in the concentration $N_U$ from the stagnant lower layer having a concentration $N_L$. The process of eddy diffusion is equivalent to the movement of equal quantities of water upward and downward, each carrying with it quantities of the nonconservative element in proportion to its concentration in the layer from which it comes. If $dx/dt$ is the rate at which water is exchanged, $N_U\, dx/dt$ and $N_L\, dx/dt$ are the rates at which $N$ is transported downward and upward respectively and $(N_L - N_U)\, dx/dt$ is the rate at which $N$ is lost from the lower layer. Under steady-state conditions the fraction of $N$ lost from the lower layer in unit time is $(N_L - N_U)\, dx/N_L \cdot h$ while the fraction of water lost is $dx/h$. Consequently, the fraction of $N$ lost in unit time equals $(N_L - N_U)/N_L$ times the fraction of water lost. Since $(N_L - N_U)/N_L$ is less than one, the fraction of $N$ lost in unit time is less than the fraction of water lost. The half-life period, $\tau$, is inversely proportional to the fraction lost in unit time. Therefore,

$$\tau_{\text{water}} = \frac{N_L - N_U}{N_L} \cdot \tau_N.$$

In the case of the Cariaco Trench the concentration of phosphorus immediately above sill depth is 1.2 mg atoms/m³ while the mean concentration below sill depth is 2.4 mg atoms/m³. If these values be taken for $N_U$ and $N_L$ respectively, $(N_L - N_U)/N_L = 0.5$. $\tau_N$ has been estimated to be 142 years and consequently $\tau_{\text{water}}$ is 71 years. In other words, the water is circulating through the basin twice as fast as is phosphorus. The estimate is not precise because of the simplification of the model but serves to illustrate the principle that accumulation occurs when a constituent circulates through a region more slowly than does the water.

### c. The relation of stagnation to stability

Equation (2) implies that for a given rate of regeneration, and under steady-state conditions, the gradient in concentration, $dN/dz$, resulting from the accumulation of a nonconservative element at depth, will be inversely proportional to the coefficient of diffusivity, $A$. This coefficient in turn tends to vary inversely with the stability of the water column, and consequently the accumulation of the products of regeneration may be expected to be greater where strong density gradients are present.

Richards and Vacarro (1956) have investigated this relation in a number of basins, using the development of anoxia, shown by the concentration of hydrogen sulfide, as a criterion of the accumulation of products of organic decomposition. Table IX shows that the expected relation obtains between the stability of the water column at sill depth, the development of anoxic conditions, and the estimated age of the water in the basin.

TABLE IX

Comparison of Stability, Accumulation of Hydrogen Sulfide and Age in Stagnant Basins (after Richards and Vaccaro, 1956)

|  | Sill depth, m | Stability at sill depth, $\Delta\sigma_t/m$ | $H_2S$-S, mg at/$m^3$ | Age estimate, years |
|---|---|---|---|---|
| Catalina Basin | 982 | $31 \times 10^{-5}$ | nil | — |
| Santa Monica Basin | 737 | 44 | nil | — |
| Santa Barbara Basin | 475 | 88 | nil | 2–20 |
| Kenoe Bay | 40–50 | 520 | 13.4 | — |
| Cariaco Trench | 146 | 440 | 30 | 100 |
| Black Sea | 40–50 | 3000 | 300 | 5600 |

### B. Seasonal Variation in Vertical Exchange

Under steady-state conditions the distribution of concentrations provides no information on the rates of the processes on which these concentrations depend. In the previous example a measurement of the rate of production of organic matter was used to introduce the factor of time into the calculations. The coefficient of diffusion could have served this purpose were its value known. Where the equilibria change with the season, the resulting changes in concentration may be employed to evaluate the rates at which the component processes take place. A study of the seasonal exchange of oxygen across the sea surface, made in the Gulf of Maine by Redfield (1948), will serve to illustrate this procedure, and, in addition, provides an example of the use of stoichiometrical relations in separating changes due to biological and physical processes.

### a. The exchange of oxygen across the sea surface

The quantity of oxygen dissolved in water beneath the sea surface depends on a balance between the rate at which it is exchanged with the atmosphere and the rate at which it is produced or consumed by biological activity, provided the effects of advection may be neglected. Changes in the quantity of dissolved oxygen, $\Delta Q_O$, during the time interval $\Delta t$ consequently result from an equilibrium represented by

$$\frac{\Delta Q_O}{\Delta t} = \frac{\Delta Q_R}{\Delta t} + \frac{\Delta Q_E}{\Delta t} \tag{3}$$

in which $\Delta Q_R/\Delta t$ is the rate of change in quantity due to biological activity and $\Delta Q_E/\Delta t$ is the rate at which oxygen is exchanged with the atmosphere across the sea surface. $\Delta Q_O/\Delta t$ was obtained from hydrographic data by integrating the quantity of dissolved oxygen underlying a square meter of sea surface to a depth of 200 m (below which the changes in oxygen concentration were negligible) at two periods separated by the interval $\Delta t$ and taking the difference. $\Delta Q_R/\Delta t$ was obtained by a similar integration of the inorganic phosphorus concentration, and multiplying the result by the oxidative ratio of phosphorus. If concentrations are expressed in mg atoms, $\Delta Q_R/\Delta t = -276\ \Delta P/\Delta t$. The mean rate of diffusion of oxygen across the sea surface, $\Delta Q_E/\Delta t$, was obtained by introducing these measurements into equation (3).

The results of the study are illustrated in Fig. 9. During the winter period oxygen entered the water at a greater rate than it was reduced by organic activity. As a result the quantity of oxygen in the water column increased.

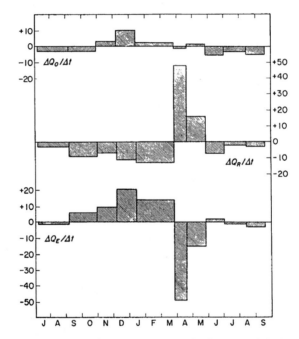

Fig. 9. Seasonal cycle in change in oxygen content of water of Gulf of Maine. $\Delta Q_O/\Delta t$, rate of change of oxygen content of water column; $\Delta Q_R/\Delta t$, rate of change attributed to biological activity; $\Delta Q_E/\Delta t$, rate of change attributed to diffusion across sea surface. Ordinates: rate of change in units of $10^4$ ml per m$^2$ × months. Abscissa: time in months. (Recalculated from data of Redfield, 1948, using oxidative ratio $\Delta O/\Delta P = -276$.)

This was possible because of the increased solubility which accompanied the cooling of the water. In the spring, when the growth of phytoplankton is intense, oxygen was produced more rapidly than it was reduced by organic activity. Oxygen escaped to the atmosphere at a nearly equivalent rate during this period with the result that there was little change in the content of the water column. During the summer the oxygen content of the water decreased at a rate which corresponded closely to the rate of consumption by biological activity. Consequently the exchange with the atmosphere was small during this period.

These estimates show that the changes in the oxygen content of the water are small compared to the processes on which they depend. Thus, during the period between 4 September and 22 March, about $84 \times 10^4$ ml/m$^2$ of oxygen entered the water. The change in the oxygen content of the water column during this time was only one-fourth of this amount, three-quarters being consumed in the decomposition of organic matter. This demonstrates the degree to which the oxygen content of the water is stabilized by exchanges with the atmosphere. The quantity of oxygen crossing the surface of the Gulf of Maine annually is equivalent to that in a layer of air about 4 m in thickness.

The rate of diffusion of oxygen across the sea surface may be expected to vary with the difference in oxygen pressure in the atmosphere and in the surface water, according to the expression $\Delta Q_E / \Delta t = E(P - p)$. $P$ is the partial pressure of oxygen in the atmosphere, and $p$ its tension in the water. $p$ was determined from the degree of saturation of samples drawn from a depth of one meter. $E$ is the exchange coefficient of oxygen.

It was found that the pressure head of oxygen $(P - p)$ was positive during the winter and negative during the remainder of the year, as required by the sign of $\Delta Q_E / \Delta t$. The estimated values of the exchange coefficient $E$ are about $20 \times 10^6$ ml per m$^2$ per month per atmosphere during the winter period and agree in magnitude with estimates of this constant based on laboratory experiments. The values for other periods are not satisfactory and suggest that water samples drawn from a depth of one meter do not give a correct indication of the tension of oxygen in the water at the surface at times when the surface water is stable.

### b. Vertical exchange as a function of depth and season

The factors on which the concentration of nonconservative elements in a vertical water column depend vary greatly with depth. In the sub-surface water of the euphotic zone the growth of phytoplankton tends to reduce the concentration of nutrient elements while at greater depths regeneration tends to increase their concentration in proportion to the quantities of organic matter sinking to decompose at each level. The motion of the water on which the vertical diffusion coefficients depend also varies with depth. The procedures which may be employed to determine the rate of biological activity at different depths and of the vertical diffusion required to maintain the observed concentrations as they varied with the season are illustrated by an investigation of these factors in Long Island Sound made by Riley (1956, 1956a).

For the purpose of the analysis the water column underlying a unit area was divided into a series of segments of which the length, $\Delta z$, was equal, except that the water immediately below the surface and above the bottom was included in volume units of one-half this length (see Fig. 10). Characteristic values for the temperature and phosphorus concentration based on hydrographic data obtained at stated intervals were assigned to the mid-depth of each segment, and to the water at the surface and bottom. These values were considered to represent the mean temperature, $T$, and mean phosphorus concentration, $N$, of the segment as a whole. It was assumed that horizontal movements of water were negligible in their effect on this distribution of properties.

In the case of the full segments the rate of change in $N$ in the interval $\Delta t$ depends on the balance between the rate of change due to biological activity, $R$, and that due to vertical diffusion across the upper and lower boundaries. Using the subscripts $z$ and $-z$ to distinguish properties of lower and upper boundaries respectively, the equilibrium may be expressed as

$$\frac{\Delta N}{\Delta t} = R + \frac{1}{\Delta z}\left(A_z \cdot \frac{\Delta N_z}{\Delta z} - A_{-z} \cdot \frac{\Delta N_{-z}}{\Delta z}\right). \tag{4}$$

In the case of the segments immediately below the surface and above the bottom, exchange takes place through only one surface and the appropriate diffusion term in (4) is omitted. Rearranging so that each term expresses a quantity rather than a concentration, (4) becomes

$$\frac{\Delta N \cdot \Delta z}{2\Delta t} = R \cdot \Delta z + A_z \cdot \frac{\Delta N_z}{\Delta z} \tag{4a}$$

for the sub-surface segment, and

$$\frac{\Delta N \cdot \Delta z}{2\Delta t} = R \cdot \Delta z - A_{-z} \cdot \frac{\Delta N_{-z}}{\Delta z} \tag{4b}$$

for the segment above the bottom.

In solving these equations for a given segment, $\Delta N/\Delta t$ is obtained from the difference in the mean concentration of $N$ in the segment determined at two times separated by the interval $\Delta t$. $\Delta N/\Delta z$ is the average of the values of the gradients in concentration across the boundary in question at these two times, each of which is given by the difference in $N$ assigned to the adjacent segments divided by the distance between their centers.

Before these equations could be used to evaluate $R$ for each segment, it was necessary to know the value of the diffusion coefficient at the depth of each bounding surface. These values were estimated from the distribution of temperature, $T$. The flux of heat across each boundary was expressed by a modification of (4b) in which $N$ is replaced by $T$, $R$ is omitted because the temperature is uninfluenced by biological activity, and $\Delta z$ is replaced by $h$, to express the length of the water column between the bounding surface in question and the bottom. Thus

$$\frac{\Delta T \cdot h}{\Delta t} = -A_{-z} \cdot \frac{\Delta T}{\Delta z}.$$

After solving this equation for the value of the diffusion coefficient $A$ at each bounding surface, the information necessary to obtain the value of $R$ characteristic of each segment was available.

As an example of the information obtained by this procedure on the processes determining the phosphate concentrations at depth, a balance sheet of the average rates of these processes in a layer of water in the euphotic zone during a summer period is given in Table X. The phosphate phosphorus

TABLE X

Balance of Factors Changing the Concentration of Phosphorus in Water of Long Island Sound between Depths of 2.5 and 7.5 m, and during Period May 21–August 19. (Data from Riley, 1956a.)

|  | mg atoms/m$^3$ per day |
| --- | --- |
| Diffusion from layer below | $4.8 \times 10^{-2}$ |
|     „        to layer above | $-1.3 \times 10^{-2}$ |
| Net change from diffusion | $3.5 \times 10^{-2}$ |
| Change due to biological action | $-3.0 \times 10^{-2}$ |
| Change in concentration in water | $0.5 \times 10^{-2}$ |

concentration in the water increased during the period from 0.51 to 1.0 mg atom/m$^3$ or at a mean rate of about $0.5 \times 10^{-2}$ mg atom/m$^3$ per day in spite of a net loss, due to the excess of photosynthesis over regeneration, at a rate of $3.0 \times 10^{-2}$ mg atom/m$^3$ per day. This was possible because of the excess rate at which regenerated phosphorus was returned to the layer by vertical diffusion. Of the phosphate diffusing into the layer from below about three-quarters remained in the layer to effect the balance while one quarter moved on into the layer above.

The rapid turnover which these exchanges produce is brought out by the consideration that the initial concentration was only 0.51 mg atom, while the biological consumption was $3.0 \times 10^{-2}$ mg atom per day. At this rate the phosphate in the layer would have been completely exhausted in 17 days had it not been renewed by vertical diffusion.

The result of Riley's estimates of the intensity of biological activity in Long Island Sound at various depths and seasons is presented in Fig. 10. They show that, in the summer period, the absorption of phosphorus by growing plankton exceeds regeneration down to a depth of 12.5 m. Below this depth regeneration is in excess and takes place mainly in the layer immediately above the bottom. In spring the total rate of absorption exceeds regeneration, while in summer this relation is reversed. During the winter the overall biological activity appears to be greatly reduced and absorption exceeds regeneration only in the layer immediately below the surface.

It should be emphasized that the procedures reveal only the net result of the balance between absorption and regeneration. In the photic zone substantial quantities of nutrients may be absorbed and returned to the water by regeneration *in situ*. These increments do not appear in the balance. They may be determined by other methods such as those discussed in Chapter 7 which measure absorption directly.

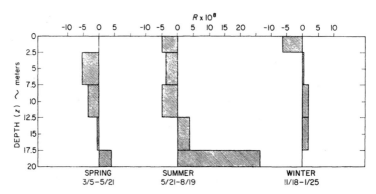

Fig. 10. Rate of change of phosphate phosphorus concentration attributed to biological activity, $R$, in layers of water at different depths and seasons in Long Island Sound. Negative values represent excess of absorption over regeneration. Units: depths, $z$, meters; $R$, $10^{-2}$ mg atom/m³/day. (Estimated from data of Riley, 1956a.)

### C. Differential Advection

Water moves with velocities which vary with depth, usually decreasing downward. Under conditions of differential advection the redistribution of nonconservative properties with respect to depth due to the sinking of organic matter leads to changes in the total quantity of the element present along the path of flow. Differential advection thus provides a mechanism by which geographical differences in distribution of nonconservative elements may be produced.

A simplified model will serve to indicate the principles and relations of the several factors involved in this process. Consider two layers of water of equal depth, $h$, moving with different velocities along the $x$-axis, separated by a surface across which a nonconservative element $N$ is transferred downward by sinking of organic matter and upward by eddy diffusion. At any position along the axis of flow, the concentration of the element is also subject to change by horizontal advection. If $-R$ represents the rate at which the mean concentration of $N$ is being reduced in the upper layer by the synthesis, sinking and decomposition of organic matter, $R$ will represent its rate of increase in the lower layer. If $A$ is the coefficient of vertical diffusion in the surface of separation and $dN/dz$ is the gradient of concentration at this boundary $(A/h) \, dN/dz$ will represent the effect of eddy diffusion on $N$ in the upper layer and $-(A/h) \, dN/dz$ its effect in the lower layer. The rate at which $N$ is altered by horizontal advection in either layer is given by $V \cdot dN/dx$, where $V$ is the velocity of flow and $dN/dx$ is the gradient in mean concentration along the $x$-axis. Under steady-state conditions the combined effect of the vertical exchanges balances the effect of advection in each layer so that

$$V_U \cdot \frac{dN_U}{dx} = -R + \frac{A}{h}\frac{dN}{dz} \tag{5a}$$

for the upper layer, and

$$V_L \cdot \frac{dN_L}{dx} = R - \frac{A}{h}\frac{dN}{dz} \tag{5b}$$

for the lower layer. The subscripts $U$ and $L$ designate the properties peculiar to the upper and lower layers respectively.

Combining these equations to eliminate the right-hand term,

$$\frac{dN_L}{dx} = -\frac{V_U}{V_L} \cdot \frac{dN_U}{dx}. \tag{6}$$

This states that under steady-state conditions the mean concentration of $N$ in the lower layer will increase along the $x$-axis as that in the upper layer decreases, and at a rate which is proportional to their relative velocities. In the case under consideration the layers are taken to be of equal depth and consequently the quantity of $N$ present is proportional to the concentration in the respective volume units. It follows that if $V_U = V_L$ the quantity of $N$ in the lower layer increases along the $x$-axis at exactly the same rate that it decreases in the upper layer. Consequently there will be no change in the total quantity of $N$ along the $x$-axis. If, as is usually the case, $V_U > V_L$, the quantity of $N$ in the lower layer will increase more rapidly than it decreases in the upper layer and the total quantity present in the two layers will increase in the direction of flow.

If the flow of the lower layer is in the opposite direction to that of the upper layer, $V_L$ is negative and the right-hand term of equation (5) becomes positive. In these circumstances the quantities of $N$ will change in both layers in the same way and will decrease in the direction in which the upper layer is flowing. Consequently countercurrent systems are particularly effective in producing changes in the distribution of nutrients along the direction of flow and lead to accumulation in the direction from which the surface current is flowing.

More complete expressions of the relations of the several factors involved in the effect of differential advection on the distribution of nonconservative properties along the axis of flow may be derived from equations (5a) and (5b). Substituting $dQ/h$ for $dN$ and $dQ_R/h$ for $R$ to express changes in the quantities of $N$ rather than concentrations, the rate of change along the $x$-axis of the total quantity of $N$ in both layers is obtained by adding the equations and rearranging the terms. Thus

$$\frac{dQ_L + dQ_U}{dx} = \frac{1}{V_U} \cdot \left(\frac{V_U}{V_L} - 1\right) \cdot \left(\frac{dQ_R}{dt} - A \cdot \frac{dN}{dz}\right). \tag{7a}$$

Similarly, the difference in the quantity of $N$ in the respective layers changes along the axis of flow according to

$$\frac{dQ_L - dQ_U}{dx} = \frac{1}{V_U} \cdot \left(\frac{V_U}{V_L} + 1\right) \cdot \left(\frac{dQ_R}{dt} - A \cdot \frac{dN}{dz}\right) \tag{7b}$$

while the rate of change in the difference in mean concentration is

$$\frac{dN_L - dN_U}{dx} = \frac{1}{V_U h_U} \cdot \left(\frac{V_U h_U}{V_L h_L} + 1\right) \cdot \left(\frac{dQ_R}{dt} - A \cdot \frac{dN}{dz}\right). \tag{7c}$$

These expressions indicate that the gradients developed in the horizontal and vertical depend on the ratio of the velocity in the two layers, and vary inversely with the absolute velocities as indicated by the term $1/V_U$. Sluggish currents are more effective than rapid ones in producing differential distribution, other things being equal. It may be noted also that the effects on the gradients in quantity depend on the velocities in the two layers, but that the effects on concentration vary with the product of the respective velocities and depths, that is with the flux in the respective layers, provided the vertical exchange remains the same.

The term $(dQ_R/dt - A \cdot dN/dz)$ indicates that the gradients will develop in proportion to the balance between the biological effect and that of vertical diffusion. As the vertical separation develops along the axis of flow, in accordance with equation (7c), $dN/dz$ will increase in value. Consequently the gradients will not develop linearly with distance along the $x$-axis but will tend to approach a limit at which the biological effect is balanced by the effect of diffusion. The coefficient of eddy diffusion, $A$, varies greatly under natural conditions, and in general in inverse proportion to the stability. Under stable conditions vertical diffusion may be small and have little effect on the gradients developed. If $A$ is large, as in the case of shallow estuaries subject to strong tidal currents, the diffusion term may be so large that gradients in the distribution of nonconservative elements do not develop.

The foregoing discussion is intended to clarify the factors which lead to a differential distribution of nonconservative elements in sea-water in situations where water moves with different velocities at various depths. It will provide a basis for discussion of the natural systems in the following pages.

## a. The estuarine circulation

In estuaries fresh water derived from the land by runoff and seepage mixes with sea-water and is carried seaward in the upper layer of the embayment. A countercurrent of sea-water moves in from the outer sea to replace that entrained in the surface outflow. The estuarine circulation is a special case of differential advection in which $V_L$ is negative. Consequently, the redistribution of nonconservative elements by the sinking of organized matter will tend to cause the concentration of $N$ to increase upstream relative to the motion of the surface layer. The estuarine circulation creates a trap in which nutrients tend to accumulate.

The Gulf of Venezuela is an example of an embayment in which the estuarine circulation is accompanied by an accumulation of phosphorus (see Fig. 11). The upper layers of the Gulf are diluted by fresh water escaping from the Maracaibo basin. The salt water entrained in the surface layer as it moves seaward is replaced by water from the Caribbean. In the deeper water at the head of the Gulf the concentration of phosphate phosphorus is 1 mg atom/m³, which is twice the quantity found in its offshore source. The total quantity of phosphorus in the water column increases from 11 mg atoms/m² at the shallow sill which divides the Gulf to 16 mg atoms/m² near the head of the Gulf. The oxygen is reduced in the deep water in the proportions expected from the $\Delta O/\Delta P$ ratio (Redfield, 1955).

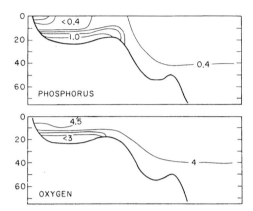

Fig. 11. Distribution of total phosphorus and oxygen in a section along axis of Gulf of Venezuela. Contour interval: Total phosphorus 0.2 mg atom/m³; oxygen 0.5 ml/l. Depths in meters. (After Redfield, 1955.)

The amount of accumulation varies greatly in different estuaries. It may be expected to increase with the rate of production of organic matter and with the length of the basin, to decrease with turbulence and the velocity of flow, and to vary with the relative depth and velocity of the surface and deep layers. Because of the variation in the accession of fresh water with the seasons, steady-state conditions are rarely observed. In shallow estuaries subject to strong tides, vertical advection is so strong that no differential distribution of nutrients can develop. Accumulation may occur, however, in small shallow estuaries if they are sufficiently isolated from tidal effects, as in the case of Great Pond, Falmouth, Massachusetts, studied by Hulburt (1956).

If the velocity of the currents is rapid the deep layer may be replaced before any notable accumulation can develop. This appears to be the case in the Strait of Juan de Fuca, where a well-developed counter current exists. In spite of the high productivity of the surface layer, no clear gradient of phosphate develops in the deeper layer along its length.

Because of the annual variation in the accessions of fresh water, and the influence of climate on the hydrography and the production of organic matter, steady-state conditions rarely exist in estuaries. The balance of the factors leading to accumulation consequently varies with the seasons. In Long Island Sound it is estimated that the countercurrent mechanism causes an accumulation of phosphorus to the order of 0.7–1.5 mg atoms/m³ in summer while comparable losses occur in the winter (Riley and Conover, 1956). Similar annual fluctuations occur in the nitrate exchange (Harris, 1959).

A striking example of the effect of seasonal variation in the influence of the countercurrent mechanism on the distribution of a nonconservative property is found in the strait connecting Lake Maracaibo with the sea. During the rainy season this strait is swept out by water escaping from the lake. The net movement is seaward at all depths. Although the deeper layers have lower velocities than the surface the strait is too short to permit any notable change in the concentrations of oxygen to develop along its length. With the advent of the dry season the outflow slackens and water of higher salinity crosses the shallow sill separating the strait from the sea, and flows lakeward beneath the outflow of surface water. Within a few months the oxygen content of the deep counter current becomes greatly reduced in the direction of its movement (see Fig. 12).

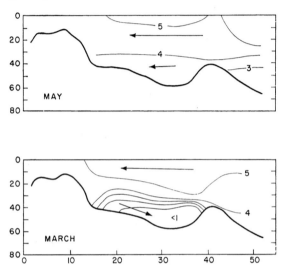

Fig. 12. Distribution of oxygen in section along Strait of Maracaibo during wet season, May, and dry season, March. Lake Maracaibo lies to the right. Arrows indicate the direction of non-tidal currents. Ordinate, depth in feet; abscissa, scale of miles; contour interval, 1 ml oxygen/m³.   (After Redfield, 1955.)

Extreme examples of the accumulation of nonconservative elements are provided by fjords. These estuaries are separated from the outer sea by sills which limit the active exchange of water to a relatively shallow layer and thus permit the deep water to stagnate. They thus represent the limiting case of differential advection in which the ratio of the velocities of the upper and lower layers is very large. The sinking of organic matter into the deep basins leads to reduction in oxygen content, the production of hydrogen sulfide and the accumulation of unusual concentrations of phosphate. In a large series of fjords examined by Ström (1936) great variations in the development of these effects are observed, varying from those in which complete anoxia does not occur to others in which as much as 40 ml/l. $H_2S$ is present. In the extreme cases as much as 10 mg atoms/m³ of phosphate phosphorus may accumulate in the deep water.

In fjords with large quantities of fresh water in the upper layers, the deep water is partially renewed regularly by advection and diffusion. In others seasonal variations in turbulence, stability and productivity lead to some renewal of the upper layers of the deep water each winter. The bottom waters, however, are renewed only periodically when sea-water crosses the sill in sufficient quantity to displace the stagnant water from the bottom upward. Ström suggests that the most general cause of renewal is the gradual warming of the deep water to the point where its density becomes less than that of the water entering over the sill. In the relatively long intervals between renewal, the effects of stagnation gradually develop. In two basins with varying degrees of stagnation, the quantity of $H_2S$ present in the bottom water increased at the rate of 1.0–1.5 ml/year.

### b. The anti-estuarine circulation

In regions where rainfall is small, evaporation from the sea surface may exceed precipitation and the additions of fresh water from rivers. Under these conditions a reversal of the currents characteristic of the estuarine circulation may be expected, since sea-water must flow in to replace water lost by evaporation from the surface, while water concentrated by evaporation will sink and flow out along the bottom. We know of no case where estuaries of this type have been well enough described to show whether the current system actually reduces the nutrient content of the water, as might be expected. However, one large sea, the Mediterranean, is well known to be low in nutrients and is characterized by an anti-estuarine circulation.

### c. The Mediterranean and Black Seas

These seas together illustrate how greatly the characteristics of the circulation can influence the distribution of the biochemically important elements in adjacent bodies of water. The Mediterranean is the most impoverished large body of water known, while in the Black Sea nutrients have accumulated to an extreme degree.

In the Mediterranean evaporation exceeds the accession of fresh water to such an extent that the salinity is increased by about 4%. The dynamics of the situation are such that a much larger volume of water flows in through the Strait of Gibraltar than is required to replace the loss by evaporation. The excess escapes through the Strait as a counter current moving seaward below the inflowing surface layer. Fig. 13 shows the situation diagrammatically. The nutrient content of Mediterranean water, as indicated by the phosphate content, is very much lower than that of the off-lying Atlantic. This is due initially to the fact that water enters the sea across a shallow sill and is thus skimmed from the surface layers of the ocean which are already greatly depleted in nutrients. This fact alone does not explain why a great accumulation of nutrients does not occur in the deep water, as it does in many other basins isolated by shallow sills.

The impoverished condition of the Mediterranean was first described by

Thomsen (1931) and is discussed by Riley (1951) in terms of the circulation pattern. The accumulation of nutrients in the deep water is limited by the exchange through the Strait of Gibraltar. A steady-state condition exists when the income of nutrients in the upper layers of the strait equals the outgo in the deeper layers. Accumulation in the deeper layers can only develop to the point where this condition is met, since any further increase in the concentration in the outgoing water will lead to a net loss of nutrients from the sea as a whole.

Fig. 13 shows that, within the Mediterranean, the concentration of phosphate

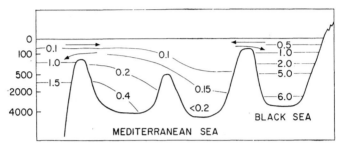

Fig. 13. Distribution of phosphorus in Black Sea, Mediterranean and off-lying Atlantic Ocean (diagramatic). Arrows indicate direction of currents in the Bosphorus and Strait of Gibraltar. Contours, phosphate phosphorus in mg atoms/m$^3$. Depths in meters.

decreases progressively in both the sub-surface and deep water as one goes eastward in the direction of flow of the surface layers. This is the relation to be expected in an anti-estuarine circulation.

The circulation of the Black Sea is similar in character to that of a fjord. It receives the discharge of the great rivers of eastern Europe of which the excess escapes through the Bosphorus as a strong surface current. Beneath this flow a counter current transports more saline water from the Mediterranean into the Black Sea. Since the Bosphorus is less than 100 m deep and the deeper part is occupied by an inflowing current, the deep water of the basin cannot escape except by mixing upward into the surface layers. Nutrients accumulated in such deep water re-enter the biochemical cycle when brought to the surface and are returned by sinking as organized matter to the depths (see Caspers, 1957).

The estuarine circulation of the Bosphorus thus produces a trap which hinders the loss of nutrients from the Black Sea and promotes their accumulation. In contrast, water from the depths of the Mediterranean may escape through the Strait of Gibraltar without re-entering the euphotic zone.

A factor which must contribute to the greater accumulation of regenerated nutrients in the Black Sea, as contrasted with the Mediterranean, is the stability of the water column. In the Black Sea the accessions of fresh water at the surface produce strong density gradients in the upper layers which retard the diffusion of deep water into the levels at which escape across the sill is possible. In contrast, the deeper water of the Mediterranean is homogeneous up to depths well above the critical level at which it can flow across the sill, and its escape is unimpeded by stability.

### d. Upwelling

Where persistent winds, such as the Trades, blow the surface water off-shore sub-surface water upwells to replace it. Upwelling takes place off the western coasts of the continents in the trade-wind zones. It is recognized by the anomalous coolness of the surface waters, their high nutrient content and the abundance of life.

A small-scale example of upwelling, and its influence on the distribution of the components of sea-water, is found in the outer part of the Gulf of Venezuela. During the winter season, when the northeast trade winds blow persistently, upwelling occurs in the lee of the peninsula of Paraguana, which forms the eastern boundary of the Gulf. Fig. 14 shows that along a section extending

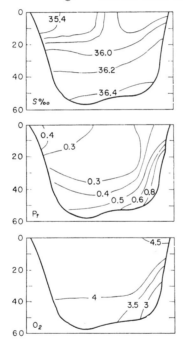

Fig. 14. Distribution of salinity ($S\%_0$), total phosphorus ($P_t$ mg atoms/m³), and oxygen ($O_2$ ml/l.) in a section across Gulf of Venezuela. Ordinate, depths in meters. (After Redfield, 1955.)

westward from the peninsula the isohalines slope upward as the coast is approached. This indicates that the more saline water at depth is being drawn landward and upward by the upwelling process to replace the surface water being blown offshore. The phosphorus and oxygen isolines slope in much the same way but more markedly. The concentrations of phosphorus are greater and those of oxygen less in deep water close to the coast than anywhere offshore. Apparently there is an accumulation of phosphorus in the deep water as it moves toward the site of upwelling, as the result of the sinking of organized matter from the surface layers as they move offshore. In support of this, the oxygen is diminished in about the proportions required for the decomposition of organic matter.

The classic examples of upwelling are found along the eastern boundaries of the oceans in the trade-wind belt. The distribution of total phosphorus in a section crossing the Benguela current off the western coast of Africa, where upwelling occurs, is shown in Fig. 15. At all depths the concentrations of phos-

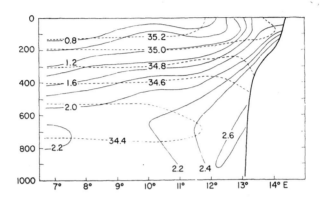

Fig. 15. Distribution of total phosphorus and salinity in a section along 24th parallel off west coast of Africa. Solid contours, total phosphorus, mg atoms/m$^3$; broken contours, salinity, ‰. Depths in meters. (Based on *Crawford* Stations 450–458, October, 1958.)

phorus increase markedly as the coast is approached. If this enrichment were due entirely to the movement of deep water into the upper layers, the isohalines should parallel the phosphorus isopleths, which is not the case. The fact that along any isohaline the phosphorus increases markedly toward the coast indicates that some process other than the physical motions of the water leads to an accumulation of this element in the upwelling area.

Upwelling can only be maintained by a flow of deep water toward the site of upwelling to replace the surface water moved offshore by the wind. Off the coasts of California (Sverdrup and Fleming, 1941) and Peru (Gunther, 1936) counter currents are present in the sub-surface waters during the periods of upwelling. It seems probable that nutrients accumulate in areas of upwelling because of the differential advection of water at different depths and in accordance with the principles which apply in estuaries. This accumulation must contribute substantially to the fertility of the water in such areas.

### D. Dynamic Equilibria in Three Dimensions

In the cases discussed above special considerations have justified a simplification of the problem by the assumption that exchanges in one or two dimensions could be neglected. Such assumptions are not justified when the distribution of properties in large bodies of water are under consideration. The treatment in such a case is much more complicated, but it has been attempted in one case, that of the distribution of oxygen, phosphate and nitrate in the Atlantic Ocean by Riley (1951). This remarkable paper, which requires and deserves careful study, develops as well a comprehensive account of the physical circulation.

Riley's procedure is to construct a mathematical model to which equation (1) can be applied.

Equation (1) is converted to finite difference form to read:

$$\delta N_0 = R + \frac{1}{\Delta x}\left(\frac{A_x}{\rho}\cdot\frac{\Delta N_x}{\Delta x} - \frac{A_{-x}}{\rho}\cdot\frac{\Delta N_{-x}}{\Delta x}\right) + \frac{1}{\Delta y}\left(\frac{A_y}{\rho}\cdot\frac{\Delta N_y}{\Delta y} - \frac{A_{-y}}{\rho}\cdot\frac{\Delta N_{-y}}{\Delta y}\right)$$

$$+ \frac{1}{\Delta z}\left(\frac{A_z}{\rho}\cdot\frac{\Delta N_z}{\Delta z} - \frac{A_{-z}}{\rho}\cdot\frac{\Delta N_{-z}}{\Delta z}\right) - \frac{1}{2}\left(V_x\cdot\frac{\Delta N_x}{\Delta x} + V_{-x}\cdot\frac{\Delta N_{-x}}{\Delta x}\right)$$

$$- \frac{1}{2}\left(V_y\cdot\frac{\Delta N_y}{\Delta y} + V_{-y}\cdot\frac{\Delta N_{-y}}{\Delta y}\right) - \frac{1}{2}\left(V_z\cdot\frac{\Delta N_z}{\Delta z} + V_{-z}\cdot\frac{\Delta N_{-z}}{\Delta z}\right). \tag{1a}$$

$N_0$ is the concentration at the center of some specified volume unit, $N_0 + \Delta N$ is the concentration at the center of the adjoining volume unit and $\Delta x$, $\Delta y$ and $\Delta z$ are the distances separating the centers along the specified axes. When the centers on any axis are equidistant, $\Delta N_x/\Delta x$, etc. represent the gradients and $A_x$, $V_x$, etc. represent average values of the physical coefficients between successive centers. A steady state is assumed, in which case the array of terms on the right-hand side of the equation equals zero.

The surface of the ocean between 63°N and 36°S is divided into rectangular areas 1000 km on a side and the water column under each area is subdivided into unit volumes separated by surfaces at intervals of 0.2 $\sigma_t$ units. To the center of each of these volumes representative values of the items—temperature, salinity, oxygen, phosphate and nitrate —are assigned, based on oceanographic data. The result is an array of data, distributed geometrically in the horizontal and so that each vertical series lies between the same $\sigma_t$ surfaces, which defines the values of the gradients $\Delta N_x/\Delta x$, etc.

The physical coefficients $A_x$, $V_x$, etc. are evaluated from the data on temperature and salinity. Relative values for the velocity of flow in each $\sigma_t$ surface are obtained by applying standard procedures of dynamic oceanography to the distribution of temperature and salinity. From these the relative transport between the $\sigma_t$ surfaces is then derived by taking account of the difference in their depths. These relative values are adjusted to accord with the principle of continuity to give absolute transports such that the total flow across the southern boundary of the array as a whole is zero. Reversing the procedure the absolute values for the velocity of flow, $V_x$, etc., are obtained. From these values the advective terms of the equation may be evaluated for each volume unit.

The coefficients of eddy diffusion, defining the exchange of water across the boundaries of the unit volumes, are obtained by applying the relaxation method to the salinity distribution. It is assumed that a steady state exists, in which case the quantity of salt entering each unit volume must equal that leaving. The boundary conditions require that, in the array as a whole, this equality shall apply also to the exchange across the southern boundary. Provisional values are assigned to the coefficients of vertical and horizontal diffusion until this condition is met within each unit volume when the advective terms are taken into account. Certain simplifying assumptions are made, including the effect of stability and shear on the vertical diffusivity, for which the original paper may be consulted.

Average values for the coefficients of horizontal and vertical diffusivity obtained are given in Table XI. These coefficients vary somewhat from one area to another.

By these procedures all the terms defining the motion of the water are determined and the value of $R$, the rate of change of any of the nonconservative properties, may be obtained for any unit volume directly from equation (1a). The average values obtained for the rate of change in oxygen concentration and of phosphorus regeneration at depth are shown in Table XI.

TABLE XI

Average Eddy-Diffusion Coefficients and Rates of Change in Oxygen and Phosphate Concentration, as Estimated by Riley (1951)

| Sigma-$t$ | Diffusion coefficients | | Rate of change in concentration | | $\Delta O/\Delta P$, mg atoms |
| | Vertical, $\dfrac{g}{cm\ sec}$ | Horizontal, $\dfrac{g}{cm\ sec}$ | Oxygen, $\dfrac{ml\ O_2}{m^3\ year}$ | Phosphate, $\dfrac{mg\ atoms\ P}{m^3\ year}$ | $\dfrac{mg\ atoms}{mg\ atoms}$ |
| --- | --- | --- | --- | --- | --- |
| 26.5 | 1.5 | $0.57 \times 10^8$ | $-0.21$ | 0.093 | $-190$ |
| 26.7 | 0.56 | 0.18 | $-0.08$ | 0.040 | $-180$ |
| 26.9 | 0.40 | 0.14 | $-0.050$ | 0.024 | $-170$ |
| 27.1 | 2.1 | 0.105 | $-0.055$ | 0.005 | $-1000$ |
| 27.3 | 1.5 | 0.025 | $-0.035$ | 0.003 | $-1000$ |
| 27.5 | 0.8 | 0.024 | $-0.013$ | 0.004 | $-280$ |
| 27.7 | 13.5 | 0.080 | $-0.005$ | 0.004 | $-100$ |
| 1500 m | 0.4 | 0.002 | $-0.0016$ | 0.003 | $-50$ |
| 4000 m | 0.5 | 0.028 | $-0.00013$ | 0.0001 | $-100$ |

The procedure is not well adapted to the analysis of the deep water because of the small gradients present. The results obtained for water below 1500 m are presented by Riley with some reserve. The values entered in Table XI indicate merely the magnitude of the items. Also, it was found impractical to deal with the surface layer by these procedures.

The description of the physical circulation obtained by the analysis of the mathematical model agrees in general with that obtained by other methods. As a check on its utility in accounting for the distribution of nonconservative elements, the results obtained were used to reconstruct the distribution of phosphate and nitrate in a north–south section of the Atlantic basin. The distribution obtained for phosphate is shown in Fig. 16. The distribution estimated from the dynamic balance of the factors evaluated from the model agree in all essential details with that observed in nature.

The estimates of oxygen consumption and phosphate regeneration provide a means of determining the oxidative ratio, $\Delta O/\Delta P$, discussed on page 32. The atomic ratio obtained for the water column as a whole, from 200 m to the bottom, is $-255$. This value differs by only 8% from the theoretical ratio of $-276$ obtained from the analysis of plankton. When the values obtained from the several $\sigma_t$ surfaces are examined, the ratios vary widely and are nearly four times greater than the theoretical ratio in the water of the intermediate layer (see Table XI).

It is difficult to believe that the character of the organic matter being oxidized differs as greatly with depth as the results indicate. Riley finds that errors in the data or analytical procedures cannot account for the discrepancy.

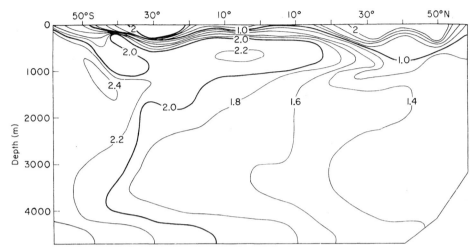

Fig. 16. Mathematical model of phosphate distribution along a north–south profile of the Atlantic Ocean estimated from analytically determined physical coefficients and assumed phosphate regeneration rates that vary in accordance with a generalized hypothesis derived from analysis. Contour interval, 0.2 mg atoms P/m³. (From Riley, 1951.)

The whole analysis is so complicated, however, that it is not possible to make a satisfactory analysis of errors. Some degree of anomaly in the oxidative ratio undoubtedly exists, but it is questionable whether the variations are as extreme as the calculations indicate.

## E. The Biochemical Circulation of the Oceans

Oxygen and phosphorus are distributed in the oceans of the world in characteristic patterns which are described by Sverdrup *et al.* (1942). In the case of the Atlantic Ocean, Riley's analysis evaluates in detail the factors on which the dynamic balance underlying the pattern depends. It is of interest to examine in a more general way the character of the circulation of the oceans to find what dominant physical features create these patterns and account for their differences.

### a. The vertical distribution of nonconservative elements

The distributions in depth of phosphate and oxygen display a similar pattern in the central basins of all oceans. Phosphate is depleted near the surface, rises to a maximum at moderate depth and then decreases somewhat and is relatively uniform in deep water (see Fig. 4). The distribution of oxygen follows an inverse pattern. The depth at which oxygen is minimal corresponds roughly, but not exactly, with that at which phosphate and nitrate are maximal and this zone will be referred to as the oxygen-minimum layer. In high latitudes the oxygen-minimum layer is absent and the vertical distributions are more uniform.

There has been general agreement that the oxygen-minimum layer arises from a dynamic equilibrium between the rate at which oxygen is removed from the respective water layers and the rate at which it is renewed by the circulation of water (see Richards, 1957). One group of investigators has emphasized the latter factor and has concluded that the oxygen content is reduced to the greatest extent where the motion of water is minimal. Seiwell (1937) and Sverdrup (1938) have pointed out that this condition is contrary to the facts in certain situations and that the observed distributions of oxygen may be accounted for by assuming various other suitable relations between the dynamic factors.

When the concentration of any nonconservative element in a moving mass of water is considered, its value will depend not only on the changes which have resulted from the balance of dynamic factors operative along its course, but also on the initial condition of the water. If the velocity of flow is great the change due to biological activity may be small and the observed concentrations may depend largely on a state established at the point of origin of the water mass. For example, a core of water of low oxygen content present in the Gulf Stream off New England may be tracked back toward its origin as far south as the Caribbean Sea (Richards and Redfield, 1955). It has been shown that the vertical distributions of oxygen and phosphorus in the Atlantic arise in substantial part from conditions existing near the sea surface at the places where the several water layers are formed, as well as from the decomposition of organic matter in the course of their flow at depth (Redfield, 1942).

The demonstration depends on separating the phosphorus present into its several fractions, as discussed on page 33. The principal motion of the water, by advection and horizontal eddy diffusion, may be assumed to be along surfaces of equal density. It is consequently instructive to plot the data as a function of $\sigma_t$ rather than depth, in which case the principal motion is in the horizontal and the continuity of motion becomes apparent. The convention is not adapted to represent conditions in the deep water with $\sigma_t > 27.8$ because the variation in $\sigma_t$ is insignificant. The conditions below 2000 m are better shown in the usual manner as a function of depth. The conditions at the depth of 200 m may be taken to represent the state of the water masses at the time of their origin. At this depth seasonal variations in the surface layers are largely eliminated.

The distribution of inorganic phosphorus along a north–south section of the Atlantic Ocean is shown in Fig. 17a, using this convention. The close relation of phosphate concentration to density distribution is striking. Maximal concentrations occur in a zone between $\sigma_t$ 27.0 and $\sigma_t$ 27.5 which meets the subsurface in the subantarctic region between 42°S and 50°S, i.e. between the approximate positions of the subtropical and antarctic convergences. This zone will be referred to as the intermediate layer. Maximum concentrations occur at about $\sigma_t$ 27.3, and their values decrease northward. This trend is interrupted in the equatorial region where a localized increase in concentration occurs.

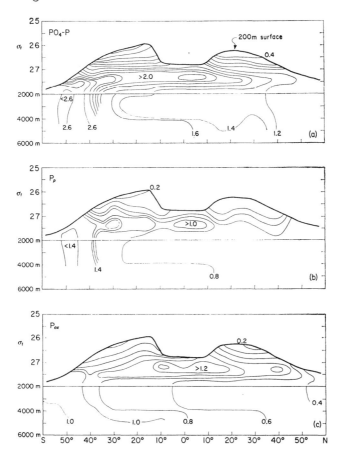

Fig. 17. Distribution of phosphorus fractions along western trough of Atlantic Ocean. (a) Phosphate phosphorus. (b) Preformed phosphorus. (c) Phosphorus of oxidative origin. Ordinates, sigma-$t$ surfaces and depths in meters. Abscissa, latitude. Contour interval, 0.2 mg atom/m$^3$. (Recalculated from data used by Redfield, 1942, and additional stations. Phosphorus corrected for salt error. Phosphorus of oxidative origin estimated using $\Delta O/\Delta P = -276$.)

The origins of the distribution of inorganic phosphorus are illuminated by the distributions of preformed phosphorus and phosphorus of oxidative origin. Preformed phosphorus also has maximal concentrations in the intermediate layer which diminish northward from the sub-surface of the subantarctic region and disappear at about 15°N (see Fig. 17b). This fraction is a conservative property. Its pattern of distribution and attenuation is very similar to that of salinity, which suggests that it is due to the intrusion of subarctic and antarctic intermediate water at an intermediate depth of the ocean. Since preformed phosphorus accounts for about one-half of the inorganic phosphorus in this layer, its northward flow must produce a substantial part of the inorganic phosphorus maximum.

The distribution of phosphorus of oxidative origin is shown in Fig. 17c. The distribution depends directly upon that of the apparent oxygen utilization from which it is estimated (see page 33). Maximum concentrations are found in the equatorial region of the intermediate layer, from which the concentrations diminish both northward and southward. This maximum must arise from the regeneration of phosphorus from organic matter derived from the surface layers of the tropical ocean.

The sub-surface water in the subantarctic region contains substantial concentrations of phosphorus of oxidative origin. The northward flow of antarctic intermediate water will carry this fraction to the intermediate layer of lower latitudes along with the preformed phosphorus. Allowance for the contribution of this fraction of subantarctic origin to the total phosphorus of oxidation can be made (see Redfield, 1942). When this is done, the occurrence of phosphorus of local oxidative origin becomes restricted in the South Atlantic to the tropical region.

The origin of the phosphate maximum is evidently complex, being due to the combined effects of fractions introduced by the transport of the water of the intermediate layer from its origin in the subantarctic regions, supplemented by local regeneration in the tropics and the subsequent redistribution of the latter by the horizontal circulation. The maximum local regeneration occurs at somewhat lesser depths than the maximum of the preformed phosphate. Accordingly the oxygen minimum lies somewhat above the phosphate maximum.

The depths between the sub-surface (200 m) and the upper boundary of the intermediate layer, in which $\sigma_t < 27.0$, are occupied by the North and South Atlantic Central Water-masses. In the equatorial region these water-masses are separated by the density distribution except for a zone of limited thickness (ca. 200 m), so that water can move between them only by passing through the sub-surface layer. In these isolated pools, which circulate as great eddies, the concentrations of the several phosphorus fractions decrease toward the surface depending on the degree of mixing of the rich water of the intermediate layer and the water of the surface layer which is impoverished by the synthesis and sinking of organized matter. The resulting patterns of distribution depend on the circulation which is peculiar to the eddies occupied by the central water-masses, and which also results in gradients in the temperature and salinity distribution.

The depths below the intermediate layer, where $\sigma_t$ is greater than 27.5, are occupied by the North Atlantic Deep Water. This water mass originates in high northern latitudes and flows southward. The concentration of inorganic phosphorus increases slowly from values of about 1 mg atom/m³ in the north to about 1.5 mg atoms/m³ at 32°S. Beyond this point the phosphate concentrations increase rapidly as the North Atlantic Deep Water mingles with deep water of antarctic origin, in which the concentrations are more than 2.6 mg atoms/m³. At depths greater than 4000 m a layer of relatively high phosphate concentration extends northward under the North Atlantic Deep Water owing to the intrusion of deep water from the Antarctic circumpolar region.

The estimated preformed phosphorus content of the North Atlantic Deep Water remains unchanged wherever its characteristic temperatures and salinities are preserved (see Table IV). The fraction of phosphorus of oxidative origin increases gradually from north to south, providing evidence that some regeneration of phosphorus due to biological activity takes place in the deep water as it drifts southward (see Fig. 17c).

From the foregoing discussion it is clear that the vertical distribution of non-conservative properties below any position on the ocean's surface is due primarily to the pattern of horizontal flow of the underlying water masses, and to characteristics which these water masses acquired at the time they were formed near the sea surface. Biological processes modify the concentrations greatly only near the sea surface, and their effects diminish rapidly with depth, as the estimates shown in Table XI indicate.

## b. The horizontal distribution of nonconservative elements

The deeper water of the Pacific and Indian Oceans is known to contain substantially more phosphate and less oxygen than that of the Atlantic and Arctic Oceans (Graham and Moberg, 1944). This fact is illustrated in Fig. 18,

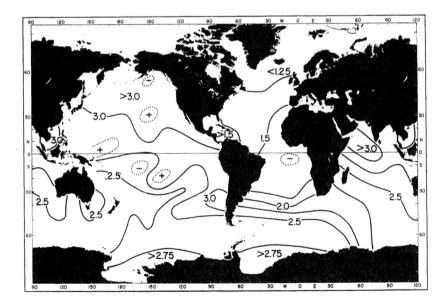

Fig. 18. Distribution of phosphorus at depth of 2000 meters in the oceans of the world. Contour interval, 0.25 mg atom/m³. (From Redfield, 1958. By courtesy of *American Scientist*.)

which shows the distribution of phosphorus at a depth of 2000 m in the oceans of the world. This depth lies below the intermediate layer in which nutrient concentrations are maximal, and the concentrations represent approximately the mean in the great mass of deep water.

In the antarctic circumpolar sea which connects the several oceanic basins the mean concentration of phosphorus is relatively uniform and varies between 2 and 2.5 mg atoms/m³. Proceeding northward in the Pacific and Indian Oceans, it increases gradually to more than 3 mg atoms/m³. In the Atlantic it decreases to less than 1.25 mg atoms/m³ and to about 1 mg atom/m³ in the Arctic basin. Thus the Pacific basin on the whole contains phosphate at about twice the concentration of the Atlantic. If extreme variation is considered, the Arctic waters of the Atlantic contain phosphate in the ratio of 1 : 3.5 when compared to the maximum concentrations present in the North Pacific, and this ratio is increased to 1 : 7 when deep Mediterranean water is compared with the richest part of the Pacific or Indian Oceans.

The differences in the nutrient content of the oceanic basins must depend on the composition of the deep water masses at the time of their origin and their subsequent modification by the balance of biological and physical factors in the course of their drift. This we have seen to be the case for the Antarctic Intermediate Water with its associated minimum oxygen layer, and for the distribution of phosphorus fractions in the North Atlantic Deep Water. The primary cause of the differences found in the deep water of the Pacific and Indian Oceans, in contrast to the Atlantic, quite clearly relates to their respective sources. The deep water of the former originates in the antarctic circumpolar sea (Southern Ocean) and flows northward with an initial concentration of phosphorus of 2 to 2.5 mg atoms/m³. In contrast the North Atlantic Deep Water is formed in the Arctic, where the concentration of phosphorus is 1–1.25 mg atoms/m³. Thus, at their origins, the deep waters of the Pacific and Indian Oceans contain twice as much phosphorus as the Atlantic. To these basic concentrations relatively small additions are made by biological effects.

Superficially, these oceans may be likened to estuaries opening in common on the antarctic circumpolar sea. In the Pacific and Indian Oceans little or no deep water is formed at the northern extremity. In consequence the circulation is estuarine, the surface waters flowing southward to be replaced by an influx of deep water drawn from the Southern Ocean. In the Atlantic the deep water is formed by sinking from the surface in high northern latitudes and is replaced by the flow of surface water northward. From the principles discussed in connection with estuaries, it might be expected that the differential advection, combined with the sinking and decomposition of organic matter at depth, would lead to the accumulation of decomposition products northward in the deep water of the estuarine Pacific and Indian Oceans and southward in that of the anti-estuarine Atlantic. Fig. 18 shows that in the direction of flow in the former oceans the increase in phosphorus concentration, which may be attributed to biological factors, amounts to about 1 mg atom/m³ while in the Atlantic the increase is about half this amount.

## References

Atkins, W. R. G., 1926. A quantitative consideration of some factors concerned with plant growth in water. Part II. *J. Cons. Explor. Mer.* **1**, 197–226.

Atkins, W. R. G., 1930. Seasonal variations in phosphate and silicate content of sea water in relation to the phytoplankton crop. Part V. *J. Mar. Biol. Assoc. U. K.*, n.s., **16**, 821–852.

Brand, T. von and N. W. Rakestraw, 1941. Decomposition and regeneration of nitrogenous organic matter in sea water. IV. *Biol. Bull.*, **81**, 63–69.

Brandhorst, W., 1959. Nitrification and denitrification in the eastern tropical North Pacific. *J. Cons. Explor. Mer*, **25**, 3–20.

Carritt, D. D., 1954. Atmospheric pressure changes and gas solubility. *Deep-Sea Res.*, **2**, 59–62.

Caspers, H., 1957. Black Sea and Sea of Azov. Treatise on Marine Ecology and Paleoecology. *Geol. Soc. Amer. Mem.* 67, Vol. 1, Chap. 25, pp. 801–889.

Cooper, L. H. N., 1933. Chemical constituents of biological importance in the English Channel. Pt. I. Phosphate, silicate, nitrate, nitrite, ammonia. *J. Mar. Biol. Assoc. U. K.*, n.s., **18**, 677–728.

Cooper, L. H. N., 1935. The rate of liberation of phosphate in the sea water by the breakdown of plankton organisms. *J. Mar. Biol. Assoc. U. K.*, n.s., **20**, 197–200.

Cooper, L. H. N., 1937. On the ratio of nitrogen to phosphorus in the sea. *J. Mar. Biol. Assoc. U.K.*, n.s., **22**, 177–204.

Cooper, L. H. N., 1938. Salt error in determinations of phosphate in sea water. *J. Mar. Biol. Assoc. U. K.*, **23**, 171–178.

Cooper, L. H. N., 1938a. Redefinition of the anomaly in the nitrate–phosphate ratio. *J. Mar. Biol. Assoc. U. K.*, **23**, 179.

Danil'chenko, P. T. and N. I. Chigirin, 1929. Notes on the chemistry of the Black Sea. Nitrogen and its compounds (in Russian). *Trudy Krym. Inst.*, **2** (2), 23–33.

Danil'chenko, P. T. and N. I. Chigirin, 1929a. Notes on the chemistry of the Black Sea. 2. On the question of the circulation of materials in the Black Sea (in Russian). *Zapiski Krym. obshch. estest.*, **11**, 5–14.

Duursma, E. K., 1960. Dissolved organic carbon, nitrogen and phosphorus in the sea. *Neth. J. Mar. Res.*, **1**, 1–147.

Fleming, R. H., 1940. The composition of plankton and units for reporting population and production. *Proc. Sixth Pacific Sci. Cong. Calif., 1939*, **3**, 535–540.

Goldberg, E. D., T. J. Walker and A. Wisenaud, 1951. Phosphate utilization by diatoms. *Biol. Bull.*, **101**, 274–284.

Graham, H. W. and E. G. Moberg, 1944. Chemical results of the last cruise of the *Carnegie*. Scientific results of Cruise VII of the *Carnegie* during 1928–29 under command of Captain J. P. Ault. Chemistry I. *Pub. Carnegie Inst. Wash.*, No. 562.

Gunther, E. R., 1936. A report on oceanographical investigations in the Peru Coastal Current. *Discovery Reps.*, **5**, 107–276.

Harris, Eugene, 1959. The nitrogen cycle in Long Island Sound. *Bull. Bingham Oceanog. Coll.*, **17**, 31–65.

Harris, Eugene and G. A. Riley, 1956. Oceanography of Long Island Sound, 1952–1954. VIII. Chemical composition of the plankton. *Bull. Bingham Oceanog. Coll.*, **15**, 315–323.

Harvey, H. W., 1926. Nitrate in the sea. *J. Mar. Biol. Assoc. U. K.*, n.s., **14**, 71–88.

Harvey, H. W., 1947. Manganese and the growth of phytoplankton. *J. Mar. Biol. Assoc. U. K.*, **26**, 562–579.

Harvey, H. W., 1947a. Fertility of the ocean. *Proc. Linnean Soc. London*, **158**, 82–85.

Harvey, H. W., 1955. *The Chemistry and Fertility of Sea Waters*. Cambridge Univ. Press, Cambridge, 224 pp.

Hulburt, E. M., 1956. The plankton of Great Pond, Massachusetts. *Biol. Bull.*, **110**, 157–168.

Ketchum, B. H., N. Corwin and D. J. Keen, 1955. The significance of organic phosphorus determinations in ocean waters. *Deep-Sea Res.*, **2**, 172–181.

Ketchum, B. H. and A. C. Redfield, 1949. Some physical and chemical characteristics of algae grown in mass culture. *J. Cell. Comp. Physiol.*, **33**, 281–300.

Ketchum, B. H., R. F. Vaccaro and N. Corwin, 1958. The annual cycle of phosphorus and nitrogen in New England coastal waters. *J. Mar. Res.*, **17**, 282–301.

Kriss, A. E., 1949. The role of micro-organisms in the accumulation of hydrogen sulfide, ammonia, and nitrogen in the depths of the Black Sea (in Russian). *Priroda*, **6**, 35–46.

Langley, H. E., Jr., 1958. Discussion of "Oxidation, reaeration, and mixing in the Thames estuary", by A. L. H. Gameson and M. J. Barrett. Tech. Rept. W-58-2. Robert A. Taft Sanitary Engineering Center, Cincinnati, Ohio, pp. 91–92.

McKinney, R. E. and R. A. Conway, 1957. Chemical oxidation in biological waste treatment. *Sewage and Indust. Wastes*, **29**, 1097–1106.

Menzel, D. W. and J. H. Ryther, 1961. Nutrients limiting the production of phytoplankton in the Sargasso Sea, with special reference to iron. *Deep-Sea Res.*, **7**, 276–281.

Rakestraw, N. W., 1947. Oxygen consumption in sea water over long periods. *J. Mar. Res.*, **6**, 259–263.

Redfield, A. C., 1934. On the proportions of organic derivatives in sea water and their relation to the composition of plankton. *James Johnstone Memorial Volume*, Liverpool, pp. 177–192.

Redfield, A. C., 1942. The processes determining the concentration of oxygen, phosphate and other organic derivatives within the depths of the Atlantic Ocean. *Papers Phys. Oceanog. Met., Mass. Inst. Tech. and Woods Hole Oceanog. Inst.*, **9** (2), 1–22.

Redfield, A. C., 1948. The exchange of oxygen across the sea surface. *J. Mar. Res.*, **7**, 347–361.

Redfield, A. C., 1955. The hydrography of the Gulf of Venezuela. *Papers Mar. Biol. Oceanog., Deep-sea Res.* Suppl. to Vol. 3, 115–133.

Redfield, A. C., 1958. The biological control of chemical factors in the environment. *Amer. Sci.*, **46**, 205–221.

Redfield, A. C. and A. B. Keys, 1938. The distribution of ammonia in the waters of the Gulf of Maine. *Biol. Bull.*, **74**, 83–92.

Redfield, A. C., H. P. Smith and B. Ketchum, 1937. The cycle of organic phosphorus in the Gulf of Maine. *Biol. Bull.*, **73**, 421–443.

Rice, T. R., 1953. Phosphorus exchange in marine phytoplankton. Fisheries Bull. 80, *Fishery Bull. Fish and Wildlife Service*, **54**, 77–89.

Richards, F. A., 1957. Oxygen in the ocean. Treatise on marine ecology and paleontology. *Geol. Soc. Amer. Mem.*, 67, Vol. 1, Chap. 9, pp. 185–238.

Richards, F. A., 1958. Dissolved silicate and related properties of some western North Atlantic and Caribbean waters. *J. Mar. Res.*, **17**, 449–465.

Richards, F. A. and B. B. Benson, 1961. Nitrogen/argon and nitrogen isotope ratios in two anaerobic environments, the Cariaco Trench in the Caribbean Sea and Dramsfjord, Norway. *Deep-Sea Res.*, **7**, 254–264.

Richards, F. A. and N. Corwin, 1956. Some oceanographic applications of recent determinations of solubility of oxygen in sea water. *Limnol. Oceanog.*, **1**, 263–267.

Richards, F. A. and A. C. Redfield, 1955. Oxygen-density relationships in the western North Atlantic. *Deep-Sea Res.*, **2**, 182–199.

Richards, F. A. and R. F. Vaccaro, 1956. The Cariaco Trench, an anaerobic basin in the Caribbean Sea. *Deep-Sea Res.*, **3**, 214–228.

Riley, G. A., 1951. Oxygen, phosphate, and nitrate in the Atlantic Ocean. *Bull. Bingham Oceanog. Coll.*, **12**, 1–126.

Riley, G. A., 1956. Oceanography of Long Island Sound 1952–1954. II. Physical oceanography. *Bull. Bingham Oceanog. Coll.*, **15**, 15–46.

Riley, G. A., 1956a. Oceanography of Long Island Sound 1952–1954. IX. Production and utilization of organic matter. *Bull. Bingham Oceanog. Coll.*, **15**, 325–344.

Riley, G. A. and S. A. McM. Conover, 1956. Oceanography of Long Island Sound 1952–1954. III. Chemical oceanography. *Bull. Bingham Oceanog. Coll.*, **15**, 47–61.

Robinson, R. J. and H. E. Wirth, 1934. Free ammonia, albuminoid nitrogen and organic nitrogen in the waters of the Pacific Ocean off the coasts of Washington and Vancouver Island. *J. Cons. Explor. Mer*, **9**, 15–27.

Ryther, J. H., 1960. Organic production by planktonic algae and its environmental control. The Pymatuning Symposium in Ecology. *Pymatuning Laboratory of Field Biology, University of Pittsburg*. Special Pub. No. 2, pp. 72–83.

Seiwell, H. R., 1937. The minimum oxygen concentration in the western basin of the North Atlantic. *Papers Phys. Oceanog. Met., Mass. Inst. Tech. and Woods Hole Oceanog. Inst.*, **5**, (2), 1–44.

Seiwell, H. R. and G. E. Seiwell, 1938. The sinking of decomposing plankton in sea water and its relationship to oxygen consumption and phosphorus liberation. *Proc. Amer. Phil. Soc.*, **78**, 465–481.

Skopintsev, B. A., 1957. Studies of redox potential in waters of the Black Sea (in Russian). *Gidrokhim. Materialy*, **24**, 21–36.

Skopintsev, B. A., F. A. Gubin, R. V. Vorob'eva, and O. A. Vershinina, 1958. Content of the main components of the salt water of the Black Sea and the problem of water exchange (in Russian). *Akad. Nauk S.S.S.R. Morskoi Gidrofiz. Inst. Trudy*, **13**, 89–112.

Ström, K. M., 1936. Land-locked waters. Hydrography and bottom deposits in badly ventilated Norwegian fjords with remarks upon sedimentation under anaerobic conditions. *Norske Vidensk. Ak. Oslo*, 1. *Math-Nat. Kl.*, No. 7, 85 pp.

Sverdrup, H. U., 1938. On the explanation of the oxygen minimum and maximum in the oceans. *J. Cons. Explor. Mer*, **13**, 163–172.

Sverdrup, H. U. and R. H. Fleming, 1941. The waters off the coast of southern California, March to July 1937. *Scripps Inst. Oceanog. Bull.*, **4**, (10), 261–378.

Sverdrup, H. U., M. W. Johnson and R. H. Fleming, 1942. *The Oceans, their physics, chemistry and general biology*. Prentice-Hall, New York.

Thomsen, H., 1931. Nitrate and phosphate contents of Mediterranean water. *Rep. Danish Oceanog. Exped. 1908–1910*, No. 10, **3**, 1–14.

Vinogradov, A. P., 1953. The elementary chemical composition of marine organisms. *Mem. Sears Found. Mar. Res.*, **2**, 647 pp.

Wattenburg, H., 1933. Uber die Titrationsalkalinitat und den kalzium karbonatgehalt des Meerwassers. *Wiss. Ergebn. Deut. Atlant. Exped. 'Meteor' 1925–27*, **8**, (2), 122–231.

Contribution No. 1113, Woods Hole Oceanographic Institution and Contribution No. 238, Department of Oceanography of the University of Washington.

# 3. ARTIFICIAL RADIOACTIVITY IN THE SEA

## Y. MIYAKE

### 1. Artificial Radioactivity in Oceanic Waters

At present the artificial radioactivity in the oceans consists mainly of the fission products of $^{235}U$, $^{239}Pu$ and $^{238}U$ from tests of nuclear weapons. In some areas a small amount of artificial radioactivity has been introduced into the sea by atomic waste disposal. The spectrum of fission products varies from one source to another according to the type of parent nuclides as well as the kinetic energy of the fission neutrons. More significant are the changes in their composition and activity with the lapse of time after the fission.

Table I gives the percentage of the activity of the principal radioisotopes produced by slow neutron bombardment on $^{235}U$ (1 kg) respectively 20 days and one year after fission. In twenty years only a few nuclides remain; $^{90}Sr$–$^{90}Y$ and $^{137}Cs$–$^{137}Ba$ occupy respectively 48% and 45% of the total activity. The rest of the activity will consist of $^{147}Pm$ and $^{151}Sm$.

In addition, radioactive nuclides produced by neutron capture or other nuclear reactions such as $^3H$, $^{14}C$, $^{35}S$, $^{57}Co$, $^{58}Co$, $^{60}Co$, $^{65}Zn$, $^{54}Mn$, $^{59}Fe$, $^{55}Fe$ have been detected in sea-water and marine products.

The artificial radioactivity from nuclear tests has been introduced into the oceans from the fallout, both immediate and delayed, and from direct injection through the surface or sub-surface tests. Since the ocean covers 71% of the earth's surface, of the stratospheric fallout, 71% falls into the oceans if the stratospheric dust is distributed evenly in the air. But this may not be true because the greater part of the stratospheric dust is accumulated in the same hemisphere in which the tests are carried out. According to the observation on the distribution of $^{90}Sr$ in the stratosphere (presentation by J. Spar at the United Nations Scientific Committee on the Effects of Atomic Radiation, New York, 1960), the total amount of $^{90}Sr$ in the Northern Hemisphere was 0.5 megacurie while it was only 0.2 megacurie in the Southern Hemisphere (Jan.–Sept., 1959). As the oceanic areas are respectively 60.6% and 81.0% in Northern and Southern Hemispheres, 66% of stratospheric fallout falls into the sea, assuming the above ratio of $^{90}Sr$ distribution. However, it is well known that the ground deposition has its maximum between 30°N and 60°N, where the oceanic area occupies only 50% of the earth's surface. The $^{90}Sr$ deposition in 1958 between 30°N and 60°N was estimated to be 43% of the total or 59% of that in the Northern Hemisphere; therefore, the stratospheric fallout in the sea might be about 61% of the total.

In 1960, the total deposition of $^{90}Sr$ on the surface of the earth is regarded to be roughly 6 megacuries, of which 3.6 megacuries of $^{90}Sr$ were introduced into the ocean from the sky, which is equivalent to 10 mc/km$^2$ on the average. If we assume the ratio of stratospheric dust is 2.5 : 1 between the Northern and Southern Hemispheres, the depositions of $^{90}Sr$ on the ocean surface in the two hemispheres is respectively 17 mc/km$^2$ and 5 mc/km$^2$. The level of activity of $^{90}Sr$ in sea-water depends largely upon the depth of the

[*MS received August, 1960*]

78

mixed layer and the rate of mixing of waters across the thermocline. Assuming that the thickness of the mixed layer is 100 m and most of the radioactive fall-out remains within the layer for a few years, the concentration of $^{90}$Sr in sea-water from fallout averages 0.17 µµc/l. in the Northern Hemisphere, while it is only 0.05 µµc/l. in the Southern Hemisphere.

TABLE I

The Percentage of Activity in Curies of the Principal Radioisotopes from Slow Neutron Fission of $^{235}$U (1 kg)

| Twenty days | $9.8 \times 10^6$ curies | | One year | $3.1 \times 10^4$ curies | |
|---|---|---|---|---|---|
| Nuclide | Activity in curies $(10^5)$ | % | Nuclide | Activity in curies $(10^3)$ | % |
| $^{140}$La | 13.6 | 13.9 | $^{144}$Ce–$^{144}$Pr | 164 | 52.8 |
| $^{140}$Ba | 11.8 | 12.0 | $^{95}$Nb | 45.6 | 14.7 |
| $^{143}$Pr | 11.8 | 12.0 | $^{95}$Zr | 22.4 | 7.2 |
| $^{141}$Ce | 9.5 | 9.7 | $^{147}$Pm | 17.7 | 5.7 |
| $^{133}$Xe | 6.2 | 6.3 | $^{91}$Y | 11.8 | 3.8 |
| $^{95}$Zr | 5.8 | 5.9 | $^{89}$Sr | 8.4 | 2.7 |
| $^{91}$Y | 5.5 | 5.6 | $^{106}$Ru–$^{106}$Rh | 15.2 | 4.9 |
| $^{131}$I | 5.5 | 5.6 | $^{90}$Sr–$^{90}$Y | 11.5 | 3.7 |
| $^{89}$Sr | 4.9 | 5.0 | $^{137}$Cs–$^{137}$Ba | 9.0 | 2.9 |
| $^{147}$Nd | 4.9 | 5.0 | $^{103}$Ru | 2.5 | 0.8 |
| $^{103}$Ru | 4.3 | 4.4 | $^{103}$Rh | 2.5 | 0.8 |
| $^{95}$Nb | 4.1 | 4.2 | | | |
| $^{144}$Ce | 2.3 | 2.3 | | | |
| $^{144}$Pr | 2.6 | 2.6 | | | |
| $^{90}$Mo | 1.3 | 1.3 | | | |
| $^{131}$I | 1.0 | 1.05 | | | |

The ratio of immediate or direct fallout of artificial radioactivity to the far-reaching tropospheric and stratospheric fallout is uncertain but it is estimated to be roughly 1 : 2 to 2 : 1. According to the estimation by the U.S. Atomic Energy Commission the total amount of $^{90}$Sr produced to date (1960) is equal to 9.2 megacuries, of which about 3 megacuries was immediate fallout. On the other hand, fifty explosions of multimegaton class were detected in Japan by our barometric oscillations up to the end of 1958. Assuming that the average energy of the explosions is five megatons and that half of the energy is produced by the fission process, the total amount of $^{90}$Sr produced so far is equal to 12.5 megacuries, which is a little larger than the estimated value of the U.S.A.E.C. Based on the measurement of radioactivity in oceanic waters in the Pacific (Miyake and Saruhashi, 1960), the total activity released directly into sea-water from the test areas was estimated to be about 3 megacuries in 1954. Since no larger surface or sub-surface tests were made subsequently in the Pacific, the total amount of $^{90}$Sr due to the direct injection and immediate fallout up to 1960 would have been roughly 4–5 megacuries. Miyake and Saruhashi (1960) estimate that the contaminated area in the North Pacific from flowout of radioactivity is about $3 \times 10^7$ km², which is equal to about 20% of the sea surface of the Northern Hemisphere.

The average activity in sea-water in the contaminated area would be about 1.5–2 μμc/l. including delayed fallout assuming the depth of thermocline as 100 m. Analytical results on $^{90}$Sr in sea-water in the Atlantic obtained by Bowen and Sugihara (1957) ranged from 0.016 to 0.2 μμc/l. during the period from 1956 to 1958. On the other hand, North Pacific waters from 1957 to 1959 showed 0.7–2.3 μμc/l. (Miyake *et al.*, 1960). These latter values are in good agreement with the above estimations.

## TABLE II

The Results of Radiochemical Analysis of Sea-Water at 12° 18′N, 161° 03′E, on June, 1954, in the North Pacific (Miyake and Sugiura, 1955). Values for December, 1954.

| Nuclide | Activity, μμc/l. | % | $^{235}$U (slow neutron fission product) (210 days) |
|---|---|---|---|
| $^{91}$Y | 260 | 37 | 10% |
| $^{144}$Ce, $^{141}$Ce | 400 | 55 | 15 |
| $^{106}$Ru, $^{103}$Ru, $^{106}$Rh | 26 | 3.8 | 8.2 |
| $^{95}$Zr | 15 | 2.2 | 39 |
| $^{90}$Sr, $^{89}$Sr | 14 | 2.0 | 7.5 |

Miyake and Sugiura (1955) carried out analyses for certain other radioactive nuclides, artificially produced in sea-water collected at 12° 18′N, 161° 03′E, about 450 km northeast of Bikini in June, 1954. The results are shown in Table II.

The fourth column of the table gives the percentage of the total activity for certain fission products 210 days after the slow neutron fission of $^{235}$U. Note that in sea-water the relative amounts of the rare earths were remarkably enhanced, while those of Zr and Sr were much lower.

Bowen and Sugihara (1957) also found an excess of $^{144}$Ce in relation to $^{90}$Sr both in the North and South Atlantic. The same tendency was confirmed in the Pacific surface waters by Harley (1956). Based on these findings, Ketchum and Bowen (1958) suggested the possibility of the retention of $^{144}$Ce in surface waters by microplankton. However, Bowen and Sugihara (*in litt.*) subsequently observed that, in the Atlantic, wherever high $^{144}$Ce:$^{90}$Sr ratios were found, the $^{144}$Ce:$^{147}$Pm (half-life, 2.6 years) ratio showed the mixture to be young. They concluded that the radioactive rare earths are removed from the surface layers of the Atlantic at higher rates than $^{90}$Sr, while in the Pacific the balance between the retention of [144]Ce by microplankton and its removal by sedimentation is in favor of the former.

Yamagata and Matsuda (1959) gave the value of 0.07 to 0.15 $\mu\mu$c/l. for radioactive cesium ([137]Cs) in the coastal waters of Japan in early 1958. Folsom (1960) reported that the average content of [137]Cs was 0.097 $\mu\mu$c/l. in coastal waters in Southern California during the period from November, 1959, to April, 1960. In surface waters of the western North Pacific [137]Cs from 0.8 to 4.8 $\mu\mu$c/l. was given by Miyake *et al.* (1961).

## 2. Enrichment of Radioactive Substances in Marine Products

It was rather astonishing to find considerable amounts of [65]Zn, [55]Fe and [59]Fe, which are not fission products but induced radionuclides, in the marine organisms collected in June, 1954, in the tropical North Pacific by Amano, Kawabata and other Japanese workers. The maximum activities of [65]Zn and [55]Fe were respectively 3.8 m$\mu$c/g and 7.7 m$\mu$c/g of wet tissue of a liver of albacore.

Since then, [54]Mn, [57]Co, [58]Co, [60]Co have been found in the internal organs of fishes caught in the western tropical region of the North Pacific. The same kinds of non-fission products were also detected in zooplankton and pelagic sediments collected in the central equatorial Pacific by Martin (1957) in 1956. The high levels of activity in the pelagic sediments are attributed to the enrichment of these radionuclides in the benthic organisms.

The [90]Sr concentration in the marine products was studied by Hiyama and others (1960) whose results showed a remarkable increase in [90]Sr from 1954 to 1959. [90]Sr in clam shells was 0.08 $\mu\mu$c/g of Ca in 1954 while it increased to

### TABLE III

Concentration Factors in Some Marine Organisms (wet weight volume basis). Values in parentheses are for analyses made on closely related genera (Ketchum and Bowen, 1958)

| Elements | Cesium | Strontium | Iron | Cobalt | Zinc | Cadmium | Iodine |
|---|---|---|---|---|---|---|---|
| Sea-water, g/ml | $5 \times 10^{-10}$ | $9 \times 10^{-6}$ | $4 \times 10^{-9}$ | $2.5 \times 10^{-10}$ | $1.4 \times 10^{-9}$ | $5 \times 10^{-11}$ | $5 \times 10^{-8}$ |
| Organism | | | | | | | |
| Limacina | (0.2)[a] | — | $(5 \times 10^4)$ | $10^4$ | $(3 \times 10^4)$ | $(7 \times 10^5)$ | $(5 \times 10^4)$ |
| Centropages | 0.1–1 | — | — | 600 | — | — | $(6 \times 10^3)$ |
| Calanus | 0.1–1 | 0.28[b] | $2.5 \times 10^4$ | 200 | — | — | 100 |
| Ommastrephes | (0.1)[a] | (0.3)[b] | $10^4$ | 200 | $(2 \times 10^4)$ | $2 \times 10^5$ | $(10^4)$ |
| Sagitta | — | 70 | $5 \times 10^4$ | $10^4$ | — | — | — |
| Euphausia | — | (0.3)[b] | — | 800 | — | — | — |
| Salpa | — | — | — | 60 | — | — | — |

[a] Based upon analyses for potassium, assuming the Cs/K ratio in the organisms to be the same as that in sea-water.

[b] Derived from Ca analysis and Sr/Ca ratio given by Thompson and Chow (1956).

4—s. II

$0.31 \sim 0.42$ μμc/g of Ca in 1959. In the fish bones $^{90}$Sr content ranged from 0.19 μμc/g of Ca to 0.7 μμc/g of Ca. In a composite sample of plankton, Nakai *et al.* (1960) reported values as high as 6 μμc of $^{90}$Sr per 1 g of Ca collected in the coastal waters of Japan in 1959. The maximum value of $^{90}$Sr in benthos was 14.3 μμc/g of Ca in Cœlenterata collected at the ocean bottom near Hachijo-jima Island. These results showed the remarkable enrichment of the dissolved or particulated forms of the radionuclide in marine products.

Observations carried out by the Japanese Bikini Expedition (1954, 1956) indicated that the gross activity was from 1000 to 10,000 times more concentrated in zooplankton than in water when compared on a weight basis.

TABLE IV

Approximate Concentration Factors of Different Elements in Members of the Marine Biosphere. The Concentration Factors are Based on a Live Weight Basis (Krumholz, Goldberg and Boroughs, 1957)

| Element | Form in sea-water | Concentration in sea-water, μg/l. | Algae (non-cal-careous) | Concentration factors | | | |
| --- | --- | --- | --- | --- | --- | --- | --- |
| | | | | Invertebrates | | Vertebrates | |
| | | | | Soft | Skeletal | Soft | Skeletal |
| Na | Ionic | $10^7$ | 1 | 0.5 | 0 | 0.07 | 1 |
| K | Ionic | 380,000 | 25 | 10 | 0 | 5 | 20 |
| Cs | Ionic | 0.5 | 1 | 10 | — | 10 | — |
| Ca | Ionic | 400,000 | 10 | 10 | 1000 | 1 | 200 |
| Sr | Ionic | 7,000 | 20 | 10 | 1000 | 1 | 200 |
| Zn | Ionic | 10 | 100 | 5000 | 1000 | 1000 | 30,000 |
| Cu | Ionic | 3 | 100 | 5000 | 5000 | 1000 | 1000 |
| Fe | Particulate | 10 | 20,000 | 10,000 | 100,000 | 1000 | 5000 |
| Ni [a] | Ionic | 2 | 500 | 200 | 200 | 100 | — |
| Mo | Ionic-particulate | 10 | 10 | 100 | — | 20 | — |
| V | ? | 1 | 1000 | 100 | — | 20 | — |
| Ti | ? | 1 | 1000 | 1000 | — | 40 | — |
| Cr | ? | 0.05 | 300 | — | — | — | — |
| P | Ionic | 70 | 10,000 | 10,000 | 10,000 | 40,000 | 2,000,000 |
| S | Ionic | 900,000 | 10 | 5 | 1 | 2 | — |
| I | Ionic | 50 | 10,000 | 100 | 50 | 10 | — |

[a] Values from Laevastu and Thompson (1956).

According to Harley (1956), the radioactivity was on the average 470 times higher in plankton than in sea-water collected west of the Bikini area. Such an enrichment in plankton provides a detection device for the presence of radioactivity in sea-water. Ketchum and Bowen (1958) suggested that organisms may modify the distribution of radionuclides in the sea once living matter has accumulated the radioactivity. Following their death, the organisms will sink toward the bottom and carry with them radionuclides within the organic or skeletal structures. While living, some of the organisms may make vertical migrations which could transport a part of the radioisotopes from a contaminated to an uncontaminated layer. The net effect of such processes depends on the concentration factors or enrichment factors of radioelements in marine products. Generally the concentration factors of radioelements cover a range from less than unity to values as large as $10^6$ (see Tables III and IV).

The concentration factors of some radionuclides such as iron, cadmium, zinc, etc. are exceedingly high (Tables III and IV). The concentration of rare-earths has been confirmed; Ketchum and Bowen (1958) gave the value of the order of $10^3$ as a concentration factor of Ce. As to the concentration of cadmium in marine organisms, Kawabata and others determined 0.9 μμc of $^{113m}$Cd/g wet tissue in the liver of albacore caught in the tropical North Pacific in 1956. The presence of $^{115m}$Cd in fish organs was confirmed by Shirai et al. (1957). The concentration factor of cesium is low, which is attributed to the hold-back action by the stable alkali elements in sea-water.

Yamagata (1957) gave the values of 0.021 μμc of $^{137}$Cs/g wet weight of the total body of a squid and 0.016 μμc/g wet weight in the flesh of a bonito both caught near Japan in the early summer of 1957. The nuclidic differentiation of fission products in plankton was observed by A. H. Seymour et al. (1957) in the sample collected near the Bikini–Eniwetok proving ground. The averaged results are given in Table V in which the relative concentrations of cerium, praseodymium and ruthenium–rhodium are observed.

TABLE V

The Nuclidic Differentiation of Fission Products in Planktonic Organisms Collected near Ujelang Island (Seymour et al., 1957)

| Isotope | Observed % | Expected % | Observed/Expected |
|---|---|---|---|
| $^{95}$Zr | 8.8 | 8.2 | 1/1 |
| Rare earths | 18.0 | 19.5 | 1/1 |
| $^{103}$Ru, $^{106}$Ru, $^{106}$Rh | 18.0 | 5.7 | 3/1 |
| $^{141}$Ce, $^{144}$Ce, $^{144}$Pr | 26.0 | 13.2 | 2/1 |
| $^{140}$Ba | 3.1 | 10.8 | 1/4 |

They also observed that there was a difference in nuclidic content of plankton collected 90 miles north of Eniwetok and plankton collected 470 miles west of Eniwetok. They reported the presence of non-fission products such as $^{57}$Co, $^{58}$Co, $^{60}$Co, $^{65}$Zn in plankton and in the liver of flying fish caught near Eniwetok.

### 3. Artificial Radioactivity as a Tracer of Water Movements in the Ocean

Up to now, artificial radioactivity, as a tracer, has not been employed to any great extent to investigate water movements in the ocean. Miyake, Sugiura and Kameda (1955) initially observed the direct evidence for the large-scale advection and diffusion of fission products in sea-water from the Bikini–Eniwetok proving ground four months after nuclear tests in March, 1954. Their observations showed significant levels of radioactivity at a distance of 2000 km from Bikini, suggesting a westward drift of about 20 cm/sec along the North Equatorial Current. Little activity was found in the North Equatorial Counter Current. Nine months later, in the spring of 1955, the *Taney* survey (U.S.A.E.C., 1956) confirmed the same magnitude of westward drift of the radioactivity, showing a significant level of activity at least 7000 km downstream from the source of contamination. The higher activity was located at that time off the coast of Luzon Island, in the Philippines.

In the summer of 1955, the maximum activity was observed in the Kuroshio region off the coast of Japan. This established the direct connection between the North Equatorial Current and the Kuroshio.

The results of the calculation of diffusion coefficients from the time change of radioactivity distribution in the North Pacific revealed values between 0.8 and $1.4 \times 10^9$ where the lateral distance from the center of diffusion was from 890 km to 3200 km (Miyake and Saruhashi, 1960).

Folsom (see Revelle *et al.*, 1955) investigated the vertical mixing in the upper layer with the aid of fission products. Artifically radioactive substances were introduced at the surface in an area where the mixed layer was about 100 m thick. The rate of downward motion of the lower boundary of the radioactive water was approximately $10^{-1}$ cm sec$^{-1}$. This motion ceased abruptly at the bottom of the mixed layer.

The vertical distribution of artificial radioactivity in the open ocean was first studied by the Japanese Bikini Expedition in June, 1954. The results showed most of the activity to be confined to the mixed layer above the thermocline. However, depth profiles showed that activity was present at some locations a few hundred meters deep. Using such data Miyoshi (1956) calculated the diffusion coefficient in the vertical direction to be 5–10 g cm$^{-1}$ sec$^{-1}$ in the tropical area of the North Pacific. In March, 1955, the activity extended to about 600 m below the surface (Harley, 1956).

Horizontal diffusion in the relatively shallow waters of the Bikini lagoon was studied by Munk, Ewing and Revelle (1949). The average value of the horizontal component of the effective diffusion coefficient, $A_h$, during three days in which the radius of the radioactive area increased to about 4 km was found to be $1.5 \times 10^5$ g cm$^{-1}$ sec$^{-1}$ and $A_h/r$ was close to 0.5 g cm$^{-2}$ sec$^{-1}$. This latter value is in good agreement with estimates of this ratio obtained by other means.

The radioisotopes useful for tracer experiments in the ocean must have a half-life compatible with the mixing rate of oceanic waters and yet short enough not to bring about a permanent hazard. Revelle *et al.* (1955) drew attention to $^{86}$Rb (19.5 d), $^{131}$I (8.0 d) and $^{140}$Ba (12.8 d). Craig (1957a) discussed the usefulness of tritium and $^{14}$C as tracers of physical processes in the sea.

## 4. Disposal of Radioactive Waste in the Ocean

In connection with the peaceful use of atomic energy, the problem of the disposal of radioactive waste in the ocean has been extensively discussed. Needless to say, waste products from nuclear reactions require special attention because they constitute hazards even in very low concentrations. In addition to the radioactive waste from nuclear tests, waste products are produced by nuclear reactors. Up to now, most of the fission products introduced into the ocean have been from weapon tests; only small quantities of radioactive waste from reactors have been put into the sea. But, in the future radioactive waste from nuclear power plants will bring about formidable problems in oceanography, if the sea is regarded as one of the possible sites of disposal.

TABLE VI

Maximum Permissible Levels of Activity in Marine Materials Adopted at Windscale, U.K.

| Type of activity | m.p.l. in water, $\mu c/ml$ | Marine material | m.p.l. in material, $\mu\mu c/g$ |
|---|---|---|---|
| Total beta | | shore sand | 25,000 |
| | | shore silt | 100,000 |
| | | sea bed | 100,000 |
| Total alpha (assumed to be $^{239}Pu$) | $3 \times 10^{-6}$ | fish | 30 |
| | | seaweed | 10 |
| Total beta (including or excluding $^{106}Ru$) | $2 \times 10^{-4}$ | fish | 2000 |
| | | seaweed | 670 |
| Ruthenium–106 | $1 \times 10^{-4}$ | fish | 1000 |
| | | seaweed | 330 |
| Strontium–90 | $8 \times 10^{-7}$ | fish | 8 |
| | | seaweed | 2.7 |

Experimental studies have been made for several years on the discharge of radioactivity from the Windscale Works of the United Kingdom Atomic Energy Authority to the coastal waters of the Irish Sea off Cumberland through a pipe-line extending about 3 km beyond the high-water mark. The results indicated that the maximum permissible discharge rate is not less than 20,000 curies per month of total beta activity. The maximum permissible levels adopted at Windscale are given in Table VI (after Dunster, 1958).

The calculation of the potential steady-state situation in which radioactive wastes are uniformly introduced into the oceans was done by Craig (1957), assuming an average residence time in the deep sea of 300 years and a constant world fission rate of $^{235}$U equal to 1000 tons per year after 100 days cooling, followed by the disposition of all the fission products into the sea. The results of calculation indicate that the total activity in the mixed layer will be equal to that from $^{40}$K. The concentration of $^{90}$Sr and $^{137}$Cs would be respectively $6.5 \times 10^{-5}$ μc and $8.4 \times 10^{-5}$ μc per liter or 160 μμc of $^{90}$Sr per 1 g of calcium and 221 μμc of $^{137}$Cs per 1 g of potassium in solution in sea-water.

# References

Bowen, V. T. and T. T. Sugihara, 1957. Strontium-90 in North Atlantic surface waters. *Proc. Nat. Acad. Sci.*, **43**, 576–580.

Craig, H., 1957. Disposal of radioactive wastes in the ocean; The fission product spectrum in the sea as a function of time and mixing characteristics. *The Biological Effects of Atomic Radiation*, Pub. No. 551, *Nat. Acad. Sci.*, Chap. 3, 34–42.

Craig, H., 1957a. Isotopic tracer techniques for measurement of physical and chemical processes in the sea and the atmosphere. *The Biological Effects of Atomic Radiation*, Pub. No. 551, *Nat. Acad. Sci.*, Chap. 11, 103–120.

Dunster, H. J., 1958. The disposal of radioactive liquid wastes into coastal waters. *Second U.N. Intern. Conf. Peaceful Uses of Atomic Energy, Geneva.* A/Conf. 15/P/297.

Folsom, T. R., 1960. Fallout cesium in surface sea water off the California coast (1959–1960) by gamma ray measurements. Private communication.

Harley, J. H. (ed.), 1956. Operation Troll, 37 pp. Joint preliminary report, U.S. Atomic Energy Commission Office of Naval Research. U.S. Atomic Energy Commission, New York.

Hiyama, Y., 1960. Radioactive contamination of marine products in Japan. Presented to United Nations Scientific Committee on the Effect of Atomic Radiation.

Hiyama, Y. *et al.*, 1957. Radiological data in Japan. United Nations Scientific Committee on the Effect of Atomic Radiation. A/AC.82/G/R.70.

Japanese Fishery Agency, 1955. Report on the investigations of the effects of radiation in the Bikini region. *Res. Dept., Jap. Fish. Agency, Tokyo*, 191 pp.

Japanese Fishing Agency, 1956. Radiological survey of western area of the dangerous zone around the Bikini-Eniwetok atolls, investigated by the Shunkotsu-Maru in 1956. Pt. 1, 143 pp.

Ketchum, B. H. and V. T. Bowen, 1958. Biological factors determining the distribution of radio-isotopes in the sea. *Contrib.* No. 968 *Woods Hole Oceanog. Inst.* (mimeographed).

Krumholz, A., E. D. Goldberg and H. Boroughs, 1957. Ecological factors involved in the uptake, accumulation and loss of radionuclides by aquatic organisms. *The Biological Effects of Atomic Radiation*, Pub. No. 55, *Nat. Acad. Sci.*, Chap. 7, 69–79.

Laevastu, T. and T. G. Thompson, 1956. The determination and occurrence of nickel in sea water, marine organisms, and sediments. *J. Cons. Explor. Mer*, **21**, 125–143.

Martin, DeC. Jr., 1957. The uptake of radioactive wastes by benthic organisms. *Proc. 9th Pacific Sci. Cong., Bangkok*, **16**, 167–169.

Miyake, Y. and K. Saruhashi, 1960. Vertical and horizontal mixing rates of radioactive material in the ocean. Report of the scientific conference on the disposal of radioactive wastes, IAEA and UNESCO, November, 1959, Monaco, 167–173.

Miyake, Y., K. Saruhashi and Y. Katsuragi, 1960. Strontium 90 in western North Pacific surface waters. *Papers Met. Geophys., Tokyo*, **11**, No. 1, 188–190.

Miyake, Y., K. Saruhashi, Y. Katsuragi and T. Kanazawa, 1961. Cesuim-137 and strontium-90 in sea water. *J. Radiation Res.*, **2**, 25–28.

Miyake, Y. and Y. Sugiura, 1955. The radiochemical analysis of radionuclides in sea water collected near Bikini Atoll. *Papers Met. Geophys., Tokyo*, **6**, 33–37.

Miyake, Y., Y. Sugiura and K. Kameda, 1955. On the distribution of radioactivity in the sea around Bikini Atoll in June 1954. *Papers Met. Geophys., Tokyo*, **5**, 253–262.

Miyoshi, H., 1956. The atoll and radiologically contaminated water. *Research in the Effect and Influences of the Nuclear Bomb Test Explosion*, Vol. 2, 964–981. Japan Society for the Promotion of Science, Tokyo.

Munk, W. H., G. C. Ewing and R. Revelle, 1949. Diffusion in Bikini Lagoon. *Trans. Amer. Geophys. Un.*, **30**, 59–66.

Nakai, Z., R. Fukai, H. Tozawa, S. Hattori, K. Okubo and T. Kidachi, 1960. Radioactivity of marine organisms and sediments in the Tokyo Bay and its southern neighbourhood. In "Radioactive contamination of marine product in Japan", U.N. Scientific Committee on the Effect of Atomic Radiation, Document A/AC.82/G/L, No. 394.

Revelle, R., T. R. Folsom, E. D. Goldberg and J. D. Isaacs, 1955. Nuclear Science and Oceanography. *First U.N. Intern. Conf. Peaceful Uses of Atomic Energy, Geneva*. A/Conf. 8/P/277, 22 pp.

Seymour, A. H., E. E. Held, F. G. Lowman, J. R. Donaldson and Dorothy J. South, 1957. *Survey of radioactivity in the sea and in pelagic marine life west of the Marshall Islands, September 1–20, 1956*. Published by Technical Information Service Extension, Oak Ridge, Tenn.

Shirai, J., M. Saiki and S. Ohno, 1957. Studies on the radioelement in the contaminating radioactive fish. III. On skipjacks caught at the Pacific Ocean in 1956 (Part 2), On the presence of [113m]Cd. *Bull. Jap. Soc. Sci. Fish.*, **22**, 651–653.

Thompson, T. G. and T. J. Chow, 1956. The strontium—calcium atom ratio in carbonate-secreting marine organisms. *Papers in Marine Biology and Oceanography*, 20–39. Pergamon Press, London.

Yamagata, N., 1957. See Hiyama, Y. *et al.*, 1957.

Yamagata, N. and S. Matsuda, 1959. Cesium-137 in the coastal waters of Japan. *Bull. Chem. Soc. Japan*, **32**, 497–502.

# 4. RADIOISOTOPES AND LARGE-SCALE OCEANIC MIXING

## W. BROECKER

## 1. Introduction

Considerable information regarding the patterns and rates of large-scale circulation can be obtained from a knowledge of the distribution of various radioactive isotopes in the sea. Both naturally produced radioisotopes with half-lives of the same order of magnitude as the time constants for oceanic mixing processes and man-made radioactive isotopes, produced in sufficient quantities by nuclear tests, are useful in this type of study. The natural isotope approach is most valuable in the study of deep water masses, where residence times of more than 100 years are involved, and the artificial isotopes for problems related to near-surface water-masses, which undergo much more rapid mixing.

As the circulation of the ocean is exceedingly complex, radioisotope data alone are not adequate to establish mixing rates and patterns. It is only when the data are combined with evidence from classical oceanographic research that any progress can be made. The most useful method of combining these two quite different types of information is to set up models of oceanic mixing based on classical observations and theory. The isotope data are used first to reject certain of these models and then to establish absolute values for various rate parameters in the remaining models. Obviously if the distribution of more than one isotope is studied, more stringent limits can be set up allowing more sophisticated models to be treated. To date the isotope approach is in its infancy; the necessary data are still sparse and the models adopted greatly oversimplified. Nevertheless, several important conclusions can be drawn from the available work.

## 2. Useful Isotopes

To be suitable for large-scale oceanic mixing studies, an isotope should have the following ideal characteristics: (1) it should be present in measurable quantities in all parts of the ocean, (2) differences in concentration well outside the limits of measurement error should exist, (3) the mode and rate of injection of the isotope into the system must be known both as a function of time and of space, (4) the isotope should move with the water acting as an infinitely soluble salt, and (5) contributions of natural and artificial production of the isotope should be distinguishable.

Only four naturally occurring isotopes, $^3$H, $^{14}$C, $^{32}$Si and $^{226}$Ra, have been given serious consideration in connection with circulation studies. Natural tritium, which potentially should be an excellent tracer, suffers from two very serious difficulties: (1) because of its low production rate ($< 2$ T atoms cm$^{-2}$ sec$^{-1}$) and short half-life ($\sim 12$ years), the abundance of tritium in the deep water masses is more than an order of magnitude below the present limits of experimental detectability, and (2) bomb tests and nuclear plant leaks have added an amount of artificial tritium probably exceeding that previously

[*MS received October, 1960*]     88

present. The exact amount added, the times of addition, and the pre-bomb background levels are subject to much uncertainty. Whether the numerous problems involved in unraveling the pre-bomb levels of natural tritium can be treated with sufficient certainty to allow valid conclusions to be drawn is unlikely. If tritium is of value in studies of ocean circulation, it will be through the use of the man-made tritium tracer.

The application of $^{32}Si$ (half-life $\sim 700$ years) recently discovered in nature by Lal *et al.* (1960) is also extremely difficult. The low activity levels ($\sim 15$ dpm/kg Si) make direct measurements in sea-water prohibitively time-consuming and expensive. Beyond this, as dissolved silica is greatly depleted in surface ocean water, $^{32}Si$ must move to a large extent independently of the water descending from surface to depth as diatom tests. Whether the biologic cycle can be sufficiently well understood to make $^{32}Si$ a valuable water tracer is somewhat doubtful.

Koczy's (1958) work on $^{226}Ra$ has already demonstrated the value of this isotope as a natural tracer. As discussed in Volume 3 release from deep-sea sediments is the dominant mode of addition of radium to the oceans. Although little detailed knowledge exists regarding the rates of release, there is some basis for assuming that the release should be relatively constant with time and with geographic location (if sufficiently large areas are considered). Once in the ocean the possibility of incorporation in particulate matter and consequent settling must be considered. Since experimental measurements suggest a universal discrimination against radium relative to calcium during the formation of carbonate materials, the absence of vertical gradients in the calcium concentration in the ocean water provides evidence that the transport of $^{226}Ra$ as calcium carbonate is negligible compared to transport in solution. The evaluation of the importance of transport in some other solid phase with a high selectivity for radium awaits a more detailed study of the vertical distribution of chemically similar elements such as strontium and barium. Of great importance is the fact that radium is not released to and transported through the atmosphere as is the case for radiocarbon.

Natural radiocarbon has received the most attention as an oceanic tracer. It is added to the oceans from the atmosphere, which has been shown to be a nearly uniform source both in space and time. Transport in solid phases can be evaluated through dissolved $O_2$ and the total $CO_2$ data, and from sedimentation rates of $CaCO_3$ and the degree of preservation of $CaCO_3$ in bottom deposits. These considerations suggest that more than 90% of the transport of radiocarbon within the ocean occurs with the water itself and that interaction with the sedimentary reservoir is negligible. One rather serious complication enters into the interpretation of radiocarbon results; whereas the transport of dissolved solids or even of water itself through the atmosphere is of little importance, that of $CO_2$ has great significance. The quantitative evaluation of this mode of transport is therefore necessary if $^{14}C$ data is to be used for water circulation rate estimates.

Of the numerous isotopes produced during nuclear tests, $^{14}C$, $^{3}H$, $^{90}Sr$ and

$^{137}$Cs seem to be the best suited for tracer studies. $^{90}$Sr (half-life 28 years) and $^{137}$Cs (half-life 30 years) have the great advantage that, prior to bomb tests, they were not present in the oceans, making corrections for natural background unnecessary. Also the availability of extensive data as to the time and space distribution of these isotopes in terrestrial soils provides a means of estimating the quantity of fallout added to the oceans as a function of geographic location and time. One potential difficulty is the possibility that $^{90}$Sr and $^{137}$Cs are incorporated into particles which move independently of the water. Although the uniform oceanic distribution of dissolved calcium (chemically similar to $^{90}$Sr) and of dissolved potassium (chemically similar to $^{137}$Cs) suggest that such transport is not important, more complete data on the depth variation in the concentrations of stable Sr and Cs in the oceans would provide a more reliable evaluation. In the absence of these data, the ratio of $^{90}$Sr to $^{137}$Cs at various levels in the ocean should provide a partial check. As $^{90}$Sr and $^{137}$Cs are dissimilar chemically they should be separated to some extent during uptake by particulate matter, whereas if movement with the water is dominant little variation in the ratio would be expected.

For bomb-produced tritium the input function is less well defined. Since $^3$H is not retained by the soils, as is the case with $^{90}$Sr and $^{137}$Cs, no permanent record of the quantity added to a given area is available. Estimates based on tritium concentrations in rainfall are not nearly so reliable because of the limited areal and time coverage. Although the amount of bomb tritium added to the surface-water systems probably exceeds that present from natural sources by a factor of at least three, the uncertainty in the natural background values will make interpretation difficult in many cases.

Although the quantity of $^{14}$C added to the carbon cycle by bomb testing is small (approximately 1% of the natural inventory), it will nevertheless prove very useful as a tracer. As a result of $^{14}$C production by tests carried out before the moratorium in the fall of 1958, the $^{14}$C concentration in surface ocean water may rise as much as 10% above pre-bomb levels. As the natural background is known to about $\pm 1\%$ in many areas and since measurements can be made to $\pm 0.5\%$ or better, the bomb $^{14}$C should be measurable in surface and near-surface masses.

### 3. Steady-State Distribution of $^{14}$C

Since at this time only natural radiocarbon data are available in sufficient quantity to warrant large-scale model calculations, the mixing rates in ocean circulation models proposed to date will be evaluated on the basis of radiocarbon data alone. Where other data are available they will be used as a cross-check.

The first question which arises in the consideration of oceanic mixing processes is whether the ocean can be approximated as a steady-state system (patterns and rates of mixing remain essentially constant with time). Unfortunately no definite answer can be given. However, since the complications

TABLE I

$^{14}C$ Results on Surface Ocean Samples

| Location | Number of samples | Range of results | Uncorrected average $\Delta^{14}C^b$ | Bomb $^{14}C$ corrected average (to 1954) | Standard deviation | Ref.[a] |
|---|---|---|---|---|---|---|
| North Atlantic 60°–80°N | 3 | −37 to −34 | −35 ± 4 | −38 ± 4 | ± 2 | 1 |
| North Atlantic 15°–40°N | 18 | −65 to −33 | −49 ± 3 | −52 ± 2 | ± 9 | 2 |
| Caribbean Sea | 8 | −76 to −46 | −56 ± 3 | −56 ± 3 | ± 9 | 2 |
| South Atlantic 0°–40°S | 16 | −76 to −39 | −57 ± 3 | −63 ± 3 | ± 10 | 2 |
| Southwest Pacific 15°–42°S | 16 | −56 to −33 | −41 ± 2 | −49 ± 2 | ± 6 | 4 |
| Eastern Pacific 20°N–35°S | 4 | −42 to −38 | −40 ± 2 | −56 ± 8 | ± 2 | 3 |
| S. Atlantic + Pacific $H_2O$ temperature 5°–11°C | 6 | −89 to −60 | −72 ± 6 | −80 ± 7 | ± 9 | 2, 3, 4 |
| Antarctic $H_2O$ temperature <3°C | 5 | −178 to −111 | −138 ± 12 | −141 ± 12 | ± 27 | 2, 4 |
| Average surface ocean | — | — | — | −58 ± 10 | — | — |

[a] 1. Fonselius and Ostlund (1959).
2. Broecker et al. (1960).
3. Bien et al. (1960).
4. Rafter and Fergusson (1958).

[b] Per mil difference from fractionation corrected steady-state atmospheric $^{14}C/^{12}C$ ratio (see Broecker and Olson, 1959).

arising from non-steady-state operation are beyond resolution, the simple models developed to date all assume steady-state operation. As the interpretation of the isotope results is strongly dependent on this assumption, discussions of the evidence supporting the steady-state hypothesis and of the consequences of non-steady-state operation are given below.

The available radiocarbon data for surface ocean water are summarized in Table I as averages for various geographic localities. The uncorrected averages are given in column 4; the averages after correction for the increase in concentration resulting from the addition of bomb $^{14}C$ in column 5. The reduction

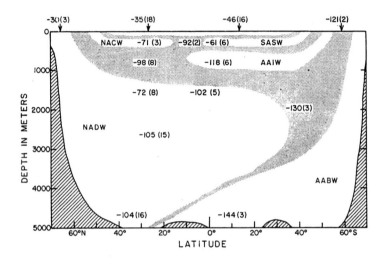

Fig. 1. Vertical profile through the western basin of the Atlantic Ocean showing idealized water-mass distribution. The numbers represent the average $^{14}C$ concentrations given as per mil difference from the age-corrected concentration of $^{14}C$ in 19th-century atmospheric $CO_2$. All the $^{14}C$ results are normalized to a common $^{13}C$ concentration in order to eliminate differences resulting from isotope fractionation. Details of this so-called $\Delta$ $^{14}C$ scale are given by Broecker and Olson (1959). The numbers in parentheses indicate the number of samples upon which the average is based.

in oceanic $^{14}C/^{12}C$ ratios resulting from $CO_2$ released by the combustion of coal and oil is assumed negligible. The corrected results are assumed to be the steady-state $^{14}C$ concentrations. The results indicate a nearly uniform $^{14}C/^{12}C$ ratio in the region from 50°N to 50°S with significantly lower values in the Antarctic and perhaps slightly higher values in the northern North Atlantic.

The results on sub-surface samples in the Atlantic Ocean are summarized in Table II and in Fig. 1. The majority of water originating in the northern North Atlantic has a fractionation-corrected $^{14}C/^{12}C$ ratio 10% below that in the steady-state atmosphere. There are three exceptions: (1) the North Atlantic Central Water has a value intermediate between that of the deeper water and that of the surface water, (2) a tongue of water in the western North Atlantic

TABLE II

$^{14}$C Results on Sub-surface Water Masses of the Atlantic Ocean[a]

| Water mass name | Sample localities | | Depth range, m | No. of samples | Average $\Delta^{14}$C | $\sigma$ | Estimated $\Delta^{14}$C for source region | Difference between source and depth value | Apparent age (yr) |
|---|---|---|---|---|---|---|---|---|---|
| | Lat. range | Long. range | | | | | | | |
| N. Atlantic Central Water | 15°–30°N | 25°–60°W | 200 to 400 | 3 | −71 ± 4 | ± 4 | −52 ± 7 | 19 ± 8 | 160 ± 70 |
| S. Atlantic Central Water | 0°–30°S | 5°E–35°W | 200 to 400 | 6 | −61 ± 5 | ± 12 | −63 ± 8 | < 10 | < 80 |
| N. Atlantic Intermediate Water | 5°–25°N | 20°–80°W | 800 to 1100 | 8 | −99 ± 4 | ± 11 | −45 ± 9 | 44 ± 10 | 400 ± 90 |
| Antarctic Intermediate Water | 0°–40°S | 5°E–35°W | 500 to 1200 | 6 | −118 ± 7 | ± 17 | −90 ± 20 | 28 ± 22 | < 350 |
| Upper North Atlantic Deep Water | 10°–40°N | 50°–70°W | 1200 to 2400 | 8 | −72 ± 6 | ± 16 | −45 ± 9 | 27 ± 9 | 350 ± 160 |
| Upper North Atlantic Deep Water | 20°S–25°N | 0°–40°W | 1400 to 2000 | 7 | −104 ± 3 | ± 8 | −45 ± 9 | 59 ± 10 | 500 ± 90 |
| S. Atlantic Deep Water | 20°–40°S | 5°E–40°W | 1500 to 2200 | 3 | −130 ± 7 | ± 12 | — | — | — |
| N. Atlantic Deep Water | 40°S–35°N | 5°E–70°W | 2500 to 4000 | 15 | −105 ± 3 | ± 11 | −45 ± 9 | 60 ± 10 | 500 ± 90 |
| N. Atlantic Deep Water | 23°–40°N | Western Basin | > 4000 | 16 | −104 ± 3 | ± 13 | −45 ± 9 | 59 ± 10 | 500 ± 90 |
| N. Atlantic Deep Water | 15°–25°N | Eastern Basin | > 4000 | 3 | −127 ± 5 | ± 8 | −45 ± 9 | 82 ± 10 | 750 ± 100 |
| Antarctic Bottom Water | 0°–35°S | Western Basin | > 4000 | 3 | −144 ± 6 | ± 10 | < −110 | < 34 | < 250 |

[a] Based on results published by Broecker et al. (1960).

### TABLE III

¹⁴C Results on Sub-surface Water Masses of the Pacific Ocean[a]

| Sample localities | | | No. of samples | Average $\Delta^{14}C$ | Standard deviation | Estimated $\Delta^{14}C$ for source region | Difference between source and depth value | Apparent age (yr) | Water mass |
|---|---|---|---|---|---|---|---|---|---|
| Lat. range | Long. range | Depth range | | | | | | | |
| 5°–25°S | 170°–180°W | 200 to 500 | 3 | −90±20 | ±35 | — | — | — | Pacific Central Water |
| 5°–30°N | 115°–130°W | 300 to 600 | 4 | −130±17 | ±33 | — | — | — | Pacific Central Water |
| 5°–40°S | 170°–180°W | 700 to 1000 | 4 | −115±21 | ±43 | — | — | — | Pacific Intermediate Water |
| 5°–25°S | 170°–180°W | 2000 to 3000 | 4 | −176±10 | ±22 | −150±15 | 26±18 | <350 | Pacific Deep Water |
| 25°–40°S | 80°–145°W | 3500 | 5 | −195±4 | ±9 | −150±15 | 45±16 | 425±150 | Pacific Deep Water |
| 15°S–30°N | 115°–135°W | 2000 to 3500 | 8 | −233±5 | ±13 | −150±15 | 83±16 | 925±150 | Pacific Deep Water |

[a] Results from western Pacific from Burling and Garner (1959) and for eastern Pacific from Bien et al. (1960).

with a $^{14}C/^{12}C$ ratio about 3% higher than the surrounding water, and (3) the deepest water in the eastern basin of the Atlantic is about 2.5% lower in $^{14}C$ than both the overlying N.A.D.W. and adjacent N.A.D.W. on the opposite side of the mid-Atlantic ridge. The A.A.I.W. and A.A.B.W., which originate in the southern South Atlantic, are significantly lower in $^{14}C$ than the deep water originating in the north. Residence time estimates given in the table are based on the assumption that the difference in activity between the water at depth and that in the corresponding source area is the result of radioactive decay ($^{14}C$ decays 1% each 80 years).

Results from the sub-surface Pacific are given in Table III. The deep water ($> 2000$ m) ranges from 0.687 to 0.850 the atmospheric value. As the Antarctic surface is deficient in $^{14}C$, residence times are similar to those for the Atlantic. As shown by Bien et al. (1960) there appears to be a south-to-north decrease in the $^{14}C/^{12}C$ ratio in the deep Pacific. Again the central waters have $^{14}C$ concentrations intermediate between those of the deep and of the surface waters.

## 4. Ocean–Atmosphere $CO_2$ Exchange Rates

As all of the water masses of the ocean simultaneously interact with one another it is necessary to consider the entire ocean as a unit before a complete understanding of the $CO_2$ cycle can be obtained. Before discussing such models, it is necessary to obtain estimates of the rate of exchange of $CO_2$ between the oceans and the atmosphere. As mentioned above, since this exchange is important to the $CO_2$ cycle and not that of dissolved solids, it must be quantitatively understood before $^{14}C$ data can yield estimates of water mixing rates. Fortunately evaluation of this exchange can be carried out without reference to oceanic mixing processes.

If it is assumed that the exchange rate is geographically uniform the following equation holds (since at steady-state the $^{14}C$ inventory in the ocean must remain constant):

$$IC_A A_{A-S} = EC_S A_{A-S} + \lambda C_O N_O,$$

where $I$ and $E$ are the invasion and evasion rates of $CO_2$ through the ocean–atmosphere interface (moles/m$^2$/yr), $C_A$, $C_S$ and $C_O$ are the $^{14}C$ concentrations in average atmospheric $CO_2$, average surface ocean $CO_2$ and average oceanic $CO_2$ respectively (moles $^{14}C$ $O_2$/mole $CO_2$), $A_{A-S}$ is the area of the ocean–atmosphere interface (m$^2$), $N_O$ is the quantity of $CO_2$ in the ocean (moles) and $\lambda$ is the decay constant of $^{14}C$ (yr$^{-1}$). Since the rate of loss of $CO_2$ to the sediments as $CaCO_3$ and the influx of bicarbonates from rivers are negligible, $E$ and $I$ should be equal at steady state. Therefore,

$$I = \frac{\lambda(C_O/C_A)N_O}{[1-(C_S/C_A)]\,A_{A-S}}$$

Thus, from a knowledge of the ratio of the average $^{14}C$ concentration in the dissolved $CO_2$ in oceans as a whole to that in atmospheric $CO_2$ ($C_O/C_A$) and of a similar ratio for average surface ocean $CO_2$ and atmospheric $CO_2$ ($C_S/C_A$), the average rate of exchange can be estimated. From existing data $C_O/C_A$ has been estimated as $0.85 \pm 0.05$ and $[1 - (C_S/C_A)]$ as $0.050 \pm 0.015$. The error in the latter stems from two sources: (1) the experimental error in the measurements, and (2) the uncertainty in the correction for the change brought about by the addition of industrial $CO_2$ to the system. The above values yield an invasion rate of $20 \pm 7$ moles m$^{-2}$ yr$^{-1}$. As a 1-m$^2$ column of air contains about 100 moles of $CO_2$ and since about 30% of the earth is covered by land, this corresponds to a mean atmospheric residence time for $CO_2$ with respect to transfer into the ocean of seven years.

Consideration must be given, however, to the possibility that the exchange is not geographically uniform. Kanwisher has demonstrated the importance of wind velocity in the rate of exchange. He points out that wind velocities are, on the average, considerably higher in the Antarctic region than over the rest of the ocean. This leads to the possibility that the exchange rate could be, on the average, as much as a factor of five higher in the Antarctic region. If this were the case, an exchange rate of 50 moles m$^{-2}$ yr$^{-1}$ would be required for the Antarctic and only 10 moles m$^{-2}$ yr$^{-1}$ for the remainder of the ocean (assuming the Antarctic comprises 12% of the ocean surface). The atmospheric residence time for $CO_2$ would rise to about 10 years. This rise occurs because the Antarctic surface water has a $^{14}C/^{12}C$ ratio averaging about 8% lower than that for the remainder of the surface ocean. As laboratory measurement of the $CO_2$ exchange rate in unstirred sea-water (unpublished work by the author) yields a value of 6 moles m$^{-2}$ yr$^{-1}$, the rate at any point in the ocean must be at least this great. In order to maintain the mixed layer exchange rate above this minimum, the contrast between the Antarctic and mixed layer of the ocean cannot exceed a factor of about eight. From the discussion which follows it will become clear that the magnitude of this geographic effect (especially as it affects the ratio of Antarctic to non-Antarctic exchange rate) is critical to the interpretation of the $^{14}C$ data.

## 5. Oceanic Mixing Models

The models considered to date are all of the box type. The ocean–atmosphere system is broken up into a series of independent, well-mixed reservoirs. Carbon dioxide with a $^{14}C/^{12}C$ ratio typical of the entire reservoir is exchanged by adjacent reservoirs. If the average $^{14}C/^{12}C$ ratio for each of the various reservoirs is fixed, the exchange rates required to maintain this distribution of $^{14}C$ can be computed.

The models considered to date can be placed in four groups: (1) the two-layer model of Arnold and Anderson (1957) and of Craig (1957), (2) the outcrop model of Craig (1958), (3) the three-box model and (4) the world-ocean model of Broecker *et al.* (1960). These models (shown diagrammatically in Figs. 2 through 8) will be referred to as models I, II, III and IV for simplicity. In each model the main body of the ocean is divided into a well-mixed surface

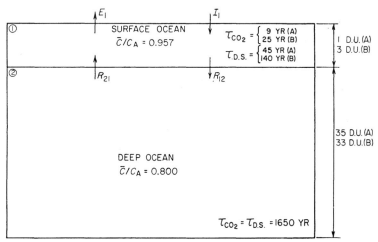

TWO LAYER MODEL (Cases IA and IB)
E1
I1
①
SURFACE OCEAN
C̄/CA = 0.957
τCO2 = { 9 YR (A) 25 YR (B)
τD.S. = { 45 YR (A) 140 YR (B)
1 D.U. (A)
3 D.U. (B)
②
R21
R12
35 D.U. (A)
33 D.U. (B)
DEEP OCEAN
C̄/CA = 0.800
τCO2 = τD.S. = 1650 YR
E1 = I1 = 22 M/M2/YR
R21 = R12 = 5 M/M2/YR

Fig. 2. Cases A and B differ in the thickness assigned to the mixed layer. The $^{14}C$ values for each reservoir are given as fractions of the pre-industrial atmospheric $^{14}C/^{12}C$ ratio and are based on the average of the available measurements. The transfer rates are computed assuming steady state. $\tau_{CO_2}$ and $\tau_{D.S.}$ refer to the mean residence times for $CO_2$ and for dissolved solids.

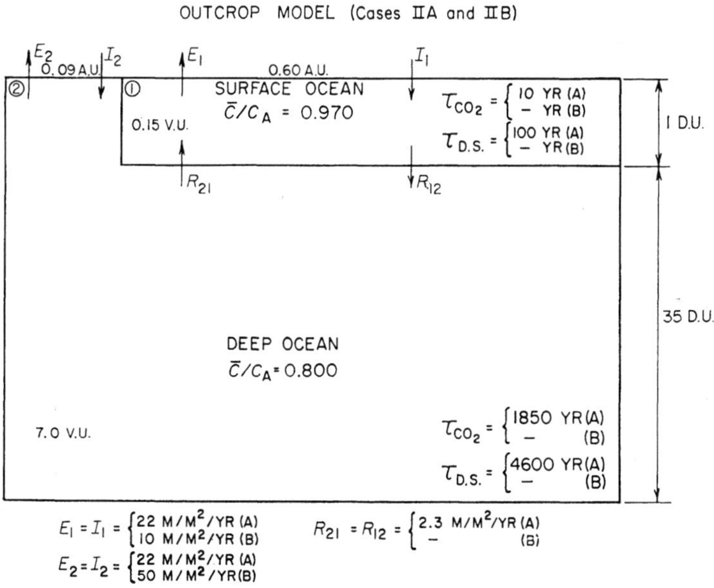

OUTCROP MODEL (Cases IIA and IIB)
E2
0.09 A.U.
I2
E1
0.60 A.U.
I1
②
①
SURFACE OCEAN
C̄/CA = 0.970
0.15 V.U.
τCO2 = { 10 YR (A) – YR (B)
τD.S. = { 100 YR (A) – YR (B)
1 D.U.
R21
R12
35 D.U.
DEEP OCEAN
C̄/CA = 0.800
7.0 V.U.
τCO2 = { 1850 YR (A) – (B)
τD.S. = { 4600 YR (A) – (B)
E1 = I1 = { 22 M/M2/YR (A) 10 M/M2/YR (B)
R21 = R12 = { 2.3 M/M2/YR (A) – (B)
E2 = I2 = { 22 M/M2/YR (A) 50 M/M2/YR (B)

Fig. 3. Cases A and B differ in the value assigned to the ratio of the ocean–atmosphere exchange rate for the polar ocean to that for the surface ocean. Case B yields negative values of $R_{12}$ and $R_{21}$ indicating that the rate of $^{14}C$ addition through the deep-water outcrop is too great to allow the low $^{14}C/^{12}C$ ratios observed in the deep sea. Thus, if this model is to be retained, either the contrast in the outcrop-mixed layer exchange rate or the area of outcrop must be considerably smaller than those adopted in Case B.

THREE RESERVOIR MODEL (Cases IIIA and IIIB)

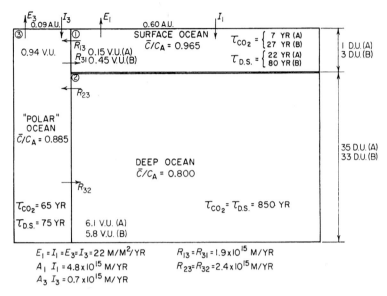

Fig. 4. Cases A and B differ in the thickness assigned to the surface layer. No direct mixing is assumed to occur between reservoirs 1 and 3.

THREE RESERVOIR MODEL (Cases IIIC and IIID)

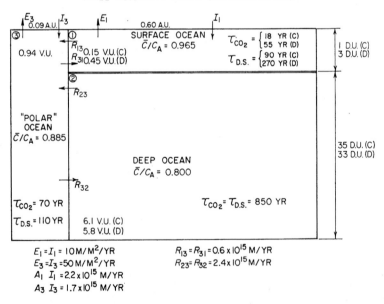

Fig. 5. Cases C and D differ in the thickness assigned to the surface layer. These two cases differ from the two in Fig. 4 only in the value assigned to the ratio of the exchange rate for the polar ocean to that for the surface ocean.

reservoir (often referred to as the "mixed layer") and a much larger deep reservoir. In order to take into account the fact that the deep water masses are fed by waters sinking in the polar regions, model II allows the deep water to outcrop at the surface and model III incorporates a third well-mixed reservoir at the end of the ocean. The world-ocean model, IV, separates the Atlantic from the Pacific plus Indian Oceans incorporating two reservoirs of vertical mixing, the Antarctic and the Arctic, at the ends of the two-layer oceans. The Atlantic and Pacific–Indian Oceans interact only through the Antarctic reservoir, the Arctic communicating only with the Atlantic (see Broecker *et al.*, 1960, for details of model IV). As each additional reservoir adds the

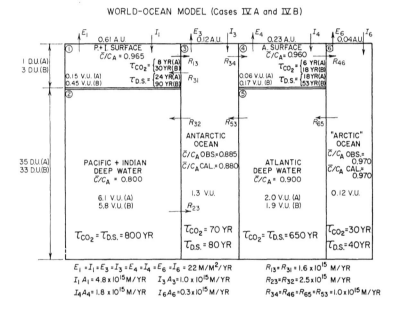

Fig. 6. Cases A and B differ in the thickness assigned to the surface ocean layer. Mixing is assumed to take place only as indicated by the arrows, i.e. a cyclic flow in the Atlantic and no direct mixing between the surface and deep reservoirs. The observed $^{14}C$ averages for the Antarctic and Arctic reservoirs, although unnecessary for the rate calculations, are given for comparison with those demanded by the model in order to satisfy the steady-state assumptions.

necessity for further assumptions and parameters, the available isotope data do not allow much more complexity than that in the present models. The numerous oversimplifications are obvious when one compares the box models with the water-mass structure of the oceans.[1]

Volumes and ocean–atmosphere interface areas have been assigned to each

---

[1] The surface-ocean $^{14}C/^{12}C$ ratios used in these models are about 10 per mil too high resulting from an industrial $CO_2$ correction applied at the time of writing (July, 1960). This correction is now known to be considerably less than that applied.

reservoir (see Figs. 2 to 8). The parameters for models I, II and III have been
chosen to approximate the Pacific plus Indian Oceans. Ratios of the fractionation-corrected $^{14}C$ concentrations in the dissolved $CO_2$ of each reservoir to that
in atmospheric $CO_2$ given in these figures are based on the data summarized
in Tables I, II and III. No data are available for the Indian Ocean.

For each model the mean residence time of both $CO_2$ and of dissolved solids
has been calculated for each reservoir. The calculations are made for mixed
layer depths of 100 m and of 300 m, and for ratios of the ocean–atmosphere

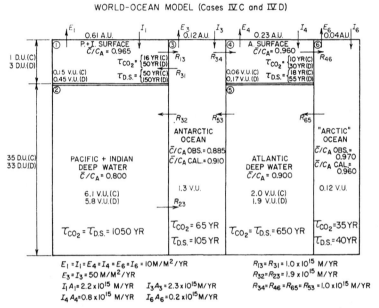

Fig. 7. Cases C and D differ in the thickness assigned to the surface ocean layers. The
cases in this figure differ from those in Fig. 6 only in the value assigned to the ratio of
Antarctic to non-Antarctic ocean–atmosphere exchange rates.

exchange rate in the Antarctic to that in the rest of the ocean of 1 and of 5,
and for various combinations of modes of interactions between the reservoirs.
For the "cyclic Atlantic" (cases A, B, C and D for the world-ocean model) five
ocean mixing-rate parameters could be determined. As only three parameters
must be fixed, the Arctic and Antarctic reservoir $^{14}C$ concentrations have been
computed for comparison with the measured values (see Figs. 6 and 7). In the
other cases such internal cross-checks are not possible.

The results of these calculations allow several important observations to be
made:

(1) Regardless of the model selected, steady-state assumptions require that
dissolved solids remain in the deep Pacific Ocean more than 1000 years before
returning to the mixed layer. The corresponding time for the Atlantic is about
two-thirds as great. The exact time is, as pointed out by Craig (1958), strongly
dependent on the rate at which $^{14}C$ can be supplied to the deep sea through

the surface of the polar seas. If, as in cases IIA, IIIC and IIID, the magnitude of this transfer is sufficiently large, the rate of transfer between the mixed layer and the deep ocean is greatly reduced leading to deep-sea residence times of several thousand years. Model IIB cannot reproduce the $^{14}C$ distribution in the oceans because the rate of supply of $^{14}C$ through the surface of the polar ocean would be too rapid to allow the low $^{14}C/^{12}C$ ratios observed in the deep Pacific.

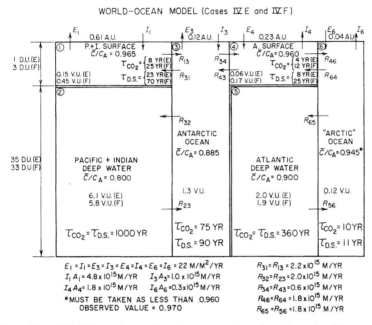

WORLD–OCEAN MODEL (Cases IV.E and IV.F)

Fig. 8. Cases E and F differ only in the thickness assigned to the mixed layer. The cases in this figure differ from those in Fig. 6 in the manner in which the Atlantic Ocean circulates. Return to the surface of Atlantic deep water is through the Arctic instead of Antarctic yielding a noncyclic rather than cyclic pattern of circulation in the Atlantic. In order to obtain non-negative mixing rates the $^{14}C$ concentration assigned to the Arctic reservoir must be intermediate between that in the surface and deep Atlantic. Contrary to this requirement, existing $^{14}C$ measurements on northern North Atlantic samples suggest that $^{14}C/^{12}C$ ratios for the Arctic reservoir may exceed those for the Atlantic surface reservoir.

(2) Models such as IIIE and IIIF requiring a return of a large fraction of the deep waters directly to the mixed layer rather than via the polar reservoir as is the case in the other models contradict the observed $^{14}C$ distribution. If such a flow pattern existed, the $^{14}C$ concentration in polar waters should be equal to or greater than that in the low latitude surface waters. This definitely is not the case for the Antarctic.

(3) The thickness chosen for the surface layer has no significant effect on the deep-water residence times. The residence time of dissolved solids in the mixed layer is, however, directly proportional to the thickness assumed for this layer.

(4) Two of the world-ocean models with a cyclic Atlantic (case A, B and C) reproduce the main features of the $^{14}C$ distribution including the high $^{14}C$ concentration in the Arctic and the intermediate value for the Antarctic. Moreover, the rate of cycling demanded for the Atlantic is consistent with oceanographic estimates of the net northward transfer of surface water in that ocean. The $^{14}C$ value for the northern North Atlantic is the key in determining whether the deep waters of the Atlantic rise at the north or south end of the ocean (the basic difference between the cyclic and noncyclic cases). Although the available data suggest a high Arctic $^{14}C$ value, and hence a dominantly cyclic Atlantic, they may not be typical of the source areas for deep water. Thus more measurements are needed to resolve this point.

## 6. Industrial $CO_2$ Effect

As shown in Table IV, the models with either the thicker mixed layer or the more rapid Antarctic exchange-rate yield predicted estimates of the magnitude of the Suess effect (decrease in the atmospheric $^{14}C/^{12}C$ ratio resulting from the combustion of fossil fuels) more consistent with the observed value of $2.4 \pm 0.7\%$ obtained from tree-ring data—1950 A.D. (Suess, 1955; Fergusson, 1958; Munnich, 1957). The predicted values are obtained by considering the industrial $CO_2$ addition to be a series of discreet $^{14}C$ withdrawals from the atmosphere at various times between the years 1850 and 1950. The magnitude of each withdrawal is estimated from fuel consumption data (see Revelle and Suess, 1957). Unfortunately, no upper limit can be placed on the mixed layer depth through consideration of the Suess effect because, as the depth increases, the predicted Suess effect approaches a limiting minimum value of $1.7\%$, a value within the range of experimental uncertainty of the tree-ring data. In all these computations exchange with the organic matter stored in soils has been considered negligible. This assumption is consistent with the radiocarbon ages of 200 to 2000 years found for soil organics (Tamm and Ostlund, 1960; Broecker and Olson, 1960).

## 7. $^{226}Ra$ Distribution

Koczy (1958) has shown that the $^{226}Ra$ concentration in surface water averages about one-half that in deep water. Assuming that radium is not removed from the surface waters by either organic or inorganic particulate matter, it is not possible to explain the radium distribution with any of the models discussed above. As shown in Table IV these models all suggest that the $^{226}Ra$ concentration in the mixed layer should be no more than $10\%$ lower than that in the deep ocean. The radium data suggest that water presently in the mixed layer has not been in the deep ocean for about one radium half-life (1600 years). In order to explain both the $^{14}C$ and $^{226}Ra$ data simultaneously, a model similar to that in Fig. 9 would have to be adopted. In this model a fourth reservoir is added between the surface and deep ocean reservoirs.

## TABLE IV

Comparison of Deep-Sea Residence Times, Suess Effect Estimates, and Surface-to-Depth Radium Concentration Ratios for the Various Models

| Case no. | Depth surface layer, m | $I_3/I_1$ | $\tau_{2-1}$ Dissolved solids, yr [a] | Predicted Suess effect, % | Predicted $\dfrac{Ra_{surface}}{Ra_{depth}}$ |
|---|---|---|---|---|---|
| *Two layer model* | | | | | |
| IA | 100 | — | 1650 | 3.9 | 0.98 |
| IB | 300 | — | 1650 | 3.0 | 0.95 |
| *Outcrop model* | | | | | |
| IIA | 100 | 1 | 4200 | 3.5 | 0.96 |
| IIB | 300 | 5 | — | — | —[b] |
| *Three reservoir model (no transfer across thermocline)* | | | | | |
| IIIA | 100 | 1 | 1850 | 3.3 | 1.00 |
| IIIB | 300 | 1 | 1850 | 2.7 | 0.99 |
| IIIC | 100 | 5 | 3700 | 3.1 | 0.99 |
| IIID | 300 | 5 | 3700 | 2.8 | 0.96 |
| *Three reservoir model (upward transfer across thermocline)* | | | | | |
| IIIE | 100 | 1 | — | — | —[b] |
| IIIF | 100 | 5 | — | — | —[b] |
| *World-ocean model (cyclic Atlantic)* | | | | | |
| IVA | 100 | 1 | 1350 | 3.1 | 0.99 (P+I) 0.95 (A) |
| IVB | 300 | 1 | 1350 | 2.6 | 0.96 (P+I) 0.95 (A) |
| IVC | 100 | 5 | 1270 | 2.8 | 0.98 (P+I) 0.95 (A) |
| IVD | 300 | 5 | 1270 | 2.6 | 0.94 (P+I) 0.95 (A) |
| *World-ocean model (noncyclic Atlantic)* | | | | | |
| IVE | 100 | 1 | 1530 | 3.0 | 0.99 (P+I) 0.96 (A) |
| IVF | 300 | 1 | 1530 | 2.7 | 0.97 (P+I) 0.96 (A) |

[a] Mean time dissolved solids remain in the deep sea before returning to mixed layer.
[b] Incompatible with [14]C distribution.

Whereas mixing between the surface and intermediate reservoir and between the polar and deep reservoir is relatively rapid, mixing between the combined surface plus intermediate reservoir and the combined polar plus deep reservoir is quite slow. The rate is controlled by the requirement that the [226]Ra concentration in the upper waters be one-half that in the deep water (i.e. $\tau_{D.s.}$ for the combined surface plus intermediate reservoir equals 1600 years). In this case

FOUR RESERVOIR MODEL (Ⅴ)

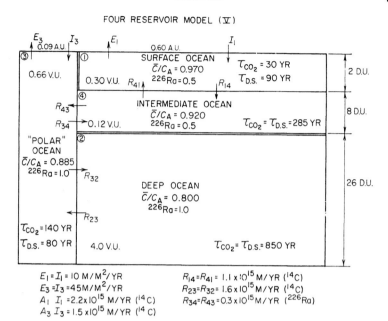

Fig. 9. An example of the type of model required to satisfy the main features of the $^{226}$Ra as well as the $^{14}$C data. The radium concentrations are given in units of $1 \times 10^{-13}$ g $^{226}$Ra per liter of water. The values of $R_{43}$ and $R_{34}$ are based on $^{14}$C data. In this case the ratio of the polar to surface ocean exchange rate is fixed by the $^{14}$C distribution.

the exchange rate of $CO_2$ across the polar ocean surface must be about four times that for the remainder of the ocean in order to supply sufficient $^{14}$C to maintain the deep waters (over 90% of the $^{14}$C decaying in the deep ocean being supplied through the polar ocean surface). Thus if the $^{226}$Ra distribution does result from radioactive decay, the oceans must be characterized by two nearly independent circulation systems separated by a relatively impermeable boundary at perhaps 1000-m depth. The intermediate reservoir is necessary in such a model in order to explain the $^{14}$C concentration observed in the surface ocean without resorting to unreasonably small ocean–atmosphere exchange rates.

## 8. $^{90}$Sr Distribution

The measurements of $^{90}$Sr at various depths in the Atlantic (Bowen and Sugihara, 1960) summarized in Table V suggest that mixing to depths of 500 m or more occurs in less than five years. When considered in conjunction with the $^{14}$C and $^{226}$Ra data, a rather serious problem arises. The $^{14}$C data for waters from 200 to 500 m fall about halfway between those for the water above and for that below. If these waters are being mixed with the surface waters as rapidly as is seemingly required by the $^{90}$Sr data, then, in order to maintain the intermediate $^{14}$C values, an equal quantity of deep water must be mixing

TABLE V

Strontium-90 in Atlantic Ocean[a]

| Station position | 15° 49'S 35° 50'W | 8° 26'S 7° 45'E | 16° 15'N 43° 37.5'W | 36° 23'N 70° 33'W | 31° 15'N 68° 10'W | 34° 30'N 65° 42'W | 34° 39'N 67° 24'W |
|---|---|---|---|---|---|---|---|
| Date collected | April 1957 | March 1957 | Nov. 1957 | July 1957 | Dec. 1957 | July 1958 | July 1958 |
| **Depth, m** | | | | | | | |
| 0 | $4.1 \pm 0.5$ | $5.0 \pm 1.0$ | $6.3 \pm 0.6$ | $10.5 \pm 0.6$ | $10.9 \pm 0.6$ | $7.6 \pm 0.6$ | $9.1 \pm 0.6$ |
| 100 | | | $5.4 \pm 0.5$ | | $7.7 \pm 0.6$ | $13.0 \pm 0.7$ | $8.2 \pm 0.6$ |
| 300 | | $2.4 \pm 1.0$ | | | | $11.8 \pm 0.9$ | $5.5 \pm 0.5$ |
| 400 | | | $2.2 \pm 0.3$ | | $3.9 \pm 0.6$ | | |
| 500 | | | | $4.1 \pm 0.6$ | | $7.3 \pm 0.4$ | $6.4 \pm 0.5$ |
| 700 | | $< 0.3$ | | | $2.3 \pm 0.4$ | $3.7 \pm 0.5$ | $3.4 \pm 0.5$ |
| 1000 | $1.5 \pm 0.6$ | | | | $< 1.0$ | $< 0.9$ | $1.7 \pm 0.3$ |
| 1200 | | | $1.6 \pm 0.8$ | | | | |

[a] Bowen and Sugihara (1960). Results are given as dis/min/100 l. of water.

upward. The rates of vertical mixing required to maintain this distribution would be an order of magnitude higher than those allowed if the surface depth contrast in $^{14}C$ concentration is to be maintained. No explanation for this anomaly is apparent. Further data, especially on pairs of fallout isotopes (for example, $^{137}Cs$ and $^{90}Sr$) and on bomb $^{14}C$, will allow a better evaluation of vertical mixing rates in the upper portion of the sea.

## 9. Non-Steady-State Conditions

Having considered the various box models it is necessary to reconsider the assumption of steady state. This assumption implies that the concentration of any one of the natural radioactive isotopes should have been constant throughout time at any point in the ocean–atmosphere system. As demonstrated by the hypothetical case in Fig. 10, cyclic mixing in the ocean would lead to time variations in the $^{14}C$ concentrations in the atmosphere and surface ocean. Furthermore, the distribution of $^{14}C$ in the ocean would not be simply related to radioactive decay as is the case in the steady-state models. Because of the backlogging of newly produced $^{14}C$ in the atmosphere and surface oceans, large differences between the $^{14}C$ concentration in the deep and the surface ocean waters could be generated quite rapidly if the rate of deep-water formation were severely reduced for a period of 20 to 100 years. Hence, estimates of isotope concentrations at times in the past would allow an evaluation of the steady-state assumption. Fortunately $^{14}C$ measurements on tree rings and historically dated samples yield such data for the atmospheric reservoir (see Fig. 11 for a summary of the available data). Although measurable variations have occurred over the past 2000 years, they are restricted to a range of slightly more than 2% either side of the mean. As such variations could equally well

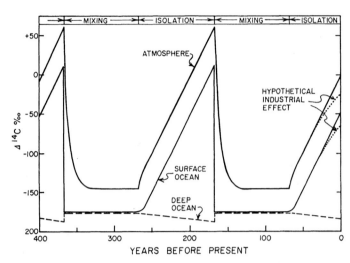

Fig. 10. A hypothetical example of the effects of non-steady-state mixing on the distribution of $^{14}C$ in the ocean–atmosphere system. One-hundred-year periods of very rapid
oceanic mixing are separated by 100-year periods during which no mixing occurs
between the upper 100 m of the ocean and the deep sea. The distribution of $^{14}C$ at
$t = 0$ corresponds roughly to that of the present (corrected for industrial $CO_2$ and
bomb $^{14}C$). Note that $^{14}C$ data are given in $\varDelta \, ^{14}C$ units (i.e. $-100$ is equivalent to
$0.900$).

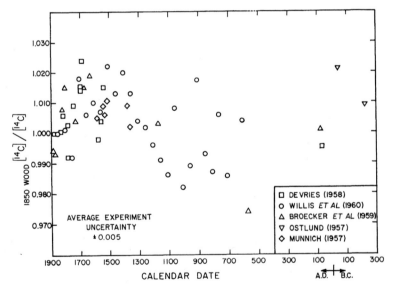

Fig. 11. Age-corrected $^{14}C$ results for tree-ring and historically dated samples of various
ages. The results place limits on the temporal variation of the $^{14}C/^{12}C$ ratio in atmospheric $CO_2$.

result from changes in the production rate of $^{14}$C, their magnitude must be considered as an upper limit on any effect resulting from variations in the rate of oceanic mixing. Consideration of model IIIA will serve to demonstrate the sensitivity of the $^{14}$C/$^{12}$C ratio in the atmosphere to changes in the rate of oceanic mixing. If the ocean–atmosphere exchange rate is assumed to remain constant and if the two water exchange rates are assumed to undergo similar fractional changes, the $^{14}$C concentrations in the four reservoirs can be shown as a function of mixing rate as in Fig. 12. It is apparent that the atmospheric

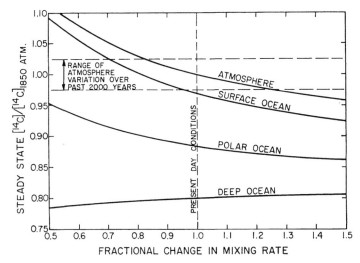

Fig. 12. Dependence of the steady-state $^{14}$C concentrations in the reservoirs of model IIIA on the rate of oceanic mixing. Although the two rates $R_{13}$ and $R_{23}$ are assumed to undergo similar fractional changes, variation in $R_{13}$ does not cause any major redistribution in $^{14}$C, $R_{23}$ being the dominant rate parameter. As the atmospheric $^{14}$C/$^{12}$C ratio has not varied outside 2.5%, its 19th-century value over the past 2000 years, the suggestion is that the rate of deep-water replacement in the ocean has not varied on a long term basis by more than 25% during this span of time.

$^{14}$C concentration is quite sensitive to variations in the rate of oceanic mixing, changes of 25% in the transfer rates being of sufficient magnitude to produce the variations observed in the tree-ring data. It should be emphasized, however, that the short-term variations in mixing rate or variations confined to a local area would not produce measurable variations in the atmospheric $^{14}$C/$^{12}$C ratio. Thus the tree-ring data can only be used to place limits on the degree of nonequilibrium in the oceans.

## 10. Conclusions

From this discussion it is clear that isotope studies offer the key to many problems involving mixing rates within the sea. To date, from $^{14}$C data, it seems probable that the residence times of water in the deep ocean may exceed 800 years in the Pacific and 500 years in the Atlantic. $^{14}$C results on tree rings suggest that the oceans have been reasonably close to steady state over the past 2000 years. $^{226}$Ra offers great promise of defining the age of surface ocean water with respect to mixing with the deep sea; and $^{90}$Sr, $^{137}$Cs and bomb $^{14}$C results should resolve the problem of the vertical mixing rates for the upper 1000 m of the ocean. Isotope data are most valuable when used as boundary conditions for oceanic circulation models derived from independent oceanographic observations.

## References

Arnold, J. R. and E. C. Anderson, 1957. The distribution of $^{14}$C in nature. *Tellus*, **9**, 28–32.

Bien, G. S., N. W. Rakestraw and H. E. Suess, 1960. Radiocarbon concentration in Pacific Ocean water. *Tellus*, **12**, 436–443.

Bowen, V. T. and T. L. Sugihara, 1960. Strontium-90 in the "Mixed Layer" of the Atlantic Ocean. *Nature*, **186**, 71–72.

Broecker, W. S., R. Gerard, W. M. Ewing and B. C. Heezen, 1960. Radiocarbon in the Atlantic Ocean. *J. Geophys. Res.*, **65**, 2903–2931.

Broecker, W. S. and E. A. Olson, 1959. Lamont radiocarbon measurements VI. *Amer. J. Sci., Radiocarbon Suppl.*, **1**, 111–132.

Broecker, W. S. and E. A. Olson, 1960. Radiocarbon from nuclear tests II. *Science*, **132**, 712–721.

Broecker, W. S., E. A. Olson and J. Bird, 1959. Radiocarbon measurements on samples of known age. *Nature*, **183**, 1582–1584.

Burling, R. W. and D. M. Garner, 1959. A section of $^{14}$C activities of sea water between 9°S and 66°S in the southwest Pacific Ocean. *N. Z. J. Geol. Geophys.*, **2**, 799–824.

Craig, H., 1957. The natural distribution of radiocarbon and the exchange time of carbon dioxide between atmosphere and sea. *Tellus*, **9**, 1–17.

Craig, H., 1958. Distribution of radiocarbon and tritium; cosmological and geological implications of isotope ratio variations. *Nat. Acad. Sci., Nat. Res. Council Pub.*, **572**, 135–147.

DeVries, Hl., 1958. Variations in concentration of radiocarbon with time and location on earth. *Koninkl. Ned. Akad. Wetenschap. Proc.*, **61B**, 1–9.

Fergusson, G. J., 1958. Reduction of atmospheric radiocarbon concentration by fossil fuel carbon dioxide and the mean life of carbon dioxide in the atmosphere. *Proc. Roy. Soc. Austral.*, **243**, 561–574.

Fonselius, S. and G. Ostlund, 1959. Natural radiocarbon measurements on surface water from the North Atlantic and Arctic Sea. *Tellus*, **11**, 77–82.

Koczy, F. F., 1958. Natural radium as a tracer in the ocean. *Proc. Second U.N. Intern. Conf. Peaceful Uses Atomic Energy*, **18**, 336–343.

Lal, D., E. D. Goldberg and M. Koide, 1960. Cosmic-ray-produced silicon-32 in nature. *Science*, **131**, 332–337.

Munnich, K. O., 1957. Heidelberg natural radiocarbon measurements I. *Science*, **126**, 194–199.

Ostlund, H. G., 1957. Stockholm natural radiocarbon measurements I. *Science*, **126**, 493–497.

Rafter, T. A. and G. J. Fergusson, 1958. Atmospheric-radiocarbon as a tracer in geophysical circulation problems. *Proc. Second U.N. Intern. Conf. Peaceful Uses Atomic Energy*, **18**, 526–532.

Revelle, R. and H. E. Suess, 1957. Carbon dioxide exchange between the atmosphere and the ocean and the question of an increase of atmospheric $CO_2$ during the past decades. *Tellus*, **9**, 18–27.

Tamm, C. O. and G. G. Ostlund, 1960. Radiocarbon dating of soil humus. *Nature*, **185**, 706–707.

Willis, E. H., H. Tauber and K. O. Munnich, 1960. Variations in the atmospheric radiocarbon concentration over the past 1300 years. *Amer. J. Sci., Radiocarbon Suppl.*, **2**, 1–4.

# 5. CHEMICAL INSTRUMENTATION

## D. E. CARRITT

## 1. Introduction

One of the relationships of chemistry to other sciences has often been likened to that of a handmaiden, ever present and ready to provide needed information and services. In oceanography, the results of applying chemical techniques and principles to obtain a better understanding of various isolated systems in the ocean show up the utilitarian aspects of chemistry and, at the same time, provide rather clear evidence of the inter-relationships and the extent of coupling between the biological, physical and geological parts of the systems functioning in the oceans.

Much of the science of oceanography is concerned with the study of processes in which there occurs a transfer of energy and of matter. Analytical and physical chemistry deal with measurements of the quantities and states of matter and of energy. It is quite natural, then, that much of the primary information which is assembled and analyzed to provide us ultimately with a model we believe to be a description of the real oceans comes from chemical measurements.

Many of the chemical measurements made in pursuit of the model building utilize procedures, techniques and instruments that are common to many other kinds of studies that depend upon chemical analyses. For example, the techniques and instruments used during the spectrographic analysis of marine sediments are essentially the same as those used during the analysis of a wide variety of samples of non-marine origin. That is, many measurements are in no way characteristically oceanographic, except for the origin of the sample.

No attempt will be made in the following discussion to review and evaluate the instrumentation associated with all of the kinds of chemical measurements that have been made in oceanography. Rather emphasis will be on the instrumentation that has been developed, and, in some cases, might be developed, to solve measurement problems in situations where the oceans place unique restrictions and requirements upon the way in which the measurements are made. Detailed descriptions of mechanisms, circuits and operating procedures will be omitted. These are adequately covered in the original literature.

The record of a symposium on oceanographic instrumentation (NAS-NRC, 1952) provides a discussion of all oceanographic instrumentation problems to that time. In the section on chemical measurements (*op. cit.*, chap. X, p. 166) it was pointed out that instrumentation problems appear in three broad categories, i.e. shipboard measurements; measurements in shore-based laboratories; and *in situ* measurements using unattended recording or telemetering buoys, as well as from a research vessel. Unique problems appear in the same categories today. Some of the problems noted in 1952 have been at least partially solved. Some of the instrumentation suggested in 1952 has been found to be inadequate or techniques not obvious at that time better suited to the problem. In short, advances have been made.

[*MS received October, 1960*]                    109

The instrumentation connected with chemical measurements made in a shore-based laboratory presents very few problems that are uniquely oceanographic. Once a sample (water, sediment or bio-material) is ashore the problems of analysis can be approached with the full compliment of analytical tools and techniques that are available for all analysis problems. It is sufficient to note here that during recent years there has been a much greater utilization of current analytical instrumentation than was obvious in the past. Significant oceanographic results have come from a wide variety of applications of mass spectrometers, X-ray techniques, neutron activation, ultraviolet, visible and infrared absorption measurement, and radioactive tracer uses, to name only a few.

As in 1952 and previous years, much recent effort has been toward improving chlorinity measurements. Present instrumentation provides an apparent improvement in precision in routine measurements of nearly an order of magnitude. The overall result of this improvement is not fully understood because the accuracy of the empirical relation between the property actually measured, conductivity, and the property sought, chlorinity, leaves an uncertainty in computed chlorinity at least an order of magnitude greater than implied by the *precision* of the conductivity measurements. Nevertheless, the development of inductive conductivity devices suggests that the instrumentation of routine, *in situ* measurements of electrical conductivity of high accuracy and precision is now possible.

## 2. Chlorinity, Salinity, Density and Related Properties

### A. Introduction

Much of what we know about the oceans, or more precisely, what we think we know about them, depends upon a belief in what has been variously called Maury's Principle, Dittmar's Hypothesis, Forchhammer's Hypothesis and, more recently, attributed to one of several other workers who studied or speculated on the chemical composition of sea-water. This hypothesis sets forth the notion that, so far as the major dissolved constituents are concerned, sea-water has constant relative composition. This is, considering sea-water to be an aqueous solution of some eleven inorganic constituents (see, for example, Lyman and Fleming, 1940, table III, p. 137), the only difference that can be found between samples is in the water content. It follows, then, that the ratios of the concentration of any one of these constituents to any other, or to any combination of others, will be constant in all samples of sea-water.

The importance of the notion of constancy of relative proportions to parts of oceanography, other than merely as a descriptive feature of the chemical composition of sea-water, is evident from an examination of the empirical relationships that have been obtained. By these relationships a variety of physical and chemical properties of sea-water are said to be predictable from measured values of concentrations of one of the major dissolved constituents, usually of chloride.

Empirical relationships in the form of interpolation formulae or tables have been constructed for the pairs of properties shown in Table I. The designations measured and computed merely indicate the prevailing use of the relationships.

TABLE I

| Measured | Computed | References |
|---|---|---|
| 1. Chlorinity | Salinity | Forch, Knudsen and Sørensen (1902) |
| 2. Chlorinity (T and P) | *In situ* density | Knudsen (1901) |
| 3. Chlorinity | The concentrations of major and several minor constituents | See for example, Lyman and Fleming (1940), Lyman (1959) |
| 4. Electrical conductivity and temperature | Chlorinity | Thomas, Thompson and Utterbach (1934) |
| 5. Refractive index (T) | Chlorinity | |
| 6. Chlorinity (T) | Vapor pressure | Arons and Kientzler (1954) |
| 7. Chlorinity (T) | Osmotic pressure | Wilson and Arons (1955) |

In each case, with the exception of 3 above, chlorinity, which is a measure of the total halide concentration but predominantly of the chloride (defined precisely below), is related to a property of sea-water that in some way involves the sum of that property contributed by all of the dissolved substances. Strictly speaking, then, the applicability of these relationships to all ocean waters depends upon the validity of the constancy notion in those waters.

At the present time the data on which judgements of this kind can be based are sparce. Dittmar's (1884) analyses of seventy-seven samples from the 1872–1876 cruise of H.M.S. *Challenger* are still the only "complete" analyses of sea-water. Since that time composition studies have been concerned primarily with the measurement of the ratio of one of the major constituents to chlorinity.

The best that can be said at present is that the relative composition of sea-water, excluding near-shore waters and shallow seas, is nearly constant. The term *nearly* must remain ill-defined until more data are available. However, Lyman (1959) showed that vertical differences between the surface and 2000 m in the Pacific, with respect to alkalinity components, nitrate, calcium, phosphate and silica, could produce a variation of 0.023‰ in salinity. In the Atlantic the difference was 0.007‰. These differences would *not* appear in salinity computed from chlorinity, even though the precision of modern chlorinity measuring methods is of the order of 0.001–0.003‰ in computed salinity.

## B. Chlorinity–Salinity Relation

Both chlorinity and salinity are properties of sea-water defined in terms of procedures that will yield a quantitative measure of the property.

The original definition of chlorinity required a knowledge of the chlorine equivalent of bromine and iodine. Chlorinity values so defined required revision with each redetermination of the atomic weights of the elements involved. A more recent, and presently accepted, definition which makes chlorinity independent of atomic weight values, given by Jacobsen and Knudsen (1940), is: "The number giving the chlorinity in per mille of a sea-water sample is by definition identical with the number giving the mass with unit gram of *Atomgewicht Silber* just necessary to precipitate the halogens in 0.3285234 kilograms of sea-water sample".

Thus, the accuracy and precision with which chlorinity can now be determined depends upon our ability to prepare atomic-weight silver, to weigh solids and solutions, and to carry out rather simple chemical operations.

The measurement of chlorinity on routine samples has been done generally by the Knudsen titration (for details see Oxner, 1920) using Copenhagen Standard Sea-Water as the primary standard. The accuracy and precision of measurements made by this procedure are of the order of $\pm 0.01\%_0$ in chlorinity.

Replacement of the colorimetric and point detection of the Knudsen titration by potentiometric methods has been reported by several investigators.

An automatic chlorinity titration, described in NAS-NRC (1952) used the potential of the cell AgCl(s) : Cl$^-$ :: Ag$^+$ : Ag for end-point detection and control of titrant delivery. Although satisfactory from the standpoint of precision and accuracy of results, excessive maintenance requirements caused abandonment of the device after several years' use.

A procedure described by Bather and Riley (1953) utilizes potentiometric end-point detection, employs relatively simple equipment and gives results of higher precision than the Knudsen titration. These authors recommend a weight titration procedure for results of high precision (standard deviation 0.001‰ in chlorinity) or a volumetric titration for somewhat lower precision (standard deviation 0.005‰ in chlorinity). With the weight titration, up to forty-five samples per day can be processed and with the volumetric procedures up to ninety samples per day.

Two methods have been used to establish the chlorinity of Primary Standard Sea-Water by the International Council for the Study of the Sea (Jacobsen and Knudsen, 1940). From 1902 to 1938 the chlorinity of new batches of Standard Water (20 different preparations were made) was established by comparison of each new batch with a previous batch. The comparisons were made using the Volhard titration as modified by S. P. L. Sørensen. In 1938 a new batch of Standard Sea-Water (labeled "Urnormal 1937") was analyzed by the usual comparison with the previous batch, and also by direct comparison with silver, according to the new definition of chlorinity.

The development of new high-precision instruments for the indirect measurement of chlorinity suggests that attention should now be paid to the methods that provide a direct measure of chlorinity, for these measurements must provide the basis for calibration and standardization of the indirect methods. The precision of the widely used Knudsen titration, $\pm 0.01$ Cl‰, is not adequate for purposes of calibrating the new conductivity bridge salinometers which have a precision equivalent to $\pm 0.001$ Cl‰ and better. The Bather and Riley (*loc. cit*) potentiometric method has a precision comparable to that of the salinometer and is being used in a fundamental study of the conductivity–chlorinity relationship of sea-water (Cox, 1960).

If we accept the idea that reference and calibration methods should have a precision and accuracy an order of magnitude greater than that in the system being studied, only the Volhard titration and the direct comparison with silver following classical atomic-weight techniques are adequate for calibration purposes and for the direct comparison with silver for establishing the value of Standard Sea-Water.

In summary, the uncertainty in chlorinity for a sample of 19.000‰ associated with various methods are: Knudsen titration $\pm 0.01$‰ Cl; Bather and Riley $\pm 0.001$‰ Cl; Volhard $\pm 0.0004$‰ Cl; and direct comparison with silver $\pm 0.0002$‰ Cl. Laboratory-type conductivity salinometers show an uncertainty equivalent to approximately $\pm 0.0005$ to $0.002$‰ Cl.

In recent years the use of so-called "salinometers," in which electrical conductivity is used to obtain a measure of chlorinity, has become widespread, and, in fact, appears now to be the preferred method in most laboratories in which large numbers of samples must be analyzed both at sea and ashore. In principle, the measurement of electrical conductivity to obtain chlorinity has a distinct advantage over the classical titration method in that conductivity measurements can be instrumented to provide a continuous measure with depth, whereas titration gives discrete values only at sampling depths. In practice, two types of salinometers have been used, one to provide continuous *in situ* measures, the other to measure samples obtained from Nansen bottles.

The Wenner Bridge (Wenner *et al.*, 1930) appears to have been the first conductivity device having the stability, sensitivity and accuracy required for routine measurements of chlorinity. The reliability of this instrument, as well as of several more recent "salinometers" of similar design, depends to a large extent on a unique feature in the arrangement of the bridge circuit. One arm of the bridge consists of a conductivity cell into which the sample to be measured is placed. Another arm of the bridge consists of a conductivity cell, as nearly identical to the sample cell as is practical, which contains either Copenhagen Standard Sea-Water or a sub-standard sea-water of known chlorinity. The conductivity cells and critical circuit components are in a thermostated bath. The associated electrical circuits provide the means of measuring the difference in conductivity between unknown and standard. Calibration is by measurement of samples of known chlorinity.

This method of comparing a sample to a standard eliminates the need for extreme precision in temperature control. For example, Paquette (1958) noted that as long as reference and unknown salinities are all at the same temperature, departures from the temperature at which calibration was performed introduce an error of approximately 0.005‰ in salinity per degree Celsius change in temperature per 1‰ difference between measured and reference salinities. In the conventional Wheatstone Bridge, temperature control to $\pm 0.003°C$ would be necessary to achieve the same precision in salinity.

The basic Wenner Bridge design has undergone at least three recent revisions and modifications. Each has been an attempt to improve the accuracy and precision of chlorinity measurements by incorporating new components, circuit features and controls into the original conductivity comparison bridge. Bradshaw and Schleicher (1956), Cox (1958) and Paquette (1958) have each constructed several conductivity bridges that have been used in their own and other laboratories. The precision (standard deviation of an individual measurement from the mean) of each of these instruments is in the range of 0.001 to 0.003‰ in salinity.

The conductance-bridge type salinometers mentioned above are laboratory-type instruments which cannot be readily modified to make *in situ* measurements. They all use classical kinds of conductivity cells in which the primary measurement is of the resistance between two metallic electrodes in contact with the samples being measured. Several *in situ* conductivity devices have been constructed using the classical-type cell but without the double cell features of the laboratory instruments.

The Woods Hole Oceanographic Institution salinity, temperature and depth recorder (S.T.D.) (Jacobson, 1948), the Johns Hopkins University Conductivity-Temperature Indicator (C.T.I.) (Carritt, 1952) and the portable temperature–chlorinity bridge described by Hamon (1956) are recent examples of devices for *in situ* measurements. Apparently all of these devices suffer from a common fault which limits their usefulness in open ocean studies. The difficulty comes from changes in cell characteristics for which there is no compensation as in the double-celled laboratory-type instrument. Even with frequent standardization and recalibration, which is difficult to do routinely at sea, the precision appears to be limited to 0.05‰ to 0.10‰ in chlorinity.

Many of the objectionable features of the classical conductivity cell have been overcome by a design first described by Esterson (1957). This cell is a toroidal transformer in which sea-water forms the equivalent of a one-turn winding linking two co-axially mounted inductors. A potential impressed on the primary element of the transformer induces a current flow in the sea-water link which, through inductive coupling, causes current to flow through the second core and associated components. The current in the sea-water path is proportional to the conductivity of the sea-water. Thus, the output signal from the second core is a measure of the conductivity of the sea-water that forms the connecting link between the two transformer elements. The transformer elements are contained in an insulating housing which maintains the relative geometry of the two elements and the sea-water path that passes co-axially through them. Since there is no direct contact of water and electrodes, polarization, fouling and poisoning of the kind encountered in classical cells are eliminated.

Field instruments constructed on the Esterson design have been used for several years by the Chesapeake Bay Institute of the Johns Hopkins University in studies in which waters ranging in conductivity from that of river-water to that of sea-water were used (Pritchard, 1959). This experience together with repeated recalibration and standardization indicates that salinometers of this type are extremely stable over long periods.

A laboratory model salinometer with a transformer-type conductivity cell has been constructed by Brown and Hamon (1961). This instrument has several features not found in other salinometers. In addition to high long-term stability, temperature-measuring and compensating circuits eliminate the need for a thermostated bath for the cell, and permit samples to be measured as quickly as they can be introduced into the cell- and bridge-balancing units adjusted. The instrument has a sensitivity of 0.0004‰ in chlorinity and a precision of 0.001‰. The electronics have a low power requirement and can be readily adapted to battery operation. The device is about the size of an ordinary brief case and weighs approximately 25 lb. The instrument is to be manufactured commercially and will sell for less than $1500.

Three of the laboratory-type salinometers mentioned above, the Schleicher, the Brown and Hamon and the Paquette instruments, were compared in a recent test in which twenty-four sea-water samples with a salinity range of 33.88‰ to 34.78‰ were analyzed by each instrument. The means of the twenty-four samples were 34.393‰, 34.394‰ and 34.39‰ respectively. The Paquette instrument will provide results significant to the third decimal place; however, because of a convention in reporting salinity by the organization that performed the analyses, the third figures were omitted in their results. The maximum deviation of the Schleicher and the Brown and Hamon results was 0.007‰ and the standard deviation 0.003‰.

The high precision capable with the conductivity-bridge salinometer provides the means of measuring a property of sea-water much more precisely than has been possible in the past. The *accuracy* of corresponding chlorinity measures may, however, not be as high as the *precision* in the relative conductivity measures. This will depend upon the constancy of the chlorinity–conductivity relation in various batches of Copenhagen Standard Sea-Water, and the method of calibration with sub-standard sea-water. Furthermore, as should be apparent from the discussion below, increased precision and accuracy in the measurement of chlorinity will not increase the accuracy of computed salinity values.

Forch, Knudsen and Sørensen (1902, p. 116) give the definition of salinity as: "Salinity (Salzmenge) is defined as follows: The weight of dissolved solid material found in 1 kilo of sea water, after the bromine has been replaced by an equivalent quantity of chlorine, all the carbonate converted to oxide, and all of the organic matter destroyed".

Forch, Knudsen and Sørensen (*loc. cit.*) measured both the chlorinity and salinity of nine samples [selected from those used to provide the primary information upon which the Knudsen Hydrographic Tables (1901) were constructed]. From the results they computed the regression equation, which is the familiar relationship:

$$S‰ = 0.03 + 1.805 \, Cl‰. \tag{1}$$

This is usually used to compute salinity values from measured values of chlorinity. Equation (1) has also been used as the definition of salinity.

It should be noted that salinity as defined is not identical with total dissolved salts. It represents a compromise between the desire to have a measure of total dissolved salts without performing analyses for each individual constituent and the limitations inherent in a less tedious set of procedures that gives a measure of a property nearly proportional to the sum of individual constituents. Lyman and Fleming (1940) discuss the difference in detail.

Equation (1) is not consistent with the notion of the constancy of relative proportions. Equation (1) predicts a salinity of 0.03% when the chlorinity is zero, whereas the constancy of relative proportions would have them both zero. Attention has been drawn to this problem several times in the past. For example, Lyman and Fleming (1940), Carritt and Carpenter (1959) and Lyman (1959) discuss various aspects of the problem, and it now seems clear that neither the concept of the constancy of relative proportion nor the salinity–chlorinity relationship, as expressed by (1), can be stated without modification or qualification.

Equation (1) was derived from the results of the analysis of nine surface samples of which only four have a chlorinity in the range of 16‰ to 22‰. The remaining five, Baltic and North Sea samples, were in the range 1‰ to 13‰. Lyman and Fleming (loc. cit., p. 143) noted that: "Brackish waters will vary according to the nature of the particular river water with which they are diluted. It happens that water from the Baltic region was selected as the standard type of diluted water in oceanography, and comprehensive tables of salinity, chlorinity and density relationships worked out. It follows from this that it is not possible, by extrapolation to zero chlorinity of measurements made in natural sea-water, to obtain the corresponding property of distilled water, although this has been attempted in the case of refractive index, dissociation constant, and other measurements. It is also incorrect to make electrical measurements in sea-water over a range of chlorinity by diluting a single sample with 'a good grade of conductivity water' although this likewise has been done. Samples diluted with distilled water will not correspond in properties to those of water of the same chlorinity from Knudsen's tables".

In terms of the physical, chemical and biological processes occurring in the oceans and around their borders, constancy of relative proportions implies, as noted by Maury (1855), that the oceans are "well shaken together". However, if, as is the case of dilution with Baltic water, a decrease in chlorinity indicates departure from "true oceanic properties" into regions where the ratios of major constituents are not constant, how then do we define "true ocean water"? Dittmar (1884) quoting Forchhammer noted: "According to Forchhammer, these ratios (ratios of the concentrations of the major dissolved constituents) are, in passing from one part of the ocean to another, subject to only very slight variations, if we omit the Caribbean, the German Ocean, the Baltic and coast water generally".

It is, then, perhaps surprising to note that, despite the warning noted above, approximately half of the samples that were analyzed for chlorinity and density and the results used in the construction of Knudsen Tables, and approximately half of the samples that form the basis for equation (1) were taken where departures from constancy of relative proportions would be expected.

With only nine measurements of salinity, made on a set of samples which most certainly are not representative of all oceans and all depths, it is natural to question the applicability of the relationships based upon those samples to all ocean waters. How different will the actual salinity be from that computed from (1) and a measured value of chlorinity? *On the average* the difference will probably be small. Lyman and Fleming (1940) computed the sum of major anions and cations from the *average* of measured element-to-chloride ratios and, after applying the necessary corrections to obtain salinity as defined, obtained a value in excellent agreement with that computed from (1). This, however, is only a part of the problem, because we are seldom concerned with *average* conditions. Often we depend upon departures from a mean to provide information about processes occurring in the ocean.

Obviously it is possible to resolve these difficulties. We need only collect a representative set of samples and measure salinity (as defined) and chlorinity. At this point, however, it seems hardly worth the time, effort and expense. Salinity is a defined property of sea-water and one that is difficult to measure with high accuracy and precision. *Furthermore, the use of computed values of salinity, even though we had reliable empirical relationships, tells us nothing more about the oceans than would be obtained by using directly the property from which we compute salinity.* For example, we can learn no more from a temperature-salinity curve or envelope than from a temperature–chlorinity envelope, and the latter is fundamentally a truer statement of the information at hand.

If it is desirable to retain the concept of salinity, that is a property which is roughly proportional to the total dissolved salts, other properties that can be readily measured with high accuracy and precision can be used. For example, the total anion or cation equivalents can be measured by ion-exchange techniques which terminate in a simple acid or base titration. These measures will be more nearly proportional to total dissolved salts than any of the properties now measured and linked empirically with chlorinity, and so to salinity. For example, conductivity and refractive index give measures of the sum of the property of each of the dissolved constituents in sea-water. However, neither of these properties is strictly analogous to salinity. The measured conductance of a sea-water sample can be written as the sum of the ion conductances of each of the dissolved constituents. However, the ion conductance of one constituent is not a simple function of the concentration of that constituent. Thus, it is possible to have significant changes in the concentrations of, say, two constituents, as in an exchange process, and so significant changes in salinity and yet insignificant changes in conductance.

### 3. Water Tagging and Tracer Techniques

Studies of the motion of sea-water encompass a very broad size spectrum bounded on the small end by molecular phenomena and on the large end by entire oceans. Regardless of the part of the size spectrum singled out for study the direct measurement of motion requires that a portion of the sea be tagged in order that changes in position with time be measurable.

The measurement of water motion has been accomplished by a variety of techniques. The measurement of currents using mechanical devices is discussed elsewhere in this volume, see, for example Chapters 13 and 14. Measurements with current meters, the GEK, drogues and buoys provide information concerning speed and direction of water motion, but little or no information on the turbulent mixing processes in the sea. The latter has been obtained by measuring the distribution of a variety of chemical substances, both those naturally produced and those introduced by man.

The measurement of natural and artificial radioactivity is discussed in Chapters 3 and 4 of this volume. Many of the problems connected with the use of artificial radioactive tracers in circulation studies have been discussed by Folsom and Vine (1957).

The recent developments of fluorimetric techniques for the measurement of the concentration of certain kinds of artificially introduced tags offer the possibility of conducting diffusion or turbulent-mixing studies on a time and size scale not practical by other methods.

An ideal tag for diffusion studies should have the following properties:

(a) Its physical, chemical and biological behavior should be identical with that of water. It is, after all, primarily water that is to be followed.

(b) It should be amenable to quantitative measurement over an extremely wide range of dilution, and the measuring techniques should be adaptable to *in situ* procedures.

(c) The cost, ease of handling, and toxicity to both the experimenter and naturally occurring marine organisms should limit neither the size nor number of experiments that can be conducted.

The fluorimetric technique mentioned above was originally developed for use in turbid, near-shore and estuarine waters. Pritchard and Carpenter (1960) described the results of several experiments in which advection and turbulent mixing were studied.

The usefulness of the new fluorimetric technique comes from a very fortunate combination of the properties of a commercially available dye, the nature of naturally occurring fluorescent substances and the availability of a fluorimeter with characteristics that make it well suited to *in situ* measurements.

Rhodamine B is a dye commercially available in ton quantities (E. I. du Pont). It is normally produced in acetic acid solution which can be adjusted with methanol to *in situ* density. It shows maximum absorption at 550 m$\mu$ (fluorescence is strongly excited by the 546 m$\mu$ mercury line) and maximum fluorescence at 575 m$\mu$. It is biologically inert and is not absorbed to any appreciable extent by bottom sediment and suspended materials. Unlike fluorescein, it is quite stable with respect to photodecomposition.

The wavelengths of maximum excitation and fluorescence permit the design of a measuring procedure that, for most practical purposes, is free from interference by naturally fluorescing and light-scattering substances.

The Turner Fluorimeter (G. K. Turner Associates, East Palo Alto), when fitted with the appropriate cut-off optical filters between exciting light source and test solution and between test solution and photomultiplier detector, shows a detection limit (1% full scale) of 0.002 parts of dye per billion parts of water. In turbid inshore waters this is reduced by background fluorescence to 0.008 parts per billion. The fluorimeter can be obtained with a flow cuvette and recorder, making continuous *in situ* measurement possible by pumping the sample through the instrument.

Carpenter (1960, unpublished MS) made a comparison of rhodamine B and radioactive tracers. Using the figures of 20,000 curies of gamma emitter, at a cost of $2.50 per curie, and a detection limit of $2 \times 10^{-9}$ curie per liter as suggested by Folsom and Vine (1959) for a deep-water tagging experiment, Carpenter estimated that 0.1 kg of dye is equivalent to one curie of gamma activity or 2 metric tons of dye for an equivalent experiment, with a direct material cost reduction of 250%. In addition the cost and hazards of handling 20,000 curies of gamma activity would be considerably greater than handling two tons of pigment.

It now appears that the use of rhodamine B with fluorimetric detection is very well suited to a variety of tagging experiments in most natural waters.

## 4. Measurement of Dissolved Gases

The Winkler Method has been the universally used procedure for measuring dissolved oxygen in sea-water. Its simplicity is both an asset and a liability. The procedure can be carried out at sea by relatively unskilled technicians. Nevertheless, the procedure involves several chemical reactions all of which are liable to interference by side reactions, if operating procedures are not strictly followed, and by reactions with a variety of dissolved substances which in analyses of sea-water are tacitly assumed to be absent. Also, the usual standardization procedure is not in terms of dissolved oxygen, but rather utilizes a standard dichromate, iodate or biniodate solution, for which an oxygen equivalent must be assumed. Furthermore, the common practice in oceanography of computing "per cent saturation" or "oxygen deficit" implies, first of all, that saturation values for given temperatures and chlorinities are well known and, secondly, that the water samples found by analysis to be under-saturated were at some previous time just saturated.

It now appears that our knowledge of the accuracy of the results of Winkler titrations, the accuracy of published saturation tables, and the validity of the assumption of previous saturation under the conditions specified in saturation tables may limit the usefulness of oxygen data as now measured.

Carritt (1954) pointed out that, whereas oxygen saturation tables provide values for saturation from an atmosphere with a pressure of 760 mm Hg, actual variations in atmospheric pressures over the sea surface may produce departures of oxygen concentration from the standard atmosphere values as high as 0.4 ml $O_2/l$. It will be noted below that this effect can be at least partially determined by measurement of the concentration of dissolved nitrogen or argon, which undergo little or no chemical change when in solution.

Procedures that involve standard solutions of potassium dichromate, potassium iodate or potassium biniodate as the means of standardization for the Winkler titration are in common use. Thompson and Robinson (1939) recommended the use of potassium biniodate in preference to the other two substances after a study of the accuracy and precision attainable with each standard. Jacobsen, Robinson and Thompson (1950) adopted this recommendation and gave detailed instructions for the entire Winkler procedure.

Nevertheless Föyn (*in litt.*) has objected to the use of potassium biniodate on the grounds that the variability in the composition of solid potassium biniodate leads to measurable variation in the normality of the thiosulfate being standardized. At least one commonly used procedure (U.S. Hydrog. Off. Pub. No. 607) recommends the use of standard potassium dichromate.

Armstrong (1959, unpublished MS) and Carritt (1959, unpublished MS) showed that air oxidation of iodide to iodine during standardization procedures can lead to low values for thiosulfate normality and so to low oxygen values in the Winkler titration. The oxidation is a photochemical reaction that is favored by high concentrations of iodide and of acid. The rate of reaction of dichromate with iodide is increased at high acid and iodide concentrations, and in some procedures increased speed has been sought by increasing the concentrations of these substances to the point where the photochemical air oxidation becomes a significant source of error.

Both iodate and biniodate react rapidly at acid and iodide concentrations where the photochemical oxidation of iodide is insignificant.

The use of an oxygen solution of known concentration obviously will eliminate many of the problems encountered with the standards named above. A simple saturation device has been constructed and tested and found to be an adequate solution to the problem (Carritt, 1959, unpublished MS). The device consists of a three-liter flask, which in use contains about a liter of water, arranged to be slowly rotated by a motor-drive in a constant temperature water bath. The speed of rotation is such that no cavitation occurs, yet a large area of continually renewed liquid film comes in contact with the atmosphere. Tests using distilled water, for which the saturation concentration–temperature relation appears to be accurately known, show that excellent agreement between experimental and tabulated values can be obtained provided that temperature is controlled or known to be constant to $\pm 0.05°C$ and that atmospheric pressure has changed less than 1 mm Hg during equilibration. Tabulated saturation values must be corrected for the difference between actual barometric pressure and the standard 760 mm Hg atmospheric pressure. One hour of equilibration produced saturated water starting with water 66% saturated.

The use of a standard oxygen water not only eliminates some of the difficulties connected with the use of previously mentioned standards, it is also the first step in obtaining a complete and unambiguous blank for the complete Winkler method.

The ratios of oxygen to nitrogen, nitrogen to argon and oxygen to argon can be measured with great precision using the mass spectrometer (see, for example, Benson, 1959; Benson and Parker, 1959, and Richards and Benson, 1959). Measurement of the ratios that involve argon offers the means of establishing the concentrations of oxygen and nitrogen in sea-water samples if the solubility of argon is accurately known and if the degree of saturation with respect to all three gases is the same during contact of sea-water with the atmosphere. In order to utilize the full potential of this technique, saturation values of all three gases need to be better known.

The dissolved oxygen electrode system described by Carritt and Kanwisher (1959) and Kanwisher (1959) offers the means of continuous measurement of dissolved oxygen in marine and fresh waters both clean and polluted and, at least in principle, it can be adapted to *in situ* measurements at all depths in the sea.

The development of infrared spectrophotometric instruments (see, for example, Smith, 1953, and Kanwisher, 1960) for the measurement of the partial pressure of $CO_2$ in a gas mixture offers the means of examining variations of $CO_2$ content in the atmosphere and gas transfer processes between atmosphere and the oceans. The fourteen papers presented in a symposium on the circulation of $CO_2$ in the atmosphere and the oceans at the General Assembly of the I.U.G.G. at Helsinki, Finland, in 1960, provide a summary of recent applications and results. Of interest from the point of view of what appears to be necessary in future instrumentation and measurement in chemical oceanography are the results presented by Hood and Ibert. A comparative study of $CO_2$ partial pressures measured with the infrared instrument and values computed from measurements of alkalinity, pH, temperature and chlorinity by the procedure of Buch *et al.* (1932) shows large differences. In the open oceans the computed values were always higher than measured and in shallow waters overlying calcareous deposits the reverse was found.

It is clear that a close examination of the details of both methods is in order. We need especially to examine the value of the constants used in computing the $CO_2$ partial pressure, as well as the relative rates of the reactions that are involved. A method of measuring pH in sea-water with higher accuracy and precision than can be done at present is needed. Complexing reactions that involve carbonate, bicarbonate and $CO_2$ should be studied and non-volatile alkalinity components—possibly short-lived organic constituents—should be sought.

## References

Arons, A. B. and C. F. Kientzler, 1954. Vapor pressure of sea-salt solutions. *Trans. Amer. Geophys. Un.*, **35**, 722.

Bather, J. M. and J. P. Riley, 1953. The precise and routine potentiometric determination of the chlorinity of sea water. *J. Cons. Explor. Mer*, **18**, 277.

Benson, B. B., 1959. Oxygen isotope fractionation during the utilization of oxygen dissolved in sea water. Paper presented at Intern. Oceanog. Congr., New York, 1959.

Benson, B. B. and P. D. M. Parker, 1959. A new technique for dissolved gas analysis with application to the study of dissolved nitrogen in aerobic ocean waters. Paper presented at Intern. Oceanog. Congr., New York, 1959.

Bradshaw, A. L. and K. E. Schleicher, 1956. A conductivity bridge for the measurement of salinity of sea water. *Woods Hole Oceanog. Inst.* Ref. 56-20 (unpublished manuscript).

Brown, N. and B. V. Hamon, 1961. An inductive salinometer. *Deep-Sea Res.*, **8**, 65–75.

Buch, K., H. W. Harvey, H. Wattenburg and S. Gripenberg, 1932. Über das Kohlensaunesystem in Meerwassen. *Rapp. Cons. Explor. Mer*, **79**, 1.

Carritt, D. E., 1952. Oceanographic instrumentation. Chap. X. *Div. Phys. Sci., Nat. Acad. Sci.-Nat. Res. Council*, Pub. 304, 166–193.

Carritt, D. E., 1954. Atmospheric pressure changes and gas solubility. *Deep-Sea Res.*, **2**, 59–62.

Carritt, D. E. and J. H. Carpenter, 1959. The composition of sea water and the salinity-chlorinity density problem. *NAS-NRC Pub.* 600, 67.

Carritt, D. E. and J. W. Kanwisher, 1959. Electrode system for measuring dissolved oxygen. *Anal. Chem.*, **31**, 5.

Cox, R. A., 1958. The thermostat salinity meter. Nat. Inst. Oceanog., Wormley, Godalming, Surrey, England. *NIO Internal Rep.*, No. C2.

Cox, R. A., 1960. Report in Symposium on Comparative Chemical Oceanography at the General Assembly of I.U.G.G. at Helsinki, Finland.

Dittmar, W., 1884. Report on researches into the composition of ocean water, collected by H.M.S. *Challenger. Challenger Reps., Physics and Chem.*, 1.

Esterson, G. L., 1957. The induction conductivity indicator (I.C.I.). A new method for conductivity measurement at sea. The Chesapeake Bay Inst., The Johns Hopkins Univ., Baltimore, Maryland. Ref. 57-3.

Folsom, T. R. and A. C. Vine, 1957. On the tagging of water masses for the study of physical processes in the oceans. *NAS-NRC Pub.* 551, chap. 12.

Forch, C., M. Knudsen and S. P. L. Sørensen, 1902. Berichte über die Konstantenbestimmugen zur Aufstellung der hydrographischen Tabellen. *Kgl. Danske Videnskab., Selskabs. Skrifter, Naturvidenskab. math.*, 6 Raekke, *Afd* XII, 1.

Hamon, B. V., 1956. A portable temperature-chlorinity bridge for estuarine investigations and sea water analysis. *J. Sci. Instrum.*, **33**, 329.

H. O. Pub. 607, 1955. *Instruction Manual for Oceanographic Observation*. Department of the Navy, U.S. Hydrographic Office, Washington, D. C.

Jacobsen, A. W., 1948. An instrument for recording continuously the salinity, temperature and depth of sea water. *Trans. Amer. Inst. Elec. Eng.*, **67**, 714–22.

Jacobsen, J. P. and M. Knudsen, 1940. Urnormal 1937 or Primary Standard Sea Water 1937. *Assoc. d'Océanog. Phys.*, U.G.G.I. Pub. Sci. No. 7.

Jacobsen, J. P., R. J. Robinson and T. G. Thompson, 1950. A review of the determination of dissolved oxygen in sea water by the Winkler method. *Assoc. d'Océanog. Phys.*, U.G.G.I. Pub. Sci. No. 11.

Kanwisher, J. W., 1959. Polarographic oxygen electrode. *Limnol. Oceanog.*, **4**, 210.

Kanwisher, J. W., 1960. pCO$_2$ in sea water and its effect on the movement of CO$_2$ in nature. *Tellus*, **12**, 209.

Knudsen, M., 1901. *Hydrographical tables*. Copenhagen.

Lyman, J., 1959. Chemical considerations. *NAS-NRC Pub.* 600.

Lyman, J. and R. H. Fleming, 1940. Composition of sea water. *J. Mar. Res.*, **3**, 134.

Maury, M. F., 1855. *Physical geography of the sea*. Harper and Bros., New York.

NAS-NRC, 1952. Oceanographic Instrumentation. (A conference held at Rancho Santa Fe, California, 21–23 June, 1952, under the sponsorship of the Office of Naval Research.) *NA S-NRC Pub.* 309.

Oxner, M., 1920. Chlorination par la méthode de Knudsen. *Bull. Comm. Intern. Expl. Sci. Mer Méditerranée*, No. 3.

Paquette, R. G., 1958. A modification of the Wenner-Smith-Soule salinity bridge for the determination of salinity in sea water. *Univ. Wash., Dept. Oceanog.*, Ref. 58-14.

Pritchard, D. W., 1959. The *in situ* measurement of "salinity" with the induction-conductivity-indicator. *N A S-N R C Pub.* 600, 146.

Pritchard, D. W. and J. H. Carpenter, 1960. Measurement of the turbulent diffusion in estuarine and inshore waters. Paper presented at the General Assembly of I.U.G.G. at Helsinki, Finland.

Richards, F. A. and B. B. Benson, 1959. Nitrogen/argon and nitrogen isotope ratios in anaerobic marine environments. Paper presented at Intern. Oceanog. Cong., New York, 1959.

Smith, V. H., 1953. A recording infrared analyzer. *Instruments*, **26**, 421.

Thomas, B. D., T. G. Thompson and C. L. Utterbach, 1934. The electrical conductivity of sea water. *J. Cons. Explor. Mer*, **9**, 28.

Thompson, T. G. and R. J. Robinson, 1939. Notes on the determination of dissolved oxygen in sea water. *J. Mar. Res.*, **2**, 1–8.

Wenner, F., E. H. Smith and F. M. Soule, 1930. Apparatus for the determination aboard ship of the salinity of sea water by the electrical conductivity method. *Bur. Stand. J. Res.*, (RP 223), **5**, 711–32.

Wilson, K. G. and A. B. Arons, 1955. Osmotic pressures of sea water solutions computed from experimental vapor pressure lowerings. *J. Mar. Res.*, **14**, 195.

[Contribution No. 1296 from the Woods Hole Oceanographic Institution.]

# 6. WATER SAMPLING AND THERMOMETERS

## H. F. P. Herdman

## 1. Sampling

The collection of subsurface water samples and the determination of an accurate temperature at the depth where the sample is taken are still important factors in oceanography, and the method of taking routine serial temperatures and samples remains essentially as described by Sverdrup, Johnson and Fleming (1942). These authors also mention two types of subsurface water-sampler commonly used (Ekman and Nansen) which are still standard equipment, but do not mention a further type which is in common use in European waters. This is the Knudsen reversing water bottle (1929), which is similar to the Nansen bottle in that it is frameless, being clamped to the wire rope at the bottom and held to the wire at the top by a catch. It differs from the Nansen bottle, however, in that it is sealed by independent hinged lids at either end. When lowering the sampler these lids are held open, against an internal spring, by a simple mechanism connected to the release catch. When the messenger hits this catch, the lids are immediately closed and the bottle falls away through 180°, pivoting on the bottom clamp. This action also reverses the thermometers, which are housed in a frame independently attached to the bottle.

None of these samplers is entirely satisfactory. In the Nansen type, for instance, constant attention is required to prevent leakage at the valves, and distortion of the cam-operated push rods and end plates of the Ekman bottle is often the cause of a poor seal in this type. The Knudsen bottle is more positive in its sealing action owing to the narrow openings and the small hinged lids but it is difficult to clean the inside of the tube. With valve-type bottles it is virtually impossible either to sight or to clean the interior. In recent years it has been found that this inability to clean the inside of a water bottle has led to deterioration of any metallic plating (commonly silver or nickel) used as a lining. This can be a serious defect when samples are being collected for the determination of dissolved oxygen, as the brass tubing normally used for the bottle is then exposed and will in time occlude oxygen. If the sample is only in the bottle for a few minutes, this effect is not appreciable, but it can be substantial for samples from the greater depths, which may be held in the water bottle for an hour or more. To overcome this difficulty, the Munro–Ekman water bottle, used by the National Institute of Oceanography in Great Britain, was first of all lined with a silver tube spun into the bottle, then with a proprietary brand of a phenol-formaldehyde plastic and, latterly, with nylon. All these linings are inert in sea-water, but nylon has been preferred since it is flexible and so not easily cracked.

More recently, Fjarlie (1953) has described a simplified water-sampling bottle with better flushing arrangements. In this type only the thermometer frame reverses, the bottle remaining clamped to the wire rope. Closure of the tube is effected by flexibly mounted lids with "mousetrap" type springs on

[*MS received July, 1960*]    124

the hinges. The weight of this bottle, with thermometer frame for three thermo-meters, is 10 lb. It has a capacity of 1.14 l.

Water-sampling bottles rigidly clamped to the wire, such as the Ekman or Fjarlie types, are preferable in rough weather since there is less chance of damaging the thermometers against the ship's side than with the free-swinging Nansen- or Knudsen-type bottles. The weight of the water bottle is, however, a very important factor. Much time can be saved on station, and many more stations worked in a given time, if a considerable number of sampling bottles can be attached in series on the wire at one time. Using 4-mm diameter wire cord with a breaking load of 20 cwt, and a weight of 100 lb at the end, the safe working load at 5000 m depth should never exceed 750 lb (approximately one-third of the breaking load). With 5000 m of wire cord of 4 mm diameter the weight will weigh, in water, approximately 630 lb, thus leaving only 120 lb for the weight of the bottles. Ekman-type bottles, with the thermometers, weigh in water 17–20 lb, depending on the country where they are made, Nansen bottles weigh 7–8 lb and the Knudsen-type weigh 10 lb. Thus, in good condi-tions, the maximum number of bottles which can be used on one deep cast is: Ekman 6, Nansen 14, Knudsen 12 and Fjarlie 12. It cannot be too strongly emphasized, however, that these are *maximum* numbers, only to be used when weather conditions are very good.

So far, all deep-sampling water bottles have been made entirely of metal—usually brass—and so, as already explained, it is necessary to line the collecting tube. A subsurface sampling bottle made almost entirely of plastic or other non-metallic materials has, however, been designed by the National Institute of Oceanography in Great Britain and should eliminate most of the problems of maintenance and corrosion. Similar to the Fjarlie bottle in that only the thermometer frame is reversed on closing, the N.I.O. sampling bottle is con-structed mainly of polypropylene and has an improved type of seal for the collecting tube which is free of all internal obstructions. The capacity of the tube is 1.3 l. and the weight of the complete bottle in water is 3 lb. This will allow the use of 20 to 25 bottles in series. In other words, in fine weather, the normal range of samples required at a station in a depth of 5000 m could be obtained in one cast.

Large uncontaminated subsurface samples at all depths are becoming in-creasingly needed for, among other things, the determination of trace elements, and also in connection with work on the productivity of the oceans. Because of the large quantities to be collected the sampler is usually of such a size that it cannot be used in series and design varies largely with individual needs, the only common factor being the necessity to have a non-metallic lining. For instance, the large sampler used by the National Institute of Oceanography in Great Britain for productivity work was adapted from a Nansen–Pettersson water-sampling bottle, and holds 4 l. It is lined with nylon but provision is not made for recording the temperature. On the other hand, a large sampler constructed in 1937 for fisheries work in Canada had a capacity of 15 gallons (*ca.* 68 l.).

The collection of surface samples and the determination of surface temperature do not always receive the care with which subsurface observations are
taken. Too often the sample is taken in a canvas bucket and the temperature
read after an appreciable time. The errors thus introduced can be considerable
and, in an effort to overcome them, Lumby (1927, 1928) has described an easily
handled sampler which can be towed alongside a ship and which automatically

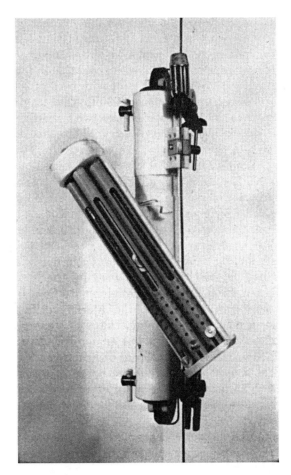

Fig. 1. N.I.O. plastic water-sampling bottle, with reversible frame for thermometers.

collects a clean sample. The thermometer cannot wholly be withdrawn when
reading it and part insulation of the sampler allows an interval of at least 4 min
for reading. More recently, Lumby and Haes (MS, 1957) have devised a simple
near-surface sampler which can be fitted in a circulating pump intake in the
smallest ship. This instrument permits a constant flow of water through
the whole apparatus. The temperature is read from a thermometer fixed in
the flow. If the surface temperature only is required this can automatically be

recorded by a distant reading thermometer or thermograph (Sverdrup *et al.*, *loc. cit*), but it must be remembered that the thermometer bulb, whether in an intake pipe or in a pocket in the hull, may be 9 ft or more below the actual surface, depending on the size of the ship. This is of little importance in the open ocean where a homogeneous surface layer normally extends to a depth greater than 5 m, but can give quite a fictitious reading in inshore waters.

## 2. Thermometry

Sverdrup, Johnson and Fleming (*loc. cit*) have given a very full account of the measurement of temperature and depth by means of protected and unprotected deep-sea thermometers. The thermometric method is still the simplest, cheapest and most accurate way of determining the depth at which a water-sampling bottle closes, even if the ship is drifting and there is considerable stray on the wire. An unprotected depth thermometer on at least alternate sampling bottles on a cast will give a more accurate figure for depth of reversal than calculations based on the length of wire and the wire angle between ship and surface.

In the course of years the deep-sea reversing thermometer has become an accepted instrument in oceanography. Nevertheless, it should still be considered a tribute to the glassblower's art rather than as a precision instrument since it is at times liable to malfunction without apparent reason. Herdman (1958) has analysed the performance of the reversing thermometers used by the Discovery Investigation Committee between 1925 and 1939, and afterwards by the National Institute of Oceanography. The percentage failure in 21,858 observations was 2.04, and there was little difference in this respect between German and British instruments. Whitney (1952) has commented on malfunctional behaviour in deep-sea thermometers and its correction and has also remarked (1955) on non-linear behaviour of unprotected reversing thermometers.

## References

Fjarlie, R. L. I., 1953. A seawater sampling bottle. *J. Mar. Res.*, **12**, 21–30.

Herdman, H. F. P., 1958. The reliability of deep-sea reversing thermometers. *Discovery Reps.*, **29**, 229–244.

Knudsen, M., 1929. A frameless reversing water-bottle. *J. Cons. Explor. Mer*, **4**, 192–193.

Lumby, J. R., 1927. The surface sampler, an apparatus for the collection of samples from the sea surface from ships in motion. *J. Cons. Explor. Mer*, **2**, 332–342.

Lumby, J. R., 1928. Modification of the surface sampler with a view to the improvement of temperature observation. *J. Cons. Explor. Mer*, **3**, 340–350.

Sverdrup, H. U., M. W. Johnson and R. H. Fleming, 1942. *The Oceans: their Physics, Chemistry and General Biology*. Prentice-Hall, New York.

Whitney, G. C., Jr., 1952. Notes on malfunctional behaviour and its correction in deep-sea reversing thermometers. *Woods Hole Oceanographic Institution Rep.* No. 52-29.

Whitney, G. C., Jr., 1955. A note on non-linear behaviour of unprotected reversing thermometers. *Woods Hole Oceanographic Institution Rep.* No. 55-62.

# II. FERTILITY OF THE OCEANS

## 7. PRODUCTIVITY, DEFINITION AND MEASUREMENT

E. STEEMANN NIELSEN

### 1. Introduction

Quantitative investigations of the marine plankton were initiated early in the history of oceanography. The most important pioneer work was done by the German scientist Victor Hensen towards the end of the 19th century. Attempts to understand the background of oceanic production were also made during this early period, for example by Hensen (1890).

Production as such was not investigated during this initial period of plankton science. It was the standing stock of organisms—plants and animals—which was measured. No methods were available at the time for measuring the production directly. We must admit that we still lack methods for measuring the production of animals directly. All estimates of animal production are based on measurements of the standing stock combined with assumptions and theoretical considerations (see, for example, Cushing, 1959).

The situation is more fortunate in the case of the primary production of organic matter by the photo-autotrophic planktonic algae. As will be shown later, the development of methods for directly measuring primary production began during the twenties of this century. During recent years the measurements of primary production have become one of the main tasks in oceanography.

A real production of organic matter takes place almost exclusively during photosynthesis; chemosynthesis is of very little importance. Organic matter is synthesized from inorganic matter, and the radiation energy from the sun is transformed into chemically bound energy. "Production" of animal organic matter is only a transfer of organic matter—or energy—from one trophic level to another. Hence we may concentrate here on discussing primary production and outline only a few aspects of "production" at higher levels.

A necessary condition for a large-scale fishery to be possible in a certain sea area is that a significant part of the primary production is within the area; but presumably this is not the only necessary condition. Thus the transfer of organic matter between the different trophic levels must be considered to be important also. However, our present knowledge is too incomplete to allow more than a general discussion.

The next trophic level following the autotrophic plants is made up in the ocean by the herbivores of the zooplankton (secondary production). A considerable part of the organic matter—and of the energy—is lost during the transfer from one trophic level to another. Ignoring at first the sinking of the

[*MS received August, 1960*]                    129

algae, we find that the losses are due partly to the respiration in the herbivores, partly to incomplete digestion of the algae used as food. Due to incomplete digestion a greater or smaller part of the organic matter is thus not transferred to the next ordinary trophic level but instead to a "side track" consisting of heterotrophic plants which assimilate the organic matter. In all probability some of these heterotrophic plants are eaten by certain herbivores. Some of the organic matter transferred into the "side track" may thus be returned to the "main track". In shallow water, incompletely digested algae in the form of "pellets" may sink to the bottom where they are used as food by herbivorous bottom animals (see Harvey, 1945, p. 151).

A part of the respiration in the herbivores supplies the necessary energy for the synthesis of various organic substances on the basis of the organic matter assimilated. These substances are used for growth and reproduction. Another part of the energy activated by respiration is used for work—in particular for the work involved in the catching of food. Our present knowledge of this subject is but slight; probably the work which the herbivores must necessarily carry out in order to collect food increases with decreasing concentrations of the algae. It is to be expected that the animals in the plankton-poor parts of the oceans must work hard in order to collect enough food. Thus a very considerable part of the organic matter assimilated by the herbivores in oligotrophic water is used to sustain the respiration operating the mechanisms necessary for the catching of food. Only a relatively small part of the assimilated matter can be used here for growth and reproduction. Accordingly, the lifetime of the herbivores in such areas must be relatively longer than in eutrophic areas. This is a very important point to be considered if we want to use the size of the standing stock of the herbivores for estimating their production. In all probability the relation between standing stock and production in the herbivores and carnivores varies to a considerable extent, being highest in oligotrophic water.

The rates of the biochemical processes taking place in the organisms need not be especially high just because the temperature is high. These processes are enzymatic and their rates are dependent on the combined effects of temperature and enzyme concentration. A low temperature may be counteracted by a high concentration of the enzymes (see Steemann Nielsen and Hansen, 1959). In fact, it has been known for a long time that in species of animals found both in the Arctic and in temperate waters, the rate of respiration at a certain temperature is highest by far in the Arctic (see Spärck, 1936). Finally, a selection takes place in nature. In oligotrophic waters herbivores able to tolerate a slow growth rate will presumably become predominant.

Few oceanic organisms are left to die from natural causes. Usually nearly all of them are devoured. However, in very eutrophic shallow coastal waters this seems not always to be true for the planktonic algae. The same is the case in areas where the spring bloom starts as an "explosion". In contrast to the open ocean, where the main bulk of planktonic algae is usually found in the photic zone (see Table I), most of the planktonic algae in very eutrophic shallow freshwater lakes are often located below the lower boundary of the

TABLE I

The Vertical Distribution of Phytoplankton (Coccolithophorides, Diatoms, Dinoflagellates) in the Central Minimum Area of the South Atlantic. [According to Steemann Nielsen and Aabye Jensen, 1957 (calculated after Hentschel)]

| Depth, m | Number of cells per litre |
|---|---|
| 0 | 2600 |
| 50 | 3600 |
| 100 | 2400 |
| 200 | 287 |
| 400 | 60 |
| 700 | 16 |
| 1000 | 13 |

TABLE II

Vertical Distribution of Chlorophyll in the Lake Wesslingsee (5 August, 1948). Depth of photic layer (according to the vertical distribution of oxygen) was about 2.5 m (Gessner, 1949)

| Depth, m | Chlorophyll, mg/l. |
|---|---|
| 0 | 33.6 |
| 2 | 76.4 |
| 3 | 33.0 |
| 5 | 33.0 |
| 6 | 76.4 |
| 7 | 96.0 |
| 8 | 75.0 |
| 10 | 75.0 |

Chlorophyll in the photic layer, 75 mg/m$^2$.
Chlorophyll below the photic layer 485 mg/m$^2$.

photic zone (see Table II). Of course all of the algae are produced in the photic zone, but they sink passively to the bottom, where they settle, die and become partly decomposed. The herbivorous zooplankton is unable to reduce effectively the sinking masses of algae by grazing. In all probability the duration of the sinking of a single alga is too short owing to the small depth. Thus an organic sediment is formed by means of the partly decomposed plankton algae. This type of locality might be characterized as an inharmonious biotype. A material part of the primary production—in many cases probably the main part by far —never reaches the next ordinary trophic level, the herbivores, but is lost as sediment. The mineral oil, which at present is utilized to such a high degree, in all probability originates from the primary production in super-eutrophic, inharmonious marine, shallow areas during ancient geological periods.

Of course planktonic algae also sink in the open ocean. However, the vertical distribution of the algae, as shown in Table I, according to the reports from the *Meteor* Expedition, clearly indicates that ordinarily nearly all of the algae are eaten either within the photic zone or during the sinking through the water-masses just below the photic zone. However, exceptions exist (see Bernard, 1958). We may characterize the ocean in general as a harmonious biotype. Most of the organic matter from the primary production is utilized by the next trophic level, the herbivores. Presumably the transfer again to the subsequent trophic level, the carnivores, is harmonious too.

With regard to the primary production, it is of extreme importance whether the conditions for the production are stable or unstable. If the development in the sea is to be harmonious it is necessary that the conditions for growth are stable at least to some extent. Catastrophes like the red tide (see page 188) occur presumably because the planktonic algae for some reason or other "grow wild". The normal "brakes", ordinarily regulating the growth of the algae, do not function. A normal "brake" of primary importance is grazing by the zooplankton. Nathansohn (1910) first called attention to the importance of zooplankton as the regulators of algal production. This early and outstanding contribution to biological oceanography has received surprisingly little attention from plankton scientists.

A provisional lack of zooplankton regulating algal production may sometimes be a perfectly normal phenomenon in the ocean. At higher latitudes, during a quiet period in spring, the production of phytoplankton often starts in an explosive way because a surface layer has been established due to heating by the sun (North Atlantic). Only a few grazing zooplankton organisms will be present provisionally. The ordinary brake for algal production is thus lacking. After a while, when a stock of zooplankton has been produced, an equilibrium will be established between zooplankton and phytoplankton. The grazing of the zooplankton can usually keep the stock of phytoplankton at just the appropriate level corresponding to the external growth conditions, such as the supply of nutrient salts. The size of the production at a higher trophic level cannot be considered to be merely a simple and uncomplicated product of the size of the production at the lower level. Production at all trophic levels must be considered as an interrelated whole.

During recent years there has been some disagreement concerning the concept of the size of the primary production in the big anticyclonic eddies in the oceans, such as the Sargasso Sea. In contrast to earlier investigations on the standing stock of plankton (see Lohmann, 1920 and Hentschel, 1933–1936), Riley *et al.* (1949) considered these ocean areas to be in fact very productive. This view was refuted by Steemann Nielsen (1952) by means of the measurements with the carbon-14 technique made during the *Galathea* Expedition in such areas—not only in the Sargasso Sea. Sverdrup (1955) did the same on a purely theoretical basis. However, the disagreement must now be considered to be finally settled. During the first International Oceanographic Congress in 1959, Menzel and Ryther (1960) were able to show that in the very neighbourhood of the Bermudas, where most of the biological Sargasso Sea work has been done by the Woods Hole Oceanographic Institute, the thermocline breaks down during winter causing vertical mixing down to a considerable depth. Nutrient salts are thus carried up into the photic zone giving rise to a considerable primary production during winter. However, at the same time they showed that the Bermudas (32°N) are situated at a borderline south of which this breaking down of the thermocline in winter does not take place. In the Sargasso Sea proper a low production rate is found throughout the whole year.

## 2. Definitions

When discussing oceanic production it is very important to make clear exactly what is meant by the different expressions used. A study of the literature reveals that different authors often interpret identical terms in very different ways; e.g. the word "production" has even been used by some authors as synonymous with "standing stock". Such severe confusions must of course be avoided. But even minor uncertainties may cause trouble.

If we consider the autotrophic plants exclusively it is relatively simple to present exact definitions, cf. the report from the "Committee on Terms and Equivalents", appointed during the Plankton Symposium at Bergen, 1957 (Anon, 1958, *Rapp. Cons. Explor. Mer*, **144**). By standing stock is meant the quantity of autotrophic plants at a given time. It is synonymous with standing crop. The term standing stock is used also for the quantity of organisms at other trophic levels, for example, for the herbivorous zooplankton organisms. The standing stock may be measured as biomass (=live weight), as dry plankton (=plankton dried to a constant weight), as dry organic matter (=dry plankton less ash), as displacement volume (=volume of fluid displaced by plankton which has been drained of water), as calculated volume (by counting and measuring the algae) or as plasma volume (=calculated volume less skeleton and vacuoles). The standing stock of phytoplankton may also be given in terms of the concentration of chlorophyll. The relationship between chlorophyll and total organic matter is variable, however. Chlorophyll should be used only with caution as a measure of the phytoplankton standing stock.

Since the term production is not unequivocal we prefer instead: gross primary production = rate of real photosynthesis, and net primary production = rate of real photosynthesis less rate of respiration by the algae. All of these quantities are given as carbon fixed (or released) per square metre or per cubic metre per unit of time. The time unit is usually 24 h. When measuring net primary production per square metre, we ordinarily account for only the respiration taking place in the photic layer (=the layer between the surface and the compensation depth, where, for a period of 24 h, the rates of gross production and respiration are identical). Ordinarily the respiration in the algae found below the photic layer must be disregarded. In the oceans this is usually relatively unimportant, because the bulk of the algae are found in the photic layer. However, in some very productive shallow localities we may find the main bulk of algae below the photic layer (see page 131). In other areas the vertical daily mixing of the water-masses extends below the lower limit of the photic zone. In the latter case the respiration in the whole mixed layer must be taken into account when calculating net primary production per square metre.

In the case of experimental methods, such as the oxygen technique to be described on page 139, it is not possible to measure the respiration of the autotrophic algae separately. The respiration of all organisms found in the experimental bottles is included. In fact, we measure instead community respiration, but unfortunately not the total community respiration. All bigger organisms ordinarily escape during the sampling of the water for the experiments.

Studies of community metabolism in the sea must be considered complicated. It is rather significant that the only detailed studies of aquatic community metabolism based entirely on observations have been made in two freshwater springs (Odum, 1957; Teal, 1957). In such localities the turnover takes place in one direction only, in contrast to ordinary marine and freshwater localities where it is cyclic, making investigations much more complicated. For investigating the community metabolism in the springs, Odum's "up-stream–down-stream" method can be used, whereby all water running through the springs is used for a compound experiment.

The only generally reliable technique by which it is possible at present to measure exclusively the respiration of the photoautotrophic planktonic algae found in the sea is a specific, rather complicated modification of the carbon-14 method (see page 145).

### 3. Methods for Measuring Primary Production

#### A. General Statements

The rate of primary production may be measured either directly or indirectly by estimating the standing stock of phytoplankton and using a conversion factor. In the first case the production is either measured experimentally by enclosing water samples in bottles or by utilizing differences in the water-masses during a certain period by measuring some property at the start and at the end of this period. All of the methods have their advantages and disadvantages. They have all contributed to our present knowledge of marine production. However, the use of several of these methods is restricted. The size of the organic production is of importance for the applicability of most of them. The existence of seasonal variations is also of importance. Some of the methods are applicable only if a typical sequence exists between a winter season without production and a summer season with production.

Finally, it must be mentioned that it is possible in broad outline to estimate theoretically the relative sizes of the primary production in oceanic areas where the hydrographical conditions are known. It is especially important to know the extent of the intermixing of the euphotic surface layers with nutrient-rich sub-surface water. Present knowledge of primary production in the ocean has shown an exceedingly remarkable correlation with hydrographic conditions. In Fig. 1 the chart shows the outlines of oceanic productivity based on the theoretical considerations exclusively. It was presented by Sverdrup (1955). A chart based on all of the measurements of the rate of primary production made during recent years would not look very different. This statement does not mean, however, that we can now cease to make real measurements of the primary production in the sea and let the physical oceanographers tell us about organic production. On the contrary, the close relationship between primary productivity and hydrography is a stimulus for the biologists to proceed in

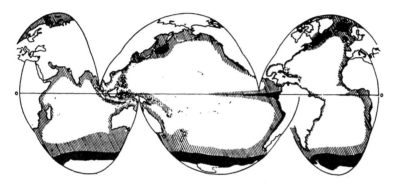

Fig. 1. Theoretical representation of relative productivity of ocean areas. Heavy shading indicates very productive areas, light shading moderately productive regions. (After Sverdrup, 1955.)

making detailed observations of the rate of organic production. Even a single individual measurement of the rate of organic production in a hydrographically complicated area does not present a random value but is a property characteristic of the water-mass in question. The rate of organic production may, to some extent, be treated in the same way as chemical or physical properties such as salinity and temperature (see page 158).

### B. Standing Stock as an Index of the Productivity of the Sea

All of the pioneer work concerning oceanic productivity was based on the standing stock. In the beginning only catches taken with fine silk nets were used for estimating phytoplankton. In the first decade of the present century, Lohmann (1908), however, showed that usually most of the phytoplankton is not retained by these nets. Silk nets may give quite erroneous results with regard to the amount of plankton found in the sea. In some instances they may catch the main part of the planktonic algae present in the water, in other cases practically nothing. In most cases the diameter of the meshes of the finest silk is much too large to retain any considerable part of the planktonic algae found in the sea.

Much to the disadvantage of plankton science, it took 50 years before the use of net methods was abandoned. However, at the Plankton Symposium at Bergen in 1957 convened by the International Council for the Exploration of the Sea (I.C.E.S.) a recommendation was unanimously adopted: "Net methods should not be employed in quantitative phytoplankton studies." Braarud (1958) has published a survey of the counting methods for the determination of the standing stock of phytoplankton. Sedimentation combined with the use of an inverted microscope is now used by most workers.

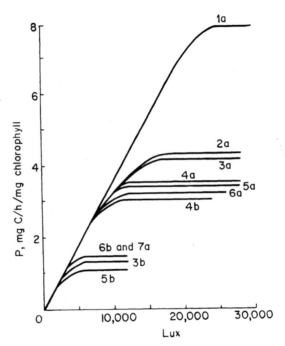

Fig. 2. Rate of gross photosynthesis per unit of chlorophyll as a function of light intensity; a = surface plankton, b = plankton from the lower part of the photic zone, 1 = tropical plankton, 2–4 = temperate summer plankton, 5 and 6 = arctic summer plankton, 7 = temperate winter plankton. (After Steemann Nielsen and Hansen.)

In recent years the concentration of the pigments active in photosynthesis—primarily chlorophyll a—has been employed, in addition to their use as an index of the standing stock of plants as a means of estimating the rate of potential photosynthesis and thus the rate of primary production (Ryther and Yentsch, 1957). However, the rate of light-saturated photosynthesis at the same concentration of chlorophyll varies within wide limits. Fig. 2 presents some typical curves showing the rate of photosynthesis as a function of the light intensity. Fortunately plankton from similar habitats gives similar curves.

Therefore, we ordinarily find a reasonably high correlation between the potential rate of photosynthesis and the concentration of chlorophyll, if we restrict ourselves to plankton samples from the same area, the same depth and the same season (see Fig. 3). Unfortunately even then, due to difficulties in measuring the concentration of chlorophyll, the results may at times turn out to be poor.

If we want to use chlorophyll as a general method for estimating primary production it will be necessary to take into consideration the shape of the different curves showing the rate of photosynthesis as a function of light (see page 136). Krey (1958) has published a survey of the chemical methods for the determination of the standing stock of phytoplankton.

When comparing the standing stock of primary producers in the oceans, the planktonic algae, with those in nearly all other biotypes, one striking

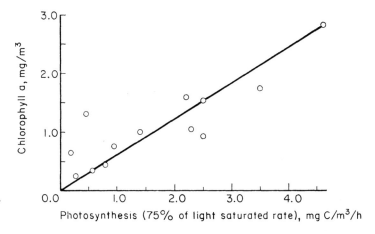

Fig. 3. Rate of photosynthesis as a function of the concentration of chlorophyll a. Arctic surface plankton, July.

difference is seen. A stock of planktonic algae is often able to produce an amount of organic matter of the same size as that originally found in the algae in the course of a day. Even if we take our fastest growing agricultural crops this is not possible. In maize during the month with maximum growth a daily increase of organic matter is found which varies between 10 and 20%. The average for the whole growth period is much less (Müller, 1948). It is easy on infertile land to find populations of perennial plants—for example lichens—for which it takes years to produce as much organic matter as is found in the plants at a given time.

In oceanic phytoplankton the relation between organic matter present in the stock and organic matter daily produced varies considerably. The light conditions are of primary importance. During the winter at higher latitudes very little, if any, organic matter is produced, thus causing a decrease in or even the virtual disappearance of the phytoplankton. A stock of planktonic algae must grow if the stock is not to be decimated due to sinking and grazing. This is, for example, in contrast to a population of terrestrial lichens. This type of population is able to survive during a long winter season without being much reduced with regard to the content of organic matter.

Our knowledge of the influence of the nutrient conditions—supply of P and N—on the growth rate of the planktonic algae is still scant. Probably minor variations in the growth rate have a considerable influence on the balance found in the sea, for example, between phytoplankton and zooplankton. According to the material from the *Galathea* Expedition, Steemann Nielsen and Aabye Jensen (1957) arrived at a phytoplankton growth rate of about 25% per day for the relatively unproductive tropical seas. Such a growth rate would compensate for the losses due to grazing and sinking and thus condition a constant, but relatively small, standing stock of planktonic algae.

### C. Consumption of $CO_2$ or Nutrient Salts as a Means of Measuring Primary Production

This method was first used by Atkins (1922) in the western English Channel. He proceeded originally by determining the loss of carbon dioxide under 1 m$^2$ of surface from the end of winter to the height of summer. As the loss in carbon dioxide equals the amount used in photosynthesis, a minimum value of the organic production in six months could be calculated. It was not possible for example to consider the exchange of $CO_2$ with the atmosphere or regeneration of $CO_2$ due to the respiration of all organisms, including also the planktonic algae. Atkins (1923) later used the loss of phosphate under 1 m$^2$ of surface in the same way and Cooper (1933) based his estimate—also for the English Channel—on the change in the content of oxygen and nitrate. The agreement between the different techniques was rather good.

These methods have been used relatively little outside the English Channel. Many difficulties may turn up. At best it is possible to obtain the order of magnitude of the value of the minimum production in a given sea area. Even for the English Channel a detailed investigation will turn out to be difficult (Cooper, 1958). Recently Steele (1958) has used the method in the northern part of the North Sea. The loss of phosphate in the photic layer during the production season was measured and a correction for the regeneration of phosphate was made by assuming the regeneration rate to be the same in the photic layer and in the water layers below, where no uptake of phosphate takes place. Simultaneous—but unfortunately rather few—measurements by means of the carbon-14 technique have shown a good agreement.

The method of using the consumption of a nutrient as a means of measuring primary production cannot be used universally. Certain conditions are strictly compulsory. First, definite seasons must exist. Winter conditions without any primary production are necessary. Secondly, no real interchange must take place with either the bottom or other water-masses. A definite thermocline thus must be found.

In Russia and in the U.S.A. (cf., e.g. Fedosov, 1958) the daily variation in the oxygen content of the water between morning and afternoon has been used as a means of estimating the organic production of marine areas. The method can be employed only in very productive areas. Even here many difficulties may present themselves (cf. Gessner, 1959).

### D. An Experimental Method using Oxygen Production

In photosynthesis oxygen is produced proportionally with the assimilation of carbon dioxide. If only carbohydrates were produced we would have the equation:

$$CO_2 + H_2O \xrightarrow{\text{light}} (H_2CO) + O_2.$$

Equivalent amounts of $CO_2$ and $O_2$ should thus be assimilated and released. However, proteins and fats are produced in addition to the carbohydrates. Usually nitrate is the source of nitrogen used. It becomes reduced before being converted into organic matter. In accordance with the basic composition of planktonic algae, like *Scenedesmus* and *Chlorella*, most authors use a $CO_2/O_2$ exchange factor of about 0.75, expressing photosynthesis measurements obtained by the oxygen technique in equivalents of carbon (Steemann Nielsen, 1952; Ryther, 1956a).

An experimental oxygen method of measuring organic production in the sea was introduced by Gaarder and Gran (1927). It has been used in slightly different modifications. In the most recommendable form, samples are collected from the various depths and siphoned into bottles with glass stoppers. Some of the bottles are used for determining the concentration of oxygen before the start of the experiment using the Winkler technique. The other bottles are again lowered to the depths from where the samples came and kept there for 24 h fixed to a line hanging down from an anchored buoy. As some of the bottles are wrapped in black material, it is possible to determine both the photosynthesis and the respiration taking place in the water samples. In respiration approximately equivalent amounts of carbon dioxide and oxygen are produced and consumed.

The oxygen content in the black bottles minus that in the initial bottle represents the combined rates of respiration of all of the organisms present, i.e. planktonic algae, animals, bacteria. The oxygen content in the clear bottles minus that in the black bottles represents the rate of photosynthesis by the plankton.

In principle the oxygen technique is very simple. However, in practice many difficulties may be encountered. It is easy to understand why so many workers have abandoned this technique after a first unsuccessful attempt; to be used properly it requires a great deal of skill and experience, and the water samples analyzed must be fairly productive. Steemann Nielsen (1958) has described the various difficulties and pitfalls, and the measures which may be taken. One of the main difficulties is the production of air bubbles in the bottles during the experiment, making the later determination of the oxygen production or uptake illusory. This difficulty is of no importance for the bottles kept at a depth of

1 m or below. But in bottles kept at the surface or in tubes on the ship deck air bubbles are ordinarily produced if no special measures are taken. The size of the bubbles varies considerably from bottle to bottle.

In the open ocean the planktonic algae are distributed within the whole photic layer, which is usually about 50–120 m deep. Although as a rule the rate of production per surface unit is by no means very low, the measurements are difficult because of the considerable vertical extension of the photic layer. Thus the amount of algae found in a single litre of water from the photic zone of the ocean is equal to the amount in the whole water column below a surface area of only about 0.1 cm². If the rate of production per square metre per day is put at 200 mg C—a rate somewhat higher than the average value for the oceans— the average production per litre per day in the photic zone is equivalent to 0.005 mg $O_2$. This means that the oxygen production per day is only about 0.1% of the oxygen ordinarily found.

The lower limit of the oxygen technique depends on the sensitivity of the Winkler technique for measuring oxygen dissolved in water. In a single titration it is impossible to determine oxygen with this technique more accurately than $\pm 0.02$ mg $O_2/l$. In many coastal waters it is even impossible to determine $O_2$ with this accuracy. The colour shift may become rather indistinct. The facts presented above tell us that the oxygen technique cannot be used in the oceans if the duration of the experiments is only 24 h.

The enclosure of a water sample in a bottle is in fact a rather severe inter-ference. In prolonged experiments the plankton may settle and thus give rise to considerable complications (difficulty in taking up $CO_2$, for example) especially if the density of plankton is high. By providing an effective stirring of the water this complication may be overcome. In plankton-poor water another difficulty turns up. Theoretically a prolonged experimental period would offer the possibility of counteracting a very low rate of photosynthesis. Unfortunately, bacteria at once start growing in large quantities if sea-water is enclosed in bottles. Due to the solid surfaces now present, organic matter, which is present in concentrations of about 1.2–2.0 mg C/l. in all sea-water, is at once attacked by the bacteria. Apparently because of the low concentration it cannot be attacked when the water is in situ. Solid surfaces seem to be a neces-sary condition (see for example, ZoBell, 1936). Jones et al. (1958) have shown that in plankton-poor oceanic water the number of bacteria may increase over a 24-h period by a factor of nearly 500. In experiments lasting three days the rate of respiration of these bacteria in ocean water is often as high as 0.2 mg $O_2/l$. per day. This rate is of an order of magnitude which is very different from that of the rate of photosynthesis of the planktonic algae in oligotrophic waters. This would be a sufficient cause for rejecting the use of experiments of long duration in plankton-poor water. Unfortunately, however, a distinct difference is often found in the oxygen contents of light and dark bottles after such pro-longed experiments (see Table III). The table shows further that the reduction of the algal stock to one-fifth—by filtration through paper—increases the oxy-gen consumption in the dark by a factor of 2.5 instead of decreasing it.

TABLE III

$O_2$ Experiments with Surface Water from the Sargasso Sea, 3–6 June, 1952. Triplicates used each titrated twice, mg $O_2$/l. $\pm 0.01$. (After Steemann Nielsen, 1958.)

| Water | Initial bottles | Clear bottles | Dark bottles |
|---|---|---|---|
| Non-filtered | 6.60 | 6.57 | 6.51 |
| Paper filtered | 6.59 | 6.61 | 6.37 |

In some way or other, light must be able to reduce the bacterial activity in the bottles containing oligotrophic ocean water. The problem seems to be rather complicated. Probably more than one cause may be found. Some species of bacteria ordinarily found together with planktonic algae are definitely influenced by even weak daylight filtered through several layers of glass (see Jørgensen and Steemann Nielsen, 1960). In light and dark bottle experiments with sea-water the results have been conflicting. Doty (1958, table IV) found, for example, that the bacteria developed much faster in the dark bottles whereas Vaccaro and Ryther (1954) did not find any difference between the two kinds of bottles. Possibly the concentration of planktonic algae is of importance, a light-induced decrease being conspicuous only in plankton-poor water.

### E. The Carbon-14 Technique

Radioactive tracers have given rise to a revolution in methods used in many parts of biology and also in the exploration of oceanic primary production. Whereas ordinary chemical methods can be used if the rates of the processes exceed a certain, often relatively high, minimum value, an isotope technique may be employed even when the rate of a process is extremely low. Carbon-14 was used for measuring the rate of primary production in the sea for the first time by the Danish *Galathea* Expedition round the world, 1950–1952. The technique was described by Steemann Nielsen (1952) and in more detail by Steemann Nielsen and Aabye Jensen (1957). Different details and minor modifications of the method have been presented for example by Doty (1958) and by Sorokin (1958).

In principle the technique is rather simple. The practical application of the carbon-14 technique in field-work does not require any expert knowledge. On the other hand, both technical skill and familiarity with tracer work is necessary when preparing $^{14}C$ ampoules to be used in field-work and when measuring radioactivity both in the ampoules and in the samples collected during the field work. In order to facilitate the work for laboratories without staff members having the necessary expert knowledge, UNESCO has provided funds for an international centre at Charlottenlund Castle, Denmark. This centre distributes standard $^{14}$C ampoules ready for use, and subsequently counts the radioactivity from the filters with the plankton.

A well-defined amount of $^{14}$CO$_2$ (as dissolved NaHCO$_3$ in sealed 1-ml ampoules) is added to the sea-water before an experiment. The content of CO$_2$ (total) in this water must be determined or estimated. If we assume that $^{14}$CO$_2$ is assimilated at the same rate as $^{12}$CO$_2$, then by determining the content of $^{14}$C in the plankton after the experiment we also determine the total amount of carbon assimilated. It is necessary only to multiply the amount of $^{14}$CO$_2$ found by a factor corresponding to the ratio between CO$_2$ (total) and $^{14}$CO$_2$ at the beginning of the experiment. The amount of $^{14}$C assimilated is determined by measuring the $\beta$-radiation from the plankton retained by a collodion filter. $^{14}$C has a half-life of 5500 years. Thus $^{14}$C ampoules will remain virtually unaltered for hundreds of years.

There is no reason for describing here all of the different details of the technique; the literature has already been cited above. One detail, however, should be mentioned. It concerns the type of Geiger–Müller tubes to be used for measuring $^{14}$C. A windowless counter is now used by most workers using this special isotope, and also by many of those working with primary production. The windowless counter has one definite advantage as compared with tubes possessing end-windows. At a definite rate of radiation from $^{14}$C the number of counts is much higher. A mica-window—even a thin one—absorbs a considerable part of the radiation from $^{14}$C.

Unfortunately the windowless counter has a definite disadvantage which has given trouble to several of the workers using the carbon-14 technique for measuring primary production (see, for example, Doty, 1959, page 43). It is difficult to determine the self-absorption curve necessary for estimating the activity of the solution found in the ampoules if a windowless counter is used. The present author has, therefore, always avoided the use of such counters.

For the purpose of counting the activity of the solution in an ampoule, carbon dioxide is precipitated as barium carbonate which is then filtered off by a membrane filter. Of primary importance is the extrapolation of the activity thus obtained to a sample thickness of 0 mg/cm$^2$ on the filter. This is absolutely necessary, because the weak radiation from $^{14}$C is readily absorbed also by the barium carbonate itself. This loss is called self-absorption. $^{14}$C emits $\beta$-rays with very varied, although always low, energies. If all $\beta$-rays had the same energy, a curve on semi-logarithmic paper showing the percentage of maximum specific activity as a function of sample thickness would be a straight line. In this case an extrapolation to zero thickness is very easy. If a relatively thick mica-window is used, only the most energy-rich radiation from $^{14}$C is able to penetrate and we obtain an almost straight self-absorption curve. In this way the number of counts registered would be considerably reduced, and consequently thin mica-windows must be used instead. Fig. 4 presents three self-absorption curves: one obtained with a windowless counter, one with an end-window counter (1.5 mg/cm$^2$) and one with an end-window counter (3.0 mg/cm$^2$). Although the last curve is nearest to the straight line, the present writer prefers the intermediate curve. The loss in counts is much less and the curve approaches a straight line sufficiently.

The difficulty in obtaining a reliable curve when using the windowless counter is due to the difficulty in preparing very thin, even samples of barium carbonate crystals. If the curve approaches a straight line, it is unnecessary to prepare such very thin samples. By means of scintillation counting Jiits and Scott (1961) have overcome the difficulty.

Another detail in the carbon-14 technique to be discussed here is the removal of any carbon in inorganic form before measuring the activity of the filters. After drying the filters, the present writer places them for 20 min in a closed container above concentrated hydrochloric acid. Experiments have shown that this procedure removes all inorganic carbon completely. Some workers instead pass a volume of very dilute hydrochloric acid through the filters by suction.

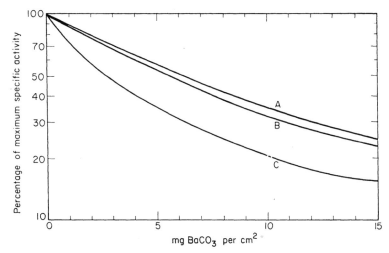

Fig. 4. Self-absorption correction curves. A = end-window 3.0 mg/cm$^2$, B = end-window 1.5 mg/cm$^2$, C = windowless counter.

When making experiments on photosynthesis with planktonic algae, it is important that no settling of the algae takes place. This is important regardless of the technique used for measuring photosynthesis. When making *in situ* measurements, where the bottles are placed on a wire hanging down from an anchored buoy, it would be very difficult to make any special arrangement for stirring the experimental water. In marine areas this is usually unnecessary anyway. The waves, the swell and the wind keep the buoy in constant motion and consequently the bottles are never at rest. Some stirring will take place.

In tank experiments on the other hand, stirring is advisable. The simplest method is to place the bottles on a rotating wheel in a water bath (see Fig. 5). The rotating speed is without importance. In addition to the stirring an even illumination of all the bottles is obtained, also an important condition neglected by some workers. Doty and Oguri (1957) have reported experiments showing the influence of agitation in tank experiments. More or less independent of the degree or the method of agitation, they found an increase of about 30 to 50% in productivity in each case. These workers used an experimental time of about 4 h. If the duration of the experiments is not too long, the effect of non-agitation may be without importance in some cases.

Fig. 5. Water-bath with rotating disk.

### F. Measurements of the Rate of Respiration by Means of the Carbon-14 Technique

There has been some disagreement concerning the explanation of what is really measured by the carbon-14 technique (see, for example, Ryther, 1954). The disagreement was settled, however, when it was found out that the duration of the experimental time was of importance (see Steemann Nielsen, 1958).

In experiments of short duration 50–70% of the $CO_2$ produced by the respiration taking place in the algae is photosynthesized again before being released by the cells (Steemann Nielsen, 1955). Fig. 6 presents experiments showing that only about one-third of previously assimilated $^{14}CO_2$ is released in the dark as compared to simultaneous experiments in the light. In a productivity experiment of long duration—24 h or more—during the greater part of the experiment, the same ratio is found between $^{14}C$ and $^{12}C$ both in the experimental water and in the algae. Under such conditions the method resembles in fact an ordinary chemical technique, where in the light we measure photosynthesis minus respiration, i.e. net-uptake of carbon dioxide. In experiments of short duration on the other hand, the carbon-14 technique measures something between net and gross production. To obtain values of either of these two quantities, it is necessary to make a correction.

A proper correction is possible only if the rate of respiration is known. A method for measuring the rate of phytoplankton respiration exclusively by means of the carbon-14 technique was described by Steemann Nielsen and Hansen (1959). A curve showing the rate of $^{14}C$-uptake as a function of the light

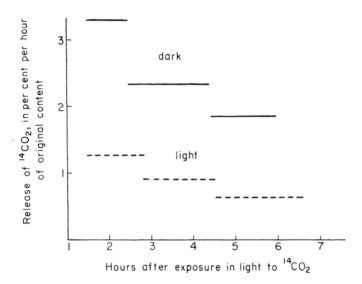

Fig. 6. The release of previously assimilated $^{14}CO_2$ in the light and in the dark. *Chlorella.* (After Steemann Nielsen.)

intensity is obtained by making experiments at different light intensities. By extrapolation the apparent rate of respiration is deduced. It is necessary, however, to correct for the interaction of photosynthesis and respiration (Fig. 7). The real respiration rate is about 10/6 of that obtained directly. The curve for the net production is found by drawing a line parallel to the original curve but intersecting the ordinate at a point representing the true respiration rate.

It would be too complicated always to determine the rate of respiration experimentally. Fortunately this is by no means necessary. Experiments with plankton collected in natural environments of all kinds have shown that the rate of respiration compared with the rate of light-saturated photosynthesis is always low—on the average about 8% of it (Steemann Nielsen and Hansen, 1959). It is thus justified to correct for the interaction of photosynthesis and respiration in measurements of photosynthesis at light saturation by arbitrarily estimating the rate of respiration to 8% of the rate of light-saturated photosynthesis, as was done by Steemann Nielsen (1952). The rate of the true net production is thus obtained by multiplying the experimentally found value by a factor of 96. In order to obtain a value of true gross production we must multiply by a factor of 106.

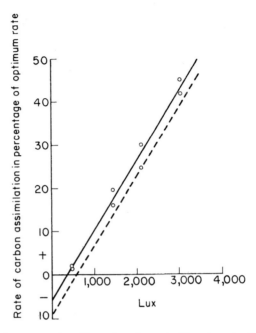

Fig. 7. Experiments for measuring the rate of respiration (see text). (After Steemann Nielsen and Hansen, 1959.)

### G. The Influence of Dark Fixation and Isotope Discrimination

It is generally known that, in animals as well as in plants, a dark exchange of $CO_2$ takes place which has nothing to do with photosynthesis. In this way labelled $CO_2$ may be introduced into the organic matter of both plants and animals. In a culture of algae the dark fixation of $^{14}CO_2$ is negligible compared to the fixation due to photosynthesis at light saturation, as was first shown by Brown, Fager and Gaffron (1949).

In ordinary $^{14}$C-experiments for measuring organic production in the sea other organisms are present in addition to the autotrophic algae. Nevertheless, in experiments lasting 4 h, the dark fixation in water from the photic layer is usually only a few (1–3) per cent of the fixation at light saturation. In water with a very low production, preferably from the lower boundary of the photic layer, dark fixation may be relatively higher (Fig. 8). The same is true in polluted water with a high stock of bacteria. Whenever possible, dark bottles should be employed in addition to the ordinary clear bottles and the dark fixation rates should be subtracted from the ordinary measurements.

In experiments of short duration, dark fixation is usually of minor importance only. In experiments of long duration, however, a very pronounced growth of bacteria takes place. It is a well known fact, as already mentioned on page 140, that a huge growth of bacteria starts as soon as sea-water is enclosed in bottles. Jones *et al.* (1958) showed that after 24 h the number of bacteria had increased by a factor of nearly 500. In oligotrophic water with few algae, the dark fixation by the bacteria may, therefore, easily exceed the fixation due to photosynthesis if the duration of the experiment is long. It cannot be repeated often enough that the enclosure of a water sample in a bottle is a severe interference. The duration of experiments must be kept short!

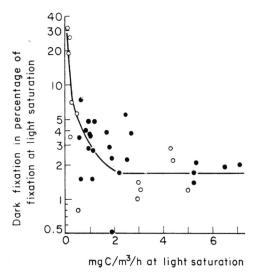

Fig. 8. Rate of dark fixation of $^{14}C$ as a function of the photosynthetic rate at light saturation. Summer plankton from the North Atlantic; filled circles from the surface, open circles from the lower boundary of the photic zone. (After Steemann Nielsen.)

In agreement with Anderson and Libby (1950), Steemann Nielsen (1955) showed that $^{14}CO_2$ is assimilated about 5% slower than is $^{12}CO_2$. This estimate was based partly on experiments made by van Norman and Brown (1952) and partly on the author's own experiments. Not all workers have taken the influence of isotope discrimination into consideration when employing $^{14}C$ for measuring primary production. Although the correction is small it should be taken into consideration.

It is likely that small amounts of assimilates are lost in most cases either by excretion during the experiments or owing to damage during the final filtration of the water sample. The experiments presented by Fogg (1958) seem to indicate that such losses are usually of no great importance.

However, in some cases the loss may be considerable (cf. McAllister *et al.*, 1961). As shown by Lasker and Holmes (1957) it is due to the fragility of the labelled cells. Many naked flagellates are destroyed beyond recognition on the surface of membrane filters after filtration. As it is difficult to predict the presence of such fragile flagellates it would be desirable always to take some precaution to minimize the loss. Among such precautions (cf. Guillard and Wangersky, 1958) may be mentioned: (a) filtration of at the maximum 100 ml (less, if the plankton concentration is very high) and (b) the use of a reduced vacuum, when using thin, fast Millipore filters.[1] If the thick, slow filters from the Membranfiltergesellschaft, Göttingen, are used, a higher vacuum may be used. Owing to the long pores, the active suction becomes much less. Duursma (1960) has shown that in certain northern waters the content of dissolved matter may vary considerably between winter and summer. This does not necessarily mean that the organic matter is primarily excreted during the growth of the algae. It is very likely that most of the dissolved organic matter originates either from algae only partly digested by the herbivores or from dying algae (cf. Guillard and Wangersky, 1958). As mentioned on page 132, during the spring outburst of algae the herbivores are often unable to eat all the algae produced.

### H. Different Modifications of the Carbon-14 Method

For the determination of the production of matter under 1 m² of surface, three different modifications of the method may be employed. In principle the best modification is the *in situ* method resembling Gaarder and Gran's oxygen method. Instead of measuring the oxygen metabolism in the bottles suspended at the different depths, the rate of photosynthesis is determined by the tracer technique. To estimate the total photosynthesis below 1 m² of surface of the sea, the measurements from the individual depths are presented graphically (Fig. 9). The area to the left of the curve represents the photosynthesis below the surface. This modification of the method is particularly suitable for measurements in coastal waters. In Danish waters, the lightships have been found ideal for such work (Steemann Nielsen, 1958). It is possible to adapt the technique in such a way that the crew on the lightship is able to do the field work adequately. A simple device for filtering the water samples is particularly important (Fig. 10).

It must be borne in mind that the conditions for the algae in the bottles are only approximately the same as in the sea, where vertical currents ordinarily prevent the algae from remaining constantly at the same depth. As very high light intensities have the effect of reducing the rate of photosynthesis, and, as it takes hours before this effect has disappeared, the lack of vertical movement of the algae during an experiment may affect the result. Rodhe (1958) has shown that in certain circumstances the yield of a 24-h experiment may be 21% or less than the sum of short-time experiments lasting a total of 24 h. The present author considers a 24-h experiment to be too prolonged for ordinary purposes (cf. page 140). Experiments from sunrise to noon or from noon to sunset must be considered more suitable.

In order to compute the rate of the production for a whole day, it is necessary

---

[1] Personal communication by Dr. G. Anderson.

only to multiply by a factor of 2. In the tropics the effect of very high light intensities is particularly conspicuous, for example the afternoon depression such as that demonstrated by Doty and Oguri (1957). In this case it is important always to use the same period for making experiments.

The *in situ* method is time-consuming even when the experiments last only from noon to sunset. It is, therefore, in most cases too expensive to use on

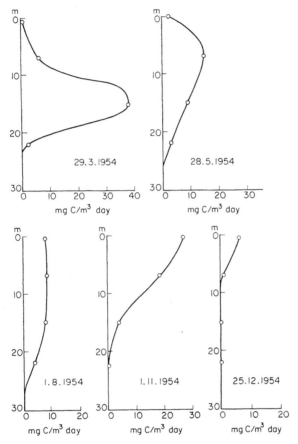

Fig. 9. *In situ* experiments in the Kattegat. Sunshine in all cases. (After Steemann Nielsen.)

expeditions with big ships. Instead a simulated *in situ* method may be used. In experiments using water from the different depths the samples are suspended in a tube with running surface water on the deck. In order to imitate the light conditions at the different depths, plates of neutral glass filters are placed above the bottles containing sub-surface water. These neutral filters are chosen in such a way that they match the light absorption in the sea; since, however, the light quality here changes with the depth the imitation is only approximate. Berge (1958) showed that the results from such simulated *in situ* experiments

(a)

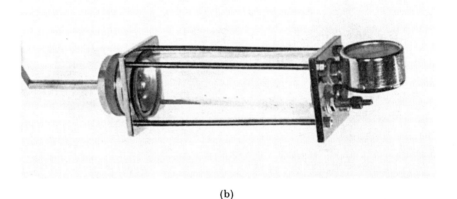

(b)

Fig. 10. An apparatus for filtering plankton by pressure.

were not significantly different from those of corresponding samples suspended
simultaneously at the normal depth in the sea. Fig. 11 presents such a simulated
*in situ* experiment. Menzel and Ryther (1960) have used this method for
measuring the annual variation of primary production near Bermuda in the
Atlantic. A simple device for agitating the bottles during such experiments has
not yet been invented. However, in areas with habitually rough weather, it is
sufficient to put glass pearls into the bottles, which are placed horizontally.

Results obtained during the S.C.O.R. intercalibration trials in Honolulu, September, 1961 (MS), indicate that the kind of neutral filters used in simulated *in situ* experiments is important. Neutral filters made of stainless steel screen or nylon net are very likely to be abandoned. It seems that the ultraviolet light penetrating such filters—in contrast to neutral filters of glass—may be harmful to the algae from the lower part of the photic zone. Inhibition of photosynthesis, besides being caused by excess light of all wavelengths, is also brought about by ultraviolet light (see Gessner, 1955). Owing to the thin walls of the experimental bottles the protection against ultraviolet light is not complete (unpublished data).

With the "tank" method $^{14}$C is added to water samples from the various depths which are then exposed to a well-defined light intensity in a water-bath at the same temperature as that found in the sea. After three to four hours the

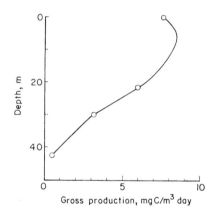

Fig. 11. A simulated *in situ* experiment in the North Atlantic (52° 00′N, 32° 42′W), 16 August, 1955. (After Steemann Nielsen.)

plankton is filtered off and the rate of photosynthesis determined in the usual way. It is the rate of potential photosynthesis which is directly measured by this modification. As mentioned on p. 143, it is essential that the bottles are agitated during the experiments. From the measurements of potential photosynthesis it is possible to estimate the rate of photosynthesis going on below a surface unit of the sea. In order to make this calculation it is necessary briefly to consider the influence of light on the photosynthesis in the algae found at the different depths.

Ryther (1959) has computed that, under ideal conditions (below saturation intensity and with total absorption of the light by the algae), 17.5% of the visible light (=about 9% of the total light) may be converted in the photosynthesis of algae. However, photosynthesis is proportional to light intensity only at low light intensities. At higher light intensities the overall process is limited by the rates of enzymatic processes. Hence a curve showing the rate of photosynthesis as a function of light intensity (Fig. 12) reaches a saturation point. At even higher light intensities the process is depressed. Under ordinary light conditions in nature, the algae found near the surface will be able to utilize the absorbed light in the most efficient way only during a short period

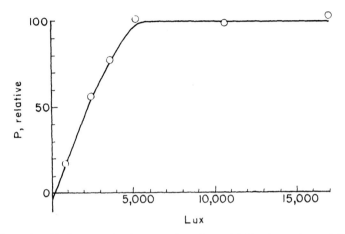

Fig. 12. The rate of photosynthesis as a function of light intensity. (After Steemann Nielsen.)

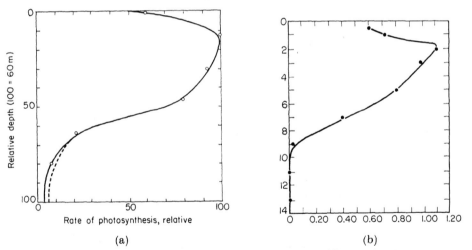

Fig. 13. Rate of photosynthesis in surface water exposed at different depths; (a) South China Sea, (b) South coast of Iceland during May. (After Steemann Nielsen.)

of the day. During the greater part of the day light saturation is reached, or still worse, light intensities prevail at which photosynthesis is depressed.

Therefore, in bright weather the highest rate of photosynthesis is found not at the very surface but at a depth where the light intensity is reduced by a factor of about 2 or 3. This is the case both in the tropics and at higher latitudes during the summer (Figs. 13a and b).

The most efficient utilization of the absorbed light takes place near the lower boundary of the photic zone. Light saturation never occurs here. It has been demonstrated (Steemann Nielsen and Hansen, 1959; Ryther and Menzel, 1959) that the algae are able to adapt those enzymes active in photosynthesis to the

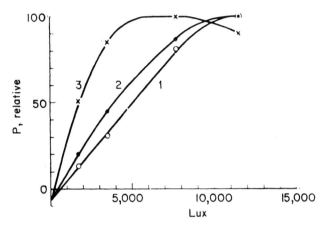

Fig. 14. Light intensity and relative rates of net photosynthesis in arctic summer plankton from three different depths; 1 = surface, 2 = 25 m, 3 = 50 m. (After Steemann Nielsen and Hansen, 1959.)

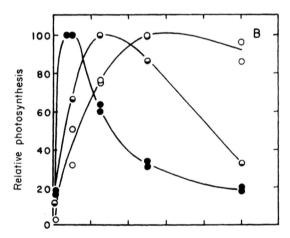

Fig. 15. Photosynthesis–light intensity curves of Sargasso Sea phytoplankton from depths to which 100% (open circles), 10% (half-filled circles) and 1% (filled circles) of the surface light penetrated. October. The whole abscissa = $12 \times 10^3$ ft-candles. (After Ryther and Menzel, 1959.)

different light intensities and temperatures found at the habitat (see Fig. 14). At the same time the light-induced decrease in photosynthesis in algae adapted to high light intensities sets in only at very high light intensities, whereas this decrease in "shade" adapted algae is found already at relatively low intensities (Ryther and Menzel, 1959; and Fig. 15).

By using the measurements of potential photosynthesis obtained with the "tank" method, it is possible to calculate the production of matter below a surface unit throughout a whole day. We must determine, first, the dependence of photosynthesis on the light intensity for the plankton from the different depths, and, secondly, the variation during the day of the light intensity at the depths from which the samples originate. By measuring the variation of light throughout the day by means of a deck-photometer and by making one determination of the light penetration into the sea, it is possible to compute exactly the relative light curves for all depths.

In practice, however, such a procedure would be too time-consuming. Formulae have therefore been developed for the calculation of primary production below a surface unit by means of tank-experiments. In contrast to Steemann Nielsen (1952) and Sorokin (1958) who tackled the problem empirically by making simultaneous experiments *in situ* and in the tank, Ryther (1956) and Rodhe *et al.* (1958) developed their formula on a more or less theoretical basis. The latter authors showed that their formula agreed with the empirical relationship used by Steemann Nielsen. Ryther (1959) has reconsidered the problem. Where he originally used a standard curve describing photosynthesis as a function of light intensity, he now takes the light adaptation into account. It must be emphasized that all formulae to be used—both those empirically found and those developed more or less theoretically—must take into consideration the special conditions at the habitat of the plankton.

Talling (1957) has presented a widely applicable expression for the total or integral photosynthesis of a population beneath a unit area of surface. The integral photosynthesis was shown to be related to the logarithm of the surface light intensity over a wide range of conditions.

## 4. The Relation between the Rate of Gross Production and Net Production Below a Surface Unit

When measuring primary production it must be emphasized that the fact most important to general oceanography is the size of the net production below a surface unit. The net production represents the potential source of organic matter—or better energy—which can be transferred to the next trophic level.

At low latitudes and during the summer at higher latitudes it is usually possible to calculate without much difficulty the approximate rates of net production below a surface unit.

By assuming the rate of respiration to be 8% of the light-saturated photosynthesis, Steemann Nielsen and Aabye Jensen (1957, p. 110) calculated the respiratory rate per 24 h of all algae found in the photic layer in the tropical ocean to be about 40% of the photosynthesis taking place daily. Ryther (1959) has tackled the problem in a general way by making a somewhat simplified assumption. Fig. 16 (after Ryther, 1959) presents relative photosynthesis, respiration and percentage of respiratory loss below a surface unit as a function of incident radiation.

The percentage of the respiratory loss ranges from 100 at radiation values of 100 cal/cm² per day or less to 28 on extremely bright, long days.

Taking into consideration the light adaptation of both the photosynthetic and respiratory rates, the curves are oversimplified. At a total radiation of e.g. 200 cal/cm² per day all values may vary considerably. The radiation in question approaches the maximum one, e.g. during autumn at higher latitudes. However, in the tropics, 200 cal/cm² per day will only be found when the weather is strongly overcast. The curves will also vary considerably depending

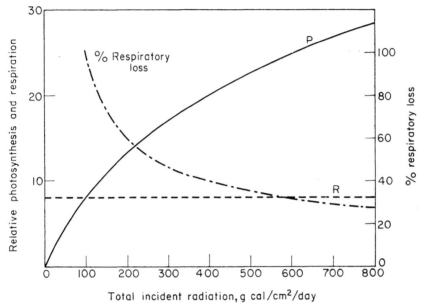

Fig. 16. Relative photosynthesis respiration, and percentage of respiratory loss as a function of incident radiation. (After Ryther, 1959.)

on whether the phytoplankton is distributed evenly within the photic layer or concentrated, for example, near the surface or near the lower boundary of the photic layer.

Calculating reliable values for the rate of the net production can sometimes be a rather time-consuming job which requires a great deal of expert knowledge. In general it must therefore be recommended that measurements of primary production are presented in terms of gross production.

## 5. Transparency of the Water versus Organic Production

A very simple method of estimating oceanic production is to measure the transparency of the sea or even more simply to estimate the colour of the sea.

Such a method can, of course, be employed only in the open sea well beyond the influence of the coast and only as a first approximation.

It must be emphasized that only the light absorbed by the photosynthetically active pigments found in the algae is used in photosynthesis. In a hypothetic ocean free of any particles and dissolved organic matter the depth of the photic layer would be about 140 m (according to Jerlov in Steemann Nielsen and Aabye Jensen, 1957). The lower boundary of the photic layer is ordinarily considered to be the depth where 1% of the surface light is found.

In the most transparent areas of the oceans—for example, in the centre of the Sargasso Sea—the depth of the photic layer is about 120 m. It is obvious that only a few algae can be present in this water, which has optical properties

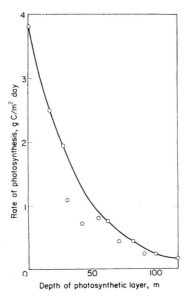

Fig. 17. Depth of photosynthetic layer and maximum rate of photosynthesis per m² surface. (After Steemann Nielsen and Aabye Jensen, 1957.)

resembling those of distilled water. Fig. 17 presents a curve showing the maximum rate of production below a surface unit as a function of the depth of the photic layer. All stations from the *Galathea* Expedition were divided into classes according to the depths of the photic layer. From every class the station showing the highest rate of production was selected.

If the rate of production per square metre is definitely lower than the curve values, the opacity of the water must be due primarily to causes other than living planktonic algae, for example minerals brought out from land, or remnants—including "yellow matter"—from a production which has taken place previously in the water. In the latter case the result is that in oceanic waters with considerable seasonal fluctuations in the size of the primary production, such as in the temperate part of the North Atlantic, highly transparent water is not found during the unproductive parts of the year. Although during the height of summer the rate of primary production is relatively low in certain areas of the north-eastern part of the North Atlantic (about 100 mg C/m² per day), the depth of the photic layer does not exceed 60 m. The transparency in the ocean far from the coast reflects not only the actual primary production at the time but also to some extent the production which has previously taken place in the water-mass.

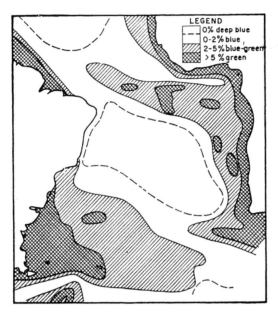

Fig. 18. Distribution of the colour of the sea. South Atlantic. (After Schott.)

Fig. 18 (Schott in Hentschel, 1933–1936), showing the distribution of the colour of the sea in the South Atlantic, represents at the same time the rate of primary production. At present it is not possible to publish a much better relative chart for this area (see Steemann Nielsen and Aabye Jensen, 1957). It agrees even in detail with the measurements of primary production—and standing stock of phytoplankton—made in the South Atlantic.

### 6. The Significance of Single Individual Measurements of Primary Production

The significance of single individual measurements of primary production must necessarily vary depending on the area from which the measurements originate. Both variations due to the sequence of seasons and to horizontal displacements of the water-masses must be considered. If a typical seasonal variation is found, as is always the case in the temperate and arctic zones, it is compulsory to make measurements throughout the productive season.

If a field "station" is situated in the centre of a large and hydrographically more or less uniform area, single individual measurements reflect the immediate situation found in this big mass of water. Berge (1958) has presented a considerable series of measurements from the Norwegian Sea made during a relatively short time. He showed that the rate of primary production may be treated in a way similar to that used for chemical and physical properties (see Fig. 19). Horizontal displacements of the water are of little importance near the centre of a big water-mass. It is only necessary to make enough measurements

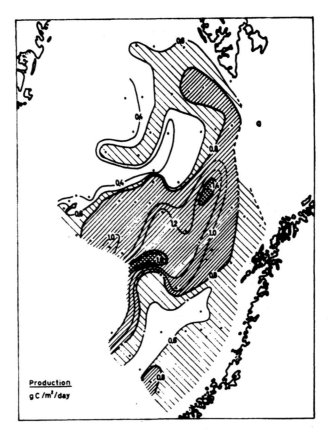

Fig. 19.  Primary production in g C/day below 1 m² of the surface. Norwegian Sea, June, 1954. (After Paasche, 1960.)

throughout the year in order to be able to follow the seasonal fluctuations. Measurements made twice a month—as on the Danish light-vessels—must ordinarily be considered sufficient. Even measurements made once a month give at least a general idea of the situation.

If, on the other hand, the station is situated near the border of two different water-masses, variations from day to day may be very pronounced depending on which water-mass predominates. During experiments at the northern entrance to the sound off Elsinore (see Steemann Nielsen, 1937), the rate of primary production was found to vary greatly depending on whether the current came from the north, carrying saline Kattegat water, or from the south, carrying brackish (ordinarily rather unproductive) Baltic water.

Rodhe (1958) has presented a series of daily measurements from a fixed station in the Swedish Lake Erken. The station was situated at the border between the main part of the lake and a much more productive bay of the lake. The variations from day to day were often very considerable depending on the direction of the wind. Jónasson and Mathiesen (1959), on the other hand, have published similar daily measurements from the Danish Lake Esrom Sø. They showed that the rate of production was nearly identical throughout the lake. Accordingly, the variations from day to day on a fixed station were found to be negligible.

In certain parts of the tropics the conditions for primary production must be fairly constant throughout the year. In other parts the monsoon shifts affect the conditions for primary production to a great extent. Even upwelling may take place during part of the year. In the big anticyclonic eddies, situated in all oceans between the latitudes of about 10°–35°, winter mixing does not reach the depths where the nutrient-rich water is found. Under these conditions primary production proceeds at a low, rather steady, rate throughout the year (see Steemann Nielsen and Aabye Jensen, 1957; Menzel and Ryther, 1959). In the western part of the North Atlantic eddy (the Sargasso Sea) the northern boundary is situated at about 32°N. North of this boundary the winter mixing reaches down into the underlying, more nutrient-rich water-masses and thus induces the formation of a considerable winter maximum in the primary production (see Menzel and Ryther, 1959).

## 7. The Influence of Weather Conditions

When making *in situ* or simulated *in situ* measurements of primary production, it is necessary to take into account the daily variations in the light energy penetrating the surface of the sea. Ordinarily, however, the influence of such variation is relatively small.

In the case of an overcast, or partly overcast, sky the radiation received is reduced compared with that on days when the sky is clear. According to daily continuous light measurements over a period of two years in Copenhagen (Romose, 1940) the average radiation measured during May–August was 67% of the average radiation measured on the four brightest days—one from each month. About the same average has been found in Washington for a whole year (Kimball and Hand, 1936). In the course of the measurements in Copenhagen it was found that 75% of the days during May–August received a solar radiation higher than 50% of that measured during the brightest day of the month, and that 94% of the days received more than 33% of that on the brightest day.

Fig. 20 presents a schematic curve for a 100 m-deep photic layer in a tropical ocean. The rate of photosynthesis is shown as a function of the depth on a

bright day. The plankton is presumed to be identical at all depths. A curve for
a day with a continuous light intensity of 50% of that on a bright day is
identical with the one which would be obtained were the curve as such moved
upwards 15 m. The rate of photosynthesis, measured per surface unit, would
now be about 80% of that measured on the bright day. By reducing the
average light intensity at the surface to one-third, the rate of photosynthesis
would become about 65% of that registered during a bright day.

It is interesting to note that, according to Bauer (1957), who during the
summer in a Bavarian lake made daily measurements of the incident light and
of the rate of primary production below a surface unit, the rate of production
decreased to about 60% on days when the incident light was about one-third

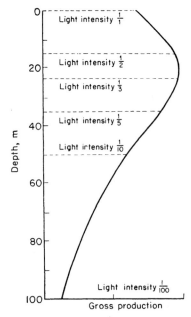

Fig. 20. Light intensity and rate of gross production at different depths. Tropical
ocean. (After Steemann Nielsen and Aabye Jensen, 1957.)

of that on a completely bright day. This is nearly the same as the value obtained
by theoretical calculations.

The relatively trivial effect of variations in the light intensity on the rate of
photosynthesis below a surface unit is due to the fact that light saturation in
planktonic algae is reached at a relatively low light intensity and that light
inhibition of photosynthesis is found at very high light intensities. At the very
surface in the summer the rate of photosynthesis per day is even higher when
the average light intensity is only half or one-third of that found during a
bright day (see Fig. 20). Days with the average light intensity reduced to one-
third are rare, as stated above. If we avoid making measurements during the
few extreme dark days, the variation in the incident light does not ordinarily

influence the measurement of primary production very much during the summer. However, an exception exists. If the bulk of the planktonic algae is concentrated near the lower boundary of the photic zone, a situation found under certain hydrographical conditions, photosynthesis will take place at the very low light intensities where, even in the shade-adapted algae, the rate of photosynthesis is directly proportional to the light intensity. Therefore, the rate of photosynthesis (gross production) per surface unit is virtually proportional to the radiation penetrating the sea surface. In the winter at higher latitudes the situation is the same. It has been found in Danish waters (Kattegat) that, in years with an exceptional number of bright days during December, a considerable stock of algae remains in the water until the New Year; they maintain a high rate of photosynthesis on the bright days. During years with relatively few bright days in winter the phytoplankton almost disappears during December (unpublished data).

In the winter at higher latitudes the light conditions prevailing on the particular days of investigation are always of decisive importance for the rate of production found. Fortunately, in the winter, the rates ordinarily found are so low that even considerable deviations from the average values would only slightly influence the rates of primary production computed for a whole year.

## References

Anderson, E. C. and W. F. Libby, 1950. World-wide distribution of natural radiocarbon. *Phys. Rev.*, **81**, 64–69.

Atkins, W. R. G., 1922. The hydrogen ion concentration of sea water in its biological relations. *J. Mar. Biol. Assoc. U.K.*, **12**, 717–753.

Atkins, W. R. G., 1923. The hydrogen ion concentration of open sea water and its variation with depth and season. *J. Mar. Biol. Assoc. U.K.*, **13**, 754–771.

Bauer, G., 1957. Die Ausnützung der Sonnenenergie durch das Phytoplankton eines eutrophen Sees. Inaugural Dissertation, München.

Berge, G., 1958. The primary production in the Norwegian Sea in June, 1954, measured by an adapted $^{14}$C technique. *Rapp. Cons. Explor. Mer*, **144**, 85–91.

Bernard, F., 1958. Données récentes sur la fertilité élémentaire en Méditerranée. *Rapp. Cons. Explor. Mer*, **144**, 103–108.

Braarud, T., 1958. Counting methods for determination of the standing crop of phytoplankton. *Rapp. Cons. Explor. Mer*, **144**, 17–19.

Brown, A. H., E. W. Fager and H. Gaffron, 1949. Kinetics of photochemical intermediate in photosynthesis. In photosynthesis in plants. *Monog. Amer. Soc. Plant Physiol.*, 403–422.

Cooper, L. H. N., 1933. Chemical constituents of biological importance in the English Channel. *J. Mar. Biol. Assoc. U.K.*, **18**, 677–753.

Cooper, L. H. N., 1958. Consumption of nutrient salts in the English Channel as a means of measuring production. *Rapp. Cons. Explor. Mer*, **144**, 35–37.

Cushing, D. H., 1959. The seasonal variation in oceanic production as a problem in population dynamics. *J. Cons. Explor. Mer*, **24**, 455–464.

Doty, M. L., 1958. Current status of carbon-14 method of assaying productivity of the ocean. University of Hawaii. MS.

Doty, M. L. and M. Oguri, 1957. Evidence for a photosynthetic daily periodicity. *Limnol. Oceanog.*, **2**, 37–40.

Doty, M. L. and M. Oguri, 1958. Selected features of the isotopic carbon primary productivity technique. *Rapp. Cons. Explor. Mer*, **144**, 47–55.

Duursma, E. K., 1960. Dissolved organic carbon, nitrogen and phosphorus in the sea.
Thesis. J. B. Wolters, Groningen.

Fedosov, M. V., 1958. Investigations of the chemical basis of productivity in sea basins.
*Rapp. Cons. Explor. Mer*, **144**, 61–64.

Fogg, G. E., 1958. Extracellular products of phytoplankton and the estimation of primary
productions. *Rapp. Cons. Explor. Mer*, **144**, 56–60.

Gaarder, T. and H. H. Gran, 1927. Investigations of the production of plankton in the
Oslo Fjord. *Rapp. Cons. Explor. Mer*, **42**, 3–31.

Gessner, F., 1949. Der Chlorophyllgehalt im See und seine photosynthetische Valenz als
geophysikalisches Problem. *Schweiz. Z. Hydrol.*, **11**, 378–410.

Gessner, F., 1955 and 1959. Hydrobotanik I and II. *Veb. Deut. Verl. Wiss. Berlin*.

Guillard, R. R. L. and P. J. Wangersky, 1958. The production of extracellular carbohydrates in some marine flagellates. *Limnol. Oceanog.*, **3**, 449–454.

Harvey, H. W., 1945. *Recent advances in the chemistry and biology of sea water*. Cambridge
University Press.

Hensen, V., 1890. Einige Ergebnisse der Plankton-Expedition der Humboldtstiftung.
*Sitz. Königl. Preuss. Akad. Wiss. Berlin, Sitz. Physik-Math.*, **17**, 1–11.

Hentschel, E., 1933–1936. Allgemeine Biologie des Südatlantischen Ozeans. *Wiss. Ergebn.
Deut. Atlant. Exped. 'Meteor'*, **11**, 1–344.

Jiits, H. R. and B. D. Scott, 1961. The determination of zero-thickness activity in Geiger
counting of $C^{14}$ solutions used in marine productivity studies. *Limnol. Oceanog.*, **6**,
116–121.

Jónasson, P. M. and H. Mathiesen, 1959. Measurements of primary production in two
Danish eutrophic lakes. *Oikos*, **10**, 137–168.

Jones, G. E., W. H. Thomas and F. T. Haxo, 1958. Preliminary studies of bacterial
growth in relation to dark and light fixation of $C^{14}O_2$ during productivity determinations. *Spec. Sci. Rep. Fish.*, **279**, 79–86.

Jørgensen, E. G. and E. Steemann Nielsen, 1960. Effect of daylight and of artificial
illumination on the growth of *Staphylococcus aureus* and some other bacteria. *Physiol.
Plantarum*, **13**, 541–545.

Kimball, H. H. and T. F. Hand, 1936. In Dugger, B.M., *Biological effects of radiation*.
McGraw-Hill, New York, **2**, 211.

Krey, J., 1958. Chemical methods of estimating standing crop of phytoplankton. *Rapp.
Cons. Explor. Mer*, **144**, 20–27.

Lasker, R. and R. W. Holmes, 1957. Variability in retention of marine phytoplankton by
membrane filters. *Nature*, **180**, 1295–1296.

Lohmann, H., 1908. Untersuchungen zur Feststellung des vollständigen Gehaltes des
Meeres an Plankton. *Wiss. Meeresuntersuch., Abt. Kiel*, N.F., **10**, 129–370.

Lohmann, H., 1920. Die Bevölkerung des Ozeans mit Plankton nach den Ergebnissen der
Zentrifugenfänge während der Ausreise der "Deutschland" 1911. Zugleich ein Beitrag
zur Biologie des Atlantischen Ozeans. *Arch. Biontologie*, **4**, 7–617.

McAllister, C. D., T. R. Parson, K. Stephens and J. D. H. Strickland, 1961. Measurement
of primary production in coastal sea-water using a large-volume plastic sphere.
*Limnol. Oceanog.*, **6**, 237–258.

Menzel, D. W. and J. H. Ryther, 1960. The annual cycle of primary production in the
Sargasso Sea off Bermuda. *Deep-Sea Res.*, **6**, 351–367.

Müller, D., 1948. *Plantefysiologi*. København.

Nathansohn, A., 1910. Beiträge zur Biologie des Planktons. *Intern. Rev. ges. Hydrobiol.
Hydrog.*, **1**, 37–72.

Norman, R. W. van and A. H. Brown, 1952. The relative rates of photosynthetic-
assimilation of isotopic forms of carbon dioxide. *Plant Physiol.*, **27**, 691–709.

Odum, H. T., 1957. Tropic structure and productivity of Silver Springs, Florida. *Ecol. Monog.*, **27**, 55–112.

Paasche, E., 1960. Phytoplankton distribution in the Norwegian Sea in June, 1954, related to hydrography and compared with primary production data. *Fiskeridirektor. Skr. Ser. Havunders*, **12**, 11, 1–77.

Riley, G. A., H. Stommel and D. F. B. Bumpus, 1949. Quantitative ecology of the plankton of the western North Atlantic. *Bull. Bingham Oceanog. Coll.*, **12**, (3), 1–109.

Rodhe, W., 1958. The primary production in lakes: some results and restrictions of the [14]C method. *Rapp. Cons. Explor. Mer*, **144**, 122–128.

Rodhe, W., R. A. Vollenweider and A. Nauwerck, 1958. The primary production and standing crop of phytoplankton. *Perspectives in Marine Biology, Scripps Inst. Oceanog.*, 1956, 299.

Romose, V., 1940. Ökologische Untersuchungen über *Homalothecium sericeum*, seine Wachstumsperioden und seine Stoffproduktion. *Dansk Bot. Ark.*, **10**, 4, 1–138.

Ryther, J. H., 1954. The ratio of photosynthesis to respiration in marine plankton algae and its effect upon the measurement of productivity. *Deep-Sea Res.*, **2**, 134–139.

Ryther, J. H., 1956. Photosynthesis in the ocean as a function of light intensity. *Limnol. Oceanog.*, **1**, 61–70.

Ryther, J. H., 1956a. The measurement of primary production. *Limnol. Oceanog.*, **1**, 72–84.

Ryther, J. H., 1959. Potential productivity of the sea. *Science*, **13**, 602–608.

Ryther, J. H. and C. S. Yentsch, 1957. The estimation of phytoplankton production in the ocean from chlorophyll and light data. *Limnol. Oceanog.*, **2**, 281–286.

Ryther, J. H. and D. W. Menzel, 1959. Light adaptation by marine phytoplankton. *Limnol. Oceanog.*, **4**, 492–497.

Sorokin, J. I., 1958. The results and prospects of using radioactive $C^{14}$ for a study of organic matter cycle in water basins. Radioisotopes in scientific research. *Proc. 1st (UNESCO) Intern. Conf.*, **4**, 633.

Spärck, J., 1936. On the relation between metabolism and temperature in some marine Lamellibranchs, and its zoogeographical significance. *Kgl. Danske Videnskab. Selskabs. Biol. Medd.*, **13**, 5.

Steele, J. H., 1958. Production studies in the Northern North Sea. *Rapp. Cons. Explor. Mer*, **144**, 79–84.

Steemann Nielsen, E., 1937. The annual amount of organic matter produced by the phytoplankton in the Sound off Helsingør. *Medd. Komm. Danm. Fisk. Havund., Ser. Plankton*, **3**, 3, 1–37.

Steemann Nielsen, E., 1952. The use of radioactive carbon ($C^{14}$) for measuring organic production in the sea. *J. Cons. Explor. Mer*, **18**, 117–140.

Steemann Nielsen, E., 1955. The interaction of photosynthesis and respiration and its importance for the determination of C-discrimination in photosynthesis. *Physiol. Plantarum*, **8**, 945–953.

Steemann Nielsen, E., 1958. Experimental methods for measuring organic production in the sea. *Rapp. Cons. Explor. Mer*, **144**, 38–46.

Steemann Nielsen, E., 1958a. A survey of recent Danish measurements of the organic productivity in the sea. *Rapp. Cons. Explor. Mer*, **144**, 92–95.

Steemann Nielsen, E. and E. Aabye Jensen, 1957. Primary oceanic production, the autotrophic production of organic matter in the oceans. *'Galathea' Rep.*, **1**, 49–120.

Steemann Nielsen, E. and V. K. Hansen, 1959. Measurements with carbon-14 technique of the respiration rates in natural populations of phytoplankton. *Deep-Sea Res.*, **5**, 222–233.

Steemann Nielsen, E. and V. K. Hansen, 1959a. Light adaptation in marine phytoplankton populations and its interrelation with temperature. *Physiol. Plantarum*, **12**, 353–370.

Sverdrup, H. U., 1955. The place of physical oceanography in oceanographic research. *J. Mar. Res.*, **14**, 287–294.

Talling, J. F., 1957. The phytoplankton population as a compound photosynthetic system. *New Phytol.*, **56**, 133–149.

Teal, J. M., 1957. Community metabolism in a temperate cold spring. *Ecol. Monog.*, **27**, 283–302.

Vaccaro, R. F. and J. H. Ryther, 1954. The bactericidal effects of sunlight in relation to light and dark bottle photosynthesis experiments. *J. Cons. Explor. Mer*, **20**, 18–24.

ZoBell, S. E. and D. Q. Andersson, 1936. Observations on the multiplication of bacteria in different volumes of stored sea water and the influence of oxygen tension and solid surfaces. *Biol. Bull.*, **71**, 324–342.

## 8. ORGANIC REGULATION OF PHYTOPLANKTON FERTILITY[1]

### L. Provasoli

## 1. Introduction [1]

### A. The Problems of "Good" and "Bad" Waters

The search for biologically active substances in sea-water is a recent approach to the age-long problem of "bad" and "good" waters. There is no doubt that water-masses differ in ability or inability to sustain growth of various organisms. Most of the interest in the properties of waters has centered on the words "good" and "bad"[2] for the organisms we harvest from the sea. In this respect, the most fertile areas of the sea are near the coasts, above shallow bottoms (banks), and in zones of water mixing (merging of different bodies of water, upwelling, etc.). Because all life in the sea depends on the primary producers of organic matter, most of our knowledge centers on the algae and we will restrict our presentation of data almost solely to them.

The early work on the physical and chemical conditions governing growth of phytoplankton in the sea was guided by the knowledge of the time that photosynthetic organisms require only mineral nutrients. Of these, nitrogen, and especially phosphorus, assumed primary importance because they are essential and are often present in waters in "limiting" quantities; hence the growth of phytoplankton, it was thought, should correlate with the N and P content of waters. This is often, but not always so: e.g. productivity along the coast of California is far less than around the British Isles, yet the phosphate content of California waters is many times higher. In the meantime, many nutritionists found that algae need, besides N and P, some organic substances present in several natural extracts (soil, peat, seaweeds, etc.) and also in sea-water (Allen, 1914). This led to the recognition that many algae need vitamins (reviewed by Provasoli, 1956; Droop, 1957; Provasoli, 1958).

The present interest in the organic components of waters, and particularly in the biologically active substances, is, however, due to the powerful arguments advanced by Lucas (1947, 1949, 1955, 1958) that some water organisms might affect the growth of other organisms by producing necessary nutrients, by removing inhibitory compounds, and by excreting inhibitory substances. The vitamin requirements of algae, what we already know of the vitamin cycle in sea-water and the production of antibiotics by marine organisms dramatically

---

[1] Aided in part by contract NR 104–202 with the U.S. Office of Naval Research and by grant G–10783 from the National Science Foundation.

[2] The utilitarian meaning attached to the terms "fertility" and "good and bad waters" is far too general and often misleading. Waters favorable for spring diatoms are not necessarily favorable for dinoflagellates, or tuna, and no water is bad enough not to support some sort of community. While it is obvious that the plants and animals living in the same biocoenosis share common requirements and resistance to a number of environmental factors, we can expect that other factors are peculiarly important for each species or taxonomic groups. Therefore, any discussion about the quality of waters should be in reference to a specific organism or group.

[MS received October, 1960]

confirm Lucas's postulates. But all this, perhaps even for the algae, is only a fraction of a bigger, more fascinating picture.

## B. Considerations on the Chemical Approach

Extracting and identifying the organic compounds in sea-water is complex because the dissolved organic C averages 2 mg/l. (maxima up to 20 mg/l.). These minimal quantities have to be separated from the 35,000 mg of inorganic salts in a liter of sea-water; the salt content of fresh water varies from 50–800 mg/l. In these few mg of organic matter we can expect to find any known and a few unknown organic compounds; some of them, like vitamins and other biologically active substances, are present in $10^{-9}$ quantities or less. The main obstacle is the separation of organic components from inorganic salts. This necessary step for any detailed analysis is by no means simple even for fresh waters. The extraction and identification of the microcomponents is further complicated by the need to process very large quantities of water. Jeffrey and Hood (1958) have reviewed and evaluated the various methods for isolation of organic matter.

To determine the effectiveness of various methods, they use a $^{14}C$-labeled test solution: a large sample of sea-water fertilized with N, P and labeled bicarbonate was kept in light for 90 days to favor algal growth, then placed in darkness for a long period to allow decay of the organic materials and filtered through an HA millipore filter. The effectiveness of the various methods was also tested on the recovery of biologically important compounds dissolved in natural sea-water.

Column absorption, electrodialysis and co-precipitation permit the isolation of almost all the dissolved organic matter. Co-precipitation with $FeCl_3$ or other metals at alkaline pH values is the most promising method. The recovery is close to 100% and the concentration factor is 10,000. However, the removal of the co-precipitating ions without altering the organic materials offers some problems; further, it is not known whether the metal ions and alkali destroy or change some of the organic compounds. Column absorption, especially with pretreated activated carbon, is very effective, but recovery from the absorbent is difficult and partial. Electrodialysis with cellulose membranes retains 97% of the organic matter and is ideal for the separation of colloidal micellae and large molecules, but inorganic sulfate is also retained and has to be eliminated by electrochromatography, and large volumes of water must be evaporated at low temperature. Electrodialysis with ion-exchange membranes employed for industrial desalting of sea-water permits work with large volumes; the recovery of organics is equal to that with electrodialysis. Electrodialysis perhaps may be improved by the use of several large cells separated by physical barriers of membranes of increasingly fine porosity; large sheets of the existing dialysing membranes of very fine and graded porosity are produced for other uses.

Ion-exchange and solvent extraction permit only a partial recovery of the dissolved organic matter. Ion-exchange resins offer no advantages: impractically large volumes of resins are necessary, huge amounts of liquid must be evaporated, and the resins give a large organic blank.

Extraction with water-immiscible solvents or with solvents after evaporation of sea-water is a simple and quick way to prepare organic concentrates without altering materially the organic compounds. Phenol, 2,4-lutidine, ethyl acetate and benzoyl alcohol remove substantial quantities of organic material. Even though any one solvent separates only a fraction of the organic matter, this method has been successful in extracting and identifying several organic substances in sea-water.

It might be interesting to try the method of "gel filtration", which acts as a molecular sieve. Like dialysis, it separates large molecules from small, but apparently is far more rapid and versatile. Columns or beds made of small grains of cross-linked dextran [Sephadex G–25 and G–50 produced by Pharmacia Laboratories (American address, Box 1010, Rochester, Minnesota)] form a gel which retards and apparently traps small molecules in the intramolecular spaces of the polymer, while large molecules do not enter the gel phase and flow freely in the interstitial fluid (i.e. between the grains of polymer molecules). This system has been employed to separate into two distinct fractions ammonium chloride from the proteins in serum (Porath and Flodin, 1959), and for the fractionation of polypeptides and proteins (Porath, 1959, 1960). It offers several advantages over dialysis: it is as rapid as filtration and can be used with the same results on a small or a large scale; high recovery is achieved in the presence of electrolytes; the polymer is inert to charged groups and stable in the pH range 2–10; and the gel beds retain their original properties even after daily use for a period of months. Two grades of cross-linkages are available, and perhaps others will be designed to trap or retard smaller molecules. Chlathrates or other systems of molecular sieves may be even better.

The direct chemical approach, hampered as it is by the inherent technical difficulties, has contributed only a few data on some of the organic substances present in relatively large quantities in sea-water. Better methods of extraction could be developed if one could direct the marine chemist to specific compounds.

The problem of the organic substances in waters and their fertility, being fundamentally biological, has also been studied by biologists. The first isolated attempts to solve the problem for algae have developed along several lines which have contributed some valuable information: the nutritional approach through culture of organisms, the biological analysis of sea-water, and the chemical analysis of the organisms and their excretions. These independent developments are all part of a comprehensive logical attack which, though mainly tested on algae, could be applied to all water organisms.

We can look at the problem in two ways: the origin and fate of the organic compounds in waters, and the nutritional needs of the organisms.

## C. Origin and Fate of Organic Substances

Obviously, almost the totality of organic substances is derived, in the present state of our planet, from the activities of living organisms.

Most water organisms are bound forever to the water environment and the cycle of organic matter is intimately dependent upon their activities. Lucas's theory of "non-predatory" relationships between water organisms includes an important part of these events because in waters the food chain is simultaneously a food pyramid and a pile of assorted excretions and metabolic products of micro-organisms, plants and animals. These organic substances fully express their biological activity because they are soluble or coupled in various ways to hydrophilic molecules and freely diffusible in the continuum of rivers, lakes and seas.

The other sources of organic matter in the sea are the rivers, which carry in the left-overs of the activities of fresh-water organisms, the elutes of rocks, and leachings of soils rich in trace metals and in the chemical residues of the activity of terrigenous microbes, plants and animals. Fluvial contributions to the open seas are minimal, and if felt at all, must be long-term; but they are significant locally, particularly in locations such as great rivers ending in relatively land-locked parts of the sea. One example is the effect of rivers and the effluent from Lake Okichobee after heavy rains on blooms of *Gymnodinium breve* (Wilson and Collier, 1955). Therefore quantitatively the plankton, especially the phytoplankton which represents the bulk of life, is the main source of organic matter in the sea, either through its secretions or decomposition.

The fate of the organic substances thus produced, and therefore the quantities of organic substances actually found in the waters, depends on the kinetics of production and consumption. Only slowly metabolizable organic substances should tend to accumulate and, being in larger quantities, may be easier to extract. Though these substances are of little or no nutritional value, they may have important functions such as the solubilization and chelation of trace metals (Shapiro, 1957; Fogg and Westlake, 1955); antibiosis, $CO_2$ and rH buffering; emulsification of fatty substances; and chemical conditioning of sexuality, spawning, moulting, etc. Such substances do accumulate in fresh waters as the humic substances, the yellow acids (Shapiro, 1957) and peptides (Domogalla, Juday and Peterson, 1925). In sea-water the situation seems to be different: the C/N ratio of the organic solutes is low (2–6) while in fresh waters it is high (>10). This difference may, however, be only apparent, because in the sea the origin of the soluble organics is almost entirely autochthonous (from organisms) while in fresh waters the effect of the surrounding soil and its leachings are very important (see discussion on page 173).

Conversely, organic substances which are in continuous demand as nutrients should be present in small quantities. Direct chemical analysis of the waters can be of little use; studies of turnover of labeled compounds seem more helpful. Granted that most of the inherent difficulties of identifying and following the fate of a variety of organic compounds in sea-water remain, this analysis is inescapable if we want to trace the important chemical events in the environment. The important thing is to know what is worth tracing.

Since we are dealing with knowns and unknowns produced by organisms, we must go to the origin, to the aseptic cultures of the organisms themselves:

chances are better there of finding larger quantities of the products of their metabolism. Besides, one can determine the conditions under which some substances are produced and released into the environment, study the chemical composition of the organisms, employ them as food organisms, or let them die and be broken down by micro-organisms. This experimental simplification does not mean that the task is simple.

In effect, what we are looking for is the chemical relationships between organisms mediated by the water environment; of the hundreds of organic substances present in the waters only a few may be important for a particular receiver species.

## D. *Nutritional Requirements*

The complementary approach to the problem is the study of the nutritional, morphogenetic and physiological requirements of the organisms. Obviously, the nutritional requirements can be identified only in the absence of any other organisms.[1] Axenic cultures are also indispensable in discerning the chemical relationships between the organisms of a given biocenosis. The lack of these exogenous contributions often results in no, or poor, growth, in abnormal morphogenesis, incomplete life cycles, mortality of larvae, difficulties in moulting, lack of copulation and ovideposition, etc. To impute these events to the artificiality of *in vitro* experimentation is to reject the opportunity of finding the cause and to miss potential bioassays.

The methodology and principles of nutritional studies are well established. Briefly, when attempting the isolation in axenic culture of the organism, it is advantageous to reproduce as far as possible the environmental conditions, both physical and chemical. Since often we do not have a precise knowledge of the environment, or we are resorting to culturing as a means to learn more about the biotic factors of the environment, we have to break new ground. We cannot afford to neglect any scrap of information or hint, experimental or observational. For instance, in dealing with an exacting alga, knowledge of the nutrition of other algae may be inadequate or irrelevant; one may have to resort to information on the nutrition of higher plants, plant-tissue culture, and animal nutrition. This applies to inorganic and organic requirements as well as to how to solve particular technical problems such as pH, rH, trace-metal buffering, emulsifiers, etc. In any case it is also advisable to try a variety of natural products and extracts rich in unknowns (soil, algal extracts; powders, infusions or hydrolysates of various organs; blood, coconut milk; etc.). If any of these proves useful, substitution of the natural products with their known major components will follow and may lead to a complete substitution with chemically defined compounds or even to a new growth factor or regulator (Hutner, Cury and Baker, 1958; Hutner, Provasoli and Baker, 1961).

Nutritional studies, in defining the factors needed for growth, indicate to the marine chemist and the ecologist which are the important parameters of the environment. The assumption is, if an organism requires absolutely a

---

[1] This is a wholly unnatural condition but so is the light-and-dark-bottle method. Somehow we seem to understand more easily the mechanism of a watch when we tear it apart, even if unable to reassemble the parts, let alone reconstitute a functional unit.

substance *in vitro*, this metabolite, or its physiological equivalent, must be in the environment: most of the algal flagellates of the littoral zone need vitamin $B_{12}$ and indeed this vitamin is measurable in sea-water. However, the nutritional studies, to have ecological meaning, should be extensive and not limited to growth factors (which are, after all, only one of many important variables).

As explained recently (Provasoli, 1958), one needs to know—besides the minimal mineral and organic requirements—the idiosyncrasies, tolerances and abilities to utilize a host of N, C, and P compounds. The success of an organism is controlled by the quantity and quality of metabolites; deficiency or excess are both detrimental.

Similarly, emphasis on the limiting factors of the environment implies belief in a simplification that can rarely be true. Just as interactions between organisms are sometimes fundamental, so are interrelations between chemical and physical factors.

Thermophilic bacteria, *Ochromonas malhamensis*, *Euglena gracilis*, etc., can be grown above the normal "lethal" temperature limit of the species if the vitamins and metals are increased manyfold; temperature is a nutritional variable (Baker *et al.*, 1955; Hutner, Baker *et al.*, 1957; Hutner, Aaronson *et al.*, 1958). Braarud (1945) noted that some boreal diatoms (*Chaetoceros* spp.) bloom in the warmer water (18°–20°C) of the Oslo Fjiord, but only in polluted areas. Temperature and salinity interact affecting the growth of marine fungi (Ritchie, 1957, 1959). Gradients of light and temperature markedly affect production and the ratios of the photosynthetic and accessory pigments of algae (Halldal, 1958). Similarly, the folic acid requirement of some organisms can be spared or substituted by *p*-amino-benzoic acid, the $B_{12}$ requirement by other cobalamins, methionine or deoxyribosides; some photosynthetic organisms can grow in darkness on exogenous carbon sources.

Axenic cultures of ecologically important organisms are the most fruitful biological approach to the problem of fertility of waters. In the course of determining nutritional needs of organisms one often detects the indispensability of organic substances which act at concentrations so low that they are not easily assayable chemically. The organism which has this need, or others more technically suitable because of their rapid and abundant growth and because of their specificity, can be employed for the quantitative determination of the active principle (see Chapter 9). Axenic cultures are also necessary, as mentioned, to determine the production of "external metabolites" and excretions which might affect growth and morphogenesis of other organisms. Their chemical composition can be analyzed giving information on the nutritional needs of their predators.

### E. Biological Analysis of Sea-Water

Another interesting approach to the fertility of waters is the so called "biological analysis". Selected organisms can serve to differentiate water-masses into "good" or "bad" by their growth responses or other measurable physiological events (hatching, setting of larvae, etc.). Once established, one can try with different treatments to see whether the "bad" waters lack some nutrients or contain inhibitory substances for the test species. When sufficient nutritional information is available for a group of organisms, we can add to freshly collected samples of waters (with their living population) a mixture of all the nutrients required, the mixture minus one of the nutrients, and each single nutrient. We have thus a means of discerning the lack of some nutrients and of evaluating by the incidence of scarcity, which parameter is more ecologically important. Other profitable uses of this approach can be found: H. Kylin (1941), employing germlings of *Ulva*, found that waters of 70-m depth were low in N, P, Fe, Zn, Co, Mn; A. Kylin (1943) determined the need of them and level of trace metals favorable for *Ulva*.

## 2. Data from Chemical Analysis

### A. Quantities of Organic Matter in the Seas

Organic matter into water is arbitrarily divided into: (*a*) particulate (living organisms + detritus = seston) and (*b*) "dissolved" organic matter (in general particles < 1 μ). Strikingly, the amount of dissolved organics exceeds by many orders of magnitude the amount of particulate: in the euphotic zone, rich in phytoplankton, it is seven to eight times greater and in deep waters, in which the plankton is scarce, it may be up to a thousand or more times greater.

Only recently has it been possible to analyze the dissolved organic carbon in sea-water. The first reliable method was developed by Krogh and Keys (1934); newer methods were developed by Kay (1954), Skopintsev (1959) and Duursma (1960). Duursma, in a comprehensive, stimulating paper, reviews methods for determining the dissolved organic C, N and P in sea-water, the data obtained and their significance. The quantities of total organic C in open seas vary from 0.2–2.7 mg C/l. Higher values are found in more landlocked areas: 3.3 in the Black Sea, 4.6 in the Baltic, 6 in the Sea of Azov, and 8 in the landlocked coastal area of the Dutch Wadden Sea (Table I). Contrary to the early data of Krogh, who found an almost homogeneous distribution from surface to deep water, wide variations were found by the other investigators. Especially significant are the data of Duursma derived from several hundred analyses in the Wadden, North and Norwegian Seas, and in the North Atlantic Ocean. He found that "there are well defined areas where maximum and minimum concentrations occur, both horizontally and along the vertical" (Duursma, *op. cit.*, p. 133). Water-masses can be characterized by their peculiar concentrations in organic nitrogen and carbon, besides the usual variations in salinity and temperature. His distribution charts and data indicate that concentrations of dissolved organics may permit the identification of smaller water-masses, especially if one considers also the C/N ratios, which vary widely and independently. In a year's study of the station L.V. *Texel* in the North Sea, Duursma found clear seasonal variation in dissolved organic matter. Since the highest concentrations of organic C are reached several weeks after phytoplankton blooms, he considers that nearly all the dissolved organic compounds are produced by breakdown of the dead phytoplankton and that the excretion

## TABLE I

### Dissolved Organic Matter in Sea-Water[a]

| Author | Locality | mg C/l. | mg N/l. |
|---|---|---|---|
| Krogh (1934) | Atlantic Ocean (Bermuda) | 2.35 ± 0.07 (along vertical) | 0.244 ± 0.08 (same) |
| Dazko (1939) | Black Sea | 2.4 | |
| Dazko (1951) | Black Sea | 2.83–3.36 (seasonal variations) | |
| Dazko (1955) | Sea of Azov | 4.63–6.02 | |
| Kay (1954) | Baltic | 2.0–4.6 (maximum in euphotic zone) | |
| Plunkett and Rakestraw (1955) | Pacific (3 stations) | 0.6–2.7 | |
| Skopintsev (1955) | North Atlantic | 1.04–1.97 | |
| Skopintsev (1959) (referring to other Russian investigators) | Atlantic Ocean | 2.40–2.48 | 0.24–0.26 |
| | Pacific | 0.98–2.68 | 0.07–0.11 |
| | Greenland Sea | 2.0–2.1 | 0.03–0.38 |
| Duursma (1960) | North Atlantic | 0.2–1.3 | 0.04–0.4 |
| | Norwegian Sea | 0.45–1.38 | 0.1–0.21 |
| | North Sea (L.V. *Texel*) | 0.5–1.8 | 0.08–0.54 |
| | Wadden Sea | 1.0–8.0 | 0.1–0.6 |

[a] Derived from Duursma's table II (1960), with added concentration ranges found by Duursma.

of dissolved organic matter by living phytoplankton is not directly demonstrable in the sea. This conclusion is contrary to the finding, under experimental conditions, that several species of algae excrete considerable amounts of organic solutes. This discrepancy deserves further work to find out if the excretions are produced by healthy cells during logarithmic growth or by resting or dying cells, and what are the conditions favoring release of external metabolites. More detailed field studies are also needed to see whether Duursma's findings are general. He determined at the microscope "the relative proportions of the number of several species" (Duursma, *op. cit.*, p. 29), but the plankton populations were not studied in detail; this could give a more precise meaning to the data on particulate matter and chlorophyll determinations.

The practical and precise method for analyzing dissolved organic carbon of Duursma allows finally a detailed and more biologically significant definition of water-masses, and already poses some new puzzles. It is interesting for instance to compare the C/N ratios found in fresh water and sea-water. Birge and Juday (1934), in grouping the data from several hundred lakes according to total organic carbon, found a C/N ratio of 12.2 for lakes containing 1.0–1.9 mg C/l. Duursma found in North Atlantic deep waters of similar C content a C/N ratio varying from 2.5–6.5 with a mean of 2.7. The surface waters (0–200 m) of the same transect (a meridional section running southward from Cape Farewell in Greenland, 60°N, to about 43°N) show a tendency toward large variations in C/N ratios (up to 20–30) at higher temperatures ($> 7°C$) whereas at lower temperatures ($< 7°C$), and especially in September, the C/N ratios are closer to the values of deep water. Living phytoplankton and zooplankton have an average C/N ratio of 5.7–6.7 with limits between 4 and 14 (Fleming, 1940), proteins of 2–3, and skeletons of invertebrates 2.9–3.3 (Vinogradov, 1953). Why the striking difference in C/N ratios between fresh waters and sea-water? Are the high values in fresh water only due to the influence of the land on lakes, or to physical conditions such as depth and size of the basins? Hutchinson (1957) has calculated from the data of Birge and Juday the amounts of the two types of dissolved organic matter found in lakes: one (allochthonous) derived from bogs, peat and soil has a brown color and an approximate C/N ratio of 45–50; the other (autochthonous) derived from the decomposition of plankton has an approximate ratio of 12. Skopintsev (1959) found that the end product of complete decomposition of marine organisms is a "water humus", a carbon–protein complex (possibly a pectin or uron) of high biochemical stability. This product has a C/N ratio of 10, remarkably close to the one calculated by Hutchinson. In the open sea we cannot expect to find significant amounts of allochthonous dissolved organic matter but we could expect to find the autochthonous, yet we find lower ratios (2–6) in some surface waters and in deep waters. The warm surface waters of the North Atlantic have variable and high ratios, while the cooler surface waters have ratios similar to the deep waters. This could be due to a slower decomposition in cool waters: the dead organisms will be decomposed only partially before they sink; in deep water the scarcity of micro-organisms and the cool temperature may account for the stability of C/N ratios of the dissolved organic portion. If so, the dissolved organic matter in deep water should still be a good nutrient and not a product resistant to bacterial attack. Many experiments with oxygen bottles in fact show that the deep waters can support a good population of micro-organisms (provided solid surfaces are available). This, however, does not explain the difference, presumably in the early phases of decomposition, between fresh water and sea-water. From the starting point of 6 (ratio in living plankton) decomposition produces in fresh waters higher ratios (i.e. N is preferentially consumed) and in sea-water lower ratios (C is preferentially consumed or high-N compounds are produced). If we assume that, during decomposition in sea-water, high-N products are released in large amounts, then the low C/N ratios found do not exclude the presence, though in low quantities, of stable products like "water humus" which have high C/N ratios. Quite likely decomposition proceeds differently in sea-water: production of amines is after all quite noticeable in sea-water food at the market; and the bacterial flora of sea-water is quite different from that of fresh waters.

## B. The Nature of Dissolved Organic Substances

Vallentyne (1957) has comprehensively reviewed the organic compounds found in lakes, oceans, sewage and soil. Data on the identification of some of the organic solutes in sea-water are summarized in Table II.

TABLE II

### Organic Compounds Identified in Sea-Water

| Substances | Quantities | Locality | Method | Authors |
|---|---|---|---|---|
| "Rhamnoside" Dehydroascorbic acid | Up to 0.1 g/l. present | Inshore waters Gulf of Mexico | Activated charcoal absorption, ethanol elution | Wangersky (1952) |
| "Carbohydrates" (as arabinose equivalents) | 0.0–20 mg/l. | Estuary, Gulf of Mexico | N-Ethyl carbazole | Collier et al. (1953) |
| "Carbohydrates" (as sucrose equivalents) | 0.14–0.45 mg/l. | Pacific Coast, U.S.A. | Anthrone and N-ethyl carbazole | Lewis and Rakestraw (1955) |
| "Carbohydrates" (as arabinose equivalents) | 0.0–2.6 mg/l. (max. of 12 mg/l. at surface, 29°N, 80°, 31′W) | South Atlantic (30°N–25°N) | N-Ethyl carbazole | Anderson and Gehringer (1958) |
| | 0.0–3.0 mg/l. (23% = 0.0; 50% = 0.2–1 mg/l.) | Continental Shelf, Gulf of Mexico (50 mg/l. in red tides of G. breve) | N-Ethyl carbazole | Collier (1958) |
| Citric acid | 0.025–0.145 mg/l. | Littoral Atlantic (French Coast) | | Creac'h (1955) |
| Malic acid Acetic and formic acids [a] | 0.028–0.277 mg/l. < 0.1 mg/l. | Northeast Pacific (surface and inshore) | Chloroform or ether extraction at pH 3; partition chromat. on silica gel column | Koyama and Thompson (1959) |
| Fatty acids (up to 20 carbons) | 0.4–0.5 mg/l. (weight of methyl esters) | Gulf of Mexico | Ethyl acetate extraction at pH 2. Gas–liquid chromatography | Slowey et al. (1962) |

TABLE II—(cont.)

| Substances | Quantities | Locality | Method | Authors |
|---|---|---|---|---|
| Amino acids (from hydrol. proteins) | Traces to 13 mg/ m³ [b] | Gulf of Mexico, Yucatan Strait, Reef (British Honduras), Caribbean | Co-precipitation of organic material with $FeCl_3 + NaOH$; acid hydrolysis; paper and ion-exchange chromat. | Tatsumoto et al. (1961) |
| Vitamin $B_{12}$ | Present | See Table XII | | |
| Plant hormones | Present | North Sea | Chloroform extraction at pH 5; ether extract of residue. Measured biologically | Bentley (1959) |

[a] Acetic, formic, lactic, and glycolic (up to 1.4 mg/l.) acids are liberated from breakdown of larger organic molecules during the long extraction procedure (4–5 weeks).

[b] 18 amino acids were found in the hydrolysates. The amounts and kind of amino acids vary widely in samples.

The data are few and limited. Aside from the "carbohydrates" estimation, based on diverse localities and depths, the other data represent successful extractions and identifications of the components of the dissolved organic matter and are limited to a very few samples of sea-water. The $N$-ethyl carbazole or anthrone methods are not strictly specific for carbohydrate. Nevertheless, this colorimetric reaction can be quantitated and serves to measure a biologically active organic fraction which affects quantitatively the pumping rate of oysters (Collier et al., 1950, 1953). This organic fraction may be of algal origin; the filtrate of cultures of several marine flagellates and diatoms contains organic substances reacting with $N$-ethyl carbazole (Collier, 1958; Guillard and Wangersky, 1958). The oyster activity of such filtrates has not been reported. Perhaps it may be possible to separate, and later analyze, the $N$-ethyl carbazole reactive fraction from the organic solutes of sea-water extracted by co-precipitation with Fe in alkali.

## 3. Organic Products of Algae and Bacteria

### A. Excretion of Carbohydrates

Several fresh-water green algae, mostly *Chlamydomonas* species, liberate soluble organic products in the culture medium. Six *Chlamydomonas* release as solutes from 10 to 45% of the total organic matter produced (Allen, 1956; measured as organic material oxidizable by dichromate). The products formed include polysaccharides, glycolic, oxalic and pyruvic acids and only traces of organic nitrogen. Fifteen species of fresh-water *Chlamydomonas*, two *Chlorosarcina* and one *Gloeocytis* liberate in the medium 10–115 mg/l. of soluble polysaccharides. This excretion constitutes 2 to 25% (*Chlamydomonas mexicana*) of the total organic matter produced (Lewin, 1956). The main components of the polysaccharides (precipitated by ethanol and hydrolyzed with $H_2SO_4$) are galactose and arabinose for all species except *C. ulvaensis* (glucose and xylose). The associated sugar moieties are fucose, rhamnose, mannose, uronic acids and several unidentified components (identification by chromatography).

Allen found that the production of extracellular compounds parallels growth, i.e. the substances are released by living and dividing cells; the same applies to *C. parvula* (Lewin, 1956) and *Anabaena cylindrica* (Fogg, 1952); however, production of mucilage around the cells continues in old cultures, reaching 40–60% of the total organic matter produced (*Chlamydomonas parvula* and *C. peterfii*) (Lewin, 1956).

Also unicellular marine algae, in bacteria-free culture, excrete large amounts of "carbohydrates" (Table III). These results were obtained with the *N*-ethyl carbazole hydrolyzing method, which gives a purple-red coloration with carbohydrates; optical densities were converted to glucose equivalents, except for *Prorocentrum* sp. (arabinose equivalent). Guillard and Wangersky (1958) found that excretion of carbohydrates does not parallel growth: it is very low and does not exceed 3 mg/l. during exponential growth. Carbohydrates accumulate in the medium at or right after maximum growth ($10^6$–$10^7$ cells/ml). This accumulation may come from lysis of dead cells or a modified metabolism when division is hampered by nutritional deficiencies but photosynthesis is active; *Navicula pelliculosa* only under these conditions makes huge amounts of capsular polysaccharides (J. C. Lewin, 1955).

Not all marine organisms behave similarly: multiplying *Katodinium dorsalisulcum* produces a polysaccharide mucilage (McLaughlin *et al.*, 1960), so much so that mucoid masses form; in stationary cultures they float to the top because of entrapped bubbles of photosynthetic oxygen. When growth has reached an optimal cell concentration ($10^4$ cells/ml), the cells keep producing polysaccharide until the medium gels (1.4–2.6 g/l. of polysaccharide are produced in a month). Hydrolysis of the polysaccharide yielded glucose, galactose and fructose. The supernatant of *Katodinium* was the only one to give a carbazole reaction typical of sugar (purple); the supernatants of *Monochrysis lutheri* and *Prymnesium parvum* gave different colors, respectively straw-green and light green. Guillard and Wangersky (1958) report for the same strains of these organisms, and with the same reagent, a good production of "carbohydrates" (Table III). They do not mention any discrepancy of the color reaction, yet they (Wangersky and Guillard, 1960) noted a different color reaction (blue) obtained with the *N*-ethyl carbazole method in *Amphidinium carteri* supernatants. This filtrate gave the usual purple carbohydrate

TABLE III

Production of "Carbohydrates" in Marine Algae[a]

| Species | Carbohydrate produced (mg/l.) | |
| --- | --- | --- |
| | Before maximal growth | Highest value (stationary phase) |
| CHLOROPHYTES | | |
| Dunaliella euchlora | 3.1 | 9.0 |
| Chlorella sp. (No. 580 Indiana Univ. Cult. Coll.) | | 9.0 |
| Chlamydomonas sp. ("Y" R. Lewin) | 2.1 | 10.6 |
| Chlorococcum sp. | | 27.0 |
| Pyramimonas inconstans | 2.8 | 5.4 |
| DIATOMS | | |
| Cyclotella sp. | | 1.5 |
| Nitzschia brevirostris | | 25.6 |
| Melosira sp. | | 60.0 |
| CHRYSOMONADS and CRYPTOMONADS | | |
| Isochrysis galbana | | 25.0 |
| Monochrysis lutheri | 1.7 | 15.7 |
| Prymnesium parvum | 5.8–15.9 | 123.0 |
| Rhodomonas sp. | 1.9 | 8.8 |
| DINOFLAGELLATES | | |
| Amphidinium carteri (interference, see text) | | > 5.0 |
| Prorocentrum sp. | | ∼ 20.0 |
| Katodinium dorsalisulcum[b] | | 0.6–2.4 g/l. |

[a] Data from Guillard and Wangersky (1958); $N$-ethyl carbamate as glucose equivalents, except for *Prorocentrum* (data from Collier, 1958; as arabinose equiv.).

[b] From McLaughlin *et al.*, 1960; for extraction of the polysaccharide, see text.

reaction after the blue-reacting material was eliminated by dialysis. The substance responsible for the blue coloration and fishy odor is apparently the end-product of the hydrolysis of an analog of acetylcholine.

Brown and red seaweeds produce large quantities of mucilaginous polysaccharides; several of them, e.g. alginic acid, agar, etc., are economically important [see papers in: Braarud and Sorensen (1956), and Lewin (1955) for polysaccharides of other marine algae]. These polysaccharides, being utilized by several bacteria as C sources, may be ecologically important; the specialized microflora which they support probably produces vitamins and other growth factors.

7—s. II

## B. Excretion of Organic Nitrogen

Extra-cellular nitrogenous products have been noted frequently in cultures of bacteria or blue-green algae fixing nitrogen, but no data were available on the nature of these substances before the work of Fogg (1952) on *Anabaena cylindrica*: liberation of extra-cellular organic nitrogen accompanies its growth. The relative amounts of the excreted nitrogen vary during growth: as much as 50% of the total nitrogen taken up is excreted during the early logarithmic growth; it decreases to a minimum (10–20%) at half growth; increases again in older but still healthy and growing cultures; and moribund material liberates large quantities of soluble organic nitrogen (but cultures in these conditions were not used by Fogg in his work). The amount excreted is not affected appreciably by cultural conditions except in the later stages of growth: it may be increased by nutritional deficiencies such as Fe, or slightly decreased by Mo deficiencies (Mo deficiencies affect nitrogen fixation and nitrite assimilation in *A. cylindrica*; Wolfe, 1954).

The nitrogenous excretion consists principally of polypeptides with lesser amounts of amide N (amides are high in young cultures and decrease as they become older). As much as 4–8 mg/l. of extra-cellular N are produced in culture. The polypeptides after acid hydrolysis yield amino acids. Chromatographic analysis suggests that the polypeptide fraction varies in composition: the polypeptide of a 28-day culture in a medium without added N contained in decreasing quantities serine, threonine, glutamic acid, glycine and tyrosine, and traces of alanine, valine, leucine; a 12-day culture in a medium with ammonium phosphate gave in decreasing quantities, glutamic acid, alanine, valine, leucine, glutamine, glycine, tyrosine, phenylalanine, aspartic acid.

Production of polypeptides is not restricted to nitrogen-fixing organisms: algae representing four different classes liberate considerable amounts of extra-cellular N and polypeptides (Fogg and Westlake, 1955). However, since far greater amounts are released in old cultures, autolysis rather than excretion is responsible for the increased amount of soluble organic N (Table IV). The production of polypeptides may be widespread amongst micro-organisms: it is common in non-nitrogen-fixing bacteria (Proom and Woiwood, 1949); several bacilli produce polypeptides with antibiotic properties like polymixin, bacitracin, valinomycin, tyrocidin and gramicidin.

The extra-cellular nitrogenous substances may or may not serve as nutrients to other organisms. The polypeptide of *Anabaena* does not give a perceptible reaction with ninhydrin, suggesting a ring structure which may not be susceptible to attack by the usual proteolytic enzymes. It has no antibiotic properties and is not utilizable as a N-source by *A. cylindrica, Chlorella* sp. and *Oscillatoria* sp.; it is not known whether bacteria can utilize them (Fogg, 1952).

Peptide nitrogen occurs dissolved in lakes in amounts ranging from 0.057 to 0.436 mg N/l. (Domogalla, Juday and Peterson, 1925). Fogg and Westlake (1955) found smaller quantities of peptide N in English lakes: the concentration

## TABLE IV

Extra-cellular Organic Nitrogen Produced by Fresh-Water Algae
(from Fogg and Westlake, 1955)

| Species | | Age of culture (days) | Cells per ml $\times 10^6$ | Extra-cellular N (mg/l.) | | |
|---|---|---|---|---|---|---|
| | | | | Total organic | Free $\alpha$-amino N | Peptide $\alpha$-amino N |
| *Chlamydomonas* | | | | | | |
| *moewusii* | (1) | 30 | 3.8 | 2.8 | 0.17 | 1.50 |
| | (2) | 24 | — | — | 0.027 | 0.2 |
| *Chlorella* | | | | | | |
| *pyrenoidosa* | (1) | 15 | — | — | 0.053 | 0.019 |
| | (2) | 31 | 7.5 | — | 0.043 | 0.000 |
| *Tribonema œquale* | | 60 | — | — | 0.041 | 0.065 |
| *Navicula pelliculosa* | | 27 | — | — | 0.089 | 0.093 |
| *Anabaena* | | | | | | |
| *cylindrica* | (1) | 30 | — | 0.51 | 0.013 | 0.073 |
| | (2) | 100 | — | 5.6 | 0.41 | 3.0 |

of peptide N shows no apparent correlation with the trophic state of the lake and the occurrence of peptide does not depend on any particular algae.

We have no data on production of extra-cellular organic nitrogen in cultures of marine algae. The high C/N ratios in the organic solutes found in sea-water indicate, as mentioned, that in the sea the situation is quite different from that in fresh waters; however, we know nothing about their nature. Furthermore, Duursma's results indicate that the release of organics in the sea occurs after the algal blooms.

### C. Growth Factors

Vitamin $B_{12}$ (cyanocobalamin), thiamine and biotin are indispensable requirements for many marine algae (Table VIII). Hence these growth factors should be present in the environment: cyanocobalamin and thiamine have been measured in sea-water. The search for the producers is going on, stimulated by the pioneer work of Lochhead and collaborators (1951, 1957) on the growth-factor relationships among the micro-organisms of the soil. Most of the work in the sea centers around vitamin $B_{12}$ and the family of cobalamins[1]; cyanoco-

---

[1] A series of cobalamins is produced by micro-organisms; these cobalamins differ one from another largely in respect to the type of nucleotide portion of the molecule or its absence (factor B). For cyanocobalamin the nucleotide is 5,6-dimethylbenzimidazole; for pseudo-$B_{12}$, adenine; for factor A, 2-methyladenine; for factor H, 2-methylhypoxanthine; for factor III (factor I), 5-hydroxybenzimidazole—other cobalamins have been obtained by biosynthesis with *Escherichia coli* (Bernhauer, 1956). The vitamin $B_{12}$-requiring organisms have characteristic patterns of response toward the cobalamins (see Table X, and chapter on bioassays). It suffices here to mention that most algae respond only to

balamin being the growth-factor needed by the majority of the auxotrophic algae.

The data indicate that bacteria are the major producers of vitamin $B_{12}$ in the sea. Ericson and Lewis (1953) isolated 34 bacteria from the Baltic Sea and epiphytic on seaweeds; 70% of them (mainly *Pseudomonas* and *Achromobacter* spp.) are cobalamin producers (*Escherichia coli* bioassay). An equal number of marine bacteria was isolated by Starr *et al.* (1957) from mud and waters from the southeast and west coast of Texas, a lagoon, laboratory tanks, and mullet intestine: 70% of them had activity for *Escherichia coli* and 30% for *Euglena gracilis*. Twenty per cent of 60 bacteria isolated from muds of the Bahia Fosforescente (Puerto Rico) had *Escherichia coli* activity (Burkholder and Burkholder, 1958). Burkholder (1959) isolated from muds and water of Long Island Sound 344 bacteria; 24% of them produced cobalamins. Marine bacteria collected near shore produce more non-$B_{12}$ cobalamins than true vitamin $B_{12}$: this is indicated by the following data. Chromatographic analysis of the filtrates show more non-$B_{12}$ cobalamins than true $B_{12}$ (Ericson and Lewis, 1953); more bacterial strains produce cobalamins active for *Escherichia coli* than for *Euglena gracilis* (Starr *et al.*, 1957). Most marine bacteria live epiphytically on particulate matter. Suspended solids and muds of Sapelo Island, Georgia coast, the Bahia Fosforescente and Long Island Sound are relatively rich in *Escherichia coli*-active cobalamins (Starr, 1956; Burkholder and Burkholder, 1956, 1958; and *in litt.*). A differential assay with *Ochromonas malhamensis* and *Escherichia coli* on the same bottom deposits of the Bahia Fosforescente shows that the quantity of true $B_{12}$ is 7–23% of the *Escherichia coli*-active cobalamins (Burkholder and Burkholder, 1958). A similar situation should exist in sea-water but data are lacking. Most determinations of "$B_{12}$" in sea-water have been done with the *Euglena gracilis*, which, as noted, is not specific for true $B_{12}$. Cowey (1956) did a few differential assays: in sea-water from Aberdeen Bay almost half the total cobalamins was $B_{12}$; in oceanic waters the non-$B_{12}$ cobalamins were present in far lower relative quantities. More work is needed to find out whether this holds generally; if it is so then we should explain the discrepancy between what is actually found in the waters, what has been produced by marine bacteria, and what is present in muds. In any case, differential assays for sea-water are indispensable because the $B_{12}$-requiring algae themselves have different specificity patterns; those of low specificity (*Escherichia coli* pattern) utilize all cobalamins, but the others able to utilize only vitamin $B_{12}$ may be discriminated against, depending on the ratio of non-$B_{12}$ cobalamins to vitamin $B_{12}$.

Many marine bacteria require vitamins for growth and they compete for

---

cyanocobalamin (*Ochromonas malhamensis* and "vertebrate" pattern); *Euglena gracilis* responds also to pseudo-$B_{12}$ and factor A, and *Escherichia coli* to all cobalamins including factor B (no nucleotide), and methionine. The *Escherichia coli* bioassay hence is equivalent to "total cobalamins", the *Ochromonas* assay only to cyanocobalamin, and the *Euglena gracilis* assay is intermediate between the *Ochromonas* and *Escherichia coli* assays.

vitamins with all the other organisms. Since their metabolism and rate of division are higher than other consumers, they may be quantitatively the most important consumers. Table V points to vitamin exchanges between some

## TABLE V

Marine Bacteria (Burkholder, in press)

| No. of cultures | Patterns of requirements | | | | No. of vitamin producers | | | |
|---|---|---|---|---|---|---|---|---|
| | Biotin | Thiamine | Nicotinic acid | $B_{12}$ | Biotin | Thiamine | Nicotinic acid | $B_{12}$ |
| 29 | + | | | | | 3 | 29 | 1 |
| 7 | | + | | | 6 | | 7 | |
| 2 | | | + | | 1 | | | |
| 3 | | | | + | 3 | 1 | 3 | |
| 3 | + | | | + | | | 3 | |
| 5 | | + | | + | 3 | | 5 | |
| 1 | | + | + | | 1 | | | |
| 1 | + | + | + | | | | | |
| 1 | + | + | | + | | | 1 | |

marine bacteria (Burkholder, in press). Here nutritional interdependency is made dramatically manifest by vitamin producers being themselves dependent on other vitamins. This chemical symbiosis supports Lucas's postulates on the importance of external metabolites in the sea.

Apparently bacteria are not the only producers of vitamins. Several unicellular algae contain appreciable amounts of vitamins (Table VI); these data refer to the content in vitamins of the cells and not of the supernatants. Therefore, we do not know whether these algae excrete the vitamins; if they do not the vitamins could become available for the consumers only after death of the cells. The vitamins are synthesized by the algae since several analyses were done on algae grown in bacteria-free culture and in media lacking the vitamin tested (see Brown *et al.*, 1955; Robbins *et al.*, 1951). Furthermore, the $B_{12}$ content of the various *Chlorella* species is of the same order for species grown under impure or aseptic conditions. However, the species of Table VI are freshwater algae and belong to the green and blue-green algae. Species belonging to these algal groups are not very abundant in the sea, except for some blue-green algae, like *Trichodesmium erythraeum*, which form blooms in warm seas.

The content in cobalamins of many seaweeds has been analyzed by Robbins *et al.* (1951), Ericson and Lewis (1953), Southcott and Tarr (1953) and Hashimoto (1954). These data and their ecological significance are discussed by Provasoli (1958a). In brief, the red seaweeds are richer in cobalamins than the brown seaweeds, and more species of reds than browns contain cobalamins. Since the algae were collected from nature, we do not know whether the

## TABLE VI

### Vitamin Content of Algae

| Species | Vitamin | Quantities per 100 g dry matter [a] | Author |
|---|---|---|---|
| *Chlorella pyrenoidosa* | thiamine | 1–4.1 mg | Combs (1952) |
| | riboflavin | 3.6–8 mg | |
| | nicotinic ac. | 12–24 mg | |
| | pyridoxine | 2.3 mg | |
| | pantothenic ac. | 0.8–2.0 mg | |
| | biotin | 14.8 µg | |
| | choline | 300 mg | |
| | $B_{12}$ (E. g.) [b] | 2.2–10 µg | |
| *Chlorella vulgaris* [d] | $B_{12}$ (E. c) [b] | 6.3 µg | Brown, Cuthbertson and Fogg (1955) |
| *Chlorella ellipsoidea* | $B_{12}$ (E. g.) [b] | 4.2–8.9 µg | Hashimoto (1954) |
| *Anabaena cylindrica* [d] | $B_{12}$ (E. c) [c] | 63- 110 µg | Brown, Cuthbertson and Fogg (1955) |
| *Microcystis* sp. (bloom) | $B_{12}$ (E. g.) [b] | 12.3 µg | Hashimoto (1954) |
| *Plectonema nostocorum* [d] | $B_{12}$ (E. g.) [b] | 6–7 µg | Robbins et al. (1951) |
| *Calothrix parietina* [d] | $B_{12}$ (E. g.) [b] | 64 µg | ,, |
| *Aphanizonemum flos-aquae* [d] | $B_{12}$ (E. g.) [b] | 28 µg | ,, |
| *Diplocystis aeruginosa* [d] | $B_{12}$ (E. g.) [b] | 24 µg | ,, |

[a] Dry weight of alga.

[b] $B_{12}$ as assayed with micro-organism: E. g. = *Euglena gracilis*, E. c. = *E. coli*.

[c] 65–70% of this is true $B_{12}$; the remainder is pseudo-$B_{12}$ and factor A.

[d] Bacteria-free.

See also Kanazawa, A., 1962. Studies on the vitamin B-complex in marine algae. I On vitamin content. *Mem. Fac. Fish. Kagoshima Univ.*, **10**, 38–69.

vitamins were synthesized by the seaweeds or absorbed from the environment (it is well known that seaweeds accumulate inorganic ions 100–1000-fold). Ericson (1952) found that *Pelvetia caniculata* has a very poor ability to concentrate radioactive $B_{12}$, but when collected in nature it contains at least 0.5 µg $B_{12}$/g; *Polysiphonia fastigiata* concentrates radioactive $B_{12}$ 90-fold, but when collected from nature is devoid of $B_{12}$. Since the quantity of cobalamins in sea-water is very low, these data favored the hypothesis that $B_{12}$ is synthesized by seaweeds. Later, Ericson and Lewis (1953) found that epiphytic bacteria of seaweeds produce $B_{12}$ and non-$B_{12}$ cobalamins: from this and other considerations they conclude that the $B_{12}$-compounds in seaweeds are probably of bacterial origin, even though true $B_{12}$ is preponderant in seaweeds and non-$B_{12}$ cobalamins are predominantly excreted by the epiphytic bacteria. The question can only be answered by the use of aseptic cultures of seaweeds; so far the red algae which have been cultured aseptically, *Goniotrichum elegans*

(Fries, 1959), *Bangia fusco-purpurea* and *Antithamnion* sp. (Provasoli and Iwasaki, unpublished), need $B_{12}$ for growth.

Folic, folinic, pantothenic and nicotinic acids were also measured in seaweeds. The content of vitamins in seaweeds varied seasonally, and in the younger and older parts of the thallus (Ericson, 1953; Ericson and Carlson, 1953). Larsen and Haug (1959) found a correlation between salinity and the nicotinic acid and biotin content of *Ascophyllum nodosum* and *Fucus vesciculosus*; vitamin maxima occur in lower salinity and do not seem to depend on pollution carried in by fresh waters.

Whatever the origin of vitamins in seaweeds, the vitamins are bound to be released in coastal waters either by excretion or decomposition of the seaweeds —indirectly, more vitamins probably are produced by the bacteria decomposing the immense quantities of dead seaweeds in the tidal zone.

Phytohormones ("X" and "Z") biologically active on *Avena* coleoptiles were found in phytoplankton samples (predominantly diatom); zooplankton samples showed high activity, on chromatography, in the indoleacetic zone (Bentley, 1958, 1959).

## D. Antibiotics

The reciprocal exchange of vitamins is one aspect of the interaction amongst organisms. Another is the production of antibiotics. Co-operation and conflict amongst organisms at the macroscopic level, especially for land plants, is an important branch of ecology. The same phenomenon, but at the chemical levels of antibiosis and growth promotion, is neglected. The possible biological and ecological importance of these phenomena is discussed boldly by Burkholder (1952) and Brian (1957). The high solubility of many biologically active substances peculiar to the water environment permits rapid exchange of metabolites but it also limits their effectiveness through dilution; therefore, only substances active at extreme dilutions ($10^{-15}$), such as vitamins, are eligible for ecological importance. The antibiotics, at least those now known, are effective at much higher concentrations ($10^{-6}$–$10^{-9}$); one wonders if they can play an important part in the *free* water environment. Their action may, however, be quite significant when the dilution factor is minimized, as in symbiosis, cohabitation and parasitism. One example is the relationship *Phaeocystis*–euphausids–Antarctic birds described by Sieburth (1958, 1959); a similar condition may be responsible for the antibiotic effect of corals (Burkholder and Burkholder, 1958a); the production of antibiotics by seaweeds (Chesters and Scott, 1956) may affect the kind of epiphytic flora normally present in seaweeds.

The sea may become a grab bag of medicinals, but the data obtained for these practical motives, e.g. measurements of pathogen-active antibiotic production by marine organisms, obviously lack ecological meaning. Screening procedures are based generally on sensitive terrigenous bacteria followed by a screening on human pathogens. To have any ecological significance, the initial screening should be based on an assay with sensitive marine bacteria, and later on the ecologically important bacterial species. But we cannot say now

TABLE VII

Vitamin $B_{12}$ Content of Marine Seaweeds

| Species | | $B_{12}$ activity as measured by | | |
| | | E. coli (Ericson and Lewis), range μg/g dry wt. | E. gracilis (Robbins et al.), range μg/g dry wt. | E. gracilis (Hashimoto), range μg/100 g [a] |
|---|---|---|---|---|
| **Red seaweeds** — High content | Laurentia pinnatifida, Rhodomela subfusca, Polysiphonia brodiaei, Ceramium rubrum,[b] C. tenuicorne | 0.20–0.63 | | |
| | Ceramium rubrum,[b] Champia parvula, Chondria tenuissima, Lomentaria baileyana | | 0.05–0.1 | 0.8–4.0 |
| | Ceramium rubrum,[b] C. tenerrimum, Martensia elegans, Acanthopeltis japonica, Gelidium amansii, Chondrococcus japonicus | | | |
| | Rhodymenia palmata, Chondrus crispus,[b] Gigartina stellata, Porphyra sp. | 0.4–0.13 | | |
| **Low content** | Agardhiella tenera, Chondrus crispus,[b] Polysiphonia variegata | | 0.01–0.045 | |
| | Batrachospermum moniliforme, Carpopeltis angusta, Meristotheca papulosa | | | 0.1–0.2 |

| Group | Species | | | |
|---|---|---|---|---|
| Brown Seaweeds | Himanthalia elongata, Alaria esculenta, Laminaria hyperborea, Sphacelaria arctica, Laminaria digitata | 0.1–0.26 | | |
| | Ascophyllum nodosum,[b] Chorda filum, Chordaria flagelliformis, Fucus spiralis, F. vesiculosus,[b] Laminaria agardhii, Mesogloia divaricata, Sargassum filipendula | | < 0.01 | |
| | Ectocarpus sp., Dictyota dichotoma, Padina arborescens, Ishige okamurai, Colpomenia sinuosa, Hydroclathrus cancellatus, Laminaria angustata, Endarachne binghamiae, Scytosiphon lomentaria, Eisenia bicyclis, Ecklonia cava, Undaria pinnatifida, Hijikia fusiformis, Sargassum enerve, S. ringgoldianum, S. horneri, S. serratifolium, S. pilabiferum, S. thunbergii | | | 0.1–1.2 |
| | Pelvetia canaliculata, Fucus vesiculosus,[b] F. serratus, Ascophyllum nodosum,[b] Laminaria saccharina | 0.04–0.075 | | |
| Green Seaweeds | Chara tomentosa, Valonia ventricosa, Ulva lactuca,[b] Enteromorpha intestinalis[b] | 0.2–0.6 | | |
| | Ulva lactuca[b] <br> Enteromorpha intestinalis[b] | | 0.01–0.045 <br> 0.05–0.1 | |
| | Enteromorpha linza, Chaetomorpha crassa, Monostroma nitidum, Letterstedtia japonica, Ulva pertusa, Caulerpa okamurai, Codium divaricatum | | | 1.4–3.9 <br> 0.1–0.8 |

[a] It is not clear whether the author refers to 100 g dry or wet weight; higher values were found in algae predigested with trypsin.
[b] Species analyzed by more than one author.

which are the ecologically important species; type culture collections are still grossly inadequate, and the species ill-identified at best.

### a. Antibiotics of bacterial origin

Two of the most important groups which produce antibiotics in the soil, the actinomycetes and the fungi imperfecti, seem scarce in sea-water. We do not know whether any truly marine representatives of them exist. As a consequence the antibiotic picture of the ocean should be quite different from the soil. The predominant flora of the sea is composed of Gram-negative and pleomorphic Gram-positive bacteria; the commonest genera are *Pseudomonas*, *Vibrio*, *Flavobacterium*, *Achromobacter*, *Bacterium*, and corynebacteria (Zobell, 1946; Wood, 1950). However, our knowledge of the bacterial flora of the seas may be entirely biased by the enrichment media employed. If some marine bacteria shared the nutritional characteristics of the marine algae, they could never be isolated with the extremely rich media in current use: the marine algae, many of which are quite euryhaline, are particularly sensitive to concentrations of organic substances (10–30 mg % of "peptones" drastically inhibit their growth). The only criterion to ensure the qualitative and quantitive appreciation of the marine flora is the matching of microscope counts with colony counts.

A few actinomycetes (*Nocardia*, *Micromonospora*, *Streptomyces*) and mycobacteria have been isolated from marine coastal sediments, nets, cordage and rotting seaweeds. Grein and Meyers (1958), after a survey, consider that the species isolated are probably of terrestrial origin: they are not different morphologically from terrestrial species; the halophilic tolerance of terrestrial actinomycetes is as good as the one of the "marine" species; both types grow better at sea-water concentrations of less than 50%. Several isolates of Grein and Meyers exhibited antibiotic production in culture media. A few marine *Bacillus* and *Micrococcus* (9 out of 58 marine species tested) had antibiotic activity against non-marine forms (Rosenfeld and Zobell, 1947). A conclusion is that marine bacteria may account for some of the observed antibacterial action of sea-water on enteric and fresh-water forms (Greenberg, 1956). The antagonism between marine micro-organisms is probable but unproven.

### b. Antibiotics of algal origin

Extracts of several marine seaweeds are antibacterial (Pratt, Mautner *et al.*, 1951; Mautner, Gardner and Pratt, 1953; Vacca and Walsh, 1954; Chesters and Scott, 1956; Allen and Dawson, 1959). This seems a widespread ability in seaweeds: antibiotics were found in green, brown and red seaweeds, but it is not known whether they are excreted in the sea or released after death. Antibiotic activity and antibacterial spectrum vary with species, and activity within each species of seaweed varies in the different months and may even be lacking (Chesters and Scott, 1956). The seaweeds tested by Allen and Dawson inhibit only Gram-positive bacteria (*Bacillus subtilis*, *Staphylococcus aureus* and *Mycobacterium smegmatis*) and not *Escherichia coli*[1] while the species tested by Chesters and Scott inhibited also Gram-negative fresh-water bacteria (*Escherichia coli* and *Pseudomonas aeroginosa*, etc.); marine bacteria are mostly Gram-negative.

Unicellular fresh-water algae also produce antibiotics: antibacterial substances are present in old cultures of *Protosiphon* and *Stichococcus* (Harder and Oppermann, 1953) and in *Chlorella* extracts (Spoehr *et al.*, 1949). Photooxidized unsaturated fatty acids are responsible for the antibacterial action of *Chlorella* and, apparently, also for *Protosiphon* and *Stichococcus* (the evidence is scanty). Unsaturated fatty acids produced and liberated upon death by *Chlamydomonas reinhardii* are apparently responsible for the antibiosis of *C. reinhardii* on *Haematococcus pluvialis* (Proctor, 1957). Since unsaturated fatty acids are produced by algae, including the seaweeds (Collyer and Fogg, 1955, for green algae, and table in Fogg, 1953), one wonders whether some antibacterial effects of seaweeds are not due to fatty-acid toxicity, especially in the alkaline range. This possibility was not considered in the work on seaweeds.

Several laboratories are now testing a variety of marine unicellular algae for antibacterial activity. The only published report is the very instructive story of *Phaeocystis* sp. Arctic birds were known to have a reduced gastro-intestinal microflora; this was confirmed for Antarctic birds by Sieburth (1959), who found that blood serum of penguins was active against Gram-positive bacteria. The antibiotic activity was traced through the food chain: from the crustacean *Euphausia superba*, which is the staple diet of penguins, to *Phaeocystis*-like algae which bloom in the Antarctic and are a food for *Euphausia* (Sieburth, 1959a). The antibacterial substance of *Phaeocystis* inhibits Gram-positives more than Gram-negatives; it is water- and ethanol-soluble and is soluble in non-polar solvents, stable to heat and alkali, but inactivated by heat-drying and mineral acids (Sieburth and Burkholder, 1959). The substance was finally identified as acrylic acid (Sieburth, 1960).

### c. Antibiotics of invertebrate origin

Water homogenates of living sponges from the temperate and subtropical zones contain substances which inhibit a variety of marine bacteria (isolated from sponges and sea-water) and the usual forms employed for antimicrobial assays (Jakowska and Nigrelli, 1960). This activity was found in species of *Microciona*, *Halichondria*, *Cliona*, *Tedania*, *Haliclona*, *Dysidea* and *Oligocera*. The active substances are heat-stable and can be selectively extracted with solvents, also from frozen and heat-dried sponges. The activity ranges from inhibition of marine bacteria to "broad-spectrum antibiosis" ("ectyonin" extracted from *Microciona prolifera*) and specific action against *Pseudomonas pyocyanea* and *Candida albicans*.

Extracts of several gorgonian (horny) corals had a strong antibiotic action against both the usual fresh-water assay species and marine bacteria. On the

[1] Gram-positive bacteria are notoriously more sensitive to antibiotics, on the whole, than are Gram-negative.

contrary, extracts of the few stony corals tested had little or no antibacterial action (Burkholder and Burkholder, 1958a). Antibacterial action was also found in extracts of unidentified "hydroids" (Allen and Dawson, 1959). Since most corals have endozoic zooxanthellae and harbor epizoic algae, it is not known whether the living tissues of the hydrozoan or the associated flora produce the antibacterial substances.

### d. Antialgal substances

Algae produce, besides antibacterial antibiotics, auto-inhibitors and substances inhibiting the growth of other algae (heteroinhibitors). The phenomenon is widespread among fresh-water algae; the literature is critically discussed by Proctor (1957); Hartman (1960) has tabulated the data on the interacting organisms, including all the species studied by Léfèvre et al. (1952). These substances are found in filtrates of old cultures and it is likely that they were mostly released by living mature cells or during autolysis, but excretion by healthy cells is not excluded. The rate of production of these inhibitory substances has not been followed experimentally and their nature is unknown; those found by Léfèvre et al. are, in general, heat-labile.[1] Presumably marine algae behave similarly.

### e. Crustacean inhibitors

Unicellular algae apparently release inhibitors for Crustacea: old cultures of Chlorella vulgaris inhibit filtering rate and growth of Daphnia magna (Ryther, 1954). Tigriopus japonicus and the brine shrimp Artemia salina utilize as food only a limited assortment of marine flagellates; others are unsuitable (Provasoli, Shiraishi and Lance, 1959). The unsuitability of these organisms as food may reflect nutritional incompleteness, but the possibility that nutritionally adequate food organisms are unsuitable because they produce inhibitory substances is not excluded. In the sea two relationships between phytoplankton and zooplankton have been observed: an inverse relationship, i.e. heavy grazing by zooplankton (Harvey, 1934) or avoidance of phytoplankton bloom by predators (Hardy, 1936). Avoidance may indicate that some species of phytoplankton produce inhibitors, or, more simply, obnoxious substances (perhaps tastes and odors, or other substances repressing the feeding reaction).

### E. Poisons

It is well known that from time to time spectacular algal blooms discolor the sea; some of them lead to mass mortality of marine invertebrates and fishes. Though commonly known as "red tides" because many of these blooms

---

[1] The blue-green algae are notorious for suppressing other algae. The fresh-water Nostoc muscorum produces in bacteria-free culture a "dihydroxy-anthraquinone" which inhibits growth of several algae (Cosmarium, Phormidium, Euglena), baker's yeast, and bacteria (Jakob, 1961).

are due to red-pigmented dinoflagellates, other discolorations (brown, yellow, green or cloudy) are also responsible for producing mass mortality; conversely some red blooms like the ones of *Trichodesmium erythreum* (a red-pigmented blue-green alga) in the Red Sea, Indian Ocean and the Sea of California are generally harmless. The poisonous blooms, though not annual, recur in certain coastal regions: in general, where fertilization occurs (upwelling, pollution, heavy soil runoffs). Maps of bloom localities are given by Hayes and Austin (1951) and Ballantine and Abbott (1957); Brongersma-Sanders (1957) gives a detailed review of the outbreaks, the causative agents, and an impressive bibliography.

Most of the mortalities are due to dinoflagellates, specifically to *Gonyaulax polyedra*, *G. monilata*, *Gymnodinium splendens*, *G. breve*, *G. mikimotoi*, *G. galatheanum*, *Cochlodinium catenatum*, *Exuviaella baltica* and *Pyrodinium phoneus*. Two other dinoflagellates, *Gonyaulax catenella* and *G. tamarensis*, do not cause mortality of marine organisms, including shellfish, but are ingested by shellfish, which concentrate the poison in the siphon or digestive glands and so become poisonous to man. Two other algae, the chloromonadine *Horniella marina* (Subrahmanyan, 1954) and the diatom *Thalassiosira decipiens* (Takano, 1956), caused, respectively, extensive mortalities off the Malabar Coast and in Tokyo Bay. Mortalities of fishes in brackish waters in Denmark (Otterstrom and Nielsen, 1939) and in fish ponds in Israel (Shilo and Aschner, 1953) are caused by the chrysomonad *Prymnesium parvum*.

Mass mortalities occur at the peak of the bloom. In bacteria-free cultures of *Gonyaulax catenella*, the toxin is localized in the cells during the logarithmic phase of growth and accumulates in the medium, when death and disintegration affect the population (Provasoli and McLaughlin, unpublished). This does not exclude some toxin excretion by actively dividing cells, i.e. concentrations below the sensitivity limit of the mouse bioassay. The ecological situation may be similar; toxic levels may be reached only when large numbers of dino-flagellates are senescent or dying.

A wealth of information on the poison of *Gonyaulax catenella* and *Prymnesium parvum* (Schantz, 1960; Burke *et al.*, 1960; Shilo and Rosenberger, 1960) and other products of marine organisms appeared in vol. 90, pp. 615–950, of the *Ann. N.Y. Acad. Sci.* devoted to "The biochemistry and pharmacology of compounds derived from marine organisms".

If the poisons of other dinoflagellates are as potent as the one of *Gonyaulax catenella*, we have another example of extremely active biological substances released in sea-water. The poison of *Gonyaulax catenella* is one of the most potent known, i.e. 1 mouse unit = 0.18 μg (1 unit is the dose killing all 20-g mice in 7–15 min); the lethal dose for man = 3–4 mg. This neurotoxin, unlike botulinum toxin, is a strongly basic nitrogeneous compound of low molecular weight (372), yielding under various oxidations and hydrolytic procedures guanidopropionic acid, guanidine, urea, ammonia and carbon dioxide (Schantz, 1960). A survey on the production of poisons by many dinoflagellates reared in bacteria-free culture is under way at the Haskins Laboratories.

## 4. Nutritional Requirements of Marine Algae

The nutritional requirements of algae have been reviewed recently (freshwater algae, Krauss, 1958; nutrition and ecology, including marine algae, Provasoli, 1958). Several marine algae have been grown bacteria-free in artificial media of known composition (Provasoli, McLaughlin and Droop, 1957). These media seem adequate for most of the algae growing in the littoral zone, including the red seaweeds (Fries, 1959; Provasoli and Iwasaki, unpublished). The photosynthetic marine algae are fundamentally photoautotrophic, i.e. synthesizing their organic carbon almost exclusively from $CO_2$. Many species, however, have acquired needs for growth factors while retaining their basic photoautotrophic abilities.

### A. Inorganic Requirements

The tolerance to salinity variations are indeed quite great; most neritic marine flagellates can grow from 12–40‰ salinity (optima between 20–24‰). Only species obtained from the Caribbean and Gulf of Mexico seem to prefer a salinity of 33‰. Tolerance to variation in ratios and concentrations of Na, K, Ca and Mg, is, in general, high; this versatility has been exploited in compounding non-precipitating artificial media low in Ca and Mg (Ca 10–20 mg %; Mg 50–70 mg %). Iron, Mn, Mo, Cu, Co and V have been demonstrated to be essential for fresh-water algae; B, Zn and S are also needed but their essentiality has not been proven. We may assume that this holds for marine algae also, even though no available data meet the rigorous requirements for purity of chemicals for these studies. In artificial media the addition of metal-buffered (chelated) trace-metal solutions is favorable for sustained growth despite the trace metals introduced as impurities of the "chemically pure" major salts.[1] Studies on single-trace-metal requirements may be misleading when metal-chelators are present in the media, because addition or removal of single heavy metals affects simultaneously the hold of the chelator on all other chelatable ions in the medium. Removal of chelators results in precipitates if the trace metals are added before sterilization; separate aseptic additions do not guarantee absence of precipitates and are cumbersome for extensive nutritional studies. However, a precise study of the trace-metal requirements is unavoidable in view of the results of Johnston (see pp. 200–201). Almost all photosynthetic marine algae utilize nitrates. Ammonia is utilized, and usually is not toxic at the average ecological levels of N, but at higher levels it becomes rapidly toxic in the alkaline range (Shilo and Shilo, 1953; McLaughlin, 1958). Ammonia toxicity is unlikely to be ecologically important and may be significant only in highly polluted zones. The species which invaded and replaced

[1] Chelators improve growth of higher plants and plant tissues (Wallace, 1960; Klein and Manos, 1960). The finding of Goldberg (1952) that the diatom *Asterionella japonica* utilizes only particulate iron has not been confirmed yet; the other species of algae so far tested grow poorly or do not grow in artificial media when precipitates occurred during sterilization. Oceanic forms may behave differently as postulated by Provasoli (1956).

the normal marine flora of Great South Bay when pollution from duck farms accumulated because of poor flushing of the bay are resistant to ammonia (1 mg atom N/l.) and utilize urea and other organic sources of N (Ryther, 1954a). Amino acids and other organic N are in general poorly utilized, if at all, by photosynthetic marine algae. An exception is *Hemiselmis virescens* which requires glycine; colorless species might be expected to utilize and even to require amino acids as do *Gyrodinium cohnii* (Provasoli and Gold, 1962) and *Oxyrrhis marina* (Droop, 1959). Inorganic phosphate and also certain forms of organic P are good P sources.

This summary of the inorganic requirements shows that only the trace metals and their status in sea-water (as particles or solutes) can be a cause of "good" or of "bad" waters, when inorganic N and P are not deficient.

## B. Organic Requirements

As mentioned, photosynthetic marine algae have slight, if any, ability to utilize organic nitrogen.

Glycerophosphoric, adenylic, cytidylic and guanylic acids and perhaps other nucleotides serve as P sources for all the marine species tested (extending the data of Chu, 1946, and Harvey, 1953). The widespread ability to utilize these organic compounds of P suggest that the P fraction of the organic solutes in sea-water should be closely measured and that P need not be mineralized to be available to phytoplankton organisms (see review in Provasoli, 1958, p. 294). This may explain why turnover of P is more rapid than expected and why in the early spring one diatom species after another can bloom so rapidly. In preliminary experiments, we noted that during the logarithmic phase of growth of *Navicula pelliculosa*, P becomes at first undetectable in the medium, but later significant amounts of P are released in the medium, and this before any portion of the population has become senescent or dying.

Organic acids and sugars are in general not utilized as carbon sources by photosynthetic marine algae. At times small concentrations of organic C are stimulatory, but the increase in growth is too little to suspect utilization as a C source. Similarly, very low concentrations (0.1–1 mg %) of amino acids and purines or pyrimidines are stimulatory; this may be due to chelation of trace metals and also to sparing of some substance along the pathways of bio-synthesis.

The above remarks on the ability to utilize organic substances are derived from data on algae living in the photic littoral zone and may not apply at all to algae living in other zones of the sea, especially the shallow and deep bottoms and deep waters. Many pennate diatoms live close to, and, in the coastal muds: 11 of 26 species of diatoms isolated from such habitats not only utilize exo-genous carbon sources but grow in darkness if supplied with glucose, acetate or lactate (Lewin and Lewin, 1960). A colorless marine diatom, *Nitzschia putrida*, has been found and is also heterotrophic (Pringsheim, 1951); it is probable that more colorless species exist in muds. This heterotrophic ability fits the need imposed by the environment: several species of diatoms were found to migrate and reproduce in the mud where no light can reach them (Moul and Mason, 1957). But the existence of diatoms in mud is not, apparently, restricted to shallow waters: large populations of *Ethmodiscus rex* were found in material collected in the Marianna Trench (Wiseman and Hendey, 1953); 14 species of diatoms were present in good number in mud samples collected during the *Galathea* Expedition of 1951 from depths of 7000–10,000 m (Ferguson Wood, 1956). This preserved diatoms material is not only composed of empty frustules, but of many cells containing protoplasm which can be stained with the usual cytological dyes. This, along with the observation that no planktonic species were found in the samples, excludes the "rain" theory (how could the protoplasm of the cells remain intact during the slow descent to the depths?), and favors the existence of autochthonous populations of living diatoms in the deep (Ferguson Wood, 1956).

Flagellates below 20 $\mu$ ( = nanoplankton = $\mu$ flagellates), endowed with photosynthetic pigments, had escaped attention until recently because they are not retained by the finest plankton nets, even though Lohmann had demonstrated their presence and abundance in the sea by centrifuging water samples. New methods for the quantitative estimation of the nanoplankton (Knight-Jones, 1951; Ballantine, 1953) reveal its great importance in the productivity of the seas. The nanoplankton in the North Sea is mainly composed of chrysomonads (*Chrysochromulina*) and other small flagellates which, like *Chromulina pusilla*, present very difficult taxonomic problems (Manton, 1959; Manton and Parke, 1960). The pioneering team of Parke and Manton is describing systematically the various species (Parke *et al.*, 1955–1959) by culturing hundreds of isolates, following the life cycles, and studying the morphology with the electron microscope. We can expect many exciting surprises, like the recent finding that *Crystalolithus hyalinus* is the motile phase of the non-motile *Coccolithus pelagicus* (Parke and Adams, 1960). In warm seas the unicellular flagellates and blue-green algae are a conspicuous part of the phytoplankton (500–4000 cells/ml, Bernard, 1958, 1959). The coccolithophorid chrysomonads, especially *Coccolithus fragilis*, are the predominant flagellates in the southern Mediterranean and constitute 75–96% of the phytoplankton volume, while dinoflagellates represent from 3–15% and naked flagellates and *Nostoc* constitute less than 1% of the volume; in the Southern Mediterranean, diatoms are rare (Bernard, 1958a). Remarkably, the richest layers for *Coccolithus* are not only in the euphotic zone (0–50 m) but also in the 250–800 m zone; furthermore, the density of cells in the layer 1000–2500 m is still very high, about two-thirds of the density found at 250–800 m (Bernard, 1958a). Growth of pigmented coccolithophorids in deep waters is not limited to the Mediterranean, but is also found in the Indian Ocean and the tropical Atlantic Ocean (off the Senegal Coast). In the Indian Ocean their maximum is found at 200 m and in the tropical Atlantic the population of the 200, 300, 400 m are between half and three-quarters of the maximum found (50 m). In fact, the growth in the column 0–200 m is several times less than below 200 m, where no photosynthesis can take place (Bernard, 1958, 1959). Daily collections in the Mediterranean show that *Coccolithus* doubles every five days (Bernard, 1958a); their growth in the non-photic layers should then depend upon organic matter. Several *Chryso-chromulina* actively ingest particles (even graphite!) and cells up to 5 µ (Parke *et al.*, 1955–1959) but none of the nanoplankton organisms has been grown in bacteria-free culture and it is not known whether they can utilize dissolved organic matter.[1] The fresh-water chrysomonad, *Ochromonas malhamensis*, grows luxuriously *in vitro* on organic solutes; it can also grow almost as well phagotrophically or photosynthetically. Even supposing that the chrysomonads living in non-euphotic zones of the sea are as versatile as *Ochromonas* nutritionally, where do they find food to enable them to divide at least once in five days? The content of organic solutes in sea-water is apparently quite low (2–20 mg/l.) and bacteria are supposed to be very scarce in deep waters!

## C. Growth Factors

The sampling of marine species in bacteria-free culture is indeed small but it is clear that many marine photosynthetic algae need vitamins (Table VIII); fresh-water algae behave similarly. A complete list of the vitamin requirements of marine and fresh-water species is given by Provasoli (1958). Since then J. C. and R. A. Lewin (1960) have studied the requirements of 26 species of marine diatoms, Droop (1959) of *Oxyrrhis marina*, McLaughlin and Zahl (1959) of two symbiotic dinoflagellates, and Fries (1959) of a red alga[2] (all included in Table IX, with new results since ms. preparation). Table IX summarizes the old and new data, including 23 species of *Volvocales* (Pringsheim and Pringsheim, 1959). The tentative conclusions reached several years ago on only two dozen species are still valid (Provasoli and Pintner, 1953). Briefly: (*a*) photosynthetic algae, like autotrophic bacteria, have species that do not and some that do need vitamins (=auxotrophs); (*b*) only three vitamins are required, alone or in combination; in order of incidence, vitamin $B_{12}$, thiamine and biotin (Table IX); (*c*) auxotrophy does not correlate with any particular environment or with the lack or presence of heterotrophic abilities; (*d*) the algae have an unexpectedly narrow and stereotyped need for only three vitamins even though the algae live in environments rich in all vitamins and many accompanying micro-organisms (bacteria and fungi) have widely different vitamin requirements.

[1] Organic carbon sources improve growth in light but cannot sustain growth of *Hymeno-monas* sp., *Pavlova gyrans*, and *Syracosphaera* sp. in darkness. *Coccolithus huxleyi* has very poor heterotrophic abilities (Pintner and Provasoli, in press). Some species of *Chryso-chromulina* are now bacteria-free.

[2] Since the writing of the MS the following species have been investigated: the diatom *Cyclotella nana* requires $B_{12}$ and *Detonula confervacea* has no vitamin requirement (Guillard and Ryther, 1962); the chrysomonads *Hymenomonas* sp., *Coccolithus huxleyi*, *Syraco-sphaera* sp., and *Ochrosphaera neapolitana* require only thiamine; *Pavolva gyrans* requires $B_{12}$ and thiamine (Pintner and Provasoli, in press); the blue-green *Synechocystis* sp. requires only $B_{12}$ (Van Baalen, 1961); the red seaweed *Nemalion multifidum* requires $B_{12}$ and perhaps pyridoxamine (Fries, 1961).

## TABLE VIII

## Vitamin Requirements of Marine Algae[a]

| Species | B$_{12}$ | Thiamine | Biotin |
|---|---|---|---|
| **CHLOROPHYCEAE** | | | |
| *Dunaliella salina, D. primolecta, D. euchlora, D. viridis, Nannochloris atomus, N. oculata, Pilinia* sp., *Platymonas* sp., *Prasiola stipitata, Stephanoptera gracilis, Stichococcus cylindricus*[b] | O | O | O |
| *Stichococcus cylindricus*[b]*, Platymonas tetrathele* | R | O | O |
| *Brachiomonas submarina, Pyramimonas inconstans* | R | R | O |
| **BACILLARIOPHYCEAE** | | | |
| *Amphora coffaeiformis,*[b] *Navicula* sp., *N. incerta, N. menisculus, Nitzschia putrida, N. angularis* var. *affinis, N. filiformis, N. frustulum,*[b] *N. hybridaeformis, N. laevis, N. curvilineata, N. marginata, N. obtusa* var. *scalpelliformis, N.* aff. *ovalis, Phaeodactylum tricornutum, Stauroneis amphoroides.* | O | O | O |
| *Achnanthes brevipes, Amphora perpusilla (coffaeiformis?) A. coffaeiformis,*[b] *A. lineolata, Cyclotella* sp., *Nitzschia frustulum,*[b] *N. ovalis, N. punctata, Synedra affinis, Skeletonema costatum, Stephanopyxis turris* | R | O | O |
| *Amphipleura rutilans, Amphora coffaeiformis,*[b] *Nitzschia closterium*[b] | R | R | O |
| *Amphiprora paludosa,* var. *duplex, Amphora coffaeiformis,*[b] *Nitzschia closterium*[b] | O | R | O |
| **CHRYSOPHYCEAE** | | | |
| *Stichochrysis immobilis* | O | O | O |
| *Hymenomonas carterae* | R | O | O |
| *Pleurochrysis scherffelii* | O | R | O |
| *Hymenomonas elongata, Isochrysis galbana, Microglena arenicola, Monochrysis lutheri, Prymnesium parvum* | R | R | O |
| **CRYPTOPHYCEAE** | | | |
| *Hemiselmis virescens, Rhodomonas* sp. (10 strains) | R | R | O |
| *Rhodomonas lens* | S | R | O |
| **DINOPHYCEAE** | | | |
| *Exuviaella cassubica, Glenodinium foliaceum, Gonyaulax polyhedra, Gymnodinium splendens, Gyrodinium californicum, G. resplendens, G. uncatenum, Peridinium balticum, P. chattoni, P. trochoideum* | R | O | O |
| *Amphidinium klebsii, A. rhynchocephalum, Gynmodinium breve, Oxyrrhis marina* | R | R | R |
| *Gyrodinium cohnii* | O | S–R [c] | R |

TABLE VIII—(cont.)

| Species | $B_{12}$ | Thiamine | Biotin |
|---|---|---|---|
| CYANOPHYCEAE | | | |
| *Phormidium persicinum* | R | O | O |
| RHODOPHYCEAE | | | |
| *Goniotrichum elegans, Bangia fusco-purpurea* | R | O | O |

[a] R = required; O = not required. Most of the data were contributed by Droop; J. C., and R. A. Lewin; Provasoli; Pintner; McLaughlin; and Gold; other contributors are A. Gibor, Ryther, and B. Sweeney (the author's references, up to 1958, are given in Provasoli, 1958).

[b] Species represented by strains with different vitamin requirements.

[c] Grows slightly but indefinitely without the thiamine; the addition of the thiamine results in a 100 or more-fold increase in growth.

The only perceptible trend is that algal groups differ in the incidence of species requiring vitamins: the Cyanophyceae, Chlorophyceae and Bacillariophyceae are algal groups in which about half or less of the species require vitamins (perhaps the non-requirers predominate); in the other algal groups the vitamin requirers predominate. The latter algal groups are the richest in animal tendencies (i.e. many species have lost the photosynthetic pigments and phagotrophy is widespread even in species with photosynthetic pigments, i.e. many chrysomonads). It has been postulated that incidence of auxotrophy may correlate with developed animal tendencies (Provasoli, 1956). For practical purposes the vitamin $B_{12}$-like cobalamins and thiamine can be considered the two most important vitamins for the phytoplankton and they may be relevant ecological factors. We have mentioned that assessing the ecological importance of $B_{12}$ is difficult because several cobalamins are present in waters and the vitamin $B_{12}$-requiring organisms have themselves different patterns of specificity. Tables X and XI summarize the situation for algae. Table XI lumps freshwater and marine algae because, as mentioned, they apparently behave similarly; unpublished data of our laboratory as well as data in collaboration with Droop are included (Droop *et al.*, 1959). Of the three algal groups most important ecologically in the sea, the diatoms (Bacillariophyceae) seem to have the widest specificity. This may be fortuitous; still, they are apparently both the most abundant producers of organic matter in the temperate and cold seas and utilize all the known cobalamins in the environment. Therefore, although dependent on $B_{12}$-like compounds, they hold the advantage over the other $B_{12}$-requirers of narrower specificity. However, some cryptomonads—a group not abundant in the sea—have wide specificity. Because of this complex situation, assays on sea-water to be meaningful should be done with at least two bioassay organisms in order to measure "total cobalamins" (*E. coli* assay) and "true $B_{12}$" [*Ochromonas* or *Thraustochytrium* assay (see Adair and Vishniac, 1958)]. Furthermore, the data of J. C. and R. A. Lewin (1960) indicate that

## TABLE IX

### Summary of Vitamin Requirements of Fresh-Water and Marine Algae

| Algal group | Number of species | No vitamins | Require vitamins | $B_{12}$ | Thiamine | Biotin | $B_{12}$ + thiamine | Biotin + thiamine | $B_{12}$ + biotin + thiamine |
|---|---|---|---|---|---|---|---|---|---|
| Chlorophyceae | 68 | 24 | 44 | 10 | 8 | | 26 | | |
| Euglenineae | 9 | 0 | 9 | 2 | 1 | | 6 | | |
| Cryptophyceae | 11 | 0 | 11 | 2 | 1 | | 7 | | |
| Dinophyceae | 17 | 1 | 16 | 11 | | | 0 | 1 | 4 |
| Chrysophyceae[a] | 22 | 1 | 21 | 2 | 5 | | 9 | 1 | 2 |
| Bacillariophyceae | 39 | 21 | 18 | 11 | 3 | | 4 | | |
| Cyanophyceae | 10 | 9 | 1 | 1 | | | | | |
| Rhodophyceae | 4 | 0 | 4 | 4 | | | | | |
| Totals | 180 | 56 | 124 | 43 | 18 | | 52 | 2 | 6 |
| Totals for single vitamins | | | | 103 | 78 | 10 | | | |

[a] Two chrysomonads require $B_{12}$ + biotin.

TABLE X

Patterns of Specificity toward Vitamin $B_{12}$-Like Compounds

| Organism | Type of nucleotide | | |
|---|---|---|---|
| | Benzimidazole ($B_{12}$, factor III) | Adenine (pseudo $B_{12}$, factor A) | No nucleotide (factor B) |
| Mammals and *Ochromonas malhamensis* | + | 0 | 0 |
| *Lactobacillus leichmannii* and *Euglena gracilis* | + | + | 0 |
| *Escherichia coli* | + | + | + |

TABLE XI

$B_{12}$ Specificity in Fresh-Water and Marine Algae

| Algal group | Number of species studied | Specificity pattern | | |
|---|---|---|---|---|
| | | mammalian | lactobacillus | coli |
| Chlorophyceae | 10 | 10 | | |
| Chrysophyceae | 8 | 7 | 1 | |
| Dinophyceae | 8 | 6 | 2 | |
| Euglenineae | 5 | 1 | 4 | |
| Cryptophyceae | 7 | 2 | | 5 |
| Bacillariophyceae | 10 | 2 | | 8 |
| Cyanophyceae | 1 | | | 1 |
| Rhodophyceae | | | | |

strains of diatoms of the *same* species may require different vitamins or not require them at all. This physiological variation among strains, which may also occur in other algal groups, obliges us to be extremely careful in extrapolating data pertaining to a strain of any species to other strains or localities. To avoid gross errors one must correlate the data on vitamin content of waters, especially $B_{12}$, with laboratory nutritional findings obtained exclusively on species isolated from the *same water samples*. Since this entails a great deal of work, such precise analysis should be limited to the history of blooms in localities where, as in Long Island Sound and off the English coast around Plymouth, the succession of forms and other ecological factors have been thoroughly studied.

It is suspected that seaweeds or other highly differentiated algae may need plant hormones for normal morphogenesis. *Ulva lactuca* produced the normal flattening and a short leafy thallus only in nutrient sea-water ($+N$, P, trace metals, vitamins) enriched with adenine and kinetin (Provasoli, 1958b).

However, this result could not be repeated with other samples of sea-water; the artificial sea-water media are inadequate for growth. *U. lactuca* responds also to indoleacetic acid and gibberellins.

## 5. Crustacea and Organic Solutes

Recent work on two Crustacea, *Tigriopus japonicus* and *Artemia salina* grown aseptically, exhumes again the hypothesis of Putter on the nutritional role of dissolved organic substances for marine animals. In feeding *Tigriopus* with a variety of aseptic algal flagellates, it was observed that *Chroomonas* sp. and *Isochrysis galbana*, fed singly, cause larval mortality and adult infertility in *Tigriopus* after supporting several normal generations: respectively, after four and eight generations. The number of generations before mortality sets in indicated that biologically potent micro-nutrients could be responsible. The addition of a vitamin mixture or glutathione to the medium restored normal growth for several additional generations (Shiraishi and Provasoli, 1959). This experiment obviously does not tell us whether vitamins added as solutes to the two-membered culture alga *Tigriopus* (a) modify the metabolism of the prey, or (b) become concentrated in the algae, or (c) are ingested directly from the medium by *Tigriopus*. It does show that vitamins in sea-waters affect, directly or indirectly, growth and fertility of the herbivores. Other effects of vitamins on Crustacea have been described. The addition of 200 mg/l. of pantothenic acid to septic cultures of *Daphnia* fed on *Chlamydomonas* tripled the life span and increased egg production tenfold (Fritsch, 1953). All the barnacles (*Balanus* sp.) in a tank when exposed to a maximal concentration of 14 µg/l. of ascorbic acid, immediately initiated copulating activities (Collier, Ray and Wilson, 1956).

The axenic culture of *Artemia* on artificial media shows that organic solutes can be utilized by Crustacea (Provasoli and Shiraishi, 1959). The main nutrients (blood serum, peptone, liver infusion, vitamins, nucleic acids, etc.) are added as solutes. *Artemia* grows to adulthood in this medium if particles (starch or cellulose) are present in abundance, but it dies, soon after the second meta-nauplius has consumed its yolk, if the particles are omitted. The nutrient solutes support growth only if enough drinking takes place; drinking in turn depends upon the feeding reaction caused by ingestion of particles; without particles they do not drink enough. The ecological significance is obvious: the soluble organic matter in sea-water is utilized, but since the quantities of nutrients dissolved are very small, the nutritional dependence of Crustacea on nucleic acids, proteins or amino acids eventually present in sea-water is minimal; only substances like vitamins and hormones, which act at extreme dilutions, can be ecologically significant.

## 6. Data from Biological Analysis of Sea-Water

The problem of "good" and "bad" waters—the fisherman's preoccupation—parallels frustrations of biologists trying to rear marine animals in the laboratory.

The change of water-masses around Plymouth caused a change in fauna and distress to the scientists of this laboratory which pioneered so much work in marine biology. During 1930 Wilson experienced great difficulties in rearing larvae of polychaetes in the local sea-water. Later it became clear through the work of Russell, Cooper, Armstrong and Harvey, at the Marine Biological Association at Plymouth, that the local planktonic and hydrographic conditions had changed. This prompted Wilson to analyze the biological properties of the new water-masses (typified by the indicator species *Sagitta setosa*) and compare it with the *S. elegans* water which previously surrounded Plymouth. Employing *Echinus esculentus*, *Ophelia bicornis* and *Sabellaria alveolata* as bioassay organisms, he found the two types of water remarkably different: in *setosa* water, eggs and larvae of the sea urchin and worms developed abnormally; in *elegans* water, they grew normally (Wilson, 1951; Wilson and Armstrong, 1952). In nature *elegans* water supports good growth of phytoplankton and zooplankton. Addition of antibiotics, filter-sterilization, variations in pH, addition of $B_{12}$, ascorbic acid, or a metal chelating agent (EDTA), and of supernatants of thick cultures of diatoms and flagellates all did little to improve bad waters for the above organisms (Wilson and Armstrong, 1954, 1958). Experiments in which eggs were allowed to develop in a mixture of the two types of water suggested that it is the presence in good water of something beneficial rather than the presence in bad water of something harmful which makes the difference. Extracts with activated carbon and acetone of the bad and good waters gave confusing results: neither supported normal growth. Wilson attributed this to the properties of the samples of waters employed for extraction: both water samples, including the good water, supported abnormal growth.

The original observations of DeValera (1940) that superficial waters of the tidal *Fucus–Ascophyllum* zone permit the normal development from zygotes of *Enteromorpha*, while the water of 30-m depth allows only stunted and slow growth, resulted in a luckier chase. H. Kylin (1941, 1943, 1946) employed germinating zygotes of the seaweeds *Enteromorpha* and *Ulva* as bioassay organisms, and counted the number of cells produced by the germinating filament at a fixed time in various enrichments of the infertile deep water: he found that these waters are poor in $NO_3$, $PO_4$, Fe and Mn. A. Kylin (1943, 1945), with the same technique, obtained normal growth with super-added Zn, Mn, Fe and Co, while Ni, Al and Cd were inert. H. Kylin (1946) concluded that fertility of the inshore waters for seaweeds is due to their relative richness in N, P, and trace metals, and that these important elements for plant growth diminish with depth and distance from the shore: waters of 70-m depth are poorer in trace metals than the 30-m depth waters. The success of the two Kylins and the lack of results of Wilson reflect the status of the knowledge of the two fields of nutrition; much was known of algal and plant nutrition and Kylin could make a more educated guess as to what might be lacking in poor waters. Wilson faced the utter unknown; very little, if anything is known about the physiology of sea urchins and marine worms; any of a thousand known substances could be responsible.

At present, with the knowledge acquired on the nutrition of marine algae, it is possible to bioassay the biological properties for the phytoplankton of different water-masses. This possibility, though not fully exploited, has given some extremely interesting results.

Sea-water samples with their natural flora of living micro-organisms and phytoplankton were enriched with nutrients and incubated in continuous light for various periods. The effect of the various enrichments was gauged either by the amount of growth elicited during a fortnight (the samples being observed periodically during this period, Thomas, 1959; Johnston, in press) or by measuring the uptake of $^{14}CO_3$ after 24 h incubation in a light of 1500 ft candles (Ryther and Guillard, 1959). The enrichments were: (a) single additions of N, P, Si, trace elements, soil extract (Thomas); (b) a complete enrichment: N + P + Si + trace elements + soil extract (Thomas); N + P + Si + thiosulphate + under-chelated trace-metal mixture + vitamins (Ryther and Guillard); N + P + Si + over-chelated trace-metal mixture (Johnston); (c) the complete enrichment minus one of the components (Thomas; Ryther and Guillard); complete enrichment minus only the chelated trace metals (Johnston); (d) an equal mixture of surface sea-water and deep water (100 or 1000 m) from the same water column (Ryther and Guillard). Though the methods differ somewhat and the samples are from very different areas, all the sea-waters supported good growth only when enriched with trace metals; the other enrichments were in-different; Si was stimulatory. Thomas treated only two samples: one from an oligotrophic area west of Baja California, and one from a eutrophic area off Central America. The oligotrophic sample had a natural population of 500 cells/l. of diatoms and dinoflagellates; none of these organisms grew in any enrich-ment. This might be due to the scant natural inoculum more than to an adverse (bad) type of sea-water. The eutrophic sample had 500,000 cells/l. of diatoms and dinoflagellates. The only effective single additions were soil extract and, in a minor way, the trace metals; early growth of diatoms followed by late growth of a *Gymnodinium* was favored by the complete enrichment and the complete enrichment minus P or N; the diatoms did not grow when Si or trace metals were lacking from the complete enrichment but the *Gymnodinium* grew. A similar contrast between the northern spring diatoms (*Skeletonema*, *Chaeto-ceros*, *Nitzschia*, *Thalassiosira*) and dinoflagellates (mainly *Ceratium*) was found also by Johnston: the type of enrichment he employed favored growth of the spring diatoms in all samples of sea-water (the amount of growth depend-ing upon the sample) while the dinoflagellates present in the natural inoculum failed to grow. Omission of the chelated trace-metal mixture from the enrich-ment resulted in no or poor growth, including the dinoflagellates, in all types of waters (Johnston employed hundreds of samples collected in different seasons and depths in the North Sea, North Atlantic, Faroe and Icelandic waters).

Ryther and Guillard treated seven samples of surface waters collected from the continental shelf to the Gulf Stream and six samples in the Sargasso Sea; omission of the trace-metal mix from the complete enrichment had the most

pronounced effect, reducing photosynthesis almost to the level of the un-
enriched controls; all other omissions, except Si, gave the same high $^{14}C$
uptake as the complete enrichment; omission of Si resulted in about half the
amount of $^{14}C$ uptake of the complete enrichment. Ryther and Guillard do not
mention the species composition of the phytoplankton present in the various
samples. Two interesting additional observations were made. The mixing of
deep waters (1000 m) with the surface waters of the Sargasso results in a photosynthesis as high as the surface water with complete enrichment, defining
experimentally the elements contributed by the deep waters which impart
renewed fertility after mixing. Even more interesting is that the surface waters
of the Sargasso are extremely poor in nitrate and phosphate yet the addition
of these two elements had no effect on photosynthesis (growth in complete
enrichment minus N or P = growth in complete enrichment). The authors
comment ". . . that the phytoplankton may be dependent upon the rate of
regeneration of these elements [N and P] and/or their presence as dissolved
organic compounds and more-or-less independent of their instantaneous concentrations as inorganic salts". It would seem necessary to measure organic P
and N of these deep waters.[1]

Johnston went a step further; he found why the addition of trace metals is
necessary. The trace-metal mixture employed by him (PI Metals, see Provasoli
et al., 1957) is over-chelated; the ratio chelator/trace metals in milliequivalents
is about 2:1, raising the possibility that either the chelated (1:1) trace metals
or the excess free chelator of the mixture could be responsible for improving
"bad" waters for phytoplankton. In fact, the addition of chelators alone, as
EDTA and DTPA, and in a lesser degree of NTA and EDDHA,[2] is as effective
as the addition of the chelated trace-metal mixture. These results fit perfectly
with the ecological situation: at the normal pH of sea-water, iron, manganese
and probably the other trace elements are extremely insoluble. The total Fe
in sea-water varies between 1–60 μg atoms/l. (see results of various authors in
Harvey, 1955, and Goldberg, 1957); in the offshore waters off the Norwegian
coast the total Fe is 3–21 μg atoms/l. (Braarud and Klem, 1931). However,
Cooper (1937) calculated that no more than $10^{-8}$ μg atoms/l. of Fe can remain
in solution at pH 8.0–8.5. Therefore the deficiency of trace metals depends
quantitatively far more on the physical status governing their availability to
the cells than on the total amount. Growth promotion by metal chelators is
due to their solubilizing power at the pH of sea-water, making available the
trace metals which are present but largely unavailable. Indeed, Johnston
found that most of the "bad" waters for phytoplankton became fertile upon
addition of chelators, and some "good" waters became poor. The latter happening may be due to very high total trace metals in the samples. The ability to
form soluble metal complexes is not at all restricted to the artificial chelates,

---

[1] See Addendum, page 210.

[2] EDTA = ethylenediamine tetraacetic acid; EDDHA = ethylenediamine-di-(o-hydroxy-
phenylacetic acid); DTPA = diethylenetriamine pentaacetic acid; NTA = nitrilotriacetic
acid.

but is shared by many organic compounds, e.g. amino acids, nucleotides and hydroxy acids, which have been found or could occur in sea-water (see Section 2 of this chapter and Chapter 9). Some components, then, of the organic matter in the sea perform the dual role of nutrients and of solubilizers of the indispensable trace metals. The experiments of Johnston show that the quality of waters for phytoplankton may depend largely on the presence or absence of trace-metal solubilizers.[1] The search for these substances is, then, of paramount importance.

Johnston, besides enriching samples of sea-water containing their living phytoplankton, did two other types of assay in his attempts to assess the biological properties of the waters. In one assay (No. 1), filtered sea-water samples were enriched with one-fifth volume of medium S36, then autoclaved, and inoculated with a bacteria-free culture of *Skeletonema costatum* (Droop's strain); medium S36 was developed for the same strain by Droop (1955a). In the second assay (No. 2), filtered sea-water samples were enriched with N, P, Si, and a chelated trace-metal mixture, autoclaved, and inoculated with unialgal cultures (i.e. with unknown bacterial flora) of *S. costatum* and *Peridinium trochoideum*. These two assays show that the quality of sea-water varies from place to place, with depth and season. However, the unialgal cultures of *Peridinium* and *Skeletonema* gave a different assessment of sea-water quality indicating that the quality of sea-water acts differently on different classes of organisms (confirming the previously noted contrast between diatoms and dinoflagellates). Hence we should always specify for which organisms waters are good or bad. The assay response obtained with bacteria-free *S. costatum* (No. 1) does not correlate with temperature, salinity, oxygen, blue fluorescence, phosphate, phytoplankton and zooplankton abundance or dominance; it hints that waters of superior quality for *S. costatum* are recently mixed oceanic and neritic waters, and that waters collected after the bloom of spring diatoms are poor. The assay response with unialgal cultures (No. 2) correlates with plankton classification.

Earlier experiments had detected a very important difference between the assay with bacteria-free and with unialgal *Skeletonema*. Bacteria-free *Skeletonema* grew very poorly or not at all in 215 samples of sea-water collected in different seasons and localities of the northern seas when vitamins were omitted from the enrichment (N, P, Si, with and without chelated trace-metal mixture). Samples of sea-water similarly enriched but inoculated with unialgal (i.e. bacterized) cultures of *Skeletonema* supported poor to good growth depending upon the sea-water samples. Evidently the bacteria of the unialgal cultures

---

[1] In fresh waters some terminal products of microbial metabolism, like the "humic substances", perform this action. The yellow organic acids extracted by Shapiro (1957) are good trace-metal chelators and so are the polypeptides produced extracellularly by blue-green algae (Fogg and Westlake, 1955). Their ecological importance is discussed in these papers and by Fogg (1958). Similar substances have not yet been extracted from sea-water, though the "Gelbstoff" of Kalle (1949), which may be similar to the yellow acids of Shapiro and the "water humus" of Skopintsev (1959), may perform the same function.

almost always produce enough vitamin $B_{12}$ to satisfy the $B_{12}$ requirement of the diatom. The addition of vitamins improved only 38% of the sea-water assayed with unialgal *Skeletonema* (Johnston, in press). This explains why Ryther and Guillard did not report any decrease in growth when the vitamins were omitted from the complete enrichment. These authors enriched sea-waters which contained their natural microflora. It is well known that when sea-water is put in glass containers the added surface of the containers promotes heavy bacterial growth (Zobell and Anderson, 1936; Jones, Thomas and Haxo, 1958) and that many marine bacteria produce vitamins.

These observations prompted Johnston to enrich sea-water samples (to be assayed with bacteria-free *Skeletonema*) with a one-fifth volume of medium S36, which contains vitamins along with N, P, Si and trace metals. With this enrichment, the growth of bacteria-free *Skeletonema* in the numerous samples of sea-water varied from substantially inferior to superior to the controls (i.e. growth obtained in undiluted medium S36). The clear need for a vitamin supplement demonstrates that *Skeletonema* requires higher levels of vitamins than are present in the water tested. This experimental evidence contradicts the assertion of Droop (1957a) that the lowest amount of $B_{12}$ found by Cowey (1956) in the North Sea—0.1 m$\mu$g/l.—should support a crop of twenty-five million cells of *Skeletonema* per liter. Johnston employed Droop's bacteria-free culture of *Skeletonema* for his assay of waters from the North Sea. The assertion of Droop was based on a calculation of the molecules of $B_{12}$ required to produce 1 $\mu^3$ of living protoplasm of $B_{12}$-requiring algae.[1] Daisley (1957) had already contested the validity of applying such data to a dynamic ecological situation. The results of Johnston are even more remarkable because *Skeletonema* can utilize, besides true $B_{12}$, the widest range of $B_{12}$-like cobalamins; in nature this species should have an advantage over other vitamin $B_{12}$ requirers with narrow specificity.

Another important result obtained with the two *Skeletonema* assays is that the enriched sea-water samples elicit "poor" to "very good growth", indicating that unknown substances, which are none of the nutrients added with the enrichment (vitamins, trace metals, N, P and Si), affect *Skeletonema* growth, else all waters should have responded alike to the enrichments. For instance, the following behavior, observed in many samples, supports the postulate of unknown beneficial factors:

Sea-water + one-fifth medium S36 = often far more growth than 100% S36 (for bacteria-free and unialgal *Skeletonema* assay).

Sea-water + vitamins + trace metals = often far more growth than 100% S36 (unialgal *Skeletonema* assay).

[1] This value was obtained by dividing the volume of a cell of *Monochrysis lutheri* by the number of cells obtained with given amounts of $B_{12}$. The value obtained is three molecules of $B_{12}$ for 1 $\mu^3$ protoplasm. The values obtained, with similar calculations, for the existing growth data for *Euglena* and *Stichococcus*, being of the same order, made him confident of the validity of this coefficient. Applying this coefficient to the volume of a *Skeletonema* cell, Droop derived the number of cells of *Skeletonema* that could be supported by 0.1 m$\mu$g of $B_{12}$. See also Droop (1961).

It is evident that comprehensive biological assays of sea-water with unialgal and bacteria-free cultures of varied organisms will be extremely useful.

## 7. Prospects

It is evident that the biological approach to the problem of "bad" and "good" waters for phytoplankton has been quite successful; two variables, vitamins and trace metals, have emerged and seem to be, with N and P, the important parameters of the environment for phytoplankton. In the absence of chemical methods, the biological approach (i.e. nutritional requirements, organics produced and excreted, and biological analysis of water) may also be the way to attack similar problems for the marine herbivores and carnivores. Preliminary work on the nutrition of Crustacea in bacteria-free culture seems promising.

The nutritional studies reveal new important cycles. One of them, the $B_{12}$ cycle, can now be roughly sketched : the main producers are the micro-organisms (mostly bacteria) though it is not excluded that photoautotrophic algae may be as important either as direct producers of vitamins or, after their death, as food for vitamin-producing micro-organisms. Micro-organisms (bacteria and unicellular algae, as far as we know) are the main consumers ; possibly animals are also important consumers. Filter-feeding organisms—if the scant knowledge on *Tigriopus* can be extrapolated—may absorb vitamins directly as solutes. Perhaps animals with extensive gill systems absorb vitamins or other organic micro-nutrients in these highly permeable organs. The non-living particles (clay, organic and inorganic micelles, detritus) absorb large quantities of vitamin $B_{12}$ and on ingestion may supply additional vitamins. How much the removal of vitamins from the particles affects vitamin cycles is unknown ; we do not know whether these particles fix the vitamins in a stable way or only transiently ; does partial elution maintain a certain level of vitamins as solutes during high consumption of vitamins by phytoplankton? Elution in deep muds might fertilize upwelling waters.

Consumption seems rarely to bring to zero the soluble $B_{12}$-like compounds in sea-water (Table XII). Though data are scarce, clearly coastal and bay waters, because of the influence of soil, are richer in cobalamins than open waters (Kashiwada *et al.*, 1957a; Droop, 1955; Lewin, 1954; Cowey, 1956). Surface waters show clearly a seasonal variation (Cowey, 1956) and growth of the spring diatoms (the dominant species is *Skeletonema costatum*, a $B_{12}$ requirer) in Long Island Sound is responsible for a sharp drop in $B_{12}$ level (Vishniac and Riley, 1959, 1961). There is more $B_{12}$ in deep waters (Kashiwada *et al.*, 1957; Daisley and Fisher, 1958).[1]

Vitamin $B_{12}$-like growth factors, therefore, behave like the other ecologically significant nutritional variables—but are the cobalamins limiting, and where? Ecological judgment is uncertain because most of the measurements have been done with bioassay organisms which are not specific for true $B_{12}$ (*E. gracilis* and

[1] See Addendum, page 210.

*L. leichmannii*) and many unicellular algae utilize only true $B_{12}$ (Table XI). But the assays for Long Island Sound (Vishniac and Riley, 1959) were done with a true $B_{12}$-specific organism. The quantities found in Long Island Sound do not differ substantially from the data of Droop and Lewin for other coastal waters; all of them show ample $B_{12}$. It is probable, then, that in inshore waters $B_{12}$ is rarely limiting and not a constraint on fertility. But when the quantities of $B_{12}$-like cobalamins are far lower, as in the open sea, the specificity of the various algae toward the different cobalamins may be decisive. For this reason much work has been done to determine the specificity of the algal species (Droop *et al.*, 1959). Since specificities vary widely (Table XI) both among the species and the ecologically important algal groups, it becomes necessary to measure the $B_{12}$-like cobalamins with several bioassay organisms and to determine the ratio true $B_{12}$/total cobalamins (as recommended by Cowey, 1956; Droop, 1957, and Provasoli, 1958a). Specificity alone is not enough!

Sensitivity of the various algal species toward true $B_{12}$ (Provasoli, 1958a) and probably to the other cobalamins varies also (Ford, 1953). While there are no reasons to doubt that the *specificity* data based on *in vitro* studies can be transferred to the ecological situations, *sensitivity* data are biased. Sensitivity, i.e. the dose–growth response, does not depend solely on the variable to be measured but is obviously influenced by all other cultural conditions, especially the composition of the medium and temperature. After all, the improvement of media and of cultural conditions and the standardization of the inoculum are the routine procedure employed to develop a bioassay, speed of growth and sensitivity being the desiderata of a workable assay (Hutner, Cury and Baker, 1959; Hutner, Provasoli and Baker, 1961). Laboratory assay basal media are, from an ecological standpoint, grotesque in the way *all* constraints on growth have been removed save for the vitamin being assayed: this permits maximum sensitivity *in the laboratory*. The routine artificial marine media, even when not tailored for a specific organism, also lack many constraints: they generally allow a growth equal or superior to the one recorded in natural blooms, even though light conditions are in general not optimal (200–400 ft-candles of fluorescent light).

Artificial media are extremely useful in detecting the needed and the utilized metabolites; this knowledge is of ecological importance but the data on the sensitivity to metabolites, acquired in artificial media, cannot be extrapolated directly to the natural environment because each sample of sea-water is in effect a different basal medium (Johnston, in press). Samples of sea-water enriched with one-fifth volume of S36 medium (which contains vitamins, N, P, Si and trace metals) inoculated with bacteria-free *Skeletonema costatum* support, depending upon the sample of sea-water, more or less growth than occurs in the undiluted S36 medium taken as a yardstick. Obviously other properties of these different sea-waters affect the growth response of equal levels of nutrients, including vitamins. Conversely, a level of vitamins and nutrients which in artificial media supports a defined number of cells may support more or less

TABLE

Quantity of Vitamin $B_{12}$-Like

| Locality and type of water | Assay organism | Method |
|---|---|---|
| Vineyard Sound, Woods Hole (coastal) | *Euglena gracilis* [a] | Dialysis of SW. |
| Northwest Arm, Halifax, N.S. (coastal, polluted) | *Stichococcus* sp.[b] | Direct measurement |
| Pier at Millport, Scotland (coastal) | *Monochrysis lutheri* [c] | Direct measurement |
| Aberdeen Bay, Scotland (coastal) | *Lactobacillus leichmannii* [d] *Ochromonas malhamensis* [e] | Phenol extract. of SW. |
| Northern North Sea | *L. leichmannii* and *O. malhamensis* [f] | Phenol extract. of SW. |
| Butt of Lewis | *L. leichmannii* and *O. malhamensis* | Phenol extract. of SW. |
| Norwegian Deeps | *L. leichmannii* and *O. malhamensis* | Phenol extract. of SW. |
| Bay of Biscay | *Euglena gracilis* [a] | Direct on diluted SW. |
| 12 stations 0–30°N along 130°E (North Pacific) | *Euglena gracilis* [a] | Dialysis of SW. |
| Kagoshima Bay (variations in depth and diurnal) | *Euglena gracilis* [a] | Dialysis of SW. |
| Bahia Fosforescente (Puerto Rico) | *Escherichia coli* [g] | Phenol extract. of SW. |
| Long Island Sound (coastal) | *Thraustochytrium globosum* [h] | Direct measurement |

[a] Responds to true $B_{12}$, pseudo-$B_{12}$ and factors A, C, $C_2$.

[b] The specificity of *Stichococcus* toward various cobalamins is unknown.

[c] Responds to true $B_{12}$, pseudo-$B_{12}$ and factors A, I and H.

[d] Responds to true $B_{12}$, pseudo-$B_{12}$, factors A, I and also desoxyribosides.

[e] Responds only to true $B_{12}$ and factor I.

## XII

Cobalamins Found in Sea-Water

| Quantity or range | Remarks | Authors |
|---|---|---|
| 30–200 mµg/l. | Variations due to length of aging in laboratory | Provasoli and Pintner (1953a) |
| 10 mµg/l. | | Lewin, R. A. (1954) |
| 5–10 mµg/l. | | Droop (1955) |
| 6 mµg/l. | Same sample assayed with | Cowey (1956) |
| 4 mµg/l. | 2 bioassay organisms | |
| 0.1 mµg/l. (Aug)–1.2 mµg/l. (Oct.) | Pronounced seasonal variations | Cowey (1956) |
| 0.4 mµg/l. (Apr.)–2 mµg/l. (Feb.) | | Cowey (1956) |
| 0. 5 mµg/l. (Aug.)–2 mµg/l. (Apr.) | | Cowey (1956) |
| 2.26 mµg/l. (mean of 34 samples) | 190–2190 m depth | Daisley and Fisher (1958) |
| 0.57 mµg/l. (mean of 7 samples) | < 190 m and > 2110 m depth | |
| 0.3–1.5 mµg/l. (range of 90% samples; min. = 0.0; max. = 20 mµg/l.) | Surface (0–100 m) | Kashiwada et al. (1957) |
| 0.5–2.5 mµg/l. (range 80% samples) | Below 200 m | |
| 7–26 mµg/l. (max. 20% samples) | | |
| 3.2 mµg/l. (aver. 34 samples; 6 samples 0.0) | Mouth of Bay | Kashiwada et al. (1957a) |
| 5.3 mµg/l. (aver. 57 samples; 7 samples 0.0) | Middle of Bay | |
| 6.7 mµg/l. (aver. 44 samples; 2 samples 0.0) | Deep inland part of Bay | |
| 3.0–3.5 mµg/l. (inside Bay) | Bahia Fosforescente, Puerto Rico | Burkholder & Burkholder (1958) |
| 1.3 mµg/l. (outside Bay) | | |
| 4.5–11.4 mµg/l. (after spring diatom bloom) | Pronounced seasonal variation | Vishniac and Riley (1959, 1961) |
| 12–14.6 mµg/l. (winter) | | |

*f* The values obtained with the two assay organisms are similar, i.e. true $B_{12}$ may be the major cobalamin (80–90%) in these waters.

*g* Responds to true $B_{12}$, pseudo-$B_{12}$, all factors including factor B and to methionine.

*h* Responds only to true $B_{12}$.

cells in different waters. Consequently, the bioassay of vitamins in *direct* assays of sea-water should be based on (Droop, 1955) or include (Daisley, 1958) "internal standards". The sample of sea-water is split in two portions equally enriched with N, P, Si and chelated trace metals. Two parallel growth curves are obtained either by diluting one portion $3 \times$ with artificial sea-water and adding to both portions graded amounts of $B_{12}$ (Droop, 1955) or by adding $B_{12}$ to one portion and diluting both portions stepwise with $H_2O$ (Daisley, 1958). One can calculate, by the interval separating the two regression curves, the amount of $B_{12}$ present in sea-water without referring to an "external standard" (i.e. the growth curve obtained with increasing amounts of vitamins in artificial media). Assays based on $B_{12}$ extracted from sea-water or separated by dialysis, do not need internal standards.

Since we cannot transfer laboratory sensitivity data directly to ecological situations, other means must be devised to judge whether vitamins are limiting. One method was found by Johnston (in press). He employed bacteria-free cultures of *Skeletonema*: no, or poor, growth occurred in over 200 samples of sea-water from the northern seas enriched with N, P, Si, with and without trace metals. Since the same types of waters enriched with S36 medium—which has vitamins—give much better growth, the content of vitamin $B_{12}$-like cobalamins was limiting in these waters. Over sixty of these samples were collected in winter. According to the data of Coway (1956) one would expect that 1–2 m$\mu$g/l. of vitamin $B_{12}$-like cobalamins were present. This quantity, according to laboratory sensitivity data (Provasoli, 1958a), is above the minimal quantity required by several marine algae. Unfortunately we have no laboratory data on the sensitivity of *Skeletonema*; however, a good density was obtained by Droop (1955a) in the second serial transfer in "no $B_{12}$" from a culture grown in 100 m$\mu$g/l. (the dilution factor for each transfer was 100). Similar assays with bacteria-free cultures of the important ecological species of algae of a given environment can tell directly whether the waters are deficient in vitamins. The assay with bacteria-free strains of the local ecological species permits a reliable judgment. The only inconvenience may be the duration of the assay, but this disadvantage is offset by an avoidance of the measurement of the quantity of vitamin (which has little meaning except for a comparison between different waters); this in turn requires differential assays and the determination of the vitamin specificity patterns of each organism of the environment. Bacterized unialgal cultures obviously cannot be used for this purpose—bacteria produce vitamins.

Trace metals are the other new parameter. The bioassays of Kylin, Thomas, Ryther and Guillard, and Johnston, extending the pioneering work of Harvey, prove that waters from the northern seas, as well as the Sargasso and tropical Pacific, benefit greatly from the addition of trace metals. As noted earlier, Johnston demonstrated that, in the northern waters, the deficiency is due not to lack of trace metals but to their unavailability; the addition of a metal chelator, because of its solubilizing power, is as effective as the addition of trace metals. Conceivably other waters, especially the tropical ones, may be deficient in total trace metals, and availability is less important; a suitable assay could tell.

Since trace-metal utilization and deficiency may depend either on total quantity, on availability of trace metals, or on both, the presence or absence in sea-water of organic solubilizers becomes a primary factor. Many organic substances (amino, hydroxy and nucleic acids) which have been detected or quite likely exist in sea-water can chelate the heavy metals in various degrees. In fresh waters the "humic" and "yellow" acids are probably the more important solubilizers quantitatively; autochthonous substances like the polypeptides produced by blue-green algae may also be important. A similar situation, though quantitatively less significant, may exist in the sea-waters exposed to the influence of soil and large rivers, and may be partially responsible for the fertility of bays, estuaries and, perhaps, banks where the winter mixing more easily enriches the surface waters with the organic substances of the muddy bottoms. Perhaps other substances of autochthonous origin, i.e. products of marine organisms, operate in the sea. Since fertility may depend in large degree on the availability of trace metals, and chemical extraction of organic substances is rather difficult, we have to resort provisionally to other means.

The C/N ratio may give some indications, particularly if substances similar to the "humic" and "yellow" acids (which have a very high C/N ratio) are present in sea-water. If we assume that plankton has a C/N ratio of about 6, higher values would indicate the presence of some substances of the humic type and lower values the presence of proteins, amino acids or amines. There is a possibility that either very low or very high C/N ratios may coincide with the chelating ability of sea-water.

Clearly, chemical or biological methods are needed for measuring the chelating power in sea-water. Perhaps chelating power may be measurable by employing one of the ions which are preferentially bound, like Cu or Hg, if a way is found to detect them when they appear in sea-water in free form.

A biological way of detecting the presence of chelators would be to measure chelation as a function of its ability to remove the toxicity caused by poisonous ions of high stability constant such as Cu and Hg. Fogg and Westlake (1955) have demonstrated the chelating power of the polypeptides produced by blue-green algae by comparing the toxic action of graded amounts of Cu in the presence or absence of polypeptides on the motility of a filamentous blue-green alga. The rate of movement of the filaments was perceptibly reduced at a concentration of Cu of 0.5 mg/l. whereas a similar effect in the presence of the polypeptide was observed at 8–16 mg Cu/l. *Euglena gracilis* grows to high densities in a medium visibly blue by the addition of Cu sulfate if it is over-chelated by EDTA.

A more complex type of assay than the one used by Johnston could be useful: samples of sea-water uniformly enriched with N, P, Si and vitamins, filter sterilized, could be enriched with aseptic additions of graded low concentrations of (a) a chelator; (b) a non-chelated trace-metal mixture of Fe,

Mn, Zn and Co; (c) the same mixture but 1:1 chelated with EDTA. It is possible that a differential assay of this type may distinguish between the gradations of the four possible combinations, i.e. waters (1) rich both in chelators and trace metals; (2) rich in trace metals and deficient in chelators; (3) deficient in trace metals but rich in chelators; (4) deficient in both trace metals and chelators. If these different combinations exist in sea-water then it is conceivable that the favorable effect of mixing may depend on mutual compensation of multiple deficiencies as well as a straight enrichment in nutrients by deeper waters.

Johnston was encouraged by his results with the addition of chelating agents to sea-water to reopen the question of artificial fertilization of sea-water, proposing to lower the cost of the operation by employing a chelator and minimal doses of N and P to be determined experimentally. Promising, inexpensive results were obtained by Buljan (1957): to 70–80 l. of sea-water were added 5 l. of commercial concentrated $H_2SO_4$, 100 kg of superphosphate, and two spades-full of garden or forest soil. The phosphates were dissolved with stirring and sea-water added up to 200 l., the whole being allowed to drain slowly into the water of the bay from a moving boat. Use of sulphuric acid permitted the solution of such cheap sources of phosphorus, as finely crushed phosphate, degreased bone meal, guano, etc.; the soil added vitamins, chelators and trace metals. A total of 37 mg of $P-PO_4$ was added in one year per ton of bay water. Blooms of surface algae and a conspicuous increase of phytobenthos resulted from the fertilization of the sheltered, shallow inlet of Valiko Jezero in the Adriatic Sea. The area became an excellent feeding ground for oysters (O. edulis). Growth of oysters was four times larger than it was for individuals of the same age during the two years preceding fertilization; the weight of oysters showed an average growth of 26 g per individual per year, and the rate of weight increase was five times larger than in the unfertilized part of the bay.

One of the most attractive features of fertilization experiments is that they provide a means of judging whether in laboratory experiments, or in analysis of sea-water for a limited number of constituents, the really important factors may have been missed. Perhaps experimental ecology will gradually reduce the "trial-and-error" procedures that have characterized fertilization experiments in the past.

## Addendum

Since the preparation of the MS a few publications worth mentioning have appeared.

Vishniac (1961) perfected a fungus bioassay for thiamine in sea-water (sensitive to 25 mμg/l.).

The occurrence of vitamin $B_{12}$ in the Sargasso Sea (Menzel and Spaeth, 1962) has been measured for one year using the diatom Cyclotella nana (for the bioassay method see Ryther and Guillard, 1962). The quantity of $B_{12}$ in waters above 50 m fluctuates from undetectable to 0.03 mμg/l. from May to October and this paucity of $B_{12}$ seems to control the species composition of the phytoplankton. The dominant organism throughout the year is *Coccolithus huxleyi*, which does not require $B_{12}$ (but requires thiamine), and the bloom of diatoms occurs in April after the level of $B_{12}$ has increased to 0.06–0.1 mμg/l. (a dozen species of diatoms isolated from the Sargasso require $B_{12}$).

A new series of enrichments of depleted Sargasso waters (at the end of the diatom bloom) containing their natural populations shows clearly that the relative proportions of various nutrients are directly responsible for species composition (Menzel, Hulburt and Ryther, in press). The initial population in the samples was dominated by *C. huxleyi* (90% numerically). One series of enrichments was done in glass carboys, the other in polyethylene bottles.

In glass carboys the enrichment with N + P caused dominancy of diatoms (70% *Skeletonema costatum* + 25% of three species of *Chaetocerus*); N + P + Fe produced a short, rich bloom of *Chaetocerus simplex* (later supplanted by unidentified flagellates). In the polyethylene bottles N + P enrichment caused a small, short-lived *Skeletonema* bloom followed by a large *Coccolithus* bloom; N + P + Fe favored preponderance of flagellates; N + P + Si favored a large *Nitzschia closterium* bloom concomitant with a smaller *Coccolithus* bloom; N + P + Si + Fe caused a dense bloom of *C. simplex*. In an environment which is almost stable for temperature and light, as the Sargasso Sea, the chemical environment seems, more than any other factor, to be responsible for the distribution and seasonal succession of forms.

Menzel and Ryther (1961) confirm and extend the results of Ryther and Guillard (1959). Iron is the most limiting factor for primary productivity of the Sargasso Sea and iron deficiency is not removed by the addition of EDTA. However, the addition of iron alone results in a short burst of growth, and sustained productivity is achieved only by adding Fe + N + P; evidently N + P are also limiting. These data from enrichments fit into the ecology of the Sargasso Sea: N and P are minimal and total Fe averages 10 μg/l. without any seasonal peak; *C. huxleyi*, the yearly dominant species, has an unusually low requirement for Fe [growth for many serial transfers is not affected by lack of Fe in the enrichment (Ryther and Kramer, 1961)].

From these results it is clear that the combining of chemical and bioassay analysis of the waters with enrichment experiments and the determination *in vitro* of the nutritional characteristics of the dominant species permits good insight into the ecological events.

## References

Adair, E. J. and H. S. Vishniac, 1958. Marine fungus requiring vitamin $B_{12}$. *Science*, **127**, 147–148.

Allen, E. J., 1914. On the culture of the plankton diatom *Thalassiosira gravida*. *J. Mar. Biol. Assoc. U.K.*, **10**, 417–439.

Allen, M. B., 1956. Excretion of organic compounds by *Chlamydomonas*. *Arch. Mikrobiol.*, **24**, 163–168.

Allen, M. B. and E. Y. Dawson, 1959. Production of antibacterial substances by benthic tropical marine algae. *J. Bact.*, **79**, 459–460.

Anderson, W. W. and J. W. Gehringer, 1958. Physical oceanographic, biological, and chemical data. South Atlantic Coast of the United States. *M/V "Theodore" N. Gill Cruise* 6 and 8. *Spec. Sci. Rep.*—Fisheries. No. 265, pp. 1–99; and No. 303, pp. 1–227.

Baker, H., S. H. Hutner and H. Sobotka, 1955. Nutritional factors in thermophily: a comparative study of bacilli and *Euglena. Ann. N.Y. Acad. Sci.*, **62**, 349–376.

Ballantine, D., 1953. Comparison of the different methods of estimating nanoplankton. *J. Mar. Biol. Assoc. U.K.*, **32**, 129–148.

Ballantine, D. and B. C. Abbott, 1957. Toxic marine flagellates, their occurrence and physiological effects on animals. *J. gen. Microbiol.*, **16**, 274–281.

Bentley, J. A., 1958. Role of plant hormones in algal metabolism and ecology. *Nature*, **181**, 1499–1502.

Bentley, J. A., 1959. Plant hormones in marine phytoplankton, zooplankton and seawater. *Preprints Intern. Oceanog. Cong. A.A.A.S.*, 910–911.

Bernard, F., 1958. Comparaison de la fertilité élémentaire entre l'Atlantique tropical africain, l'océan Indien et la Méditerranée. *C.R. Acad. Sci. Paris*, **247**, 2045–2048.

Bernard, F., 1958a. Données recentes sur la fertilité élémentaire en Méditerranée. *Rapp. Cons. Explor. Mer*, **144**, 103–108.

Bernard, F., 1959. Elementary fertility in the Mediterranean from 0 to 1000 meters compared with the Indian Ocean and the Atlantic off Senegal. *Preprints Intern. Oceanog. Cong. A.A.A.S.*, 830–832.

Bernhauer, K., 1956. Über die Biosynthese von Vitaminen der $B_{12}$-Gruppe. *Zntbl. Bakteriol. Parasitenk.*, *II*, **109**, 325–329.

Birge, E. A. and C. Juday, 1934. Particulate and dissolved organic matter in inland lakes. *Ecol. Monog.*, **4**, 440–474.

Braarud, T., 1945. A phytoplankton survey of the polluted waters of the inner Oslo Fjord. *Norske Vidensk. Akad. Hvalrad.* Skr. No. 28, 1–142.

Braarud, T. and A. Klem, 1931. Hydrographical and chemical investigations in the coastal waters off Møre. *Norske Vidensk. Akad. Hvalrad.* Skr. No. 1, 88 pp.

Braarud, T. and N. A. Sorensen (ed.), 1956. *Second International Seaweed Symposium.* Pergamon Press, 220 pp.

Brian, P. W., 1957. The ecological significance of antibiotic production. *Microbial Ecology.* Cambridge Univ. Press, 168–188.

Brongersma-Sanders, M., 1957. Mass mortality in the sea in "Treatise on marine ecology and paleoecology" Vol. I. Ecology (edit. I. W. Hedgpeth). *Geol. Soc. Amer. Mem.*, **67**, 941–1010.

Brown, F., W. F. J. Cuthbertson and G. E. Fogg, 1955. Vitamin $B_{12}$ activity of *Chlorella vulgaris*, Beij. and *Anabaena cylindrica*, Lemm. *Nature*, **177**, 188.

Buljan, M., 1957. Report on the results obtained by a new method of fertilization experimented in the marine bay "Mljetska Jezera". *Acta Adriatica*, **6**, No. 6, 44 pp.

Burke, J. M., J. Marchisotto, J. J. A. McLaughlin and L. Provasoli, 1960. Analysis of the toxin produced by *Gonyaulax catenella* in axenic culture. *Ann. N.Y. Acad. Sci.*, **90**, 837–42.

Burkholder, P. R., 1952. Cooperation and conflict among primitive organisms. *Amer. Sci.*, **40**, 601–631.

Burkholder, P. R., 1959, Vitamin-producing bacteria in the sea. *Preprints. Intern. Oceanog. Cong. A.A.A.S.*, 912–913.

Burkholder, P. R., in press. Some nutritional relationships among microbes of the sea sediments and waters. In *Symp. Mar. Microbiol.* (edit. C. H. Oppenheimer). Thomas Co., Springfield, Ill.

Burkholder, P. R. and L. M. Burkholder, 1956. Vitamin $B_{12}$ in suspended solids and marsh muds collected along the coast of Georgia. *Limnol. Oceanog.*, **1**, 202–208.

Burkholder, P. R. and L. M. Burkholder, 1958. Studies on B vitamins in relation to the productivity of the Bahia Fosforescente, Puerto Rico. *Bull. Mar. Sci. Gulf and Caribbean*, **8**, 201–223.

Burkholder, P. R. and L. M. Burkholder, 1958a. Antimicrobial activity of horny corals. *Science*, **127**, 1174–1175.

Chesters, C. G. C. and J. A. Scott, 1956. The production of antibiotic substances by seaweeds. *Second International Seaweed Symposium* (edit. Braarud and Sorensen). Pergamon Press, 49–54.

Chu, S. P., 1946. The utilization of organic phosphorus by phytoplankton. *J. Mar. Biol. Assoc. U.K.*, **26**, 285–295.

Collier, A., 1958. Some biochemical aspects of red tides and related oceanographic problems. *Limnol. Oceanog.*, **3**, 33–39.

Collier, A., S. Ray and W. Magnitzky, 1950. A preliminary note on naturally occurring organic substances in sea water affecting the feeding of oysters. *Science*, **111**, 151–152.

Collier, A., S. M. Ray, A. W. Magnitzky and J. O. Bell, 1953. Effect of dissolved organic substances on oysters. *U.S. Dept. Int. Fish and Wildlife Service, Fish Bull.*, **84**, 167–185.

Collier, A., S. M. Ray and W. B. Wilson, 1956. Some effects of specific organic compounds on marine organisms. *Science*, **124**, 220.

Collyer, D. M. and G. E. Fogg, 1955. Studies on fat accumulation by algae. *J. Exp. Bot.*, **6**, 256–275.

Combs, C. F., 1952. Algae (*Chlorella*) as a source of nutrients for the chick. *Science*, **116**, 453–454.

Cooper, L. H. N., 1937. Some conditions governing the solubility of iron. *Proc. Roy. Soc. London*, **B124**, 299.

Cowey, C. B., 1956. A preliminary investigation of the variation of Vitamin $B_{12}$ in oceanic and coastal waters. *J. Mar. Biol. Assoc. U.K.*, **35**, 609–620.

Creac'h, P., 1955. Sur la présence des acides citrique et malique dans les eaux marines littorales. *C.R. Acad. Sci. Paris*, **240**, 2551–2553.

Daisley, K. W., 1957. Vitamin $B_{12}$ in marine ecology. *Nature*, **180**, 1042–1043.

Daisley, K. W., 1958. A method for the measurement of vitamin $B_{12}$ concentration in sea water. *J. Mar. Biol. Assoc. U.K.*, **37**, 673–681.

Daisley, K. W. and L. R. Fisher, 1958. Vertical distribution of vitamin $B_{12}$ in the sea. *J. Mar. Biol. Assoc. U.K.*, **37**, 683–686.

Dazko, V. G., 1939. Organic matter of certain seas. *Doklady Akad. Nauk S.S.S.R.*, **24**, 294–297.

Dazko, V. G., 1951. Vertical distribution of organic matter in the Black Sea. *Doklady Akad. Nauk S.S.S.R.*, **77**, 1059–1062. (Russian).

Dazko, V. G., 1955. On the concentration of organic matter of the waters of the Sea of Azov, before the regulation of the Don. *Akad. Nauk S.S.S.R., Hydrotechn. Inst. Novocherkask, Hydrochem., Mat.*, **23**, 1–10 (Russian).

DeValera, M., 1940. Note on the difference in growth of *Enteromorpha* species in various culture media. *Kgl. Fysiog. Sällsk. Lund Förhandl.*, **10**, 52–58.

Domogalla, B. P., C. Juday and W. H. Peterson, 1925. The forms of nitrogen found in certain lake waters. *J. Biol. Chem.*, **63**, 269–285.

Droop, M. R., 1955. A suggested method for the assay of vitamin $B_{12}$ in sea water. *J. Mar. Biol. Assoc. U.K.*, **34**, 435–440.

Droop, M. R., 1955a. A pelagic marine diatom requiring cobalamin. *J. Mar. Biol. Assoc. U.K.*, **34**, 229–231.

Droop, M. R., 1957. Auxotrophy and organic compounds in the nutrition of marine phytoplankton. *J. gen. Microbiol.*, **16**, 286–293.

Droop, M. R., 1957a. Vitamin $B_{12}$ in marine ecology. *Nature*, **180**, 1041–1042.

Droop, M. R., 1959. Water-soluble factors in the nutrition of *Oxyrrhis marina*. *J. Mar. Biol. Assoc. U.K.*, **38**, 605–620.

Droop, M. R., 1961. Vitamin $B_{12}$ and marine ecology: The response of *Monochrysis lutheri*. *J. Mar. Biol. Assoc. U.K.*, **41**, 69–76.

Droop, M. R., J. J. A. McLaughlin, I. J. Pintner and L. Provasoli, 1959. Specificity of some protophytes toward vitamin $B_{12}$-like compounds. *Preprints Intern. Oceanog. Cong. A.A.A.S.*, 916–918.

Duursma, E. K., 1960. Dissolved organic carbon, nitrogen, and phosphorus in the sea. *Netherl. J. Mar. Res.*, **1**, 1–148.

Ericson, L. E., 1952. Uptake of radioactive cobalt and vitamin $B_{12}$ by some marine algae. *Chem. & Ind. London*, **34**, 829–830.

Ericson, L. E., 1953. Further studies on growth factors for *Streptococcus faecalis* and *Leuconostoc citrivorum* in marine algae. *Arkiv Kemi*, **6**, 503–510.

Ericson, L. E. and B. Carlson, 1953. Studies on the occurrence of amino acids, niacin and pantothenic acid in marine algae. *Arkiv Kemi*, **6**, 511–522.

Ericson, L. E. and L. Lewis, 1953. On the occurrence of vitamin $B_{12}$-factors in marine algae. *Arkiv Kemi*, **6**, 427–442.

Fleming, R. H., 1940. The composition of plankton and units for reporting population and production. *Proc. Sixth Pacific Sci. Cong. Calif.*, **3**, 535–540.

Fogg, G. E., 1952. The production of extracellular nitrogenous substances by a blue-green alga. *Proc. Roy. Soc. London*, **B139**, 372–397.

Fogg, G. E., 1953. *The Metabolism of Algae*. Methuen, London, 149 pp.

Fogg, G. E., 1958. Extracellular products of phytoplankton and estimation of primary production. *Rapp. Cons. Explor. Mer*, **144**, 56–60.

Fogg, G. E. and D. F. Westlake, 1955. The importance of extracellular products of algae in freshwater. *Proc. Intern. Assoc. Theor. Appl. Limnol.*, **12**, 219–232.

Ford, J. E., 1953. The microbiological assay of "Vitamin $B_{12}$". The specificity of the requirement of *Ochromonas malhamensis* for cyanocobalamin. *Brit. J. Nutrit.*, **7**, 299–306.

Fries, L., 1959. *Goniotrichum elegans*, a marine red alga requiring vitamin $B_{12}$. *Nature*, **183**, 558–559.

Fries, L., 1961. Vitamin requirements of *Nemalion multifidum*. *Experientia*, **17**, 75.

Fritsch, R. H., 1953. Die Lebensdauer von *Daphnia* spec. bei verschiedener Ernährung, besonders bei Zugabe von Pantothensäure. *Z. wiss. Zool.*, **157**, 35.

Goldberg, E. D., 1952. Iron assimilation by marine diatoms. *Biol. Bull.*, **102**, 243–248.

Goldberg, E. D., 1957. Biogeochemistry of trace metals, in "Treatise on marine ecology and paleoecology". Vol. I, Ecology (edit. J. W. Hedgpeth). *Geol. Soc. Amer. Mem.*, **67**, 345–358.

Greenberg, A. E., 1956. Survival of enteric organisms in sea water. *Public Health Reps. U.S.*, **71**, 77–86.

Grein, A. and S. P. Meyers, 1958. Growth characteristics and antibiotic production of actinomycetes isolated from littoral sediments and material suspended in sea water. *J. Bact.*, **76**, 457–463.

Guillard, R. R. L. and J. H. Ryther, 1962. Studies on marine planktonic diatoms. I. *Cyclotella nana* Hustedt, and *Detonula confervacea* (cleve) gran. *Can. J. Microbiol.*, **8**, 229–239.

Guillard, R. R. L. and P. J. Wangersky, 1958. The production of extracellular carbohydrates by some marine flagellates. *Limnol. Oceanog.*, **3**, 449–454.

Halldal, P., 1958. Pigment formation and growth in blue-green algae in crossed gradients of light intensity and temperature. *Plant Physiol.*, **11**, 401–420.

Harder, R. and A. Oppermann, 1953. Über antibiotische Stoffe bei den Grünalgen *Stichococcus bacillaris* und *Protosyphon botryoides*. *Arch. Mikrobiol.*, **19**, 398–401.

Hardy, A. C., 1936. Plankton ecology and the hypothesis of animal exclusion. *Proc. Linnean Soc.*, 148th Session, Pt. II, 64–70.

Hartman, R. T., 1960. Algae and metabolites of natural waters. *Pymatuning Symp. Ecol. Spec. Pub.*, Pittsburg Univ. No. 2, 38–55.

Harvey, H. W., 1934. Annual variation of plankton vegetation. *J. Mar. Biol. Assoc. U.K.*, **19**, 775–792.

Harvey, H. W., 1953. Note on the absorption of organic phosphorus compounds by *Nitzschia closterium* in the dark. *J. Mar. Biol. Assoc. U.K.*, **31**, 475–476.

Harvey, H. W., 1955. *The Chemistry and Fertility of Sea Waters*. Cambridge Univ. Press, 24 pp.

Hashimoto, Y., 1954. Vitamin $B_{12}$ in marine and freshwater algae. *J. Vitaminol.*, **1**, 49–54.

Hayes, H. L. and T. S. Austin, 1951. The distribution of discolored sea water. *Tex. J. Sci.*, **3**, 530–541.

Hutchinson, G. E., 1957. *A Treatise on Limnology.* Vol. I, *Geography, Physics and Chemistry.* J. Wiley & Son. (See pp. 882–886.)

Hutner, S. H., S. Aaronson, H. A. Nathan, S. Baker, S. Scher and A. Cury, 1958. Trace elements in microorganisms; the temperature factor approach. In *Trace Elements* (edit. C. A. Lamb, O. G. Bentley and J. M. Beattie). Academic Press, 47–65.

Hutner, S. H., H. Baker, S. Aaronson, H. A. Nathan, E. Rodriguez, S. Lockwood, M. Sanders and R. A. Petersen, 1957. Growing *Ochromonas malhamensis* above 35°C. *J. Protozool.*, **4**, 259–269.

Hutner, S. H., A. Cury and H. Baker, 1958. Microbiological assays. *Analyt. Chem.*, **30**, 849–867.

Hutner, S. H., L. Provasoli and H. Baker, 1961. Development of microbiological assays for biochemical, oceanographic and clinical uses. *Microchem. J. Symp. Ser.* **1**, 95–113.

Jakob, H., 1961. Compatibilités, antagonismes et antibioses entre quelques algues du sol. Thèse No. 4485, Fac. Sciences, Univ. de Paris.

Jakowska, S. and R. F. Nigrelli, 1960. Antimicrobial substances from sponges. *Ann. N.Y. Acad. Sci.*, **90**, 913–916.

Jeffrey, L. M. and D. W. Hood, 1958. Organic matter in sea water; an evaluation of various methods for isolation. *J. Mar. Res.*, **17**, 203–224.

Johnston, R., in press. Sea-water, the natural medium for phytoplankton. Part I. General aspects. Part II. Trace metals and chelation. *J. Mar. Biol. Assoc. U.K.*

Jones, G. E., W. H. Thomas and F. T. Haxo, 1958. Preliminary studies of bacterial growth in relation to dark and light fixation of $C^{14}O_2$ during productivity determinations. *U.S. Fish and Wildlife Service, Spec. Sci. Rep.*, Fish. No. 279, 79–86.

Kalle, K., 1949. Fluoreszenz und Gelbstoff im Bottnischen und Finnischen Meerbusen. *Deut. Hydrog. Z.*, **2**, 117–124.

Kashiwada, K., D. Kakimoto, T. Morita, A. Kanazawa and K. Kawagoe, 1957. Studies on vitamin $B_{12}$ in sea water. II. On the assay method and the distribution of this vitamin $B_{12}$ in the ocean. *Bull. Jap. Soc. Sci. Fisheries*, **22**, 637–640.

Kashiwada, K., D. Kakimoto and K. Kawagoe, 1957a. Studies on vitamin $B_{12}$ in sea water. III. On the diurnal fluctuation of vitamin $B_{12}$ in the sea and its vertical distribution in the lake. *Bull. Jap. Soc. Sci. Fisheries*, **23**, 450–453.

Kay, H., 1954. Eine Mikromethode zur chemischen Bestimmung der organischen Substanz im Meerwasser. *Kiel. Meeresforsch.*, **10**, 26–53.

Klein, R. M. and G. E. Manos, 1960. Use of metal chelates for plant tissue culture. *Ann. N.Y. Acad. Sci.*, **88**, 416–425.

Knight-Jones, E. W., 1951. Preliminary studies of nanoplankton and ultraplankton systematics and abundance by a quantitative culture method. *J. Cons. Explor. Mer*, **17**, 140–155.

Koyama, T. and T. G. Thompson, 1959. Organic acids of sea water. *Preprints Intern. Oceanog. Cong. A.A.A.S.*, 925–926.

Krauss, R. W., 1958. Physiology of the fresh-water algae. *Ann. Rev. Plant Physiol.*, **9**, 207–244.

Krogh, A., 1934. Conditions of life in the oceans. *Ecol. Monog.*, **4**, 421–439.

Krogh, A. and A. Keys, 1934. Methods for the determination of dissolved organic carbon and nitrogen in sea water. *Biol. Bull.*, **67**, 132–144.

Kylin, A., 1941. Biologische Analyse des Meerwassers. *Kgl. Fysiog. Sällsk. Lund Förhandl.*, **11**, (21), 217–232.

Kylin, A., 1943. The influence of trace elements on the growth of *Ulva lactuca*. *Kgl. Fysiog. Sällsk. Lund Förhandl.*, **13**, (19), 185–192.

Kylin, A., 1943. Über die Ernährung von *Ulva lactuca*. *Kgl. Fysiog. Sällsk. Lund Förhandl.*, **13**, (21), 202–214.

Kylin, A., 1945. The nitrogen sources and the influence of manganese on the nitrogen assimilation of *Ulva lactuca*. *Kgl. Fysiog. Sällsk. Lund Förhandl.*, **15**, (4), 27–35.

Kylin, A., 1946. Über den Zuwachs der Keimlinge von *Ulva lactuca* in verschiedenen Nährflüssigkeiten. *Kgl. Fysiog. Sällsk. Lund Förhandl.*, **16**, (23), 225–229.

Larsen, B. A. and A. Haug, 1959. The influence of habitat on the niacin and biotin content of some marine Fucaceae. *Preprints Intern. Oceanog. Cong.*, *A.A.A.S.*, 927–928.

Léfèvre, M., H. Jakob and N. Nisbet, 1952. Auto- et heteroantagonisme chez les algues d'eau douce. *Ann. Stat. Centr. Hydrobiol. Appl.*, **4**, 197 pp.

Lewin, J. C., 1955. The capsule of the diatom *Navicula pelliculosa*. *J. gen. Microbiol.*, **13**, 162–169.

Lewin, J. C. and R. A. Lewin, 1960. Auxotrophy and heterotrophy in marine littoral diatoms. *Can. J. Microbiol.*, **6**, 127–134.

Lewin, R. A., 1954. A marine *Stichococcus* sp. which requires vitamin $B_{12}$. *J. gen. Microbiol.*, **10**, 93–96.

Lewin, R. A., 1955. Polysaccharides of marine algae. (Abstract). *Biol. Bull.*, **109**, 373.

Lewin, R. A., 1956. Extracellular polysaccharides of green algae. *Can. J. Microbiol.*, **2**, 665–672.

Lewis, G. J. and N. W. Rakestraw, 1955. Carbohydrates in sea water. *J. Mar. Res.*, **14**, 253–258.

Lochhead, A. G. and M. O. Burton, 1957. Specific vitamin requirements of the predominant bacterial flora. *Can. J. Microbiol.*, **3**, 35–42.

Lochhead, A. G. and R. H. Thexton, 1951. Vitamin $B_{12}$ as a factor for soil bacteria. *Nature*, **167**, 1934–1935.

Lucas, C. E., 1947. The ecological effects of external metabolites. *Biol. Rev.*, **22**, 270–295.

Lucas, C. E., 1949. External metabolites and ecological adaptations. *Symp. Soc. Exp. Biol.*, **3**, 336–356.

Lucas, C. E., 1955. External metabolites in the sea. *Deep-Sea Res.*, **3** (suppl.), 139–148.

Lucas, C. E., 1958. External metabolites and productivity. *Rapp. Cons. Explor. Mer*, **144**, 155–158.

McLaughlin, J. J. A., 1958. Euryhaline chrysomonads: nutrition and toxigenesis in *Prymnesium parvum*, with notes on *Isochrysis galbana* and *Monochrysis lutheri*. *J. Protozool.*, **5**, 75–81.

McLaughlin, J. J. A. and P. A. Zahl, 1959. Vitamin requirements in symbiotic algae. *Preprints Intern. Oceanog. Cong. A.A.A.S.*, 930–931.

McLaughlin, J. J. A., P. A. Zahl, A. Novak, J. Marchisotto and J. Prager, 1960. Mass cultivation of some phytoplanktons. *Ann. N.Y. Acad. Sci.*, **90**, 856–865.

Manton, I., 1959. Electron microscopical observations on a very small flagellate: the problem of *Chromulina pusilla*. *J. Mar. Biol. Assoc. U.K.*, **38**, 319–333.

Manton, I. and M. Parke, 1960. Further observations on small green flagellates with special reference to possible relatives of *Chromulina pusilla*. *J. Mar. Biol. Assoc. U.K.*, **39**, 275–298.

Mautner, H., G. M. B. Gardner and R. Pratt, 1953. Antibiotic activity of seaweed extract II. *Rhodomela larix*. *J. Amer. Pharm. Assoc.*, (*Sci. Ed.*), **42**, 294–296.

Menzel, D. W., E. M. Hulburt and J. H. Ryther, in press. The effects of enriching Sargasso sea-water on the production and species composition of the phytoplankton.

Menzel, D. W. and J. H. Ryther, 1961. Nutrients limiting the production of phytoplankton in the Sargasso Sea, with special reference to iron. *Deep-Sea Res.*, **7**, 276–281.

Menzel, D. W. and J. P. Spaeth, 1962. Occurrence of vitamin $B_{12}$ in the Sagasso Sea. *Limnol. Oceanog.*, **7**, 151–154.

Moul, E. T. and D. Mason, 1957. Study of diatoms populations on sand and mud flats in the Woods Hole area. *Biol. Bull.*, **113**, 351.

Otterstrom, D. V. and E. S. Nielsen, 1939. To Tilfaelde af omfattende Dodelighed hos Fisk foraarsaget af Flagellaten *Prymnesium parvum*. *Rep. Danish Biol. Sta.*, **44**, 1–24.

Parke, M. and I. Adams, 1960. The motile (*Crystallolithus hyalinus*) and non-motile phases in the life-history of *Coccolithus pelagicus*. *J. Mar. Biol. Assoc. U.K.*, **39**, 263–274.

Parke, M., I. Manton and B. Clarke, 1955–59. Studies on marine flagellates: II. Three new species of *Chrysochromulina*; III. Three further species of *Chrysochromulina*; IV. Morphology and microanatomy of a new species of *Chrysochromulina*; V. Morphology and microanatomy of *Chrysochromulina strobilus* sp. nov. *J. Mar. Biol. Assoc. U.K.*, **34**, 579–609; **35**, 387–414; **37**, 209–228; **38**, 169–188.

Pintner, I. J. and L. Provasoli, in press. Nutritional characteristics of some Chrysomonads. In *Symp. Mar. Microbiol.* (edit. C. H. Oppenheimer). Thomas Co., Springfield, Ill.

Plunkett, M. A. and N. W. Rakestraw, 1955. Dissolved organic matter in the sea. *Deep-Sea Res*, **3** (suppl.), 12–14.

Porath, J., 1959. Fractionation of polypeptides and proteins on dextran gels. *Clin. Chim. Acta*, **4**, 776–778.

Porath, J., 1960. Gel filtration of proteins, peptides and amino acids. *Biochim. Biophys. Acta*, **39**, 193–207.

Porath, J. and P. Flodin, 1959. Gel filtration: a method for desalting and group separation. *Nature*, **183**, 1657–1659.

Pratt, R., H. Mautner, G. M. G. Gardner, Sha Yi-Hsien and J. Dufrenay, 1951. Antibiotic activity of seaweed extract. *J. Amer. Pharm. Assoc. (Sci. Ed.)*, **40**, 575–579.

Pringsheim, E. G., 1951. Über farblose Diatomeen. *Arch. Mikrobiol.*, **16**, 18–27.

Pringsheim, E. G. and O. Pringsheim, 1959. Die Ernährung Koloniebildender Volvocales. *Biol. Zntrbl.*, **78**, 937–971.

Proctor, V. W., 1957. Studies of algal antibiosis using *Haematococcus* and *Chlamydomonas*. *Limnol. Oceanog.*, **2**, 125–139.

Proom, H. and A. J. Woiwood, 1949. Examination, by paper chromatography, of the nitrogen metabolism of bacteria. *J. gen. Microbiol.*, **3**, 319–327.

Provasoli, L., 1956. Alcune considerazioni sui caratteri morfologici e fisiologici delle Alghe. *Boll. Zool. Agrar. et Bachicolt.*, **22**, 143–188.

Provasoli, L., 1958. Nutrition and ecology of protozoa and algae. *Ann. Rev. Microbiol.*, **12**, 279–308.

Provasoli, L., 1958a. Growth factors in unicellular marine algae. In *Perspectives in Marine Biology* (edit. Buzzati-Traverso). Univ. Calif. Press, 385–403.

Provasoli, L., 1958b. Effect of plant hormones on *Ulva*. *Biol. Bull.*, **114**, 375–384.

Provasoli, L. and K. Gold. 1962. Nutrition of the American strain of *Gyrodinium cohnii*. *Arch. Mikrobiol.*, **42**, 196–203.

Provasoli, L., J. J. A. McLaughlin and M. R. Droop, 1957. The development of artificial media for marine algae. *Arch. Mikrobiol.*, **25**, 392–428.

Provasoli, L. and I. J. Pintner, 1953. Ecological implications of *in vitro* nutritional requirements of algal flagellates. *Ann. N.Y. Acad. Sci.*, **56**, 839–851.

Provasoli, L. and I. J. Pintner, 1953a. Assay of vitamin B$_{12}$ in seawater. *Proc. Soc. Protozool.*, **4**, 10.

Provasoli, L. and K. Shiraishi, 1959. Axenic cultivation of the brine shrimp *Artemia salina*. *Biol. Bull.*, **117**, 347–355.

Provasoli, L., K. Shiraishi and J. R. Lance, 1959. Nutritional idiosyncrasies of *Artemia* and *Tigriopus* in monoxenic culture. *Ann. N.Y. Acad. Sci.*, **77**, 250–261.

Ritchie, D., 1957. Salinity optima for marine fungi affected by temperature. *Amer. J. Bot.*, **44**, 870–874.

Ritchie, D., 1959. The effect of salinity and temperature on marine and other fungi from various climates. *Bull. Torrey Bot. Club.*, **86**, 367–373.

Robbins, W. J., A. Hervey and M. Stebbins, 1951. Further observations on *Euglena* and B$_{12}$. *Bull. Torrey Bot. Club*, **78**, 363–375.

Rosenfeld, W. D. and C. E. ZoBell, 1947. Antibiotic production by marine microorganisms. *J. Bact.*, **54**, 393–398.

Ryther, J. H., 1954. Inhibitory effects of phytoplankton upon feeding of *Daphnia magna* with reference to growth, reproduction, and survival. *Ecology*, **35**, 522–533.

Ryther, J. H., 1954a. The ecology of phytoplankton blooms in Moriches Bay and Great South Bay L.I. N.Y. *Biol. Bull.*, **106**, 198–209.

Ryther, J. H. and R. R. L. Guillard, 1959. Enrichment experiments as a means of studying nutrients limiting to phytoplankton production. *Deep-Sea Res.*, **6**, 65–69.

Ryther, J. H. and R. R. L. Guillard, 1962. Studies of marine planktonic diatoms. II. Use of *Cyclotella nana* for assays of vitamin $B_{12}$ in seawater. *Can. J. Microbiol.*, **8**, 437–446.

Ryther, J. H. and D. D. Kramer, 1961. Relative iron requirement of some coastal and off-shore planktonic algae. *Ecology*, **42**, 444–446.

Schantz, E. J., 1960. Biochemical studies on paralytic shellfish poisons. *Ann. N.Y. Acad. Sci.*, **90**, 843–855.

Shapiro, J., 1957. Chemical and biological studies on the yellow organic acids of lake water. *Limnol. Oceanog.*, **2**, 161–179.

Shilo (Shelubsky), M. and M. Aschner, 1953. Factors governing the toxicity of the phyto-flagellate *Prymnesium parvum*. *J. gen. Microbiol.*, **8**, 333–343.

Shilo, M. and M. Shilo, 1953. Conditions which determine the efficiency of ammonium sulphate in the control of *Prymnesium parvum* in fish breeding ponds. *Appl. Microbiol.*, **1**, 330–333.

Shilo, M. and R. F. Rosenberger, 1960. Studies on the toxic principles formed by the chrysomonad *Prymnesium parvum* Carter. *Ann. N.Y. Acad. Sci.*, **90**, 866–876.

Shiraishi, K. and L. Provasoli, 1959. Growth factors as supplement to inadequate algal foods for *Tigriopus japonicus*. *Tohoku J. Agr. Res.*, **10**, 89–96. (Abstract in: *Preprints Intern. Oceanog. Cong. A.A.A.S.*, 951–952.)

Sieburth, J. McN., 1958. Antarctic microbiology. *Amer. Inst. Biol. Sci. Bull.*, **8**, 10–12.

Sieburth, J. McN., 1959. Gastrointestinal microflora of antarctic birds. *J. Bact.*, **77**, 521–531.

Sieburth, J. McN., 1959a. Antibacterial activity of antarctic phytoplankton. *Limnol. Oceanog.*, **4**, 419–423.

Sieburth, J. McN., 1960. Acrylic acid, and "antibiotic" principle in *Phaeocystis* blooms in antarctic waters. *Science*, **132**, 676–677.

Sieburth, J. McN. and P. R. Burkholder, 1959. Antibiotic activity of antarctic phytoplankton. *Preprints Intern. Oceanog. Cong. A.A.A.S.*, 933–934.

Skopintsev, B. A., 1959. Organic matter of sea water. *Preprints Intern. Oceanog. Cong. A.A.A.S.*, 953–954.

Slowey, J. F., L. M. Jeffrey and D. W. Hood, 1962. The fatty acid content of ocean water. *Geochim. et Cosmochim. Acta*, **26** 607–616.

Southcott, B. A. and H. L. A. Tarr, 1953. Vitamin $B_{12}$ in marine invertebrates and sea-weeds. *Prog. Rep. Pacific Coast Sta.; Fish. Res. Bd. Canada*, No. 95, 45–47.

Spoehr, H. A., J. Smith, H. Strain, H. Milner and G. Hardin, 1949. Fatty acids anti-bacterials from plants. *Pub. Carnegie. Inst. Wash.*, 586, 67 pp.

Starr, T. J., 1956. Relative amounts of vitamin $B_{12}$ in detritus from oceanic and estuarine environments near Sapelo Island, Georgia. *Ecology*, **37**, 658–664.

Starr, T. J., M. E. Jones and D. Martinez, 1957. The production of vitamin-$B_{12}$ active substances by marine bacteria. *Limnol. Oceanog.*, **2**, 114–119.

Subrahmanyan, R., 1954. On the life-history and ecology of *Horniella marina* gen. et sp. nov. (Chloromonadineae), causing green discoloration of the sea and mortality among marine organisms off the Malabar coast. *Ind. J. Fish.*, **1**, 182–203.

Takano, H., 1956. Harmful blooming of minute cells of *Thalassiosira decipiens* in coastal water in Tokyo Bay. *J. Oceanog. Soc. Jap.*, **12**, 63–67.

Tatsumoto, M., W. T. Williams, J. M. Prescott and D. W. Hood, 1961. On the amino acids in samples of surface seawater. *J. Mar. Res.*, **19**, 89–96.

Thomas, W. H., 1959. The culture of tropical oceanic phytoplankton. *Preprints Intern. Oceanog. Cong. A.A.A.S.*, 207–208.

Vacca, D. D. and R. A. Walsh, 1954. The antibacterial activity of an extract obtained from *Ascophyllum nodosum. J. Amer. Pharm. Assoc. (Sci. Ed.)*, **43**, 24–26.

Vallentyne, J. R., 1957. The molecular nature of organic matter in lakes and oceans, with lesser reference to sewage and terrestrial soils. *J. Fish. Res. Bd. Canada*, **14**, 33–82.

Van Baalen, C., 1961. Vitamin $B_{12}$ requirement of a marine blue-green alga. *Science*, **133**, 1922–23.

Vinogradov, A. P., 1953. The elementary chemical composition of marine organisms. *Mem. Sears Fdn. Mar. Res.*, no. 3, 647 pp.

Vishniac, H. S., 1961. A biological assay for Thiamine in seawater. *Limnol. Oceanog.*, **6**, 31–35.

Vishniac, H. S. and G. A. Riley, 1959. $B_{12}$ and thiamine in Long Island Sound : patterns of distribution and ecological significance. *Preprints Intern. Oceanog. Cong. A.A.A.S.*, 942–943.

Vishniac, H. S. and G. A. Riley, 1961. Cobalamin and thiamine in the Long Island Sound : patterns of distribution and ecological significance. *Limnol. Oceanog.*, **6**, 36–41.

Wallace, A., 1960. Use of synthetic chelating agents in plant nutrition and some of their effects on carboxylating enzymes in plants. *Ann. N.Y. Acad. Sci.*, **88**, 361–377.

Wangersky, P. J., 1952. Isolation of ascorbic acid and rhamnosides from seawater. *Science*, **115**, 685.

Wangersky, P. J. and P. R. L. Guillard, 1960. Low molecular weight organic base from the dinoflagellate *Amphidinium carteri. Nature*, **185**, 689–690.

Wilson, D. P., 1951. A biological difference between natural seawaters. *J. Mar. Biol. Assoc. U.K.*, **30**, 1–26.

Wilson, D. P. and F. A. J. Armstrong, 1952. Further experiments on biological differences between natural seawaters. *J. Mar. Biol. Assoc. U.K.*, **31**, 335–349.

Wilson, D. P. and F. A. J. Armstrong, 1954. Biological differences between seawaters : experiments in 1953. *J. Mar. Biol. Assoc. U.K.*, **33**, 347–360.

Wilson, D. P. and F. A. J. Armstrong, 1958. Biological differences between seawaters : experiments in 1954 and 1955. *J. Mar. Biol. Assoc. U.K.*, **37**, 331–348.

Wilson, W. G. and A. Collier, 1955. Preliminary notes on the culturing of *Gymnodinium brevis. Science*, **121**, 394–395.

Wiseman, J. D. H. and N. I. Hendey, 1953. The significance and diatom content of a deep sea floor sample from the neighbourhood of the greatest oceanic depth. *Deep-Sea Res.*, **1**, 47–59.

Wolfe, M., 1954. The effect of molybdenum upon the nitrogen metabolisms of *Anabaena cylindrica. Ann. Bot.*, **18**, 299–325.

Wood, E. J. Ferguson, 1950. Bacteria in marine environments. *Indo-Pacific Fish. Council, Proc. 2nd meeting*, 69–71.

Wood, E. J. Ferguson, 1956. Diatoms in the ocean deeps. *Pacific Sci.*, **10**, 377–381.

ZoBell, C. E., 1946. *Marine Microbiology.* Chronica Botanica Company, Waltham, Mass.

ZoBell, C. E. and D. Q. Anderson, 1936. Observations on the multiplication of bacteria in different volumes of stored seawater and the influence of oxygen tension and solid surfaces. *Biol. Bull.*, **71**, 324–342.

# 9. BIOASSAY OF TRACE SUBSTANCES

W. L. Belser

## 1. Introduction

One of the truly fascinating problems in the study of the oceans is the fact that animal and plant life is not everywhere abundant. Ancient maps show sea monsters on every quarter, with fish leaping all over the surface of the ocean. Today, we know that this is not a true picture. Some areas of the ocean are barren of life (aquatic deserts); others possess a profusion of plant and animal species.

Redfield, Ketchum and Richards in Chapter 2 have approached the problem of oceanic fertility, dealing with two of the major constituents of sea-water—phosphates and nitrates. They point out that these two chemicals profoundly influence the biology of the sea, and, conversely, that the biology of the sea has a marked influence upon the concentrations of phosphate and nitrate. It would be a tidy solution if we were able to ascribe all biological variability to these two constituents. The observation that adjacent areas of very similar chemical composition in respect to phosphate and nitrate may vary from a dense plankton bloom to very few organisms, suggests, however, that there is not a simple relationship between nitrogen–phosphorus concentrations and biological activity. The sporadic distribution of living organisms presents us with evidence that the ocean is not a uniform environment for biological activity, yet the two major constituents of sea-water, which are known to be involved closely with marine life, do not seem to be the limiting factors in all cases. We are thus presented with the problem that marine organisms can detect differences in the physical and chemical make-up of sea-water which we have not properly identified, and that, in our search for some of the factors responsible for these differences, we have detected some substances whose role in the biochemistry of the seas cannot yet be understood.

Interest in microconstituents of sea-water possessing biological activity stems from a publication by Pütter (1907), but gained its greatest impetus as a result of two subsequent papers by Lucas (1947, 1955). These papers drew together a considerable body of evidence dealing with the effects of extracellular and extraorganismal substances upon the life and death of those particular organisms and their neighbors. Most of the evidence favored rather small molecules of organic composition of varying degrees of complexity. Thus we can have an amino acid or vitamin functioning as a regulator in a physiological process, or as an essential nutrient. There are innumerable instances cited in the literature (see Lucas, 1947, for a review) implicating soluble extracellular organic materials in feeding, homing, reproductive and other physiological responses. Lucas has applied the name "ectocrine" to these substances. Significantly, these subtle differences in sea-water were in each case observed because the organisms could detect them. A striking case in point arises from the work of D. P. Wilson (1954, 1958) in which he carefully checked the physical

[*MS received August, 1960*]

parameters of two areas, and found them almost identical. Yet, the invertebrate populations with which he worked could distinguish between them. If the sand from these two areas was cleaned with hot sulfuric acid to remove organic materials, the organisms could no longer distinguish between them. He attributed the ability of the invertebrate larvae to distinguish between the two areas to an organic film (non-living) which coated the sand in one area, while the other area was relatively clean. Clearly, then, living organisms are able to distinguish differences in environments which our tests do not show. It follows that, if we can find out what substances the organisms respond to, we can then use the organisms to test for the presence of specific materials. This is the basis for the bioassay technique.

The bioassay technique for organic materials in sea-water enjoys certain advantages over conventional chemical tests. This latter approach has been used by Johnston (1955) with some success, but there are certain basic problems involved which make this approach impractical. Many of the standard methods require processing of large quantities of sea-water to recover the trace substances in sufficient concentration for analysis. One commonly used method for the extraction of organic materials from sea-water involves adsorption on activated charcoal which selectively removes 65 to 100% of the total soluble carbon-containing materials from the water (Jeffrey and Hood, 1960). The removal of the organic matter may be accomplished by eluting with a variety of organic solvents. Either cold or hot extractions may alter some of the organic compounds; if a recycled hot-extraction method is used, to obtain quantitative recovery, one might reasonably expect considerable alteration of many of the trace organic materials. Following the elution step, the organic solvents are generally removed by cold evaporation under vacuum. This probably introduces no further changes in the organic products, but quite a bit of inorganic salt may remain after evaporation is complete. Depending upon the analytical tests to be run and the complexity of the mixture of organic materials, it may be desirable to remove these salts. This can be accomplished by dissolving the extract in distilled water and dialyzing against flowing distilled water. The inorganic salts migrate through the dialysis tubing while the organic materials are retained. If such a step is not carried out, chromotography and electrophoresis, two common methods for resolving individual compounds from a complex mixture, yield often indeterminate results.

As an example of the quantitative aspect of this problem of isolation and identification, Daisley and Fisher (1958) have reported samples of sea-water to contain 0.03 $\mu\mu g$ vitamin $B_{12}$ per ml. For the extraction technique, if one processed 50 l. of such a water sample with 100% recovery, the yield of vitamin $B_{12}$ would be about 15 m$\mu$g. This is a small quantity indeed to attempt to characterize chemically. Yet *Euglena gracilis*, the bioassay organism, is capable of detecting one-fiftieth of that quantity, with no extraction or concentration required.

As has been previously stated, the "reagent" for detecting chemical substances in the bioassay is a living organism. An example of a typical bioassay system will be used to illustrate the basic concepts of the technique. One organism which has been employed widely in sea-water bioassay is *E. gracilis*. This alga will not grow in a medium devoid of vitamin $B_{12}$ (Provasoli and Pinter, 1953). If vitamin $B_{12}$ is added to the basal medium, the alga grows, and the amount of growth is directly proportional to the quantity of vitamin $B_{12}$ added to the medium within certain concentration limits.

Thus, if one wishes to test an unknown sample of sea-water for the presence of vitamin $B_{12}$, it is a simple matter to add an aliquot of the sample to the basal medium, inoculate with *Euglena* and wait until terminal growth is reached. If growth occurs, $B_{12}$ is present, and the amount of growth (i.e. increase in biomass) will be proportional to the amount of $B_{12}$ present.

With slight modifications in the method and using a variety of organisms as the reagents, it is presently possible to assay for a large number of organic compounds. In the following sections, these organic compounds will be considered in their natural groupings; the organisms involved, the sensitivity with which they can be detected, and any special conditions or techniques required to detect them, will be discussed.

## 2. Vitamins

The group of organic compounds which have been searched for most diligently by biologists, using the bioassay technique in sea-water samples, is the vitamins. Certain of the vitamins have been implicated in the growth of marine algae (Lewin, 1959) and, for this reason, have been screened for much more stringently than most other compounds.

One vitamin which has occupied the attention of a great many investigators is vitamin $B_{12}$. This compound is highly potent biologically in millimicrogram quantities, and the assays for it are extremely sensitive (Hutner *et al.*, 1958). Sea-water is being routinely assayed for vitamin $B_{12}$ in Great Britain (Daisley and Fisher, 1958; Droop, 1955), the United States (Provosali and Pinter, 1953), Japan (Kashiwada, Kakimoto and Kanazawa, 1959) and Norway (Larsen and Haug, 1959). The assay for vitamin $B_{12}$ is a rather complicated one, inasmuch as vitamin $B_{12}$ activity resides in quite a number of similar chemical compounds. These analogues have been characterized chemically and biologically in some cases, and only biologically in others. In Table I the vitamin activity of the various compounds is characterized biologically.

The type of assay used will be largely determined by whether one is interested in vitamin $B_{12}$ itself (cyanocobalamin) or in vitamin $B_{12}$ activity, since many of the analogues may be used by organisms in place of the vitamin itself. It has been concluded by a number of people (Daisley, 1959; Droop, 1955) in the field that a good bioassay for this vitamin should include a number of organisms, so that full information regarding the vitamin $B_{12}$ activity of a sample may be obtained. Table II lists some of the organisms which are currently used, and

TABLE I

Some Properties of Naturally Occurring $B_{12}$ Vitamins.
(Modified from Coates and Ford, 1955.)

| Name of compound | Base of nucleotide | Microbiological activity [a] | | | | |
|---|---|---|---|---|---|---|
| | | Bact. coli (plate test) | Bact. coli (tube test) | Lb. leich-mannii (tube test) | Euglena gracilis (tube test) | Ochro-monas malhamen-sis (tube test) [b] |
| Cyanoco-balamin | 5, 6-Dimethyl-benziminazole | + + + | + + + | + + + | + + + | + + + |
| Pseudovitamin $B_{12}$ | Adenine | + + + | + | + + | + + | − |
| Factor A | 2-Methyladenine | + + + | + + | + + | + + | − |
| Factor B | No nucleotide present | + + + | + | − | − | − |
| Factor $C_1$ | Not known | + + + | + | + | + | − |
| Factor $C_2$ | Not known | + + + | + | + | + | − |
| Factor D | Not known | − | − | − | − | − |
| Factor E | Not known | + + + | − | − | | − |
| Factor F | Not known | + + + | + + | | | + (?) |
| Factor G | Hypoxanthine | + + + [c] | | + + [c] | | − [c] |
| Factor H | 2-Methyl-hypoxanthine | + + + [c] | + + [c] | + + [c] | | − [c] |
| Factor I (vitamin $B_{12}$III) | Not known | + + + | + + | + + | | + + |

[a] + + + denotes "fully active" (as cyanocobalamin).
   + + denotes activity of the order of 50% that of cyanocobalamin.
   + denotes activity of the order of 10% that of cyanocobalamin.
   − denotes activity of < 1% that of cyanocobalamin.
[b] The *Ochromonas* response is denoted mammalian type.
[c] Brown *et al.* (1955).

summarizes the times involved in the assay, the type response (see Table I) and the sensitivity of the assay.

As Table II shows, the organisms vary in the types of compounds to which they respond. By judicious selection of the proper combination of organisms, more or less complete characterization of the sample's vitamin $B_{12}$ activity may be obtained. This is, perhaps, the most complex situation in the bioassay of organic micronutrients.

Assays for vitamin $B_{12}$ in the North Atlantic and northern North Sea show a seasonal variation in concentration of the vitamin in surface samples. Cowey

## TABLE II

### Vitamin $B_{12}$ Assay Organisms and Their Response Patterns

| Assay organism | Sensitivity lower limits | Time required for complete assay | Type of response[a] | Ref.[b] |
|---|---|---|---|---|
| *Thraustochrytrium globosum* | 1.0 μμg/ml | 9 days | Mammalian | Vishniac |
| *Monochrysis lutheri* | 0.1 μμg/ml | 21 days | *Lb. leichmannii* | Droop |
| *Euglena gracilis* | 0.1 μμg/ml | 7–12 days | *Lb. leichmannii* | Provasoli |
| *Ochromonas malhamensis* | 0.1 μμg/ml | ? | Mammalian | Provasoli |
| *Escherichia coli* | 0.5 μμg/ml | 20 h | *E. coli* | Kakimoto |
| *Lactobacillus leichmannii* | 0.5 μμg/ml | 24–36 h | *Lb. leichmannii* | Johnston |

[a] Mammalian = Response to cyanocobalamin only. *Lb. leichmannii* = Response to cyanocobalamin, pseudo $B_{12}$, factors A, $C_1$ and $C_2$. *E. coli* = Response to most known analogues.

[b] This table was prepared by the individuals listed in the right column during a round table discussion of Vitamin $B_{12}$ bioassay at the 1st International Oceanographic Congress, 1959, New York.

(1956) reports values from 0.1 μμg/ml in summer to 0.5–1.0 μμg/ml in autumn. Daisley and Fisher (1958) examined variation in concentration with depth, in the Bay of Biscay, and found marked variability with depth, with the lowest values in the upper euphotic zone. They report a concentration range from 0.3–4 μμg/ml in a 3600-m vertical cast, using a *Euglena gracilis* assay. Vishniac and Riley (1959) have reported similar seasonal variability in $B_{12}$ in Long Island Sound, ranging from 4.5 μμg/ml to greater than 12 μμg/ml over a period of one year. Their assays were carried out with *Thraustochrytrium globosum*. Kashiwada *et al.* (1959) also found variability in concentration of $B_{12}$ both horizontally and vertically in sea-water samples. They report values ranging from 0.0 to 26.3 μμg/ml.

Practically every sample subjected to assay by any of these techniques has proven to contain some vitamin $B_{12}$ activity. It is interesting to note that characteristic of each of these investigations is a discontinuous distribution in concentration of vitamin $B_{12}$ both horizontally and vertically with marked seasonal variability. This characteristic seems to hold for most other organic micronutrients that have been studied.

Other vitamins that have been assayed in sea-water samples include thiamine, niacin and biotin. Provasoli and Gold (1959) have used one organism *Gyrodinium cohnii* to bioassay for both biotin and thiamine. This alga shows a linear response to biotin between 3 and 40 μμg/ml and to thiamine between 100 and 4000 μμg/ml. The organism requires a rather complex mixture of trace metals and organic materials, including glycerophosphate, betaine, adenylic acid, glucose and acetate which, although not a problem of assays in the laboratory, might constitute a limitation on the use of the assay at sea. This disadvantage could be overcome by returning samples to the laboratory for assay. Addition of volatile preservative (Hutner and Bjerknes, 1948) would maintain the sterility of the sample. These authors have not as yet reported any data on the application of this assay to sea-water samples.

Burkholder (1959) has applied a rather different method to the assay of biotin, thiamine and niacin. The preceding techniques are all carried out in liquid culture, and the quantitation of the assays done by densitometry or some method to estimate increase in cell mass. Burkholder uses an agar-plate method which should lend itself well to studies at sea, as well as in the laboratory. His assay organisms are bacteria isolated from the marine environment. An agar plate containing a basal medium, but lacking the vitamin to be assayed, is seeded with the assay bacterium. The sample to be tested is applied to a small sterile filter pad, which is placed on the surface of the seeded plate. The growth stimulation is estimated by measuring the diameter of the zone surrounding the pad. The major disadvantage of this technique is its relative insensitivity. The responses of the three assay organisms are: niacin, 15 m$\mu$g–1.0 $\mu$g/ml; thiamine, 4 m$\mu$g–1.0 $\mu$g/ml; and biotin, 1 m$\mu$g–0.5 $\mu$g/ml. Although no specific data are available, Burkholder asserts that physiologically significant quantities of all three of these vitamins have been found in sea-water samples and natural marine sediments.

Vishniac and Riley (1959) have assayed thiamine using a non-filamentous fungus which is as yet unidentified. The organism responds to concentrations from 25 $\mu\mu$g to 200 $\mu\mu$g/ml of thiamine. They report that detectable concentrations (i.e. greater than 20 $\mu\mu$g/ml) were rarely encountered in Long Island Sound, although near-shore samples, subject to enrichment by land drainage, showed relatively high concentrations (e.g. 63–65 $\mu\mu$g/ml at a river outfall and 30–40 $\mu\mu$g/ml near Charles Island).

Niacin and biotin in sea-water are also being assayed by Belser (1959) utilizing organisms with artifically produced gene mutations. In all of the preceding cases, the assay organisms are naturally occurring auxotrophs; that is, they have evolved in such a way that they lack the synthetic capacity for the compounds they are used to assay. Belser takes advantage of an observation from microbial genetics that gene mutation in bacteria sometimes resulted in alteration or loss of enzymes. When such a mutation occurs, and an enzyme is no longer manufactured, the organism cannot synthesize the end-product of the biosynthetic chain in which the enzyme occurs.

Belser employs *Serratia marinorubra*, a red-pigmented coccobacillus (ZoBell and Upham, 1944), which can synthesize every single compound it requires for growth, given a mixture of inorganic salts and with glycerol as the sole source of carbon. By inducing mutations with ultraviolet light, biochemically deficient strains have been isolated, each of which is incapable of synthesizing one specific end-product necessary for maintaining life. If the end-product in question is present in the medium, the organisms can grow, but if absent, no growth occurs. Two vitamin mutants have been obtained so far. These require niacin and biotin respectively. Other cases of mutants will be discussed in the sections that follow.

These mutant strains are subject to the same limitation as the marine bacteria used by Burkholder in regard to their sensitivity. The biotin mutant shows a linear response (densitometry) in the concentration range from 2 mμg/ ml to 40 mμg/ml while the niacin mutant, although not completely characterized, can detect amounts less than 1 mμg/ml. Interest thus far has been primarily centered upon the presence and distribution of organic micronutrients rather than in quantitative aspects. Therefore, only rough estimations of concentration can be made. It is significant to note, however, that of 29 separate samples of sea-water from one area which were assayed for biotin 16 (55%) showed detectable quantities of the vitamin (Belser, 1958). Fifty per cent of some more recent near-shore samples contained detectable quantities of niacin. Consistent with the observations on distribution of vitamin $B_{12}$, the biotin assays show the same variabilities in concentration both horizontally and vertically, and much the same sort of random discontinuity.

## 3. Amino Acids

Although many biologists have not attached particular significance to amino acids in sea-water (as evidenced by the emphasis on vitamin assays and the absence of amino-acid assays), the program of mutant isolation in *Serratia marinorubra* has provided strains with which some of them can be assayed. At present it is possible to assay for ten different amino acids. The sensitivity of response of the mutants is several orders of magnitude less than the vitamin responses, with the lower limits detectable in the neighborhood of 1 μg/ml. Over the last year and a half, twice-weekly samples have been taken at a station off the Scripps Institution of Oceanography pier. These samples have been subjected to assay, and conform with the vitamin studies in being characterized by marked variability. The one amino acid which has been present in 50% or more of the samples is isoleucine. Some of the others have occurred at rare intervals. Threonine, tryptophan, glycine, methionine, histidine and arginine have all appeared sporadically. None of the remaining three—cystine, proline and leucine—have occurred in any of these samples. Pelagic water samples obtained on Zig Pac cruise of SIO (see Fig. 1) contained only isoleucine, tryptophan and glycine. A few near-shore sediment samples have been analyzed, and were found to contain several amino acids and niacin.

## 4. Purines and Pyrimidines

This group of compounds has not been examined extensively. With the exception of the *Serratia* program, no assays are being run for purines or pyrimidines at this time. The group comprises adenine, xanthine, guanine, hypoxanthine, cytosine, thymine and uracil as the seven bases and their ribosides and ribotides. Several of the vitamins contain nucleic acid bases as an integral part of their structure, and a few organisms are known to have requirements for one or another base. Potentially, these compounds could have great biological significance. The mutant isolation program has yielded three mutants with requirements in this group. Two of these respond to single compounds, namely the purine adenine and the pyrimidine uracil. The third mutant is apparently blocked at an early step in purine synthesis, and will grow when supplied with any one of the four purine bases (adenine, guanine, xanthine or hypoxanthine). This has been described as a "purine" mutant. Assays of sea-water samples have detected uracil and "purine", but never adenine. The routinely obtained pier samples show uracil in about 30% of the samples, with only about 2% occurrence of "purine". In open-ocean pelagic samples, uracil has appeared in about 25% of the samples on two separate cruises, while "purine" has occurred only once in some 45 samples. The sensitivity of these mutants is in the same order of magnitude as the amino-acid mutants, 1 µg/ml.

Two significant points to be derived from the preceding discussions are: (1) that when bioassays have been run for each of the groups (i.e. vitamins, amino acids, nucleic acid bases) some representatives of each have been detected in every case, and (2) that the occurrence, distribution and concentration of each of the groups is highly variable. One might expect, in retrospect, to find representatives of each of the building blocks of biological material in sea-water. But until a very few years ago, little was known of the composition of the organic material in the sea. It is only with the advent of sensitive and specific assay methods that a clear picture of the true composition is emerging.

## 5. Perspectives

It seems worth considering at this point two directions of development for the bioassay technique. They are the elaboration and expansion in terms of the groups of materials to be assayed and the application of the bioassay to specific problems. In the discussion of vitamins, amino acids, purines and pyrimidines, it was pointed out that the ultimate aim is to be able to assay for all of the known representatives of each of these groups. It is neither necessary nor desirable, however, to confine the bioassay technique to these areas. Provasoli in Chapter 8 has discussed growth-promoting and inhibiting substances. Some recent studies indicate that these substances are not wholly confined to the preceding groups. Bentley (1959) has demonstrated the existence in sea-water of organic micronutrients possessing hormonal activity for marine algae. She believes these are related to the indole auxins found in terrestrial plants. Johnston (1959) has examined sea-water for antimetabolites and has found that they do exist, and, further, that they compete in certain specific metabolic processes. Collier et al. (1950) have demonstrated that carbohydrates may be directly involved in the filtering rate in oysters, and Loomis (1955) showed that glutathione triggers a feeding response in hydra.

It seems quite reasonable that all of these compounds might be bioassayed in the future. The basis for description of genus and species in bacteria resides in differential utilization of carbohydrates. Why should it not be possible to apply a spectrum of bacteria with different carbohydrate patterns to sea-water in order to characterize the carbohydrate components of the organic material ?

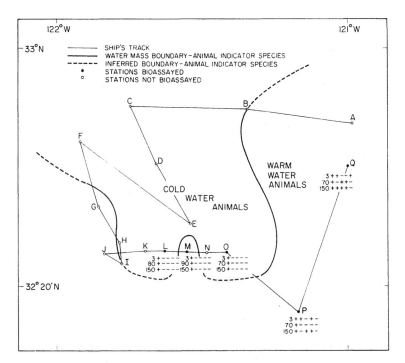

Fig. 1. Plot of Zig Pac cruise data showing water-mass boundaries as defined by animal indicator species. At the stations where bioassays were run, the results appear below the station. + denotes presence and − denotes absence of specific biochemicals. The number indicates depth of sample in meters, and the biochemicals are reported in the order: isoleucine, glycine, tryptophane, uracil and peptone. The animal indicator data are from unpublished work of L. Berner.

The mutant isolation program with *Serratia* has produced five separate mutants which respond to unidentified components of peptone, but to none of the single amino acids. It is suspected that they cannot synthesize short chain peptides, and efforts are being made to identify their requirements. Since glutathione is a tri-peptide, it seems reasonable to suppose that one might be able in the future to have a specific assay for this compound.

Vishniac (1955) has devised an assay for steroids using *Labyrinthula*. This might perhaps open a whole new area of bioassay for more complex substances in sea-water, including the steroid vitamin D.

Bernhard (1955) showed that development in *Dendraster* eggs is highly sensitive to trace concentrations of heavy metals and Arnon (1958) has studied trace-element requirements in algae. These two investigations suggest that highly specific and sensitive bioassays for trace amounts of metallic ions might also be practical.

Another perspective is the application of bioassay techniques to such specific problems in the ocean as bloom succession, sporadic plankton distribution and ecological relationships. But there are other problems of interest, possibly amenable to attack by these techniques. Some results obtained on the Zig Pac cruise (Belser, unpublished work, 1959) suggest that it might be possible to distinguish water-masses on the basis of their organic content, and to describe the boundaries between them with some accuracy. In Fig. 1 are shown some data from the Zig Pac cruise. The boundary between the warm and cold water-masses is defined by use of animal indicator species. These data have been found to correlate well with temperature–salinity studies of boundary conditions. The data from the bioassay, with the exception of station M, correlate quite well with the animal indicator data, and suggest that the bioassay will provide still another valuable parameter in the study of oceanography.

Since land drainage by rivers is often characterized by a high and rather representative organic content (Jerlov, 1955), it might also be possible to ascertain the fate of that water once it reaches the sea. Bioassays from a river effluent out into the ocean would show rates and direction of flow and rate of dilution. It might also be possible to assess the potential contribution of such drainage to productivity.

An extension of the bioassay technique which increases the sensitivity of detection is that of bioauxanography. In cases where the unknown organic materials can be separated by paper chromatography, the assay organisms may be used to detect the position of specific fractions on the paper. This is accomplished by cutting strips from the chromatogram parallel to the direction of solvent flow. These strips are placed on the surface of agar plates seeded with the assay organisms. The portion of the paper which contains the substrate for the assay organism will be surrounded by a zone of growth, thus indicating its position. This technique has been used in studies of the vitamin content of algae (Ericson and Carlson, 1953; Larson and Haug, 1959), and for various specific trace organic substances from animal body fluids (Baker *et al.*, 1959). An excellent review of microbiological assays, describing equipment, sources of organisms and application for general purposes has been published by Hutner *et al.* (1958). In this paper are described some of the non-marine problems which might conceivably be attacked by bioassay techniques.

Although the bioassay technique is relatively new in its application to oceanography, there is little doubt that it represents a potentially valuable tool. Used in combination with chemical, physical and classical biological methods, the bioassay may well extend our knowledge of the biochemical cycles in the ocean.

The author gratefully acknowledges support from the National Science

Foundation (Grant No. G 9556) and from the Marine Life Resources program, Scripps Institution of Oceanography component of the California Cooperative Oceanic Fisheries Investigations.

# References

Arnon, D. I., 1958. Some functional aspects of inorganic micronutrients in the metabolism of green plants. *Perspectives in Marine Biology*, 351–378.

Baker, H. *et al.*, 1959. Assay of thiamine in biologic fluids. *Clinic. Chem.*, **5**, 13–17.

Belser, W. L., 1958. Possible application of a bacterial bioassay in productivity studies. *Spec. Sci. Reps. Fisheries*, **279**, 55–58.

Belser, W. L., 1959. Bioassay of organic micronutrients in the sea. *Proc. Nat. Acad. Sci.*, **45**, 1533–1542.

Bentley, J. A., 1959. Plant hormones in marine phytoplankton, zooplankton and sea water. *Intern. Oceanog. Cong. N.Y., 1959*, 910–912.

Bernhard, M., 1955. Die Kultur von Seeigellarven (Arbacia lixula L.) in künstlichem und natürlichem Meerwasser mit Hilfe von Ionenaustauschsubstanzen und Komplexbildnern. *Pub. Staz. Zool. Napoli*, **29**, 80–95.

Brown, F. B. *et al.*, 1955. The vitamin $B_{12}$ group. Presence of 2-methyl purines in factors A and H and isolation of new factors. *Biochem. J.*, **59**, 82–86.

Burkholder, P. R., 1959. Production of B vitamins by bacteria of the sea. *Intern. Oceanog. Cong. N.Y., 1959*, 912–913.

Coates, M. E. and J. E. Ford, 1955. Methods of measurement of vitamin $B_{12}$. *Biochem. Soc. Symp.*, **3**, 36–51.

Collier, A. *et al.*, 1950. A preliminary note on naturally occurring organic substances in sea water affecting the feeding of oysters. *Science*, **111**, 151–152.

Cowey, C. B., 1956. A preliminary investigation of the variation of vitamin $B_{12}$ in oceanic and coastal waters. *J. Mar. Biol. Assoc. U.K.*, **35**, 609–620.

Daisley, K. W., 1959. Vitamin $B_{12}$ in sea water. *Intern. Oceanog. Cong. N.Y., 1959*, 914–915.

Daisley, K. W. and L. D. Fisher, 1958. Vertical distribution of vitamin $B_{12}$ in the sea. *J. Mar. Biol. Assoc. U.K.*, **37**, 683–686.

Droop, M. R., 1955. A suggested method for the assay of vitamin $B_{12}$ in sea-water. *J. Mar. Biol. Assoc. U.K.*, **34**, 435–440.

Ericson, L. E. and B. Carlson, 1953. Studies on the occurrence of amino acids, niacin and pantothenic acid in marine algae. *Arkiv Kemi*, **6** (49), 511–522.

Hutner, S. H. and C. A. Bjerknes, 1948. Volatile preservatives for culture media. *Proc. Soc. Exp. Biol., N.Y.*, **67**, 393–397.

Hutner, S. H., *et al.*, 1958. Microbiological assays. *Anal. Chem.*, **30**, 849–867.

Jeffrey, L. M. and D. W. Hood, 1960. Organic matter in sea-water; an evaluation of various methods for isolation. *J. Mar. Res.*, **17**, 247–271.

Jerlov, N. G., 1955. Factors influencing the transparency of Baltic waters. *K. Vet. O. Vitterh. Samh. Handl.*, F6, Ser. B, Bd. 6, No. 14.

Johnston, R., 1955. Biologically active compounds in the sea. *J. Mar. Biol. Assoc. U.K.*, **34**, 185–195.

Johnston, R., 1959. Preliminary studies on the responses of marine algae to antimetabolites. *Intern. Oceanog. Cong. N.Y., 1959*, 918–920.

Kashiwada, K., D. Kakimoto and A. Kanazawa, 1959. Studies on vitamin $B_{12}$ in natural water. *Intern. Oceanog. Cong. N.Y., 1959*, 924.

Larsen, B. A. and A. Haug, 1959. The influence of habitat on the niacin and biotin content of some marine fucaceae. *Intern. Oceanog. Cong. N.Y., 1959*, 927–928.

Lewin, R. A., 1959. Phytoflagellates and algae. *Handb. Pflanzenphysiol.*, **14**, 401–417.

Loomis, W. F., 1955. Glutathione control of the specific feeding reactions of hydra. *Ann. N.Y. Acad. Sci.*, **62** (9), 209–228.

Lucas, C. E., 1947. The ecological effects of external metabolites. *Biol. Rev.*, **22**, 270–295.

Lucas, C. E., 1955. External metabolites in the sea. *Deep-Sea Res.*, **3**, Suppl., 139–149.

Provasoli, L. and K. Gold, 1959. *Gyrodinium cohnii*, a bioassay organism for biotin and thiamine in sea-water. *Intern. Oceanog. Cong. N.Y.*, *1959*, 931–933.

Provasoli, L. and I. J. Pintner, 1953. Assay of vitamin $B_{12}$ in sea-water. *Proc. Soc. Protozool.*, **4**, 10.

Pütter, A., 1907. Der Stoffhaushalt des Meeres. *Z. allgem. Physiol.*, **7**, 321–368.

Vishniac, H. S., 1955. Marine mycology. *Trans. N.Y. Acad. Sci.*, **17**, 352–360.

Vishniac, H. S. and G. Riley, 1959. $B_{12}$ and thiamine in Long Island Sound: patterns of distribution and ecological significance. *Intern. Oceanog. Cong. N.Y.*, *1959*, 942–944.

Wilson, D. P., 1954. The attractive factor in the settlement of *Ophelia bicornus* Savigny. *J. Mar. Biol. Assoc. U.K.*, **33**, 361–380.

Wilson, D. P., 1958. Some problems in larval ecology related to the localized distribution of bottom animals. *Perspectives in Marine Biology*, 87–99.

ZoBell, C. E. and H. Upham, 1944. A list of marine bacteria including descriptions of sixty new species. *Bull. Scripps Inst. Oceanog. Univ. Calif.*, **5**, 239.

# Comparative and Descriptive Oceanography

# III. CURRENTS

## 10. EQUATORIAL CURRENT SYSTEMS

J. A. Knauss

### 1. Methods of Study of Ocean Currents

A descriptive knowledge of ocean currents has been gained in four general ways, all of which have made substantial contributions to the description of the equatorial circulation.

#### A. *Ship Drift*

The oldest method is the use of ship-drift observations (this method gives surface currents only, while the others describe sub-surface as well as surface currents). If a ship finds herself off course after making due allowance for the wind, the most likely explanation is that she has been moved by a surface current. A picture of the surface circulation can be made by a careful study of ships' logs. Accuracy is limited by any errors in allowance for the wind and errors in navigation.

The equatorial currents generally run from east to west, and the detection of such drifts requires a knowledge of longitude. Therefore, discovery of the detailed equatorial circulation depended upon the invention of the chronometer, which was first used by Captain Cook on his voyages of exploration, 1768–1779. The first mention of the Pacific Equatorial Countercurrent was by Captain Freycinet in 1817. By the middle 1850's both the French hydrographer de Kerhallet (1869) and the English geographer Findlay (1853) had published charts of the currents of the different world oceans which were qualitatively correct in most respects. The current charts of the American hydrographer Maury (1859), published during the same period, were not as good as the others. However, it was Maury who organized the systematic collection of ship-drift data which made possible many of the future improvements in these current charts. One of the most notable improvements was that of Puls (1895), who presented charts of surface currents and temperature by months for the entire equatorial Pacific. Few refinements have been made on these charts of Puls.

#### B. *Distribution of Properties*

A second method has been to infer the circulation from the distribution of properties. Water-mass analysis, the "Kernschichtmethode" of Wüst (1935) and isentropic analysis of Montgomery (1938) are three examples. These methods are similar in that the circulation is inferred from the gradient of certain properties such as temperature and salinity on possible paths along

[*MS received January, 1961*]
235

which the water particles may move. Although the method is indirect, it is still responsible for most of the little knowledge available on the deep circulation. More particularly, Defant (1936) and Montgomery (1938) used this method in describing the circulation in the equatorial Atlantic.

### C. Geostrophic Currents

The third method is the calculation of velocity from the pressure field using the geostrophic approximation. Apparently the first use of the geostrophic calculation in the description of the equatorial currents was made by Sverdrup (1932) using the Carnegie data. It has since been used by many investigators to describe the equatorial circulation and has recently been used to find a new equatorial current, a South Equatorial Countercurrent in the Pacific, which flows eastward between about 5°–10°S and is most pronounced at a depth of about 400 m (Reid, 1959).

Because an important horizontal component of the Coriolis force becomes zero at the equator, there has been some question in the past as to how close to the equator the geostrophic approximation is valid. It has recently been shown that geostrophic balance exists in the Pacific Equatorial Undercurrent (the Cromwell Current) to within half a degree of the equator where the value of the Coriolis acceleration was 1 to $2 \times 10^{-4}$ cm/sec² (Knauss, 1960).

### D. Velocity Measurements

The fourth method is the use of direct current measurements. It seems the obvious way to study the circulation, but reliable measurements in the open ocean have until recently been difficult and time-consuming. Accuracy and reliability are still problems. The discovery and further measurements of the large sub-surface current at the equator in the Pacific, the Cromwell Current, were accomplished by direct current measurements (Cromwell, Montgomery and Stroup, 1954; Knauss, 1960).

### 2. Gross Circulation Pattern

The great equatorial currents are zonal. Except where they infringe upon continents they move almost due east and west. Unlike the western boundary currents, there is no change of water-mass, water color or marked change in surface temperature as one moves out of one current and into the next. At least in the Pacific there is no surface phenomenon of any kind that gives indirect but conclusive evidence of entering or leaving one of the equatorial currents.

The classic picture of the equatorial circulation has been that of three currents: the westward-flowing North and South Equatorial Currents, and between them, the eastward-flowing Equatorial Countercurrent. These currents correspond closely to the analogous wind system—the northeast and southeast trades separated by a zone of weak and variable winds, the doldrums. The real situation is more complicated and care must be taken in extrapolating the conditions in one area to that of another. What Clarence Palmer has called "the fallacious generalization from a given longitude" is as dangerous in tropical oceanography as it is in tropical meteorology.

## A. Wind Stress

There is marked seasonal and longitudinal variation in the winds and the wind stress. A region of light variable winds, separating strong easterly trades, is to be found only in the Atlantic and in a 2000-mile band in the central Pacific. This picture does not hold true for 70% of the equatorial oceans, namely, the Indian Ocean and the Pacific Ocean east of 120°W and west of 155°W (Fig. 1).

It is, therefore, to be expected that the great zonal currents are not constant (Fig. 2). In both the Atlantic and Pacific Oceans, the South Equatorial Current

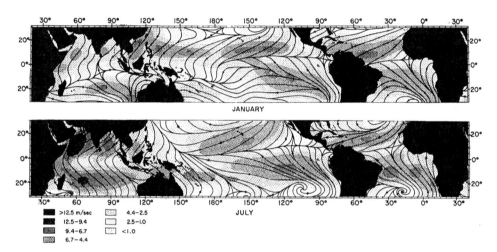

Fig. 1. Direction and magnitude of the mean surface wind in January and July. (After Mintz and Dean, 1952.)

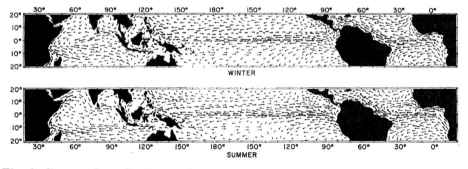

Fig. 2. Current chart for the tropical ocean areas. "Winter" and "summer" refer to northern winter and summer respectively. The chart is derived mainly from that of Schott (1943). The only major changes are those of the equatorial undercurrents in the Atlantic and Pacific (wiggly lines) and the Pacific South Equatorial Counter-current.

is stronger than the North Equatorial Current. In both, the South Equatorial Current and the Countercurrent are stronger in the northern summer than in winter, and in fact there is considerable question about the existence of a continuous countercurrent during some of the winter months. The *Deutsche Seewarte* current chart of the world oceans (from which Fig. 2 is derived; Schott, 1943) shows no eastward flow in the Atlantic east of 20°W during northern winter, and the concept of a continual countercurrent in the Pacific during the winter months has been questioned by Cromwell (1958) and Knauss (1958).

The Indian Ocean is quite different from the Atlantic and Pacific. The wind system is dominated by the monsoons. There is always a South Equatorial Current (stronger in northern summer than in winter), but the current north of the equator reverses with the season.

### B. "Sverdrup" Transport

In the light of the observed seasonal variations in both the equatorial currents and the wind stress, it is of interest to consider to what extent the observed changes in the current can be accounted for by seasonal change in the winds. Sverdrup's integrated mass transport equations have come closest to describing the observed transport in the equatorial Pacific (Sverdrup, 1947; Reid, 1948). In Sverdrup's theory there is a net north–south flow, proportional to the curl of the wind stress, $\zeta$, and inversely proportional to $\beta$, the variation in the Coriolis parameter ($\beta = df/dy$; $f = 2\omega \sin \varphi$; $\omega$ is the angular velocity of the earth; $\varphi$ is the latitude, and $y$ is positive to the north). The great zonal transports, $M_x$, are a consequence of continuity requirements resulting from the meridional changes in the curl of wind stress (see Fofonoff, Vol. I, Section III, page 344):

$$M_x = - \int_{x_0}^{x} \frac{1}{\beta} \left( \frac{\partial \zeta}{\partial y} \right) dx.$$

Wind velocities were read from charts similar to Fig. 1 for January and July. Wind stress was calculated from the data of Van Dorn (1953), and the Sverdrup transport was calculated from the above equation (Fig. 3). Except in the vicinity of the equator there is rather good agreement with the surface current charts of Schott. Specifically, the Sverdrup equations show that the North Equatorial Current is stronger in January than in July and is strongest in the western Pacific in both months. The countercurrent is farther to the north in July than in January, is stronger in July than in January and is inclined slightly to the east-northeast.

The lack of agreement in the region of the equator is not surprising. First of all, the comparison has been between surface currents and vertically integrated mass transport, and it has been tacitly assumed that the speed of the surface current is an index of its transport. This assumption is probably rather good for most of the equatorial zone since the currents are generally shallow. However, it is not a good assumption within two degrees of the equator where the eastward-flowing Cromwell Current flows below the South Equatorial Current. Secondly, the existence of the Cromwell Current strongly suggests that the assumptions of the Sverdrup model (namely the dropping of all the nonlinear terms) are not justified in the vicinity of the equator.

There are two other areas where the transport calculations in Fig. 3 disagree with the surface current in Fig. 1. The first is the position of the northern edge of the North Equatorial Current in January. Although the *Deutsche Seewarte* chart shows the northern edge of this current farther south in winter than in summer, the southward movement is to a latitude closer to 20–22°N rather

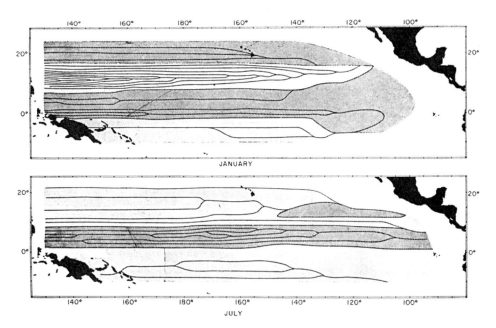

Fig. 3. "Sverdrup" transport in the tropical Pacific for January and July. Dark areas indicate eastward transport, light areas westward transport. Each line represents $3 \times 10^6$ tons/sec.

than 17°N as indicated in Fig. 3. The second area of disagreement is the indication of a weak eastward transport at about 15°N and east of 140°W during July. Although it is doubtful that such a reverse flow extends as far west as 140°W, there appears to be a growing body of evidence (as yet unpublished) of an eastward transport east of 120°W.

### 3. A Two-Layer Ocean—The Thermocline

The equatorial oceans are characterized by a layer of warm, well-mixed water separated from the colder intermediate and deep water by a narrow zone of remarkably high stability[1] (a thermocline or *sprungschicht*). The density gradient is usually greatest immediately below the mixed layer and Reid (1948a) has shown that a density *vs.* depth curve in the zone of the thermocline can be approximated by an exponential function. The thickness of the well-mixed surface layer varies from 10 to 150 m and in both the Atlantic and Pacific this mixed layer becomes thicker toward the west. The characteristics of the mixed layer and the thermocline in the equatorial Atlantic have been described in great detail by Defant (1936).

## A. Stability

The intensity of the density gradient is not the same throughout the tropical regions. The maximum values are usually found in the vicinity of the counter-current; hence, there is on the average greater stability north of the equator

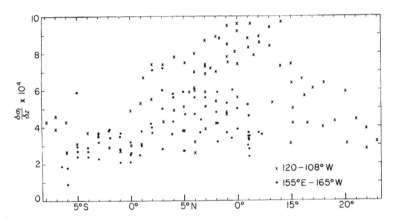

Fig. 4. Maximum "stability" in the Pacific as a function of latitude. Each value is averaged over 50-m depth. The difference between these values and true stability is less than 1%. In other words true stability also varies between $1$–$10 \times 10^{-7}$/cm as shown in this figure. (After Knauss, 1961.)

than south of it. Furthermore the stability appears to be greater in the eastern than the western halves of both the Atlantic and Pacific, although the difference is more marked in the Pacific (Fig. 4); perhaps because in the equatorial regions, the Pacific is three times the width of the Atlantic. In both the Atlantic and the Pacific there is a marked stability minimum at the equator (for the Atlantic, see in particular, Defant, 1936, pl. 40). In the Pacific, this minimum has been associated with the Cromwell Current, and it is likely that a similar current is associated with the minimum in the Atlantic. The maximum stability in the Pacific appears to be centered along the north edge of the countercurrent (Fig. 4). Defant's analysis of the *Meteor* data does not show this feature. This

[1] Stability is defined as $\left[\left(\dfrac{\partial \rho}{\partial z}\right) - \left(\dfrac{\partial \rho}{\partial z}\right)_A\right]\Big/\rho$, where $\rho$ is density and the subscript $A$ refers to the adiabatic density gradient.

may be a real difference in the two oceans; however, considering the scatter of the individual values in Fig. 4, it might mean that a similar feature would be found in the Atlantic if there were sufficient data to be averaged.

The stability values from the Indian Ocean are of the same order as those of the Atlantic and Pacific.

### B. Slope of Sea Surface

If one assumes a level isobaric surface at some depth like 1000 m, the sea surface slopes up to the west along the equator in both the Atlantic and the Pacific.[1] The slope in the Atlantic is about $4 \times 10^{-8}$ (Montgomery and Palmén, 1940); in the Pacific it is about $5 \times 10^{-8}$ (Austin, 1958). The slope is completely accounted for by the depth of the mixed layer and the distribution of density within the thermocline (the mixed layer becomes deeper toward the west). Hence, there is little or no isobaric gradient along the equator at 500 m. It is believed that the slope of the sea surface toward the west results from the mean wind stress of the easterly trades, and Montgomery and Palmén have shown that the mean slope is approximately that which would be expected for the mean wind stress. The slope of the sea surface is not constant across the Pacific; and, as noted previously, neither are the winds. It can be shown that the zone of greatest slope ($6.5 \times 10^{-8}$ between $120°$ and $170°$W) is also the zone of maximum easterly wind stress, using the data of Hidaka (1958) (Fig. 5). Austin (1958) has suggested that there is a real difference in the slope between January to June and July to December; however, this difference is not apparent in Fig. 5.

As might be expected, the slope in the Indian Ocean is quite different from the Atlantic and Pacific. The summer monsoon winds so dominate the circulation pattern that the mean wind stress along the equator, averaged over the year, is westerly rather than easterly (Hidaka, 1958). Likewise, the sea surface slopes to the west rather than to the east (Fig. 6). There might be expected to be marked seasonal variation in the slope of the sea surface, and the limited data available suggest that this is the case although the variation appears to be limited to eastern sections of the Indian Ocean. The fact that the Indian Ocean is divided down the center to $1°$S along $73°$E by the Maldive Islands perhaps complicates the slope as well as the circulation.

### C. Oxygen Minimum

Because the thermocline is a zone of high stability, reduced mixing takes place across it. The water in the mixed layer above the thermocline is high in oxygen and low in phosphate and silicate. Below the thermocline the water is low in oxygen and rich in phosphate and silicate (Fig. 7) (Montgomery, 1954;

---

[1] Because the meridional pressure gradients are very small near the equator, the zonal gradient along the equator is usually considered to be a good estimate of the mean slope of the equatorial waters. All data on which the statements in this section are based came from stations within 80 miles of the equator.

9—S. II

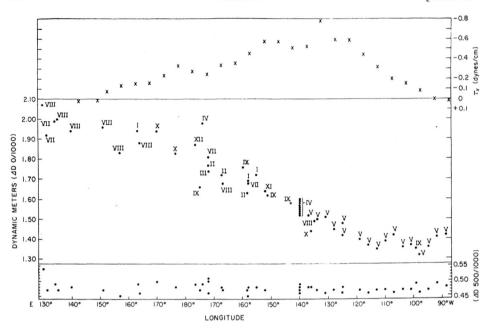

Fig. 5. The slope of the sea surface along the equator in the Pacific relative to the 1000-db surface. The 500-db relative to the 1000-db surface is also plotted for comparison. All observations made within one degree of the equator. Roman numerals refer to the month of the observation. At the top of the figure is the mean east–west wind stress. Minus values refer to an easterly wind. Wind stress values from Hidaka (1958).

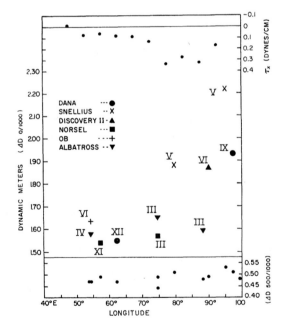

Fig. 6. The slope of the sea surface along the equator in the Indian Ocean relative to the 1000-db surface. The 500-db surface relative to the 1000-db surface is also plotted for comparison. All observations made within 80 miles of the equator. Roman numerals refer to the month of the observation. At the top of the figure is the mean east–west wind stress. Minus values refer to an easterly wind. Wind stress values from Hidaka (1958).

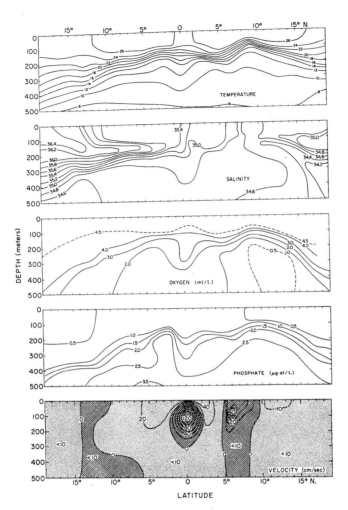

Fig. 7. Composite cross-sections of temperature, salinity, dissolved oxygen, inorganic phosphate and velocity for the central Pacific. Curves based on data taken by the U.S. Fish and Wildlife Pacific Oceanic Fisheries Investigation cruises 5 and 35 of the *Hugh M. Smith*, summer 1950 and 1956, respectively. Velocities are geostrophic velocities except near the equator where the values are based on direct observation (Knauss, 1960). One thousand meters was assumed for the "level of no motion". Darker areas indicate eastward velocity, lighter areas westward velocity.

Wooster and Cromwell, 1958). The shallow thermocline is a permanent feature in the tropics. Primary productivity is probably dependent upon mixing nutrients through the thermocline into the surface zone, where photosynthesis can take place. Regions of high productivity in the tropics are limited to zones like the equator where there is upwelling, or areas such as those off Costa Rica where the thermocline is very close to the surface.

The oxygen minimum, found immediately below the thermocline, is a particularly well-marked feature in both the eastern Atlantic and the Pacific. In both the Atlantic and Pacific the minimum is less marked at the equator than on either side. The fact that the minimum is less pronounced at the equator has been attributed to the upwelling and/or mixing that takes place across the thermocline at the equator.

The oxygen minimum zone of the eastern North Pacific includes a greater body of almost oxygen-free water than any other region in the world oceans. There is a layer of water whose oxygen content is less than 0.25 ml/l. in the upper one thousand meters which extends out from the coast of Central America some 3000 miles and at places is as much as 800 m thick (Fig. 8). The layer is not destitute of life, and some animals probably live their entire lives in this zone (Kanwisher and Ebeling, 1957).

The northern boundary of this oxygen minimum is very sharp and on any given section the boundary corresponds closely with the demarcation between what Sverdrup has called the Pacific Equatorial water and the "transition

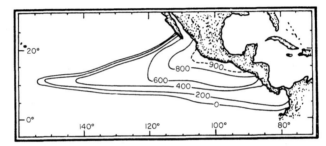

Fig. 8. The thickness of the layer of water (contours are in meters) in which the dissolved oxygen is less than 0.25 ml/l. The entire layer lies above 1000 m.

region" of the California Current (Sverdrup et al., 1942, fig. 209a). There is also a sharp faunal division at this line (see Johnson and Brinton, Chapter 18, page 388). The southern edge of this layer is less well marked, but the thickness of the layer begins to diminish at the north edge of the countercurrent.

The steady-state oxygen distribution below the euphotic zone represents a balance between oxygen consumption and the replenishment by advection and diffusion (see Sverdrup et al., 1942, p. 160). The factors controlling this balance need not be the same for all areas, and an inspection of the data indicates that the oxygen minima cannot be simply related to a single variable. For instance, the extreme minimum off the coast of Mexico does not correspond with either the zone of maximum vertical stability or maximum biological productivity. Both of these maxima occur somewhat to the south of the central oxygen minimum.

## 4. Zonal Flow—Geostrophic Currents

The great zonal currents of the equatorial waters are in approximate geostrophic balance. Charts of dynamic topography showing the geostrophic flow at the surface, such as those constructed for the Atlantic by Defant (1941), for the Pacific by Reid (1961) and for the Indian Ocean by Lacombe (1951), indicate good first-order agreement with the charts of currents based primarily on ship-drift data published by the maritime services of the United States, Great Britain, Germany and the Netherlands. Likewise, there is agreement between the geostrophic currents and the flow pattern as deduced by the distribution of properties using such methods as isentropic analysis (Montgomery, 1938).

The analysis of dynamic height observations show that there is a marked current shear at the thermocline, and that the flow beneath the thermocline is slow (Fig. 7). Direct observations in the North Equatorial Countercurrent also indicate a marked current shear in the region of the thermocline. In addition, these observations show the current maximum to be somewhat below the surface as in Fig. 7 (Knauss, 1961). In general this section (Fig. 7) agrees with the mean current chart (Fig. 2) in the position of the North Equatorial Countercurrent, in the fact that the highest speeds associated with the North and South Equatorial Currents are to be found at their southern and northern edges respectively, and in the fact that the South Equatorial Current has a higher speed than the North Equatorial Current. The South Equatorial Countercurrent, described by Reid (1959), can clearly be seen (albeit weakly) between 10° and 14°S. The Cromwell Current is indicated between 2°S and 2°N.

### A. Cromwell Current

This sub-surface, eastward flow along the equator (first described by Cromwell, Montgomery and Stroup, 1954) is a major feature of the ocean circulation. Recent measurements (Knauss, 1960) have shown that it is a fast, thin current. As defined by the 25 cm/sec contour, it is 300 km wide and about two-tenths of a kilometer thick; at its core the speed is 100–150 cm/sec. It is symmetrical about the equator. The transport of this current is remarkably high, about $40 \times 10^6$ m³/sec. It is very likely the largest current in the equatorial Pacific.

The Cromwell Current is associated with the thermocline. As the depth of the mixed layer shoals toward the east, so does the depth of the core (Fig. 9). Hydrographic measurements suggest that the flow is in geostrophic equilibrium to within half a degree of the equator. The water in the region of the equator appears to be better mixed vertically than the water on either side of it. In the vicinity of the current, there is a marked reduction in the gradients of properties associated with the thermocline. The temperature gradient is weaker. Water with high oxygen content is mixed down from the surface to depths of at least 300 m and near-surface values are lower than those on either side. Similarly, water low in phosphate appears to be mixed downward through the thermocline. The sharp tongue of high salinity water, which moves up from the south along the thermocline, is dissipated at the equator (Figs. 7 and 10).

The major features of the Cromwell Current (including its general shape, the depth of the core and the upward slope of the core to the east) are consistent with a geostrophic flow in response to the horizontal pressure gradient which would result from mixing across the thermocline at the equator (Knauss, 1960).

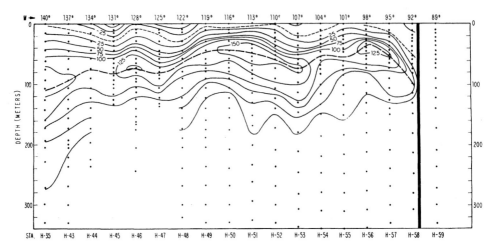

Fig. 9. East–west velocity components along the equator in cm/sec. Positive values are eastward velocities. The thick dashed line indicates the core of the Cromwell Current. The thick column at 91°W represents Isabella, one of the Galápagos Islands. Dots represent points at which velocity measurements were made. (After Knauss, 1960.)

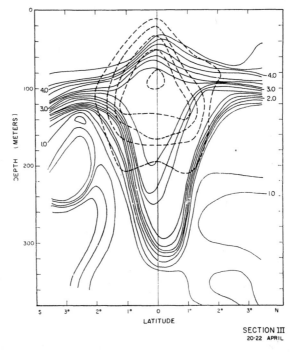

Fig. 10. Velocity cross-section of the Cromwell Current superimposed on an oxygen section made at the same time. Speed contours all in intervals of 25 cm/sec from 25 to 125 cm/sec with the highest velocity in the center. Oxygen values are in ml/l. (After Knauss, 1960.)

Using the above criteria as indicators of the current, it can be shown that, although the Cromwell Current appears to be best developed in the central eastern Pacific, it can at times be traced as far west as 150°–160°E. On the basis of the *Meteor* work (Defant, 1936; Wattenberg, 1936) a similar current can be traced along the equator in the Atlantic.[1] The observations from the Indian Ocean are not yet sufficient to indicate from the distribution of properties that a similar current exists there. The mechanisms which control this current are not yet clear; however, most attempts to date require a zonal pressure gradient commensurate with an upward slope of the sea surface to the west (Fofonoff and Montgomery, 1955; see also several articles in *Deep-Sea Res.*, **6**, no. 4, 1960). As noted before, the sea surface slopes up to the west in the Atlantic and Pacific, but slopes up to the east in the Indian Ocean (Figs. 5 and 6), which would indicate, assuming these theories are correct, that there should not be a similar current in the Indian Ocean.

### B. On the Accuracy of the Geostrophic Equation in Equatorial Waters

Because the Coriolis force approaches zero at the equator, it has long been of interest to oceanographers to know how close to the equator the geostrophic equation can be used. The core of the Cromwell Current, at a depth of 100 m, appears to be in approximate geostrophic balance even to within half a degree of the equator. More recently a comparison of direct current measurements with hydrographic observations has shown at least partial geostrophic balance in the region below the thermocline in the Pacific North Equatorial Countercurrent (Knauss, 1961). In the Cromwell Current speeds were in the range of 50–150 cm/sec; in the deep countercurrent speeds ranged from 5–15 cm/sec. In both cases the pressure and Coriolis forces which were balanced had values of $1-2 \times 10^{-4}$ dyne/g. In terms of dynamic heights, this gradient is equivalent to a change of one to two dynamic centimeters per 100 km. Because of short period changes in the density structure caused by internal waves, small scale advection or other reasons, it is difficult in practice to measure gradients so small, except by averaging a large number of observations.

### C. Seasonal Variations in the Pacific North Equatorial Countercurrent

Since the horizontal pressure gradients in the equatorial region are primarily determined by the temperature distribution, the geostrophic currents can be inferred from the slope of the thermocline. For an observer facing downstream, the thermocline would slope up to the left in the Northern Hemisphere and up to the right in the Southern Hemisphere; the steeper the slope the stronger the current. As an example, the position of the countercurrent, as shown in Fig. 7, is well defined by the temperature structure alone as being between 5°–10°N. The westward-flowing North and South Equatorial Currents are to the north and south of the countercurrent.

---

[1] The equatorial undercurrent in the Atlantic has now been observed (Metcalf, Voorhis and Stalcup, 1962).

As noted previously there are marked seasonal changes in the equatorial currents, and it is of interest to know to what extent these seasonal changes are reflected in the distribution of mass. Using the slope of the thermocline as a criterion for the presence and strength of the Pacific Equatorial Countercurrent, Cromwell (1958) and Knauss (1958) have shown that there is qualitative agreement with the seasonal change in the countercurrent and the change in the slope of the thermocline. From their analysis it is not possible to tell whether or not there is any appreciable lag in the response of the mass distribution to the changing currents, or whether the response is complete; that is, whether the countercurrent is always in geostrophic balance. This question is of some interest in view of the recent analysis of Veronis and Stommel (1956) on the response of a baroclinic ocean to a variable wind stress. The Atlantic and Pacific, however, are not the best areas to study this problem. It will be easier and more clear-cut to determine a phase lag (if, indeed, a measurable one exists) in the Indian Ocean, where the currents north of the equator reverse with the changing monsoon conditions.

## 5. Meridional Flow—Upwelling

Except at the ocean boundaries (where they are more properly treated as eastern or western boundary currents), the north–south flow in equatorial waters is weak. The presence of meridional flow has most often been described on the basis of either (a) requirements in a theory of oceanic circulation or (b) inferences from the distribution of properties (Defant, 1935, 1936; Montgomery, 1938; Sverdrup, 1947; Reid, 1948; Cromwell, 1953; Tchernia et al., 1958).

Cromwell (loc. cit.) discusses the evidence for upwelling at the equator and the lack of such evidence at the north edge of the countercurrent in the central Pacific. The latter was suggested for the Atlantic by Defant (loc. cit). Cromwell further suggests that there is a northward transport in the region of the equator in the surface layer and a convergence at about 2°–3°N. His picture of the meridional flow contrasts with the requirements of the Sverdrup theory which requires a southward transport from about 5°N to 5°S, and in the region of 2°–3°N (which is in the boundary region between the South Equatorial Current and the countercurrent) where there is neither divergence nor convergence (Reid, 1948). It should be noted, however, that the Sverdrup theory is concerned only with the integrated mass transport over a vertical water column, and, since there is the possibility of opposing currents within the column, these differences may not be irreconcilable.

### A. Geostrophic Flow in a Meridional Plane

The possibility of a weak geostrophic flow in the meridional plane should not be overlooked, and the observational data are not inconsistent with this possibility. As previously noted, the sea surface slopes upward to the west in the vicinity of the equator in both the Atlantic and Pacific (Fig. 6). This slope is equivalent to a pressure force of $5 \times 10^{-5}$ dyne/g. The balancing geostrophic currents would converge toward the equator. This slope is maintained by the wind stress, and, because of the importance of the vertical stress term in the surface layer, there may not be a geostrophic balance in the region of the mixed layer; but immediately below the mixed layer, in the upper part of the thermocline, it seems possible that geostrophic balance might be achieved. In other words, the high salinity tongues (Fig. 7) which converge toward the equator along the top of the thermocline from both sides of the Atlantic and Pacific might be explained by geostrophic currents resulting from the zonal tilt of the sea surface caused by the easterly trades. The fact that these tongues are absent in the Indian Ocean in the vicinity of the equator (Tchernia *et al.*, 1958) would be consistent with this idea since the Indian Ocean appears to slope upward toward the east (Fig. 6). In these circumstances, there should be a divergence at the top of the thermocline. An alternate explanation of the observed salinity distribution, based on an eastward zonal flow of less saline water from the Timor Sea, has been suggested by Rochford (1958).

However, both Defant and Cromwell suggest that this high salinity tongue in the South Atlantic and Pacific crosses the equator and can be found in the Northern Hemisphere. Unless this is accomplished completely by horizontal mixing, this transport is inconsistent with geostrophic flow. Fofonoff and Montgomery (1955) in their theory of the Pacific Equatorial Undercurrent (the Cromwell Current) considered such a geostrophic current. They calculated an equatorward flow of 4 cm/sec at 3° latitude.

## B. Oceanic Fronts

One of the arguments in Cromwell's picture of the meridional circulation is the occasional appearance of a well-marked convergence zone north of the equator—an oceanic front. Although such fronts, or convergence zones, have been reported in other areas, they are a particularly striking phenomenon in the equatorial Pacific where horizontal temperature gradients are usually so small. So far as this writer is aware, they have only been reported in the Pacific in the zone 5°S–5°N.[1] They are an intermittent phenomenon which makes an observational program difficult, but of the many unsolved problems concerning the dynamics of the equatorial circulation, they represent one of the most interesting.

Fronts in the equatorial Pacific have been described by Beebe (1926, chap. 2), Cromwell (1953), Cromwell and Reid (1956) and Knauss (1957). In the Northern Hemisphere they are characterized by a layer of cold water to the south infringing on warmer water to the north. At the surface the front is usually marked by a foam line, a color contrast on either side, a sharp increase in the fauna and, apparently, a marked difference in the fauna on either side of the front. The water appears to be sinking on the south, or cold, side of the front.

On the one occasion where surface current observations have been made,

[1] The records of fronts between the equator and 5°S are apparently not very well substantiated (T. S. Austin, *in litt.*).

the water to the south of the front was moving northwest at right angles to the frontal line at a speed of two knots (100 cm/sec). The estimated speed of the front, normal to itself over a 14-h period, was one knot. The cold water (which was also more saline) was overrunning the warmer water and plunging downward (Fig. 11). The temperature gradient was 3°C in 60 m (Knauss, 1957).

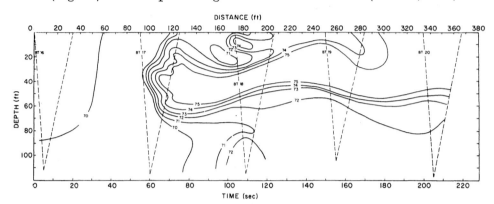

Fig. 11. An equatorial "front" plotted with no vertical exaggeration. Temperature contours, shown in degrees Fahrenheit, are based on five bathythermograph observations made while the ship was drifting across the frontal edge. (After Knauss, 1957.)

## 6. Conclusions and Speculations

This discussion of the equatorial circulation has been almost entirely descriptive and has noted only in passing a few of the many attempts to understand the equatorial circulation on a theoretical basis. There has been in the past a wide gap between the theorists and those who try to understand the oceanic circulation by observing it. Although this gap is obviously easy to understand, it most certainly has been to the disadvantage of all concerned. There are signs that this communication barrier is breaking down. With the advent of high-speed computers, it seems possible that theorists can phrase their problems in ways which do not divorce their solutions so much from reality. On the other hand, it is to be hoped that the marked increase in the oceanic observational program and the steady but slow (sometimes painfully slow) improvement in oceanographic instrumentation will permit descriptive and experimental results which are more meaningful and less ambiguous.

It would seem that the equatorial region might be a particularly fruitful region in which to study oceanic circulation. For one thing, the circulation is apparently but little influenced by the continental boundaries. This is particularly true in the Pacific where the continental boundaries are 8000 miles apart. The gross equatorial circulation pattern in the Atlantic and Pacific are similar, although there are significant differences. The circulation in the Indian Ocean, on the other hand, is very different from that in the Atlantic and Pacific. Furthermore, it is subject to a much more marked seasonal variation than that in either the Atlantic or the Pacific. A careful delineation of the differences as well as the similarities in the circulation should be of some help in any critical evaluation of a theory of oceanic circulation, just as a careful analysis of the consequences of any theory under varying conditions should point the way to the types of observation needed for verification.

## References

Austin, T. S., 1958. Variations with depth of oceanographic properties along the equator in the Pacific. *Trans. Amer. Geophys. Un.*, **39**, 1055–1063.

Austin, T. S., 1960. Oceanography of the east central equatorial Pacific as observed during Expedition Eastropic. *Fish. Bull.*, **60**, 168; *U.S. Dep. Intern. Fish Wildlife Serv.*, 257–282.

Beebe, W., 1926. *The Arcturus Adventure.* G. P. Putnam and Sons, New York and London, 493 pp.

Cromwell, T., 1953. Circulation in a meridional plane in the Central Equatorial Pacific. *J. Mar. Res.*, **12**, 196–213.

Cromwell, T., 1958. Thermocline topography, horizontal currents and "ridging" in the eastern tropical Pacific. *Bull. Inter-Amer. Trop. Tuna Comm.*, **3**, 135–164.

Cromwell, T., R. B. Montgomery and E. D. Stroup, 1954. Equatorial undercurrent in Pacific Ocean revealed by new methods. *Science*, **119**, 648–649.

Cromwell, T. and J. L. Reid, Jr., 1956. A study of oceanic fronts. *Tellus*, **8**, 94–101.

Defant, A., 1935. Der äquatoriale Gegenstrom. *Sitz. Preuss. Akad. Wiss., Phys.-Math. Kl.*, **14**, 450–472.

Defant, A., 1936. Schichtung und Zirkulation des Atlantischen Ozeans. Die Troposphäre des Atlantischen Ozeans. *Wiss. Ergebn. Deut. Atlant. Exped. 'Meteor', 1925–1927*, **6**, Tl. 1, 3.

Defant, A., 1941. Quantitative Untersuchungen zur Statik und Dynamik des Atlantischen Ozeans. Die absolute Topographie des physikalischen Meeresniveaus und der Druckflächen, sowie die Wasserbewegungen im Atlantischen Ozean. *Wiss. Ergebn. Deut. Atlant. Exped. 'Meteor', 1925–1927*, **6**, Tl. 2, 5.

Findlay, A. G., 1853. Oceanic currents and their connection with the proposed Central-American canals. *J. Roy. Geog. Soc.*, **23**, 218–240.

Fofonoff, N. P. and R. B. Montgomery, 1955. The Equatorial Undercurrent in the light of the vorticity equation. *Tellus*, **7**, 518–521.

Hidaka, K., 1958. Computation of the wind stresses over the oceans. *Recs. Oceanog. Works Japan*, **4**, 77–123.

Kanwisher, J. and A. Ebeling, 1957. Composition of the swim-bladder gas in bathypelagic fishes. *Deep-Sea Res.*, **4**, 211–217.

Kerhallet, C. P. de, 1869. *General examination of the Pacific Ocean.* (English translation from the second French Edition) U.S. Navy Hydrographic Office Publication No. 5, Government Printing Office, Washington 25, D.C.

Knauss, J. A., 1957. An observation of an oceanic front. *Tellus*, **9**, 234–237.

Knauss, J. A., 1958. The Equatorial Countercurrent as shown by the Equapac observations. *Proc. Ninth Pacific Sci. Cong., 1957*, **16**, 228–232.

Knauss, J. A., 1960. Measurements of the Cromwell Current. *Deep-Sea Res.*, **6**, 265–286.

Knauss, J. A., 1961. The structure of the Pacific Equatorial Countercurrent. *J. Geophys. Res.*, **66**, 143–155.

Lacombe, H., 1951. Application de la méthode dynamique à la circulation dans l'océan Indien et dans l'océan Antarctique. *Serv. Hydrog. Mar. Com. Cent. Océanog. d'études des Côtes*, **3**, 459–479.

Maury, M. F., 1859. *The Physical Geography of the Sea.* Sixth ed., Harper and Brothers, New York.

Metcalf, W. G., A. D. Voorhis and N. C. Stalcup, 1962. The Atlantic equatorial under-current. *J. Geophys. Res.*, **67**, 2499–2508.

Mintz, Y. and G. Dean, 1952. The observed mean field of motion of the atmosphere. *Geophys. Res. Papers.* No. 17. Cambridge, Mass., 65 pp.

Montgomery, R. B., 1938. Circulation in upper layers of southern North Atlantic deduced with use of isentropic analysis. *Papers Phys. Oceanog. Met., Mass. Inst. Tech. and Woods Hole Oceanog. Inst.*, **6**, 1–53.

Montgomery, R. B., 1954. Analysis of a Hugh M. Smith oceanographic section from Honolulu southward across the equator. *J. Mar. Res.*, **13**, 67–75.

Montgomery, R. B. and E. Palmén, 1940. Contribution to the question of the equatorial counter current. *J. Mar. Res.*, **3**, 112–133.

Puls, C., 1895. Oberflächentemperaturen und Strömungsverhaltnisse des Aequatorial-gürtels des Stillen Ozeans. Doctoral dissertation, Universität Marburg, 38 pp., 4 tabs.

Reid, J. L., Jr., 1959. Evidence of a South Equatorial Countercurrent in the Pacific Ocean. *Nature*, **184**, 209–210.

Reid, J. L., Jr., 1961. On the geostrophic flow at the surface of the Pacific Ocean with respect to the 1000-decibar surface. *Tellus*, **13**, 489–502.

Reid, R. O., 1948. The equatorial currents of the eastern Pacific as maintained by the stress of the wind. *J. Mar. Res.*, **7**, 74–99.

Reid, R. O., 1948a. A model of the vertical structure of mass in equatorial wind-driven currents of a baroclinic ocean. *J. Mar. Res.*, **7** (Sverdrup Anniversary Volume), 304–312.

Rochford, D. J., 1958. Characteristics and flow paths of the intermediate depth waters of the southeast Indian Ocean. *J. Mar. Res.*, **17**, 483–504.

Schott, G., 1943. Weltkarte zur Übersicht der Meeresströmungen. *Ann. Hydrog. Marit. Met.*, Tafel 22.

Sverdrup, H. U., 1932. Arbeider i luft- og havforskning. *Beretn. fra Chr. Michelsens Inst. Videns. Aandsf.*, **2**, 3–20.

Sverdrup, H. U., 1947. Wind-driven currents in a baroclinic ocean; with application to the equatorial currents of the eastern Pacific. *Proc. Nat. Acad. Sci.*, **33**, 318–326.

Sverdrup, H. U., M. W. Johnson and R. H. Fleming. 1942. *The Oceans: their physics chemistry and general biology.* Prentice-Hall, New York. 1987 pp.

Tchernia, P., H. Lacombe and P. Guibout, 1958. Sur quelques nouvelles observations hydrologiques relatives à la region equatorial de l'océan Indien. *Serv. Hydrog. Mar., Com. Cent. Océanog. d'études des Côtes*, **10**, 115–143.

Van Dorn, W. G., 1953. Wind stress on an artificial pond. *J. Mar. Res.*, **12**, 249–276.

Veronis, G. and H. Stommel, 1956. The action of variable wind stresses on a stratified ocean. *J. Mar. Res.*, **15**, 43–75.

Wattenberg, H., 1936. Die Verteilung des Sauerstoffs im Atlantischen Ozean. *Wiss. Ergebn. Deut. Atlant. Exped. 'Meteor', 1925–1927*, **9**, Tl. 1, Atlas, 132 pp., 61 pl.

Wooster, W. S. and T. Cromwell, 1958. An oceanographic description of the eastern tropical Pacific. *Bull. Scripps Inst. Oceanog. Univ. Calif.*, **7**, 169–282.

Wüst, G., 1935. Schichtung und Zirculation des Atlantischen Ozeans. Die Stratosphare des Atlantischen Ozeans. *Wiss. Ergebn. Deut. Atlant. Exped. 'Meteor', 1925–1927*, **6**, Tl. 1, 2.

# 11. EASTERN BOUNDARY CURRENTS[1]

## W. S. Wooster and J. L. Reid, Jr.[2]

## 1. Introduction

The general scheme of circulation of surface waters of the world ocean has been gradually revealed since the first compilation of sailing-vessel drift observations early in the 19th century. An important part of this circulation is found in subtropical gyres, which extend from the doldrums to the mid-latitude belts of prevailing westerly winds. Within these gyres, surface flow is predominantly anticyclonic, with zonal currents at the north and south, strong persistent poleward currents on the western side, and compensating equatorward drifts in the central and eastern portions. Along the eastern boundary there occur variable equatorward currents. At higher latitudes, especially in the Pacific, cyclonic subpolar gyres are observed with eastern currents, which in this case are directed polewards. It is the eastern boundary currents of the subtropical gyres with which this chapter is primarily concerned, although mention will also be made of the poleward eastern currents.

The scientific investigation of the eastern boundary currents did not take place until well into the 20th century. By the time of World War II (and the completion of *The Oceans*) two of these features, the California and Peru Currents, had been studied in a descriptive way in some detail, the former by cruises of *Bluefin* and *E. W. Scripps* (Sverdrup and Fleming, 1941; Tibby, 1941), the latter by an extensive cruise of *William Scoresby* (Gunther, 1936) and by the *Carnegie* Expedition (Sverdrup, 1930). Several crossings of the Canary and Benguela Currents were made by the *Meteor* Expedition, and the general features of the latter current were discussed by Defant (1936). Conditions off the west coast of Australia were not studied and remain little known to this day.[3]

Since the end of the war, studies of the ocean have been greatly intensified. A general theory of the wind-driven circulation has been presented (Munk, 1950) in which some of the important features of the eastern boundary currents are included. Major theoretical investigations have concentrated on the western boundary currents, but some work has been done on the process of coastal upwelling, a basic feature of the eastern currents, by Defant (1952), Hidaka (1954),

[1] Contribution from the Scripps Institution of Oceanography, University of California. The work was supported in part by the Office of Naval Research and the Bureau of Commercial Fisheries through contracts with the University of California, and in part by the Marine Life Research Program, the Scripps Institution's part of the California Co-operative Oceanic Fisheries Investigations.

[2] This paper was prepared jointly, the contributions of the two authors being equivalent. In general the work was divided by hemisphere, with Reid assembling material for the North Pacific and Atlantic, Wooster for the South Pacific and Atlantic, and the Indian Ocean. We are greatly indebted to Dr. Ronald I. Currie who furnished us with proofs prior to publication of his paper with Dr. T. John Hart on the Benguela Current, and to our colleagues who read the manuscript in draft form.

[3] Since the submission of this chapter we have seen a manuscript dealing with part of this area (Wyrtki, in press).

[MS received July, 1960]

Yoshida (1955) and Yoshida and Mao (1957). The descriptive knowledge of the eastern parts of oceans has been greatly advanced: in the North Pacific by the monthly surveys of the California Cooperative Oceanic Fisheries Investigations (Reid, Roden and Wyllie, 1958), and by cruises of the University of Washington and the Pacific Oceanographic Group of Canada; in the South Pacific by occasional studies off the Peruvian coast (Posner, 1957; Wooster and Cromwell, 1958); and in the Atlantic off the South African coast by *William Scoresby* (Hart and Currie, 1960) and by repeated cruises of the South African Division of Fisheries (e.g., Buys, 1959).

Although the descriptive data are still fragmentary and the theoretical studies incomplete, it seems desirable to synthesize on a comparative basis what is known about the common features of this important class of ocean currents. Knowledge of the similarities and differences between these analogous phenomena may serve to stimulate theoretical investigations and should facilitate the design of more efficient observational programs.

## 2. Common Features of Eastern Boundary Currents

If the eastern boundary currents had nothing more in common than their location, it would be of little interest to discuss them as a group. In fact, however, they all have certain important characteristics which make it profitable to examine them on a comparative basis. These common characteristics pertain to their source waters and surface characteristics, the nature of their flow, and the phenomena resulting from the local atmospheric circulation and the presence of boundaries imposed by the continental land-masses.

### A. Source Waters

Where the eastern currents are well developed, they are fed by waters carried eastward in the west wind drifts. In the North Pacific, for example, the North American continent is so located as to block completely the west wind drift, a portion of which turns north to feed the Alaskan Current, the remainder turning south into the California Current. In the South Pacific and South Atlantic the continents extend far enough south to intercept at least a part of the flow of the west wind drift, which turns north in the Peru and Benguela Currents, respectively.

In the Indian Ocean, on the other hand, the west wind drift lies well south of the Australian continent, a fact which may be responsible for the apparent absence of a well-developed eastern boundary current in that ocean (Schott, 1933). In the North Atlantic the principal eastern boundary in the subtropical gyre is the northwest coast of Africa along which flows the Canary Current. Farther north the flow into the Norwegian Sea can be considered analogous to the Alaskan Current.

### B. Surface Characteristics

Since the eastern boundary currents flow equatorward from relatively high latitudes, it is to be expected that their waters will be at lower temperatures

than those of comparable latitudes in the central region of the ocean. Seasonally, additional cold water is introduced along the coast by upwelling (see
page 266), and, as a general rule, surface isotherms run more or less meridionally
along eastern coasts in contrast to their zonal configuration farther west
(Fig. 1). The manner in which the annual variation of surface temperature is
affected by coastal boundary processes is discussed on page 269.

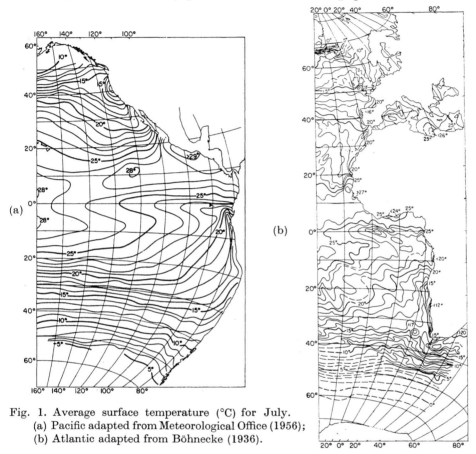

Fig. 1. Average surface temperature (°C) for July.
(a) Pacific adapted from Meteorological Office (1956);
(b) Atlantic adapted from Böhnecke (1936).

In some cases (California Current, Peru Current off Chile) the high latitudes
from which these currents flow are regions where precipitation exceeds evaporation (see Dietrich and Kalle, 1957, fig. 73) so that surface waters are relatively
fresh. Evaporation and mixing downstream then tend to increase surface
salinity. In other cases (such as the Canary Current) evaporation exceeds
precipitation in the source region, and surface salinities are high.

Average surface salinities of the various currents are as follows:

| | |
|---|---|
| California Current | less than 33.0 to 34.0‰ |
| Peru Current (off Peru) | 33.5 to 35.0‰ |
| Peru Current (off Chile) | less than 34.5‰ |
| Benguela Current | less than 35.0 to 35.5‰ |
| Canary Current | less than 35.5 to 36.5‰ |

These differences reflect the surface salinities of the source regions in the various oceans. It should also be noted that coastal upwelling seasonally tends to increase or decrease surface salinity depending on whether salinity increases or decreases with depth. For example, in the California Current, upwelling causes an increase in surface salinity whereas in the Peru and Benguela Currents its effect is to decrease surface salinity.

## C. Flow

Surface ocean flow is predominantly in the sense of the prevailing winds, which are anticyclonic about the mid-ocean highs and cyclonic about the sub-polar lows. This similarity between atmosphere and ocean circulation is, of course, more than coincidence, and many of the basic features of the eastern currents are contained in the general theory of the wind-driven circulation

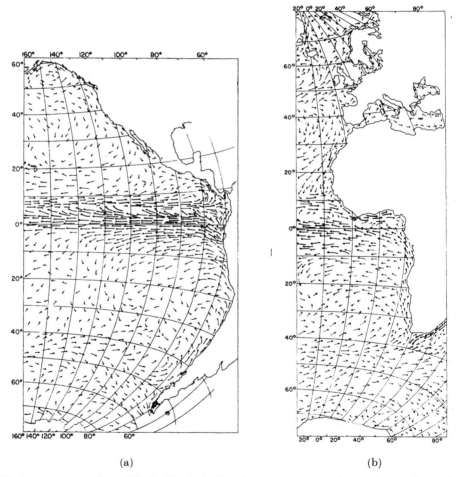

(a)                                                    (b)

Fig. 2. Average surface drift for Northern-Hemisphere summer. (Adapted from *Deutschen Seewarte*, 1942, and U.S.S.R. Ministerstvo Oborony, 1953.)

(Munk, 1950). This theory suggests that, on the eastern sides of oceans, the character of the circulation depends on the vorticity of the meridional winds, the balance being between wind stress vorticity and planetary vorticity, with lateral stress vorticity playing a minor part.

Accordingly, the eastern boundary currents are slow, broad and shallow with relatively small transport, in contrast to the intensified western currents which are narrow and swift, extend to great depths and transport vast quantities of water.

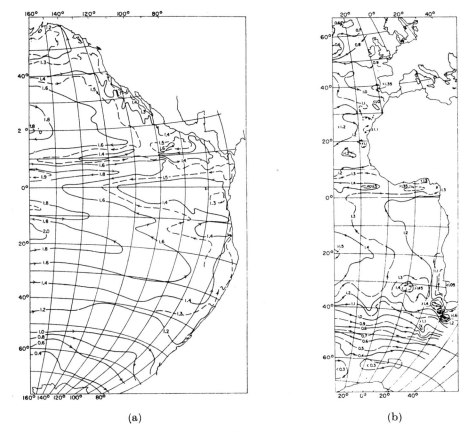

(a)                                                                    (b)

Fig. 3. Surface geostrophic circulation (dynamic height anomalies, 0 over 1000 db). (Pacific after Reid, 1961; Atlantic adapted from Defant, 1941, Teil. XIV.)

The direction of flow is evident from charts of surface drift (Fig. 2) and the surface dynamic topography (Fig. 3). Direct current measurements are scarce, but available GEK measurements (e.g. Fig. 4) support the picture of equatorward flow. However, it is more satisfactory to speak of a prevailing equatorward component, since both the direct measurements and the dynamic charts indicate the presence of numerous eddies and other irregularities of flow.

Furthermore, the existence of coastal countercurrents, either at the surface or below, is well established (Reid, 1960, 1962; Wooster and Gilmartin, 1961).

The speed of the eastern boundary currents can be estimated from drift charts or from dynamic charts; in the California and Peru Currents, GEK measurements are available. These estimates indicate that the average speed is half a knot or less, in contrast to the extreme speeds of four knots or more in the western boundary currents (Stommel, 1958).

The depth to which eastern boundary currents extend has not been determined by direct measurements. Indications are available, however, from the

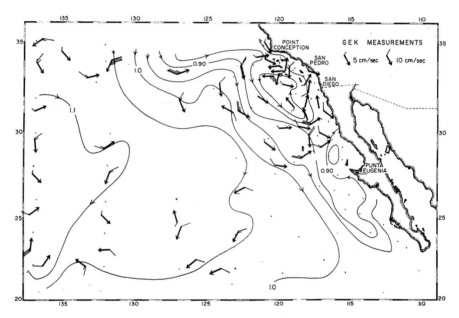

Fig. 4. Measurements of surface velocity by geomagnetic electro-kinetograph, January–March, 1954, superposed on contours of dynamic height anomalies (0 to 500 db) for same period. GEK measurements averaged over 24 h west of 123°W, over 12 h east of 123°W. (Prepared by Reid from data of Scripps Institution of Oceanography, California Department of Fish and Game, and Pacific Oceanic Fishery Investigations of the Bureau of Commercial Fisheries.)

sub-surface distribution of density or temperature, which closely parallels density (Figs. 6a–10a). In general the offshore slope of the isopycnals decreases significantly below about 500 m. Furthermore, transport calculations relative to the 1000-db or 2000-db surfaces are little different, suggesting either that there is no motion at these depths, or that motion is the same at both levels. The presence of coastal undercurrents, especially in the Pacific (see page 273), is additional evidence that surface flow is restricted to a relatively shallow layer. It seems clear that the principal equatorward flow is above 1000 m, and in most places is above 500 m. In contrast, in the Gulf Stream, sub-surface isotherms still slope strongly at depths of several thousand meters (Fig. 5), and even at 2000 m geostrophic flow is significant (Worthington, 1954; Stommel, 1958).

The width of eastern boundary currents is difficult to determine since there is no abrupt change of velocity or western "edge" to the flow. Both Gunther (1936) and Hart and Currie (1960), in the Peru and Benguela Currents respectively, have distinguished between a narrow coastal current and a broad offshore oceanic current flowing in the same general direction. In both cases the distinction is based on the contrast between the coastal regime, where boundary processes predominate, and the offshore region, which is governed by the trade winds. This point of view is in contrast with our picture of the eastern boundary currents as the principal meridional circulation on the eastern sides of oceans, the near-shore characteristics of which are modified

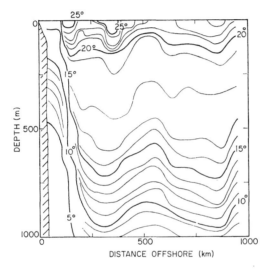

Fig. 5. Vertical temperature structure across Gulf Stream, June, 1955. (After Fuglister, 1960).

by the boundary processes. Both surface drift charts and the observed distribution of properties indicate that the bulk of equatorward transport occurs within 1000 km of the coast with an appropriately directed pressure gradient throughout this belt. The principal flow of the Gulf Stream or Kuroshio, on the other hand, is confined to a band 100 km wide or less (Stommel, 1958; NORPAC Committee, 1960).

The difference in steric level across the Gulf Stream is of the order of one meter. Across the eastern boundary currents, the difference is only about a third as large (Table I), and since it is expressed across a distance ten times as great, the resulting slope of the sea surface is very small. The average geostrophic surface velocity associated with this slope across the entire flow should be about one-thirtieth that of the Gulf Stream, a figure comparable with that observed.

Transports of the various eastern boundary currents have been calculated from available hydrographic data (Table I). Choice of the reference level below 1000 db seems to make little difference to the final computed values. Offshore stations for the calculation were selected from consideration of the offshore increase in geopotential anomaly and were usually the stations where the sea surface reached a more or less steady level (at about 1000 km offshore). The data are too scarce to permit meaningful comparison of the several currents. They show that the average transport is of the order of $15 \times 10^6$ m$^3$/sec, less than a third of the transports of the western boundary currents.

TABLE I

Mass Transport and Change in Steric Level

| Name of current | Av. latitude, ° | Ref. level, db | $\Delta$ level, cm | Transport, $10^6$ m$^3$/sec | Stations | Ref. |
|---|---|---|---|---|---|---|
| California | 32 | 1000 | 46 | 13.2 | C18 70.60, 70.265 | Scripps Inst. (1960) |
| | 34 | 1200 | 39 | 11.6 | Np 63, 68 | NORPAC Comm. (1960) |
| | 32 | 1200 | 35 | 11.6 | Np 70, 90 | NORPAC Comm. (1960) |
| | 32 | 1000 | 24 | 13.1 | *Serrano* 354, 357 | Scripps Inst. (1957) |
| | 33 | 2000 | 38 | 10.1 | *Carnegie* 130, 133 | Fleming *et al.* (1945) |
| Peru | 15 | 2000 | 27 | 17.4 | *Carnegie* 76, 70 | Fleming *et al.* (1945) |
| | 32 | 2000 | 37 | 19.0 | *Ob* 436, 444 | Inst. Okeanol., IGY World Data Center A |
| Benguela | 28.5 | 1000 | 30 | 15.7 | *Meteor* 22, 26 | Defant and Wüst (1938) |
| | 25 | 1000 | 26 | 14.9 | *Meteor* 26, 183 | Defant and Wüst (1938) |

Transport: 10.1 to $19.0 \times 10^6$ m$^3$/sec.
$\Delta$ level:    24 to 46 cm, $\overline{\Delta} = 34$ cm.

## D. Sub-surface Characteristics

The sub-surface distributions of properties of the various eastern boundary currents are shown in a series of profiles roughly normal to the coasts (Figs. 6–11). The following features are noteworthy:

## a. Temperature

Although temperatures throughout the water column are higher in the North Atlantic than elsewhere in the ocean (Reid, 1961a), the thermal structure is similar on most of the profiles in that near-surface temperatures are lower nearshore with shallow isotherms ascending toward the coast. The

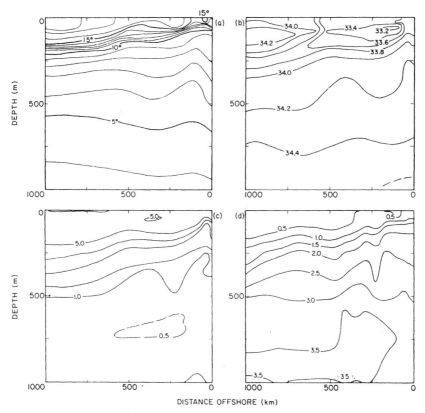

Fig. 6. Profiles across California Current at 32.5°N, September, 1955 (phosphate from August–September, 1950). (a) Temperature, °C; (b) salinity, ‰; (c) dissolved oxygen, ml/l.; (d) inorganic phosphorus, μg-atoms/l. (After NORPAC Committee, 1960, and Scripps Institution, 1960.)

general coastwards rise of the thermocline is indicative of the distribution of mass associated with the equatorward geostrophic circulation. In some cases (Figs. 6a, 7a, 9a) the vertical thermal gradient decreases close to the boundary, with the upper isotherms rising and the deeper isotherms sinking toward shore. Where higher temperatures are thus found at depth nearshore, the presence of a poleward undercurrent is indicated (see page 273). Hart and Currie (1960) suggest that, when isotherms are observed to ascend toward the coast but a strong vertical thermal gradient is still present close inshore, this

condition should be considered a relic of previous upwelling, rather than
evidence of active upwelling at the time of observation.

## b. Salinity

In all but one of the eastern boundary currents, salinity appears to decrease
with depth to minimum values at depths of several hundred to a thousand
meters. Thus, the effect of coastal upwelling is to decrease surface salinity.

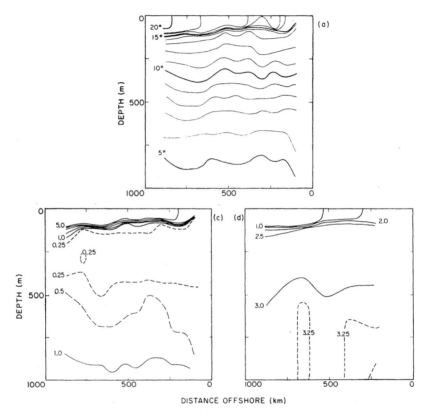

Fig. 7. Profiles across Peru Current at 14°S, July, 1952. (a), (c), (d) as in Fig. 6. (MS
    data from Scripps Institution of Oceanography.)

Only in the case of the California Current (Fig. 6b) does salinity generally increase with depth nearshore, where upwelling then causes an increase of surface
salinity. Offshore there is a complicated interdigitation of salinity minima and
maxima, reasons for which have been presented by Reid, Roden and Wyllie
(1958).

## c. Density

The vertical density structure closely parallels that of temperature, with isopycnals ascending toward the coast. Where the vertical density gradient is

decreased near the coast by the descent of isopycnals below a hundred meters or so, this structure is apparently associated with a poleward undercurrent (see page 273). Comparison of the density structure in the Kuroshio and California Currents (NORPAC Committee, 1960) shows the following differences:

1. Greater width of the region with sloping isopycnals in the eastern boundary currents.

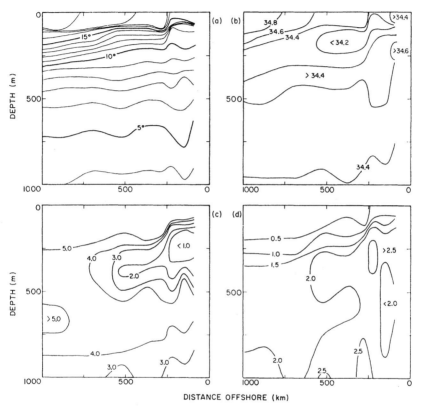

Fig. 8. Profiles across Peru Current at 33°S, May, 1958. (a)–(d) as in Fig. 6. (From data of *Ob* Cruise 3, Institut Okeanologii; MS data from I.G.Y. World Data Center A.)

2. Greater slope of isopycnals in western boundary currents.
3. Greater depth to which sloping isopycnals occur in the western boundary currents.

These differences in the density field are reflected in the character of the geostrophic flow in the two classes of currents.

## d. Oxygen

In most parts of the ocean, dissolved oxygen decreases with depth from values near saturation at the surface to relatively low values at an oxygen minimum which lies at intermediate depths. In the eastern boundary currents, isopleths of dissolved oxygen ascend toward the coast, where the oxygen minimum is shoalest and has the lowest oxygen content. Thus, in sub-surface charts of the horizontal distribution of dissolved oxygen, wedges of low oxygen extend westward from the regions of eastern boundary currents (e.g. Wattenberg, 1938, fig. 5). This condition appears to be more strongly developed over a larger area in the Pacific than in the Atlantic Ocean, although the extensive

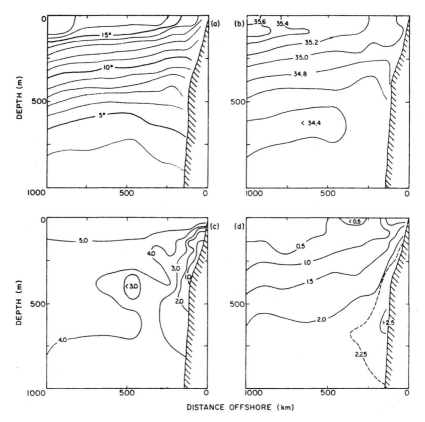

Fig. 9. Profiles across Benguela Current at 24°S, October, 1958. (a)–(d) as in Fig. 6. (After Fuglister, 1960.)

development of the anaerobic and "azoic" conditions on the sea floor near Walvis Bay (Hart and Currie, 1960) is apparently not equalled in the Pacific. At times coastal upwelling is so intense that nearshore surface waters are significantly undersaturated. Thus, surface saturation values as low as 36% and 50% have been reported off the Peruvian coast (Posner, 1957; Discovery Committee, 1949), 70% off the Canary coast (Fuglister, 1960) and even 6% off Walvis Bay on the southwest coast of Africa (Hart and Currie, 1960). Such low oxygen values do not occur off the California coast.

### e. Phosphate

In most parts of the ocean, dissolved inorganic phosphorus (usually referred to as "phosphate") increases with depth from very low values near the surface to maximum values at depths somewhat greater than those of the oxygen minima. Values at this maximum are higher in the North Pacific than elsewhere in the world ocean (Redfield, 1958). In the eastern boundary currents isopleths

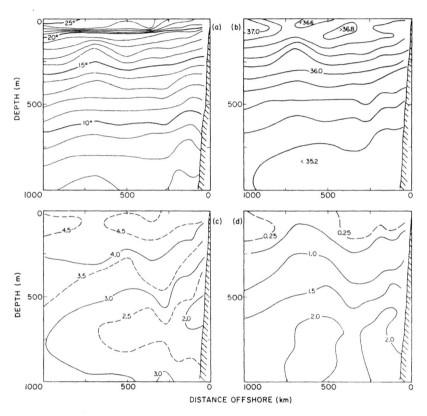

Fig. 10. Profiles across Canary Current at 24°N, October, 1957. (a)–(d) as in Fig. 6. (After Fuglister, 1960.)

of phosphate ascend toward the coast, where the highest surface values are usually found. These coastal surface values in the Peru and Benguela regions are commonly as high as 1–2 μg-atoms/l. (Posner, 1957; Wooster and Cromwell, 1958; Hart and Currie, 1960) in contrast to the usual mid- and low-latitude oceanic surface concentrations of 0.2 μg-atom/l. or less. To these unusually high surface nutrient concentrations is attributed the extraordinary productive capacity of these regions (see page 276).

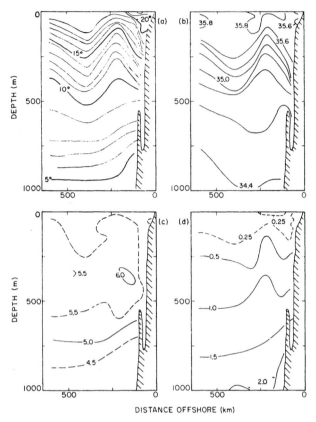

Fig. 11. Profiles west of Australia at 32°S, January, 1951. (a)–(d) as in Fig. 6. (After National Institute of Oceanography, 1957.)

## 3. Coastal Upwelling

Although the low surface temperatures of the eastern boundary currents were at first attributed entirely to their origin in high latitudes (e.g. see Gunther, 1936), it gradually became evident that temperatures did not increase monotonically downstream, and, thus, that some other process must be operating. The distributions of other properties, such as salinity, oxygen and phosphate, also suggested that vertical motion was present. There evolved a model of coastal upwelling from depths of a few hundred meters, with wind stress parallel to the coast causing an offshore transport of surface water and a compensatory replenishment with deeper lying waters (Sverdrup, Johnson and Fleming, 1942, pp. 500–503). Recent studies (Defant, 1952; Hidaka, 1954; Yoshida, 1955, 1958 and 1958a; Yoshida and Mao, 1957) have attempted to evaluate the model quantitatively. We have examined the time and space variations of upwelling on the basis of a simple relationship between wind stress and the orientation of the coastline.

### A. Ekman Transport Calculations and Upwelling

The field of mean wind stress in the world ocean has been computed for five-degree squares by Hidaka (1958), using shipboard observations as summarized in Pilot Charts of the U.S. Navy Hydrographic Office. The accuracy of this representation of the field of mean wind stress suffers because of the paucity of data from large regions of the ocean and because of inadequate knowledge of the drag coefficient at various wind speeds. However, it seems likely that the gross features of the wind-stress field are revealed in these values.

Using the wind-stress data, we have computed the Ekman transport of surface water away from the coast, according to the following equation:

$$M_n = \tau_p/f,$$

where $M_n$ is the Ekman transport normal to the coast, $\tau_p$ the wind stress parallel to the coast and $f$ the Coriolis parameter. An attempt was made to select the five-degree square lying nearest to the coast, although the squares containing adequate data were not always ideally located. The average orientation of the coastline was determined for each square and values of $M_n$ computed for each season. These values were then examined as possible crude indices of the seasonal and geographical variation of coastal upwelling.

An important deficiency of the offshore Ekman transport as an index of upwelling should be noted. The overall transport consists of both Ekman and geostrophic components. The distinction can be visualized by assuming that the Ekman transport first brings about a redistribution of mass, the resulting pressure gradient then being balanced by the Coriolis term to give the geostrophic transport. Thus the Ekman transport should be considered only the initial step in the process which, at equilibrium, gives an essentially geostrophic boundary current.

Another difficulty is that the average wind stress in a five-degree square adjacent to the coast may be a poor approximation to the actual stress operating at the boundary. This problem is mentioned by Hart and Currie (1960) who state that on the southwest coast of Africa diurnally variable coastal winds prevail in a belt extending seaward for about 80 miles from the coast.

Nonetheless, the seasonal and geographical variations in the index do appear to bear some relation to corresponding variations in coastal upwelling. Values of the index are plotted as functions of latitude and season (Fig. 12). Some of the important features of the behavior of the index are summarized below:

1. In both hemispheres maximum values of the index are usually observed in the spring or summer. Only in the Peru Current north of 30°S is a distinct winter maximum present.

2. As a rule values of the index are much larger in the Atlantic and Indian Oceans than in the Pacific. Again, winter values off northern South America are exceptional.

3. Negative values of the index are observed in the North Atlantic (south of 10°N and north of 50°N), Indian Ocean (north of 20°S), North Pacific (north of 40°N) and South Pacific (south of 45°S). In both the North and South Pacific

minimal values of the index (at 30°–35°N and 20°–25°S respectively) appear to separate two seasonally different upwelling regimes.

4. In the Canary, Benguela and California Currents the latitude of the index maximum migrates seasonally, being farther south in spring than in summer.

Unfortunately no direct measurements of vertical velocity are available, but indirect manifestations of the upwelling process can be used to investigate the validity of the index as an indicator of upwelling.

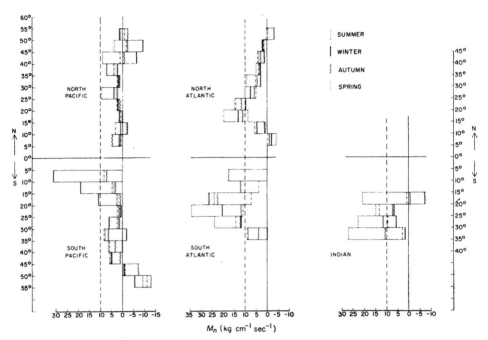

Fig. 12. Offshore Ekman transport, $M_n$, computed from mean wind-stress values (Hidaka, 1958) and average coastline orientations for five-degree squares along the eastern sides of oceans.

## a. Estimates of the speed of vertical motion

Various estimates of vertical speed have been based on the observed vertical movement of isopleths or on theoretical analyses of several models of the process. These estimates range from 10–20 m/month (McEwen, 1929) to 80 m/month (Saito, 1951; Hidaka, 1954). It is possible to use the magnitude of offshore Ekman transport to compute the speed of the associated vertical motion at the boundary. If one assumes a steady-state offshore Ekman transport of 10 kg cm⁻¹ sec⁻¹ and a coastal upwelling band 50 km wide, the compensating vertical motion is about 50 m/month, a result compatible with previous estimates.

## b. Annual range of surface temperature

Since, in the process of upwelling, sub-surface water of relatively low temperature ascends to the surface, it can be assumed that, during the season of most intense upwelling, surface temperatures will be lower than normal for that latitude. At the same time the annual variation of net incoming solar energy increases from the equator toward higher latitudes (Haurwitz and Austin, 1944). Thus, at mid-latitudes a summer maximum of upwelling will

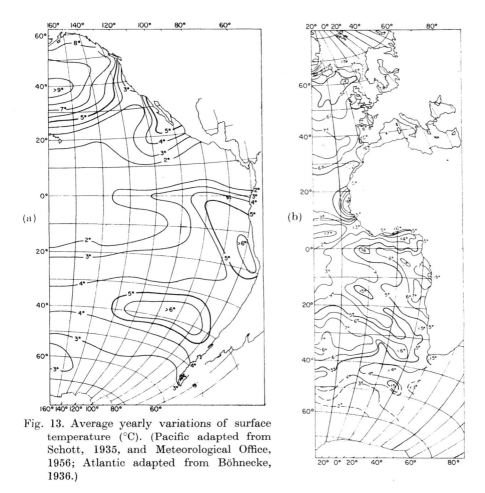

Fig. 13. Average yearly variations of surface temperature (°C). (Pacific adapted from Schott, 1935, and Meteorological Office, 1956; Atlantic adapted from Böhnecke, 1936.)

tend to suppress the annual range of surface temperature, whereas a winter maximum will tend to increase this range, relative to comparable latitudes farther offshore. At low latitudes a pronounced seasonal change in the intensity of upwelling will tend to increase an otherwise small annual range of surface temperature.

In examination of the annual range of surface temperature the following sources were inspected: Atlantic Ocean, Böhnecke (1936); Pacific Ocean, Meteorological Office (1956) and Reid, Roden and Wyllie (1958); Indian Ocean, Koninklijk Nederlands Meteorologisch Instituut (1952) and Schott (1935). The important features are summarized in Fig. 13 and Table II. It is

Table II

Annual Range of Surface Temperature (°C).

Observed Inshore and Offshore Range Compared with Predicted Inshore
Range

| Region | Predicted range | Inshore range | Offshore range | Ref. |
|---|---|---|---|---|
| North Atlantic |  |  |  |  |
| 15°–20° N | large | > 8° | 4°–5° | Böhnecke (1936) |
| South Atlantic |  |  |  |  |
| < 20°S | large | > 7° | 3°–4° | ,, |
| > 20°S | small | < 5° | > 7° | ,, |
| North Pacific |  |  |  |  |
| 33°–50°N | small | < 3° | > 7° | Reid, Roden & Wyllie (1958) |
|  |  |  |  | Meteorological Office (1956) |
| 20°–35°N | medium | > 5° | < 3° | ,, |
| < 20°N | small | < 2° | < 3° | ,, |
| South Pacific |  |  |  |  |
| < 20°S | large | > 6° | < 4° | ,, |
| > 30°S | small | 4°–5° | > 6° | ,, |
| Indian Ocean |  |  |  |  |
| 15°–20°S | large | > 5° | 3°–4° | Schott (1935) |
| > 20°S | small | 3°–4° | > 7° | ,, |

evident that the difference between inshore and mid-ocean ranges is in qualitative agreement with the principles stated above, indicating that the general picture of seasonal and geographical variation of upwelling suggested by the index is correct.

### c. Reported variations of upwelling

Descriptions of variations in upwelling in various regions of eastern boundary currents can be compared with the behavior of the index.

#### (i) California Current

Equatorward winds, more or less parallel to the coast, are strongest off Baja California and southern California in the spring and off northern California in the summer, and the location of most intense upwelling changes accordingly (Reid, Roden and Wyllie, 1958). In the fall and winter a coastal countercurrent, the antithesis of upwelling, is present at the surface (see page 273). The distribution of oceanographic properties (see their fig. 4) is consistent with this picture of seasonal change. Non-seasonal variations in the strength of upwelling also appear to be related to changes in the wind (Reid, 1960).

#### (ii) Peru Current

Although systematic oceanographic observations in the Peru Current are not yet available, a study of the average surface temperature in one-degree squares along the Peruvian coast from 4°S to 17°S (Wooster, 1960) showed a marked seasonal change in surface temperature, with highest values in summer and lowest in winter. There are too few oceanographic observations to permit a study of the seasonal variations of upwelling, and the various reported upwelling centers along the coast (Schott, 1931; Gunther, 1936) are too small to be revealed by the five-degree scale of our analysis. Negative values of the index south of 45°S are consistent with the distribution of surface isotherms which at all seasons are either zonal or dip to the south near the coast (Meteorological Office, 1956).

### (iii) Benguela Current

The seasonal variation of upwelling is described by Marchand (1953) who says:

"The influence of the Benguela Current on coastal upwelling increases in summer while in winter it is decreased. This decrease occurs because in the winter north to southwesterly winds [through west] tend to blow, and northwesterly winds in particular have the effect of piling water up on the west coast; no upwelling takes place from this cause, and the inshore countercurrent is increased but this does not induce upwelling ... thus, maximum salinity occurs in August [winter, Southern Hemisphere]. This is to be expected as upwelling, which produces lower salinity in the upper layers, is at a minimum in winter. Conversely, minimum salinity occurs in the months of January and February. These are summer months with high frequencies of southeasterly winds, and hence with considerable upwelling of water of lower salinity."

This picture of seasonal variation is generally consistent with that suggested by the index. However, Hart and Currie (1960) report an interesting diversity of opinion as to the seasons of maximum and minimum upwelling, quoting the following authorities:

| Reference | Max. upwelling | Min. upwelling |
|-----------|----------------|----------------|
| Schott (1902) | August | November |
| Franz (1920) | August | February |
| Bobzin (1922) | October–December | June |
| Defant (1936) | May–June | December–January |

Another indication of time and space variation in the intensity of coastal upwelling comes from Böhnecke's (1936) surface temperature anomaly charts (with respect to five-degree zones of the Atlantic) for the months of January, April, July and October (Fig. 14). These suggest that upwelling occurs throughout the year, is most intense in summer and fall, and migrates northward from summer to winter.

*(iv) Canary Current*

We have been unable to find any studies of time and space variations of up-welling in the Canary Current. Some indication is available from temperature anomaly charts (Fig. 14), which suggest that upwelling occurs throughout the year, is most intense in spring and summer, and migrates northward from winter to summer.

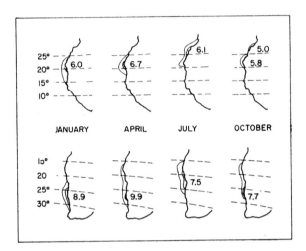

Fig. 14. Seasonal migration of upwelling along Canary and Benguela coasts, as indicated by surface-temperature anomaly with respect to five-degree zonal averages in the Atlantic Ocean for the months of January, April, July and October. Isopleth indicates −5°C anomaly and number is extreme value of anomaly inshore from isopleth. (After Böhnecke, 1936, Beil. XX–XXIII.)

*(v) West Australia*

There seems to be no clear-cut evidence of upwelling along the western coast of Australia (Fig. 11), contrary to what is suggested by the index. Surface iso-therms in the region remain essentially zonal throughout the year, and tem-perature anomaly charts give no indication, either in summer or winter, of any cooling near the coastal boundary (Schott, 1935). Although southerly winds prevail in summer, so that the index predicts strong coastal upwelling, there is no thermal evidence of upwelling, and currents during this season are weak and variable (Koninklijk Nederlands Meteorologisch Instituut, 1952). We have no explanation for the apparent failure of the index in this little-known region.

Thus, with the exception of the Indian Ocean, behavior of the index is in qualitative agreement with what is known about seasonal and geographical variations of coastal upwelling. This suggests that a simple model of dependence of vertical motion on the component of wind stress parallel to the coast and the resulting offshore Ekman transport is applicable. With better wind data it might be possible to make a more sophisticated analysis of the relationship which might then have some predictive value for particular seasons and years.

## 4. Poleward Eastern Boundary Currents

### A. Subtropical Coastal Countercurrents and Undercurrents

In regions of the better-known eastern boundary currents, there are occasional indications of surface poleward countercurrents close inshore. There are reports of such flow in the Benguela region (Hart and Currie, 1960), and the El Niño condition off northern Peru may represent another such situation (Wooster, 1960). The best known of these countercurrents is present along the North American coast during winter from near the tip of Baja California to 45°N and is known north of 35°N as the Davidson Current. There is also northwestward flow from about 32°N to 35°N inside the Channel Islands throughout the year. This has been explained by Munk (1950) as due to the fact that the northerly winds reach their maximum speed some distance from shore. Therefore, on approaching the coast, the wind-stress curl changes sign as does the meridional Sverdrup transport. An analysis of the nearshore wind stress in other eastern boundary regions may show the same feature.[1]

There is also evidence that coastal undercurrents are a common feature of the circulation on the eastern sides of oceans. As has been noted before, cross-stream density profiles frequently show a nearshore weakening of the vertical density gradient with a deepening of the isopycnals below 100–200 m. A similar nearshore deepening of isopleths of other properties occurs. The picture is reminiscent of the distribution of properties near the equator (Wooster and Cromwell, 1958) which has been related to the Equatorial Undercurrent (Wooster and Jennings, 1955; Knauss, 1960).

Such a distribution of mass indicates a poleward geostrophic flow which is apparent on dynamic charts of the 200-db surface (Fig. 15). Off the California coast, such an undercurrent has been discussed by Reid, Roden and Wyllie (1958). This current, which is present throughout the year, flows to the northwest along the coast from Baja California to at least 40°N, bringing warmer, more saline water great distances along the coast. Gunther (1936) describes an undercurrent off the Chilean coast, identifying it by the presence of waters of high salinity and temperature, and of low oxygen content. Recently Brandhorst (1959) has attributed important effects on the Chilean hake fishery to this "Gunther Current". The feature is also evident on the 200-db dynamic chart and, presumably, persists throughout the year.

Hart and Currie (1960) observed poleward flow along the southwest African coast on dynamic charts of the 200-db surface (their fig. 34) which they consider a compensatory movement connected with the process of upwelling. (This is not evident on our 200-db chart, which for the most part is based on *Meteor* data.) They consider the presence of waters of low oxygen content close to

[1] The Sverdrup (1947) equation refers to vertically integrated transport rather than to surface velocity and thus includes the possibility of sub-surface countercurrents. The width of such currents, where present, seems to be of the order of one or two hundred kilometers, so that Hidaka's (1958) wind-stress values for five-degree squares are spaced on too coarse a grid to be used in detailed analysis.

shore to be further evidence of sub-surface flow from the north. Similarly, evidence of sub-surface poleward flow off the Canary coast was inferred by Montgomery (1938) from the distribution of properties and can be traced as far as 20°N on the 200-db chart (Fig. 15).

The relation between coastal upwelling and counterflow at depth has been discussed by Yoshida and collaborators (Yoshida, 1955; Yoshida and Mao, 1957; Yoshida and Tsuchiya, 1957; Yoshida, 1958), but the exact nature of the boundary processes is not yet well understood.

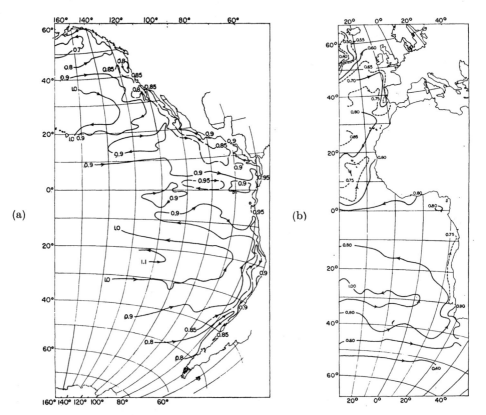

Fig. 15. Geostrophic circulation at 200 m (dynamic height anomalies, 200 over 1000 db.) (Prepared by Reid from a variety of Pacific data; Atlantic derived from Defant, 1941.)

## B. Subpolar Eastern Currents

Where the continents extend to sufficiently high latitudes, they form a barrier to the west-wind drift and cause some portion of the waters to turn poleward. These poleward currents form the eastern limbs of the subpolar gyres and are observed in the North and South Pacific and in the North Atlantic. They are similar to the subtropical eastern currents in that they are broad and slow, and transport significantly less water than the western boundary currents. In several important respects, however, they differ from the equatorward currents.

Since they flow toward higher latitudes, their surface waters are relatively warm. The isopycnals and other isopleths tend downward toward the coast (Figs. 16 and 17), so that, in accordance with the geostrophic approximation,

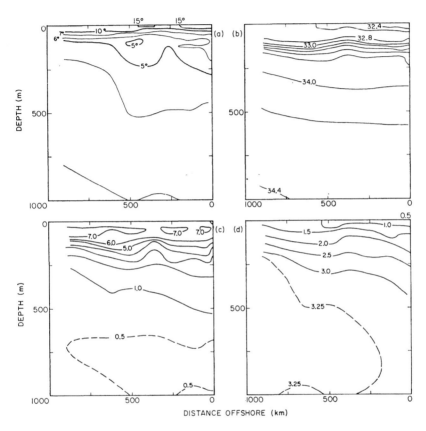

Fig. 16. Profiles across Alaska Current at 53.5°N, July–August, 1957. (a)–(d) as in Fig. 6. (After Pacific Oceanographic Group, 1957.)

the sea surface stands highest next to the coast. As shown by negative values of the offshore Ekman transport (see page 268) in these regions, the surface drift is convergent with the coast, so that light surface waters pile up at the boundary. Coastal upwelling is not possible in such circumstances, and coastal surface waters do not show the influence of this upward vertical motion. The distributions of properties (Figs. 16 and 17) suggest that poleward flow extends to depths greater than 1000 m in accordance with the conclusion of Bennett (1959) that, in the Gulf of Alaska, northward flow extends to at least 2000 m.

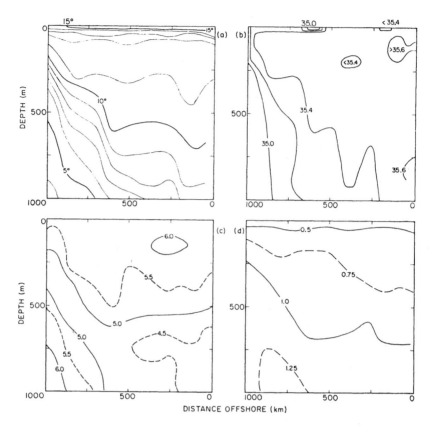

Fig. 17. Profiles across North Atlantic Current at 51.5°N, August, 1958. (a)–(d) as in Fig. 6. (After Fuglister, 1960.)

## 5. Biological Implications

Production of organic matter in the sea is limited to the surface layers where the supply of radiant energy is sufficient to permit photosynthesis. Since the vertical distribution of the nutrient elements on which the process also depends usually shows marked depletion in the surface layers and much higher concentrations in and below the pycnocline, regions of high productivity are found where the upward vertical transfer from this reservoir of nutrients is most effective. The exchange between near-surface and deeper waters takes place most commonly (1) in high latitudes, (2) along the equator, and (3) in coastal regions, particularly on the eastern sides of oceans (Wooster and Ketchum, 1957). Such exchange along the eastern boundary coasts is attributed to the coastal upwelling process discussed above. High productivity is also favored by the shallow thermocline along such coasts, since the mixed layer is usually shallower than the "critical depth" (which depends on the amount of incoming radiation, transparency of the water and the energy level at the compensation depth), a condition essential for maximal development of phytoplankton populations (Sverdrup, 1953).

Quantitative studies of the rate of production of organic matter in the oceans by the carbon-14 method have been reported by Steeman Nielsen and Jensen (1957). Over large parts of the ocean the average rate of carbon fixation was found to be less than 0.2 g $C/m^2/day$. In the regions of eastern boundary currents the following ranges of values were measured:

> Canary Current      0.11–0.67 g $C/m^2/day$
> Benguela Current    0.46–2.5 (3.8 in Walvis Bay) g $C/m^2/day$
> California Current   0.24–0.9 g $C/m^2/day$

Recently on a station near the northern limits of the Peru Current, a rate of 1.02 g $C/m^2/day$ was measured (Holmes, Schaefer and Shimada, 1957). In the same region extremely high values of surface chlorophyll "a" (up to 2.0 $mg/m^3$) were reported.

Standing crops of zooplankton are also large in eastern boundary current regions, with values in nearshore waters as high as the maximum values in high latitudes. Friedrich (1950) has prepared a chart of zooplankton concentrations in the Atlantic (using *Meteor* and other data) on which the highest values are found beyond 50°N and 50°S, and in small areas off the southwest coast of Africa, with somewhat smaller values in the Canary Current. Detailed charts of the zooplankton distribution in the Pacific (NORPAC Committee, 1960; Holmes, Schaefer and Shimada, 1957; Reid, 1962a) show similar features, with highest values beyond 40°N and 50°S and in the regions of the California and Peru Currents. Extreme values of more than 1600 parts of zooplankton per billion ($10^9$) parts of water are found in small areas of the Bering Sea, and the California and Peru Currents, diminishing to less than 25 parts/$10^9$ over most of the subtropical anticyclonic gyres.

The principal economic value to man of these regions lies in the large mid- and low-latitude concentrations of fish of commercial importance. A significant element of these resources consists of clupeoid fishes with short food chains (as the California sardine, and the Peruvian anchovy) and their predators (such as bonita, yellowfin tuna, and cormorant and other producers of guano) (Walford, 1958). An impression of the magnitude of these resources can be gained from estimates of the anchovy population along the Peruvian coast. During 1956 the production of Peruvian bird guano was approximately 330,000 tons. The principal food of the guano birds is anchovy (*Engraulis ringens*); and if one compares the phosphorus content of guano and of the fish (Hutchinson, 1950; Goldberg, *in litt.*), it appears that the minimum conversion factor is about 13 to 1. Thus at least $4.3 \times 10^6$ tons of fish were consumed to produce this much guano. At the same time other predators, including man, are attacking the same species; and the commercial catch of anchovy in 1959 was nearly two million tons (Popovici, *in litt.*). If one estimates that at least four million tons of anchovy can be removed in a year from the inshore water of Peru, by birds and man, this is equivalent to about one-seventh of the annual world landings of fish (Food and Agricultural Organization, 1958). Yet these fish are caught in a coastal strip less than 800 miles long and 30 miles wide, or only about 0.02% of the surface of the world ocean. Similar figures could be cited for other eastern boundary current regions to emphasize the great fertility of their waters.

# References

Bennett, E. B., 1959. Some oceanographic features of the northeast Pacific Ocean during August 1955. *J. Fish. Res. Bd. Canada*, **16**, 565–633.

Bobzin, E., 1922. Vergleichende Betrachtung des Klimas und der Kalten Auftriebströmungen an der Südwestafrikanischen und Südarabischen Küste. *Deut. übers, met. Beob.*, (23), H 1–18.

Böhnecke, G., 1936. Temperatur, Salzgehalt und Dichte an der Oberfläche des Atlantischen Ozeans. *Atlas. "Meteor" Rep.*, **5**, 74 charts.

Brandhorst, W., 1959. Relationship between the hake fishery and a southerly subsurface return flow below the Peru Current off the Chilean Coast. *Nature*, **183**, 1832–1833.

Buys, M. E. L., 1959. Hydrographical environment and the commercial catches, 1950–57. *Un. S. Africa, Dep. Comm. Indus., Div. Fish. Invest.*, Rep. 37, 176 pp.

Defant, A., 1936. Das Kaltwasserauftriebsgebiet vor der Küste Südwestafrikas. *Landerkdl. Forsch., Festschr. N. Krebs*, 52–66.

Defant, A., 1941. Die relative Topographie einzelner Druckflächen im Atlantischen Ozean. *"Meteor" Rep.*, **6** (2 : 4), 183–190 + Beil. X–XVIII.

Defant, A., 1952. Theoretische Überlegungen zum Phänomen des Windstaus und des Auftriebes an ozeanischen Küsten. *Deut. Hydrog. Z.*, **5**, 69–80.

Defant, A. and G. Wüst, 1938. Die dynamischen Werte für die Standardhorizonte an den Beobachtungsstationen. *"Meteor" Rep.*, **6** (2 : 3), 105–181.

Deutschen Seewarte, 1942. Weltkarte zur Übersicht der Meeresströmungen. *Deut. Seewarte*, Chart No. 2802.

Dietrich, G. and K. Kalle, 1957. *Allgemeine Meereskunde; Eine Einführung in die Ozeanographie*. Borntraeger, Berlin-Nikolassee, viii + 492 pp.

Discovery Committee, 1949. Station list R.R.S. *"William Scoresby"*, 1931–1938. *Discovery Reps.*, **25**, 143–280.

Fleming, J. A., C. C. Ennis, H. U. Sverdrup, S. L. Seaton and W. C. Hendrix, 1945. Oceanography I–B. Observations and results in physical oceanography. *Carnegie Rep.*, I–B, 315 pp.

Food and Agricultural Organization, 1958. *Yearbook of Fishery Statistics*. Rome, **9**.

Franz, A., 1920. Beiträge zur Ozeanographie und Klimatologie der Deutsch Sud-Westafrikanischen Küste nach Beobachtungen von S.M.S. "Möwe". *Arch. deut. Seewarte*, **38** (1). Also *Ann. Hydrog.*, **49**, 139–140 (1921).

Friedrich, H., 1950. Versuch einer Darstellung der relativen Besiedlungsdichte in den Oberflächenschichten des Atlantischen Ozeans. *Kieler Meeresforsch.*, **7**, 108–121.

Fuglister, F. C., 1960. *Atlantic Ocean atlas of temperature and salinity profiles and data from the International Geophysical Year, 1957–58*. Vol. 1, 209 pp., Woods Hole Oceanog. Inst., Mass.

Gunther, E. R., 1936. A report on oceanographical investigations in the Peru Coastal Current. *Discovery Reps.*, **13**, 107–276.

Hart, T. J. and R. I. Currie, 1960. The Benguela Current. *Discovery Reps.*, **31**, 123–298.

Haurwitz, B. and J. M. Austin, 1944. *Climatology*. New York, xi + 410 pp.

Hidaka, K., 1954. A contribution to the theory of upwelling and coastal currents. *Trans. Amer. Geophys. Un.*, **35**, 431–444.

Hidaka, K., 1958. Computation of the wind stresses over the oceans. *Rec. Oceanog. Wks. Japan*, **4** (2), 77–123.

Holmes, R. W., M. B. Schaefer and B. M. Shimada, 1957. Primary production, chlorophyll, and zooplankton volumes in the tropical eastern Pacific Ocean. *Bull. Inter-Amer. Trop. Tuna Comm.*, **2** (4), 129–156.

Hutchinson, G. E., 1950. Survey of existing knowledge of biogeochemistry. The biogeochemistry of vertebrate excretion. *Bull. Amer. Mus. Nat. Hist.*, **96**, xviii + 554 pp.

Knauss, J., 1960. Measurements of the Cromwell Current. *Deep-Sea Res.*, **6**, 265–286.

Koninklijk Nederlands Meteorologisch Instituut, 1952. Indian Ocean oceanographic and meteorological data. *K.N.M.I.* No. 135, 2nd Ed., 24 charts.

Marchand, J. M., 1953. Twenty-third annual report for the year ended December, 1951. *Un. S. Africa, Dep. Comm. Indus., Div. Fish.*, 181 pp.

McEwen, G. F., 1929. A mathematical theory of the vertical distribution of temperature and salinity in water under the action of radiation, conduction, evaporation and mixing due to the resulting convection. *Bull. Scripps. Inst. Oceanog. Univ. Calif.*, **2**, 197–306.

Meteorological Office, 1956. *Monthly Meteorological Charts of the Eastern Pacific Ocean.* London, H.M.S.O., M. O. 518. 122 pp.

Montgomery, R. B., 1938. Circulation in upper layers of southern North Atlantic deduced with use of isentropic analysis. *Papers Phys. Oceanog. Met., Mass. Inst. Tech. and Woods Hole Oceanog. Inst.*, **6** (2), 1–55.

Munk, W. H., 1950. On the wind-driven ocean circulation. *J. Met.*, **7**, 79–93.

National Institute of Oceanography, 1957. Station list 1950–1951. *Discovery Reps.*, **28**, 299–398.

NORPAC Committee, 1960. *Oceanic Observations of the Pacific: 1955, The NORPAC Data.* University of California Press and University of Tokyo Press, 123 plates.

Pacific Oceanographic Group, 1957. Physical, chemical and plankton data record, North Pacific survey, July 23 to August 30, 1957. *Fish. Res. Bd. Canada*, Mss. Rep. Ser. No. 4, 103 pp.

Posner, G. S., 1957. The Peru Current. *Bull. Bingham Oceanog. Coll.*, **16** (2), 106–155.

Redfield, A. C., 1958. The biological control of chemical factors in the environment. *Amer. Sci.*, **46**, 205–221.

Reid, J. L., Jr., 1960. Oceanography of the northeastern Pacific during the last ten years. *Calif. Coop. Ocean. Fish. Invest., Rep.*, **7**, 77–90.

Reid, J. L., Jr., 1961. On the geostrophic flow at the surface of the Pacific Ocean with respect to the 1000-decibar surface. *Tellus*, **13**, 489–502.

Reid, J. L., Jr., 1961a. On the temperature, salinity and density differences between the Atlantic and Pacific Oceans in their upper kilometer. *Deep-Sea Res.*, **7**, No. 4, 265–275.

Reid, J. L., Jr., 1962. Measurements of the California Countercurrent at a depth of 250 meters. *J. Mar. Res.*, **20**, 134–137.

Reid, J. L., Jr., 1962a. On the circulation, the phosphate-phosphorus content, and the zooplankton volumes in the upper part of the Pacific Ocean. *Limnol. Oceanog.*, **7**, 287–306.

Reid, J. L., Jr., G. I. Roden and J. G. Wyllie, 1958. Studies of the California Current System. *Prog. Rep., Calif. Coop. Ocean. Fish. Invest.*, 1 July, 1956 to 1 January, 1958. 27–57.

Saito, Y., 1951. On the velocity of the vertical flow in the ocean. *J. Inst. Polytech.*, Osaka City Univ., Ser. B, **2**, 1–4.

Schott, G., 1902. Die Auftriebzone an der Küste von Sudwestafrikas. *Wiss. Ergebn. "Valdivia"*, **1**, 124 et seq.

Schott, G., 1931. Der Peru-Strom und seine nordlichen Nachbargebiete in normaler und anormaler Ausbildung. *Ann. Hydrog. mar. Met.*, **59**, 161–169, 200–213, 240–252.

Schott, G., 1933. Auftriebwasser an den australischen Westküsten? Ja und Nein! *Ann. Hydrog. mar. Met.*, **61**, 225–233.

Schott, G., 1935. *Geographie des Indischen und Stillen Ozeans.* C. Boysen, Hamburg, xx + 413 pp., 37 plates.

Scripps Institution of Oceanography, 1957. *Oceanic Observations of the Pacific: 1949.* University of California Press, Berkeley and Los Angeles, 363 pp.

Scripps Institution of Oceanography, 1960. *Oceanic Observations of the Pacific: 1950.* University of California Press, Berkeley and Los Angeles, xxviii + 508 pp.

Steeman Nielsen, E. and E. A. Jensen, 1957. Primary oceanic production. The autotrophic production of organic matter in the oceans. *"Galathea" Rep.*, **1**, 49–136.

Stommel, H., 1958. *The Gulf Stream, a Physical and Dynamical Description.* University of California Press. Berkeley and Los Angeles, xiii + 202 pp.

Sverdrup, H. U., 1930. Some oceanographic results of the Carnegie's work in the Pacific—the Peruvian Current. *Trans. Amer. Geophys. Un.* (Eleventh Ann. Meet.), 257–264 [reprinted in *Hydrog. Rev.*, **8**, 240–244 (1931)].

Sverdrup, H. U., 1947. Wind-driven currents in a baroclinic ocean; with application to the equatorial currents of the eastern Pacific. *Proc. Nat. Acad. Sci.*, **33**, 318–326.

Sverdrup, H. U., 1953. On conditions for the vernal blooming of phytoplankton. *J. Cons. Explor. Mer*, **18**, 287–295.

Sverdrup, H. U. and R. H. Fleming, 1941. The waters off the coast of Southern California, March to July 1937. *Bull. Scripps. Inst. Oceanog.*, **4**, 261–378.

Sverdrup, H. U., M. W. Johnson and R. H. Fleming, 1942. *The Oceans, their Physics, Chemistry and General Biology.* Prentice-Hall, New York, 1087 pp.

Tibby, R. B., 1941. The water masses off the west coast of North America. *J. Mar. Res.*, **4**, 112–121.

USSR Ministerstvo Oborony, 1953. *Morskoi Atlas.* Tom. II. *Fiziko-geograficheskii.* Moscow. Izdaniye glavnogo shtaba voenno-morskikh sil. 76 charts.

Walford, L. A., 1958. *Living Resources of the Sea.* Ronald Press, New York, xv + 321 pp.

Wattenberg, H., 1938. Die Verteilung des Sauerstoffs im Atlantischen Ozean. *"Meteor" Rep.*, **9**, 1–132.

Wooster, W. S., 1960. El Niño. *Calif. Coop. Ocean. Fish. Invest., Rep.*, **7**.

Wooster, W. S. and T. Cromwell, 1958. An oceanographic description of the eastern tropical Pacific. *Bull. Scripps Inst. Oceanog.*, **7**, 169–282.

Wooster, W. S. and M. Gilmartin, 1961. The Peru-Chile Undercurrent. *J. Mar. Res.*, **19** (3), 97–112.

Wooster, W. S. and F. Jennings, 1955. Exploratory oceanographic observations in the eastern tropical Pacific, January to March, 1953. *Calif. Fish Game*, **41**, 79–90.

Wooster, W. S. and H. H. Ketchum, 1957. Transport and dispersal of radioactive elements in the sea. In *The Effects of Atomic Radiation on Oceanography and Fisheries.* Washington, National Academy of Sciences—National Research Council Pub. No. 551. Ch. 4, pp. 43–51.

Worthington, L. V., 1954. Three detailed cross-sections of the Gulf Stream. *Tellus*, **6**, 116–123.

Wyrtki, K., in press. The upwelling in the region between Java and Australia during the south-east monsoon.

Yoshida, K., 1955. Coastal upwelling off the California coast. *Rec. Oceanog. Wks. Japan*, **2**, 13 pp.

Yoshida, K., 1958. Coastal upwelling, coastal currents, and their variations. *Rec. Oceanog. Wks. Japan*, Sp. No. 2, 85–87.

Yoshida, K., 1958a. A study on upwelling. *Rec. Oceanog. Wks. Japan*, **4**, 186–192; [also *Geophys. Notes*, **11**, No. 13].

Yoshida, K. and H. L. Mao, 1957. A theory of upwelling of large horizontal extent. *J. Mar. Res.*, **16**, 40–54.

Yoshida, K. and M. Tsuchiya, 1957. Northward flow in lower layers as an indicator of coastal upwelling. *Rec. Oceanog. Wks. Japan*, **4**, 14–22; [also *Geophys. Notes*, **11**, Contr. No. 17].

# 12. THE SOUTHERN OCEAN

## G. E. R. Deacon

Writers dealing with the circumpolar ring of ocean round the Antarctic continent soon feel the need for a comprehensive name to refer to the whole of it. Southern Ocean, Antarctic Ocean, Southern Seas and South Polar Seas are some of the names used. The U.S. Hydrographic Office insists that there is no need for a comprehensive name, but its recent *Oceanographic Atlas* seems to follow Admiral Wilkes in using South Polar Seas. British explorers and geographers have used Southern Ocean, notably the *Discovery* Investigations which has made most physical and biological observations there. It is not a name that needs to be defined rigidly, but one that can be used without much risk of ambiguity to refer to the ring of ocean. It is a remarkably uniform ocean: differences between one sector and another are small compared with differences between one latitude and another. A chronological list of expeditions that have worked there is given in the *Antarctic Pilot* issued by the British Hydrographic Department and in *Sailing Directions for Antarctica*—including the ocean and off-lying islands—compiled by the U.S. Hydrographic Office. New work is reported in journals such as the *Polar Record, I.G.Y. Annals*, and journals dealing with oceanography, geography and geophysics.

## 1. Topography

The distance across the Drake Passage between the southernmost tip of South America and the northernmost extremity of Antarctica in Graham Land (British) or Palmer Peninsula (U.S.) is less than 500 miles. Between the southern tip of Africa and the continent it is 2200, and south of Tasmania and New Zealand about 1400 miles. Most of the ocean is deeper than 2000 fathoms but only a small proportion more than 3000 fathoms. The greatest depth—in the South Sandwich Trench—is 4519 fathoms. There is a prominent obstruction across the circumpolar channel in the submarine ridge which bends east from Cape Horn to South Georgia, south through the South Sandwich Islands and west to the South Orkney Islands and Graham Land. Along most of this arc the ridge rises to within 1000 fathoms of the surface. There are other prominent north–south ridges between Kerguelen and Gaussberg, Macquarie Island and the Balleny Islands, and a gentler rise running northeast from the Balleny Islands right across the South Pacific Ocean. The mid-ocean ridges which run southwards in the Atlantic and Indian Oceans turn eastwards at about 50°S and enforce some degree of separation at the greatest depths between the basins near the continent and those round the northern fringe. The deepest channels from the Antarctic to the northern oceans are those along the west sides of the mid-ocean ridges.

The most recent bathymetric chart is No. 2592, Antarctica, issued by the U.S. Hydrographic Office in 1958. Others published by the International Hydrographic Bureau (1952–1955), Department of External Affairs, Australia (1956) and by Kosack (Geographisch-Kartographische Anstalt, Gotha, 1955)

[*MS received June, 1960*] 281

have been discussed by Herdman (1957). There is a useful relief chart in the
U.S. Hydrographic Office Publication No. 705, Oceanographic Atlas of the Polar
Seas, Part 1, Antarctic, 1957. Most world atlases give fairly detailed bathy-
metric maps, that of the Morskoi Atlas being outstanding.

## 2. Oceanographic Data

The physical and chemical observations likely to be found most useful are
contained in the report of the German Atlantic Expedition (Brennecke, 1921),
the *Meteor* Expedition (Wüst, 1932), the Norwegian Antarctic Expedition
(Mosby, 1934), *Discovery* Investigations (Station Lists, 1929 to 1957), French
Antarctic Expeditions (C.O.E.C. Bulletin, 1951, No. 10), British, Australian
and New Zealand Antarctic and Research Expedition (Howard, 1940), German
Antarctic Expedition 1938–1939 (Model, 1958), Norwegian *Brategg* Expedition
(Midttun and Natvig, 1957), Operation Deep Freeze II (U.S.H.O. Technical
Report 29), U.S. Navy I.G.Y. Programme (Lyman, 1958), Expedition of the
Soviet Research Ship *Ob* 1955–1956 (U.S.S.R. Academy of Sciences, 1958), and
Japanese expeditions in recent numbers of the *Oceanographical Magazine* and
*Records of Oceanic Works in Japan*.

## 3. Wind Zones

The almost circular outline of the continent and the continuous ring of water
favour the development of relatively simple wind and current systems. A low-
pressure belt extends round the continent in about 65°S, though it is a few
degrees farther north in the Atlantic Ocean and a few degrees farther south in
the Pacific Ocean. North of the low-pressure belt the prevailing wind is west
and the water moves east with some northward component. The average speed
of the eastward current judged mainly from old drift-bottle records (see
Krümmel, 1911, Bd. 2, 677) is about eight miles a day, which is 2 to 3% of
the mean wind speed. South of the low-pressure trough the winds are rather
more variable but the current appears to set mainly to the west following the
trend of the Antarctic coastline. The east wind will cause some southward as
well as eastward movement, and the low-pressure region between the west- and
east-wind drifts is a region of diverging flow in which deep water tends to up-
well so that there is more mixing between the deep and surface waters. There
are not enough current measurements to produce any kind of chart, but there
is enough evidence to show that the coastal winds and currents are disturbed
near prominent topographical features. There are northward currents on the
west sides of the Ross Sea and Weddell Sea, and smaller northward deflections
in other places, near the Kerguelen-Gaussberg ridge and near Peter I Island
for example.

## 4. The Water-Masses

Examination of temperature profiles and the distributions of temperature
and salinity in vertical sections across the west-wind drift shows a natural

division into two zones which can be reasonably described as antarctic and sub-
antarctic. Surface charts of mean temperature like those of Vowinckel (1957)
and U.S.H.O. (1957) based on relatively few observations collected over many
years show an inner zone in which the surface temperature increases slowly
towards the north and an outer zone where the increase is more rapid but still
more or less uniform. Ships fitted with continuously recording thermographs
obtain a sharper picture. The surface temperature increases gradually towards
the north, away from the continent, till it is about 1 to 2°C in winter or 3 to 4°C
in summer, and then rises approximately 2°C in a very short distance. It is
shown very clearly in the surface temperature charts which Mackintosh (1946)
has based on thermograph recordings. This sharp change of temperature was

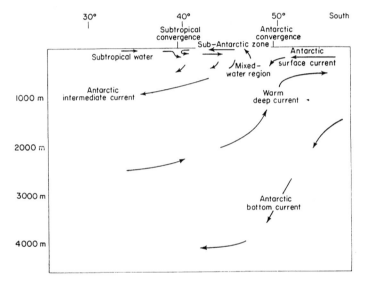

Fig. 1. The circulation of water in a vertical section from 30° to 60°S in the Atlantic Ocean.

first described by Meinardus (1923) in the meteorological report of the German
South Polar Expedition. He explained it as the line along which the ice-water
spreading northwards from the Antarctic sinks below the surface. He gave a
table showing its latitude in the Atlantic Ocean and the neighbouring parts of
the Pacific and Indian Oceans. It was next described by Schott (1926) who
called it the Meinardus Line. Wüst (1928) and Defant (1928) used the name
Oceanic Polar Front. The circumpolar voyage made by the *Discovery II* in 1932
gave a clear indication that it was continuous all round the continent, and in
subsequent publications, such as Deacon (1937) and Mackintosh (1946), the
name Antarctic Convergence has been used.

Fig. 1 shows the probable circulation of water in a vertical section across
the west-wind drift based on temperature and salinity observations in 30°W.
The deep current contains some water which has sunk from the surface in
the North Atlantic Ocean, in the boundary region between Atlantic and Arctic

currents. Water which sinks there, joined by some from the Mediterranean Sea, spreads southwards at depths of 1500 to 3000 m across the equator into high southern latitudes. Its course is more or less horizontal as far as about 50°S, where it slopes upwards above colder water formed in the Antarctic that spreads northwards along the bottom in the opposite direction. This Antarctic bottom water can be traced well into the Northern Hemisphere. Above the warm deep water at the southern end of the section there is a relatively shallow layer of poorly saline surface water. Except for a thin surface stratum that may be warmed to 2 or 3°C in summer, it is colder than the deep water, but it remains above it because of its low salinity. The action of the wind makes it move north as well as east, and its relatively high density compared with that of the warmer surface waters farther north must also make it drain slowly to the north above the more saline deep water. Temperature and salinity observations all round the continent show that it sinks rather abruptly below warmer surface water as soon as it passes the latitude where the deep layer climbs in the opposite direction.

## A. The Warm Deep Water

The presence of a warm layer between the cold and bottom layers all round the continent is evidence of a general southward transport in this layer, and isohalines in vertical sections running southwards from the Atlantic, Indian and Pacific Oceans indicate that it is supplied from the deep layers of these oceans. The source of highly saline water in the North Atlantic Ocean seems to play a predominant part. A layer of highly saline water can be seen extending southwards from the North Atlantic Ocean and Straits of Gibraltar, eastwards round the Southern Ocean and northwards in the Indian and Pacific Oceans, getting less saline and deeper all the way. There is another and apparently much smaller source of highly saline deep water in the Arabian Sea. Its high salinity, temperature and low oxygen content distinguish it clearly as far as 20°S ; south of this there is a second kind of highly saline water some 2°C colder and 2 cc/l. richer in dissolved oxygen which must be derived largely from the Atlantic Ocean.

In the Pacific Ocean the deep and bottom layers appear everywhere to be separated from highly saline surface or sub-surface water by a layer with relatively low salinity derived from surface waters that sink in the Antarctic and Arctic regions ; there seems to be no supply of highly saline water to the deep and bottom layers in addition to that derived from the eastward flow south of Australia and New Zealand. Further work may reveal some enrichment of the deep layer from a source within the complicated system of island ridges and basins north-west of New Zealand, but it seems rather unlikely.

The warm deep layer between the cold Antarctic surface and bottom waters is not, however, filled with water straight from the North Atlantic Ocean. Much of it must be formed by mixing between northward and southward movements in low and middle latitudes all round the circumpolar ocean and in the ocean itself, especially perhaps where the levels of the deep and bottom currents change abruptly in 40° to 50°S. In the Atlantic Ocean, Wüst (1936) shows that the percentage of North Atlantic water in the high-salinity layer decreases rapidly where the deep current turns eastwards into the Southern Ocean. In 40° to 50°S it is probably not more than 10%. We shall be far from a complete understanding of the general circulation of the Southern Ocean till we know more about the relative transports of the zonal and meridional movements and about the exchanges between the intermediate deep and bottom layers in the oceans outside. Till then we must think of a more or less circumpolar flow, probably lessened by the Drake Passage and the adjoining submarine arc, and fed by deep water from the oceans outside, especially from the Atlantic

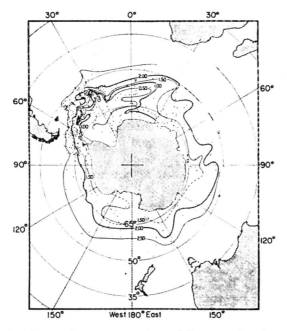

Fig. 2. Chart showing the maximum temperature of the warm deep layer in the Antarctic zone.

Ocean. In most of the circumpolar region of east winds, close to the continent, the slope of the isotherms and isohalines suggests that the deep water moves east in spite of the wind, but in the Atlantic sector and in the eastern approaches to the Ross Sea there is clear evidence that it flows west (see Fig. 2). In the Greenwich meridian the maximum temperature in the warm deep layer falls to less than 0.5°C between 58° and 63°S but rises again to more than 1°C in 64° to 66°S. The cold region coincides more or less with the trough of low atmospheric pressure and the divergence between the west- and east-wind drifts. The warm deep layer to the south must be supplied to a large extent from the west Indian Ocean sector. Its temperature and salinity fall as it flows round the south and west sides of the Weddell Sea, and, as is shown on page 287, it must eventually be the source of most of the Antarctic bottom water. North-east of the Ross Sea the divergence occurs in about 65° to 70°S, and the cyclonic movement is on a much smaller scale. More observations close to the continent may show similar eddying deep movements during periods of strong east winds.

## B. Antarctic Bottom Water

It is evident from meridional sections of temperature and salinity distribution that the cold water found at the bottom of much of the Atlantic, Indian and Pacific Oceans comes from the Antarctic, but it is apparently not formed all round the continent. It must be formed by the cooling of the warm deep water, which is the only general southward movement towards the continent, but the cooling is of a special kind. The bottom water is 1° or 2°C colder than the deep water, but only 0·05 parts per thousand less saline. The deep water is, therefore, cooled strongly, but diluted very little. As far as we know it is never exposed at the surface, and there are only very few places where the cold, less saline water always found above it becomes saline enough to cool it sufficiently by mixing without diluting it to a salinity less than that of the bottom water.

Using observations made some 30 years ago in the Weddell Sea by the *Deutsch-land*, *Norwegia* and *Discovery II*, Fofonoff (1956) has shown that the temperature and salinity of the much weakened deep current can be taken as 0.6°C and 34.70‰. Taking account of the changes in volume and density on mixing, he shows that water cooled to freezing point must have a salinity greater than 34.51‰ before it can form mixtures with the deep water that are heavier than the deep water itself. As soon as freezing water round the coast of Antarctica reaches such a high salinity, it will, therefore, be likely to form mixtures with the deep water that will be heavy enough to sink down the continental slope and feed the bottom layer. Fofonoff further showed that if the freezing water reaches a salinity as high as 34.63‰, it will already be heavier than any mixture it could form with the deep water and will sink directly down the continental slope if not confined in a depression on the shelf. Water with a salinity less than 34.51‰, even though it is cooled to its freezing point, cannot form mixtures heavier than the deep water and must float above it in the Antarctic surface layer.

Most of the Antarctic bottom water appears to be formed in the Weddell Sea. The surface water in the Antarctic coastal current is warmed and diluted in summer till its temperature is well above freezing point and its salinity less than 34‰. In winter it is cooled, and its salinity rises as ice separates out, drainage from the land ceases, snow lies unmelted above the sea-ice, the cold air over relatively warm water leads to increased evaporation, and vertical mixing is easier because the density gradients are weaker. Within a month after midsummer the salinity of the coastal water flowing westwards in about 70°S across the Greenwich meridian is already greater than 34‰ and rising rapidly. On the broad continental shelf farther south and west in the Weddell Sea it feels the full effect of winter and reaches salinities of 34.75‰ or more. There are no observations in the southernmost part of the sea in winter, but those made in summer (Brennecke, 1921; Lyman, 1958) reveal water with salinities as high as 34.60 to 34.75‰ and cooled to freezing point lying in depressions on the continental shelf. Surface observations at the northern end of Graham Land (Deacon and Marshall, 1954) indicate a similar winter maximum. There are no vertical series of temperature and salinity observations near the continental slope in winter among which we can look for evidence of cold, highly saline water sinking from the shelf, but there can be little doubt that it takes place on a large scale. It may also take place along the east coast of Graham Land, though the observations of the *Deutschland* as she drifted northwards through the middle of the sea during the winter of 1912 show a continuous though much weakened warm deep layer. The closest indication of complete convection from top to bottom observed so far is in the region between the northern tip of Graham Land and the South Orkney Islands (Deacon, 1937, p. 108).

From the Weddell Sea the temperature of the bottom water increases towards the east and north all round the continent and the increase is accompanied by a small increase of salinity and decrease in dissolved oxygen content. The ocean bottom temperatures off the Pacific side of the Graham Land peninsula are 1°C higher than those on the Atlantic side. Very cold, highly saline water is found on other parts of the Antarctic shelf, but as far as we know it is confined in depressions on the shelf and has little or no influence on the deep layers of the ocean outside. The regions where it has been found are the Ross Sea, where the salinity reaches at least 34.87‰ in winter (Deacon, 1939 and 1939a; Lyman, 1958), and Adélie Land, where summer observations show cold water with a salinity of 34.70‰ at the bottom of a depression on the shelf. Winter observations made by the *Gauss* on the shelf in 90°E (Drygalski, 1926) show that the salinity of the surface water was never as high as 34.51‰, but later observations by the *Ob* indicate that this salinity is exceeded a little farther east. More observations, especially in winter, are needed for a complete picture, but those available so far indicate that the Weddell Sea is by far the greatest source of Antarctic bottom water, and they suggest that its bottom water is an essential constituent of the mixtures which spread northwards in the Pacific and Indian Oceans as well as in the Atlantic Ocean. The temperature and salinity characteristics of the newly formed bottom water are not likely to vary since they are determined primarily by those of the deep water, but it is reasonable to expect that much more is produced in winter. This is a very interesting question to pursue. Stommel and Arons (1960) estimate that the flow from the Weddell Sea must be at least as great as that of the highly saline water from the North Atlantic Ocean. There is some evidence from the behaviour of sounding lines and dredges, from the movements of plankton and the hardness of the sea floor that the bottom current is unusually strong. Radiocarbon "age" determinations south of New Zealand have been interpreted as evidence of very slow movement, but there are many uncertainties. It is doubtful whether there is time for the carbon dioxide of the shelf water to get into equilibrium with that of the atmosphere, and we know that the shelf water mixes with a large proportion of deep water in the Weddell Sea and with more on its way round the continent. The result is water with a very complex history.

## C. Antarctic Surface Water

Conditions that produce water with a salinity less than 34.51‰ and mixtures light enough to float above the warm deep layer occur all round the continent. The southward movement in the warm deep layer is, as far as we know, the only flow towards the continent, and this, mixed with the surface water, cooled by radiation and exchanges with the atmosphere, and diluted by melting ice and snow and drainage from the land, produces cold, poorly saline water which lies in a shallow well-defined layer above the warm deep water. It has a depth of 150 to 250 m, deepest near the Antarctic convergence and near the continent, and shallowest in the divergence region between the west- and east-wind drifts. It is separated from the warm deep layer by a sharp discontinuity in which the temperature and salinity increase with depth and the dissolved oxygen content decreases. In winter it is almost uniform with depth, near its freezing point in the south, and only 1° or 2°C warmer in the north; its salinity is highest in winter. In summer it generally has a warmed and diluted surface stratum but its lower half is still for the most part well below 0°C and not warmed very much above its winter condition. The seasonal changes seem to be least in the divergence region, presumably because the formation of the warm dilute surface strata found farther south as well as farther north in summer is hindered by upwelling and some south and north transport at the surface. The great volume of poorly saline water that spreads northwards well beyond the equator in all oceans and can be fed only from the Antarctic surface currents is unmistakable evidence of their large overall flow away from the Antarctic. We have little information on which to base an estimate of the relative strengths of the circumpolar and northward transports, but it seems likely that most of the water in the surface layer spends several years in high latitudes before it escapes to the north.

## D. The Antarctic Convergence

The latitude where the convergence occurs is marked almost as clearly by changes in level of the deep and bottom currents as by the sinking of the Antarctic surface water below the warmer sub-Antarctic water. The isotherms and isohalines at great depths slope as steeply as those near the surface. Conditions in the ocean seem to be paralleled to some extent in the atmosphere, for, although the data are scanty, there are enough to indicate that the meridional atmospheric pressure gradient and the west wind are strongest in the same latitude. It might be argued that the atmospheric circulation and the wind determine the latitude of the frontal region in the ocean, but it can also be maintained that the latitude of the north-to-south gradients in the atmosphere is determined by some kind of balance between the deep and bottom layers of the ocean. The convergence plotted in Fig. 3 is not everywhere in the same latitude; it is north of 50°S in the Atlantic Ocean and south of 60°S in the eastern Pacific sector. There are very good reasons why the Antarctic bottom water should extend far north in great volume in the Atlantic Ocean since it is mainly formed there and the configuration of the land and activity of the meridional circulation in the Atlantic Ocean favour its advance to the north. There seems to be just as good reason why it should be much weaker in the east Pacific sector after travelling all round the continent, draining away to the north as it goes. Such factors may control the balance between the bottom and deep

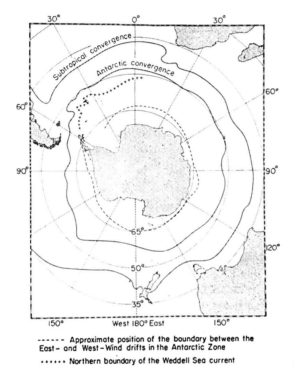

------ Approximate position of the boundary between the East- and West-Wind drifts in the Antarctic Zone
•••••• Northern boundary of the Weddell Sea current

Fig. 3. The Antarctic and subtropical convergences; the approximate position of the boundary between the west- and east-wind drifts, and the northern boundary of the cold current from the Weddell Sea.

currents, and determine the latitude of the frontal region in the ocean, the meridional gradient of temperature at the surface and the north-to-south gradient of mean atmospheric pressure.

The most remarkable feature of the Antarctic convergence is that it is practically stationary. There is some variation in reported positions but most of them seem to fall within 60 miles or so of a mean position. Eady (1951) has suggested that we should think of the mass and velocity distribution as being similar to that of an atmospheric front where unstable waves grow into cyclones and anticyclones and warm air flows round tongues of cold air which eventually descend and are replaced by warm air. The west wind should cause northward transport on the north side of the convergence as well as on the south side, but there is probably a balance between wind transport and density considerations which leads to turbulent southward progress in the sub-Antarctic water. There is some evidence of eddies 40 or 50 miles across. A ship steaming north or south across the convergence often finds a single transition from Antarctic to sub-Antarctic water, but if she is on a slanting course she often finds patches of warm and cold water. Wexler (1959), arguing from two detailed bathythermograph sections down to only 200 m, suggests that such cold patches are due to upwelling and regards them as evidence of divergence in the boundary region, but further bathythermograph sections discussed by Garner (1958) do not support this view, and simultaneous salinity and deep water observations would no doubt confirm that the observed irregularities are better regarded as minor variations introduced by local meteorological conditions into a picture basically controlled by deep-water movements. In some longitudes the convergence seems generally sharper than others and this seems to occur where a low bottom temperature and a steep slope of the deep isotherms indicate a strong flow of bottom water and deep sharp frontal conditions. The steadiness of the mean position of the convergence seems to be a reliable indication that it is not directly dependent on the surface winds.

### E. The Sub-Antarctic Zone

North of the Antarctic convergence there is a region of well-mixed water attributable to mixing processes associated with the sinking Antarctic water. The descent of this water can be traced by the weak temperature minimum it enforces below the sub-Antarctic water.[1] In 30°W, the meridian used in Fig. 1, the temperature minimum sinks below 500 m within 100 miles of the convergence. At first there is little or no decrease in salinity with depth, but as we go north there is soon evidence of a southward movement of more saline water in the sub-surface layer, and the underlying layer occupied by the descending Antarctic and mixed water is clearly distinguished by its low salinity. This salinity minimum can be traced into the Northern Hemisphere and the northward flow that it demonstrates has been called the Antarctic Intermediate

[1] It is here, close to the convergence, that Sir James Clark Ross plotted a circumpolar circle of uniform temperature below which he thought the temperature of the ocean was 4°C from top to bottom. Following Dumont Durville, he supposed that sea-water like fresh water had its maximum density at 4°C, and that the ocean depths were filled with such water of maximum density. They believed that the surface water farther south was cooler and lighter than water at 4°C because of excess of outgoing radiation and that farther north it was warmer and lighter because it was warmed by the sun. We know that their observations were faulty because their thermometers, not sufficiently protected against pressure, read 1°C too high for each 1000 m they were lowered, and their ideas were wrong because ocean water does not have a point of maximum density above its freezing point. But the surface temperature of 4°C and a correction of 1°C for each 1000 m produce much the same temperature structure as we find today in the latitudes quoted by Ross, and the agreement suggests that the convergence has not shifted in the past 100 years (Deacon, 1954).

Current. At the surface, the transport, still well within the region of prevailing west wind, must be north as well as east, but higher salinities in the sub-surface layer indicate southward flow. This is the first sign of a southward penetration of near-surface water into the Antarctic regime. It must contribute fairly largely to the mixed water north of the Antarctic convergence. It seems to be particularly strong in the Indian Ocean. There is generally a sharpening of the rise in surface temperature where the increase of salinity becomes noticeable.

### F. The Subtropical Convergence

Some 10 degrees north of the Antarctic convergence there is another sharp rise of surface temperature towards the north—from about 10° to 14°C in winter and 14° to 18°C in summer. There is a sharp rise of salinity from about 34.3‰ to 34.9‰. Details of the water circulation can only be inferred from the temperature and salinity distribution, but the sharp change seems to be caused by convergence of sub-Antarctic water moving northwards with sub-tropical water moving southwards. The convergence is farthest south on the western sides of the oceans, where warm water is carried southwards in the Brazil Current, the Agulhas Current and the East Australian Current. Its position, indicated in Fig. 2, is much more variable than that of the Antarctic convergence, especially on the eastern side of the oceans, and it probably depends largely on wind transport. It may be more appropriate to think of it as a fairly wide region over which sharp frontal conditions move. South of the Brazil, Agulhas and East Australian Currents there are generally isolated patches of eddies of warm water separated from the main currents. The German research vessel *Meteor* reported a very well defined crossing of the convergence in 41°S, 22°E: the boundary appeared as a line of current disturbance at the surface and the temperature rose 5.6°C in 1 mile and a total of 9.1°C in 5 miles.

### 5. Climatology and Zoogeography

D. D. John gave a graphic description of the significance of the Antarctic convergence in a lecture to the Royal Geographical Society in 1934. After the *Discovery II* had made observations all round the Antarctic continent, he said: "It is a physical boundary very easily and precisely detected with a thermometer by the sharp change in temperature as one passes from one zone to another. It can be detected as easily if not so precisely by a zoologist with a tow-net, because each of the two waters has a distinctive fauna of floating animal life. The zoologist need only know the species of prawns of the genus *Euphausia* to which *E. superba*, whale food, belongs. They are so numerous in the surface that his net will always catch some. If, in the neighbourhood of the convergence, he takes *E. vallentini* or *E. longirostris* he is in sub-Antarctic water. He will have crossed the convergence and be in the Antarctic when his net brings back *E. frigida* and not *vallentini* nor *longirostris*. But we, whether sailors or scientists, know and will remember the convergence best in another way: as the line to the north of which we felt one day, at the right season, after months in the Antarctic, genial air again and soft rain like English rain in the spring. I can remember a number of those days vividly. It was like passing at one step from winter into spring. In the southernmost lands in the sub-Antarctic, the islands about Cape Horn, the earth smells as earth should smell and as it never does in the Antarctic. It is, no doubt, the north-easterly course of the convergence between the longitudes of Cape Horn and South Georgia, so that the former is left far to the north and the latter to the south, that accounts for the vast difference in the climate of two islands which are in precisely the same latitude and only 1000 miles apart. The lower slopes of Staten Island are clothed with beech trees with so rich an undergrowth that it is difficult to push through. Darwin compared the richness of the region to that of a tropical forest. South Georgia, the other island, is a true Antarctic land. The snow-line of South Georgia is lower than the tree-line of Tierra del Fuego."

Many of the zoologists who have examined the *Discovery* collections have remarked on the significance of the convergence as a biological boundary. Among the floating and drifting plants and animals there are many species which are typical of either Antarctic or sub-Antarctic water, and sufficiently rare on the other side of the convergence to be regarded as intruders if found there. Others are common to both sides. Among the free-swimming animals the coastal fishes were shown by Tate-Regan (1914) to be best divided into Antarctic and sub-Antarctic species by the 6°C isotherm, which follows the Antarctic convergence very closely, and Norman (1938), using all the new material, based his classification on the position of the convergence, the result being almost the same. An example of its significance for bottom-living animals is afforded by the non-abyssal Polyzoa for whose distribution Hastings (1943) found the division into Antarctic and sub-Antarctic species to be the most conspicuous feature. The great majority were found in one region or the other but not in both. Murphy in his book on the oceanic birds of South America (1936) gives a list of fifteen birds typical of the Antarctic zone, twenty-nine of the sub-Antarctic zone and eleven common to both.

There is some interrelation between the position of the convergence and the boundaries of different marine sediments at the bottom of the ocean, and evidence provided by the stratification of sediments in certain areas has been used by W. Schott (1939) to demonstrate the probably greater extent of the Antarctic zone during the Glacial Period. Round the Antarctic continent there is a wide belt of glacial mud, but the proportion of this deposit decreases towards the north, and in the northern part of the Antarctic zone, where the growth of diatoms is particularly abundant, the sediment consists mainly of the siliceous skeletons of diatoms. Farther north, in the sub-Antarctic and subtropical regions, the most typical sediment is calcareous, mainly *Globigerina* ooze composed chiefly of the skeletons of pelagic Foraminifera. The chief exceptions are deep-water areas of red clay occurring where the calcareous material has been dissolved, and Wüst (1934) has shown that the extent of these areas is closely correlated with the northward movement of the Antarctic bottom current as well as the depth, the Antarctic bottom water being a very effective solvent for calcareous material.

The boundary between the diatom ooze and *Globigerina* ooze lies just north of the Antarctic convergence and approximately parallel to it, but in certain areas from which cores of the bottom sediments have been obtained, particularly in the neighbourhood of the Crozet Islands, diatom ooze typical of the Antarctic zone is found underlying the *Globigerina* ooze typical of warmer water. It is believed that the superposition of *Globigerina* ooze occurred during the retreat of the convergence from an advanced position held during the Glacial Period. If the convergence was farther north it should be argued that there must have been simultaneously a greater spreading of the Antarctic bottom current, and this conclusion is supported by the occurrence of *Globigerina* ooze superposed on red clay. In addition, the finding of diatom ooze superposed on glacial sediments in the northern part of the Antarctic zone points to a simultaneous decrease in the amount of glacial material carried into the sea from the Antarctic continent.

Like the Antarctic convergence, the subtropical convergence has been found to limit the distribution of some species of drifting plants and animals, and of some fishes and bottom-living animals. Certain species are typical of sub-Antarctic water and some of subtropical water. The two halves of the sub-Antarctic zone, the northern stratified water with a marked highly saline sub-surface layer, and the southern well mixed part, also appear to favour some different species, for example the Polychaeta (Tebble, 1960, p. 237).

The convergences are not boundaries but probably mark the normal limits of circulatory systems that are favourable to the breeding and growth of species that have become adapted to them. Most species, if not all, have circumpolar distributions, but there are variations in abundance from sector to sector. Marr (1956), dealing with *Euphausia superba*, the food of the Antarctic whales, emphasizes the effect of the cold current from the Weddell Sea in extending the distribution and increasing the abundance of the species in the Atlantic Ocean.

### 6. Nutrient Salts

The work of Clowes (1938) and others shows that the amounts of phosphate, nitrate and silicate in the Antarctic surface water rarely fall below the winter maxima of temperate regions, so that Hart (1934) attributes the falling off of phytoplankton after mid season to other factors. The surface water of the sub-Antarctic zone has less nutrients, but at a deeper level, between the northward movement in the Antarctic intermediate layer and the southward movement in the more saline deep water, phosphate and nitrate appear to be regenerated, possibly because of a large mortality of sinking phytoplankton as indicated by abundant deposits of diatom ooze below this region. Oxygen consumption appears to be highest in the same boundary layer between the sinking northward and rising southward flows. The decomposition enriches the south-going deep water and, in the Antarctic zone, the highest phosphate and nitrate concentrations are found in the warm deep layer. Silicate is most abundant in the bottom water, as though most of it were regenerated from the solution of the skeletons of diatoms near the bottom. There is generally a very sharp fall in silicate from south to north across the Antarctic convergence.

Further study of the patterns of phosphate and silicate distribution is likely to give useful indications about the history and circulation of the water-masses, but there appear to be large fluctuations with variations in growth and mortality, and many more observations are needed.

## 7. Future Needs

Kidson (1932), discussing the problems of Antarctic meteorology, said that the predominance of ocean- over land-surface in the Southern Hemisphere renders the mechanism of the general circulation simpler and more regular than in the Northern, and added that it should, therefore, be possible to arrive at sound generalizations from fewer data. This is undoubtedly true, but the theoretical treatment of the circulation nevertheless seems to present at least as much difficulty as the other oceans. If treated as a purely zonal current driven by the prevailing west wind, its velocity becomes much too large unless restricted by large arbitrary eddy viscosities which seem out of keeping with persistence of well defined deep and bottom layers, and warm and cold surface and sub-surface strata. Proudman (1952, p. 190) shows that this difficulty is removed as soon as we allow appreciable meridional movements in the different layers and take account of accelerations relative to the earth as they move towards higher and lower latitudes. Stommel (1958) also shows that better agreement with observations is obtained when we recognize that it is not a zonal current after all. Assuming restriction of the zonal movement by South America, Graham Land and the connecting submarine arc, he sets up theoretical models which face the difficulty of assuming a net meridional flow and produce a dynamic topography with striking resemblances to the observed pattern.

Current measurements at all depths are urgently needed, but many must be made, with as much guidance as possible from theoretical models, before much advance can be made. It is reasonable to expect considerable seasonal changes in transport: more Antarctic bottom water is likely to be formed in winter, and there is probably a greater northward flow of Antarctic surface water in summer when ice is melting. The southward flow in the deep layer must be affected by changes in both layers. It would be very useful if natural radio-active tracers, such as $^{14}C$, could be used, but their interpretation will be uncertain till there is a better background of physical measurements.

It is difficult to emphasize the need for new work without some risk of appearing to underestimate the value of what is being done. Some of the new oceanographical techniques, such as underwater photography, bottom corers, greatly improved echo-sounders and seismic methods, are being put to good use, but very few deep temperature and salinity observations and no current measurements are being made. It is amazing how little is done when account is taken of the large numbers of ships that have gone to the Antarctic in recent years. Though it lacks the fascination of adventure, the pressure of politics and prestige, and perhaps the animus of prospecting, the better understanding of the Southern Ocean and the oceanic and atmospheric circulation of neighbouring areas is, in one way and another, far more urgent than the problems of the continent.

# References

Brennecke, W., 1921. Die ozeanographischen Arbeiten der deutschen antarktischen Expedition 1911–12. *Arch. deut. Seewarte*, **39**, 1–216.

Clowes, A. J., 1938. Phosphate and silicate in the Southern Ocean. *Discovery Reps.*, **19**, 1–120.

Deacon, G. E. R., 1937. The hydrology of the Southern Ocean. *Discovery Reps.*, **15**, 1–124.

Deacon, G. E. R., 1939. The Antarctic voyages of R.R.S. *Discovery II* and R.R.S. *William Scoresby*, 1935–37. *Geog. J.*, **93**, 185–209.

Deacon, G. E. R., 1939a. The work of the *"Discovery"* Committee in the South Pacific Ocean. *Proc. Sixth Pacific Sci. Cong.*, **3**, 139–141.

Deacon, G. E. R., 1954. Exploration of the deep sea. *J. Inst. Navig.*, **7**, 165–174.

Deacon, G. E. R. and N. B. Marshall, 1954. Salinity of the Antarctic coastal current. *The Challenger Soc. Ann. Rep. and Abstracts of Papers*, **3**, 27–28.

Defant, A., 1928. Die systematische Erforschung des Weltmeeres. *Z. ges. Erdk. Jubiläums Sonderband*, 459–505.

Discovery Reports, Station Lists. *Discovery Reps.*, **1**, 1–140; **3**, 1–132; **4**, 1–232; **21**, 1–226; **22**, 1–196; **24**, 1–422; **25**, 143–280; **26**, 211–258; **28**, 299–398.

Drygalski, E. van, 1926. Ozean und Antarktis. *Deut. Südpol Exped.*, **7**, 391–602.

Eady, E. T., 1951. Circulation of water in the oceans. *Nature*, **167**, 543–545.

Fofonoff, N. P., 1956. Some properties of sea-water influencing the formation of Antarctic bottom water. *Deep-Sea Res.*, **4**, 32–35.

Garner, D. M., 1958. The Antarctic convergence south of New Zealand. *N.Z. J. Geophys.*, **1**, 577–589.

Hart, T. J., 1934. On the phytoplankton of the South-West Atlantic and the Bellingshausen Sea. *Discovery Reps.*, **8**, 1–268.

Hastings, A. B., 1943. Polyzoa (Bryozoa). *Discovery Reps.*, **22**, 301–510.

Herdman, H. F. P., 1957. Recent bathymetric charts and maps of the Southern Ocean and waters around Antarctica. *Deep-Sea Res.*, **4**, 130–137.

Howard, A., 1940. The programme and record of observations. *Brit. Austral. N.Z. Antarct. Res. Exped.*, **3**, Pt. 2, Sec. 2, 24–86.

John, D. D., 1934. The second Antarctic commission of the R.R.S. *Discovery II. Geog. J.*, **83**, 381–398.

Kidson, E., 1932. Some problems of modern meteorology. No. 8: Problems of Antarctic Meteorology. *Q.J. Roy. Met. Soc.*, **58**, 219–226.

Krümmel, O., 1911. *Handbuch der Ozeanographie*, Bd. 1 and 2. Engelhorns Nachf., Stuttgart.

Lyman, J., 1958. The U.S. Navy International Geophysical Year program in oceanography. *Intern. Hydrog. Rev.*, **35**, 111–126.

Mackintosh, N. A., 1946. The Antarctic convergence and the distribution of surface temperatures in Antarctic waters. *Discovery Reps.*, **23**, 177–212.

Marr, J. W. S., 1956. *Euphausia superba* and the Antarctic surface currents. *Norsk Hvalfangsttid.*, **45**, 127–134.

Meinardus, W., 1923. Die zonale Verteilung der Luft und Wassertemperatur. *Deut. Südpol Exped.*, **3**, 528–546.

Midttun, L. and J. Natvig, 1957. Pacific Antarctic water. *Sci. Res. Norweg. "Brategg" Exped.* 1947–48, No. 3, 1–130.

Model, F., 1958. Ein Beitrag zur regionalen Ozeanographie der Weddellsee. *Wiss. Ergebn. Deut. Antarkt. Exped., 1938–39*, **2**, 63–96.

Mosby, H., 1934. The waters of the Atlantic Antarctic Ocean. *Sci. Res. Norweg. Antarct. Exped. 1927–28*, **1**, No. 11, 1–131.

Murphy, R. C., 1936. *Oceanic Birds of South America*, **1** and **2**. American Museum of Natural History, New York.

Norman, J. R., 1938. Coast fishes. Part 3: The Antarctic zone. *Discovery Reps.*, **18**, 1–105.

Ob, 1958. Hydrological, hydrochemical, geological and biological studies. *Reps. Antarct. Exped. U.S.S.R.* 1955–56, Leningrad.

Proudman, J., 1952. *Dynamical Oceanography*. Methuen, London and New York.

Schott, G., 1926. *Geographie des atlantischen Ozeans*. Boysen, Hamburg.

Schott, W., 1939. Deep-sea sediments of the Indian Ocean. *Recent Marine Sediments*, 396–408. Amer. Assoc. Petrol. Geol., Tulsa, Okla. and London.

Stommel, H., 1958. A survey of ocean current theory. *Deep-Sea Res.*, **4**, 149–184.

Stommel, H. and A. B. Arons, 1960. On the abyssal circulation of the world ocean. 2: An idealized model of the circulation pattern and amplitude in oceanic basins. *Deep-Sea Res.*, **6**, 217–233.

Tate-Regan, C., 1914. Fishes, British Antarctic (*Terra Nova*) Exped., 1910. *Zool.*, **1**, 1–54.

Tebble, N., 1960. The distribution of pelagic polychaetes in the South Atlantic Ocean. *Discovery Reps.*, **30**, 161–300.

Vowinckel, E., 1957. Climate of the Antarctic ocean. *Met. Antarct.*, 91–108. Weather Bureau, Pretoria.

Wexler, H., 1959. The Antarctic convergence—or divergence. *The Atmosphere and Sea in Motion (Rossby Memorial Volume)*, 107–120. New York.

Wüst, G., 1928. Der Ursprung der atlantischen Tiefenwasser. *Z. ges. Erdk. Jubiläums Sonderband*, 506–534.

Wüst, G., 1932. Das ozeanographische Beobachtungsmaterial. Serienmessung. *Wiss. Ergebn. Deut. Atlant. Exped. 'Meteor', 1925–27*, **4**, 1–290.

Wüst, G., 1934. Anzeichen von Beziehungen zwischen Bodenstrom und Relief in der Tiefsee des indischen Ozeans. *Naturwissenschaften*, **22**, 241–244.

Wüst, G., 1936. Die Stratosphäre des atlantischen Ozeans. *Wiss. Ergebn. Deut. Atlant. Exped. 'Meteor', 1925–27*, **6**, 109–288.

# 13. DEEP-CURRENT MEASUREMENTS USING NEUTRALLY BUOYANT FLOATS

## G. H. Volkmann

## 1. Introduction

Our knowledge of the patterns of circulation in the deep oceans has come largely through tracing particular properties such as temperature, salinity and oxygen as they are modified with increasing distance from their source. Water of Antarctic origin can be traced to the area south of the Grand Banks in the Atlantic. The lack of a source of deep or bottom water in the North Pacific may be deduced. Wüst (1935) has used a more refined technique to trace quantitatively the spreading of Mediterranean water into the North Atlantic (and more recently, the ultimate source area of this water within the Mediterranean itself). This method remains the most powerful one for the deduction of the general features of circulation.

It is, however, not appropriate for the deduction of current velocities and transports since they will be dependent on the mixing rates along the path of the current, and the process of mixing in the ocean is, at best, incompletely understood.

With few exceptions the velocities and transports associated with the circulation in the ocean have been deduced from the geostrophic equation, in which the horizontal pressure gradients associated with a variable density field are balanced by the Coriolis force.

The most satisfactory method of measuring currents is to observe them directly. However, this is a time-consuming process and, until recently, devices for such measurements have been somewhat marginal. Pillsbury's measurements in the Straits of Florida are the outstanding example of this technique and also give a fair indication of the difficulties involved.

In 1955 J. C. Swallow of the National Institute of Oceanography developed an instrument for the measurement of deep ocean currents. Coincident with this has been a renewed interest in the deep circulation of the oceans. Swallow and others are applying this device to the problem, and a picture of the motions at depth is slowly emerging.

## 2. The Pinger

Swallow's instrument, called the pinger, consists of a small acoustic source enclosed in a pressure case which can be set to float with neutral buoyancy at a predetermined depth without attachment to surface or bottom. Being naturally buoyant at that depth it will move more or less with the current at that depth and a series of fixes made by taking bearings on its acoustic source will determine its track.

The problem may be divided into three parts: (1) the construction and calibration of the pinger itself, (2) the design of a properly matched listening system, and (3) navigation.

[*MS received November, 1960*]

The pingers in use to date are essentially identical to those first developed and described by Swallow (1955). They consist of a relaxation oscillator timed to discharge a relatively high energy capacitor through a cold cathode trigger tube into a 10-kc magneto-strictive toroid. A pulse length of about 2 msec duration and a pulse rate of a ping about every 2–3 sec has been found optimum for this system. While a longer pulse would permit narrower filtering and increased signal-to-noise ratio, the present method of measuring the time delay visually in the two-hydrophone listening system and power considerations made the 2 msec pulse a good compromise.

The instrument case is made of extruded aluminum tube closed by "O" ring sealed-end caps through one of which are led the wires for the external magneto-

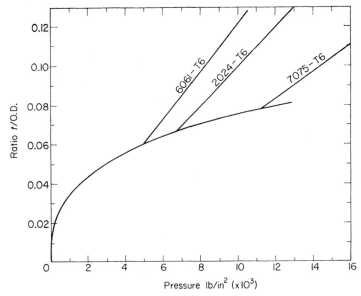

Fig. 1. Collapse pressure for cylinders of various aluminum alloys against the ratio of wall thickness to outside diameter. The curved section represents buckling of the cylinder, the straight sections represent metallic yield. (By courtesy of L. D. Hoadley.)

strictive sound source. Aluminum, while not unique, has two properties essential to the operation of the pinger. First, it is light enough to carry a pay load to deep depths without collapsing. The figure (Fig. 1) shows the collapse pressure for tubes of various alloys against the ratio of wall thickness to outside diameter. The bigger this ratio, of course, the less the flotation per foot, and for aluminum a ratio of about 0.106 gives no flotation. Secondly, the aluminum tube is less compressible than water and if made slightly heavy at the surface will gain buoyancy as it sinks. By adjusting the density of the pinger before release it can be made neutrally buoyant at any desired depth. The density of sea-water is well known and the change in volume of the aluminum cylinder, owing to temperature and pressure changes, can be computed from standard engineering formulae. While aluminum is not the only material with these two properties, it is the easiest to machine and to obtain.

In the present pinger circuits, no back discharge can occur through the gas trigger tube and the length of the pulse is governed by the mechanical resonance of the ring itself. The output is in the vicinity of 80 dB/μb at 1 m. Higher efficiency and longer pulses may be obtained from tuned oscillator circuits.

Equally important in increasing listening range is a properly designed listening system. While it would appear that a filter sharply tuned to the pinger frequency would be best, the relatively short (1–2 msec) pulse from the pinger places an upper limit of about 1000 c/s on the band width. Another important feature of the system is that it has strong rejection in the lower frequencies, since it is here that most of the ship noises occur. The receiving hydrophones should have a sensitivity of about −75 dB referred to 1 μb/V in order to get down to ambient noise on a quiet day. Ranges of 3 to 4 miles are standard with the present system and ranges of 6 and 7 miles have been obtained on quiet days. For the shallower pingers (less than 700 m), near-surface refraction of sound by sea-water can limit the range to an abrupt 2 miles or less.

Positions of the pingers are obtained by taking a series of three or more cross-bearings. Under most conditions the bearing accuracy is about ± 3 degrees. This means that the accuracy of the fixes is limited by the accuracy of the navigation. In limited areas, Loran and Decca are available. In other areas anchored buoys and bottom topography have been successfully used.

At great depths the gradient of the density of sea-water is almost solely a function of the pressure and the pinger may be expected to follow an isobar fairly closely. By fortuitous circumstances the temperature coefficients of water and aluminum are quite similar so that at shallower depths, where the properties of the water are changed more rapidly by mixing, the pinger may be expected to follow some isopleth which is a function largely of pressure and salinity. Since the motion of the pinger will be affected by the turbulent mixing motions themselves, exact solution of its path in terms of the changing properties should not be anticipated.

Other devices using similar principles can be imagined. For example, an enclosed volume of fresh water will have the same coefficients of temperature and pressure and could be ballasted to follow a given isohaline.

While to date pingers have been used only for measuring currents, the same principle can be used to float a variety of sensing elements. The measurements could either be telemetered to the surface acoustically or a release mechanism could be used to bring internally recorded data to the surface.

### 3. Measurements

A number of measurements with pingers have been reported in the literature and they suggest that deep currents are appreciably faster than had heretofore been suspected. It is, however, measurements just completed in the spring and summer of 1960 that form the main body of deep-current measurements and will provide us with new concepts of deep currents. These measurements have not yet appeared in the literature and, indeed, have not yet been fully analyzed. It has become fashionable in this, the day of rapid scientific advancement, to assure the reader that our ideas are in a state of flux, and that is the case with deep circulation in the oceans.

The first measurements (Swallow, 1955) were made off Spain in June, 1955, when two floats were followed successfully. A lunar semidiurnal tidal oscillation was detected in the movement of the floats. It was at a depth of $600 \pm 200$ m in a water depth of 5330 m.

In 1957, Swallow reports two more sets of measurements, one group again off Spain, the other in the Faroe–Shetland Channel and Norwegian Sea. One of the floats off Spain was very near the core of the Mediterranean outflow but showed a northerly drift. The others in this area were quite variable, apparently the result of the nearby bottom topography. In the Norwegian Sea an attempt was made to correlate the measured drift with gradient currents computed from a hydrographic section made at the same time. While the currents agreed in direction, subsequent measurements made further north in the Norwegian Sea were quite variable and the lack of agreement between the measured and computed velocities in the channel was not surprising.

Swallow and Hamon (1960) report another series of measurements off Spain, this time over the flat abyssal plain. They found that the currents did not decrease uniformly with depth although there was moderately good agreement between pingers at different depths and geostrophic shear in at least one instance. They also report considerable variability in the measurements with one float of particular interest which was followed for 48 days. It both speeded up and changed direction markedly during its trajectory.

In the Pacific, Knauss has reported measurements with pingers in both the Equatorial Countercurrent (Knauss and Pepin, 1959) and the Cromwell Current (Knauss, 1960). In both cases, propeller-type current meters were used for the near-surface measurements while pingers were used at the deeper depths. In the case where the two measurements overlapped in depth the agreement was good.

On the basis of theoretical consideration of thermohaline circulations, H. Stommel suggested that a southward-flowing deep current existed along the western boundaries of the oceans. [These suggestions were reported later by Stommel (1958) and stated more formally by Stommel and Arons (1960, 1960a).] In 1957 a joint cruise was undertaken by the National Institute of Oceanography with *Discovery II* and by Woods Hole Oceanographic Institution with *Atlantis* to an area off Charleston along the coast of the United States. Swallow aboard the *Discovery II* made the measurements at places determined from the deep temperature field as measured by Worthington aboard *Atlantis* (Swallow and Worthington, 1961). The current at the deep depths was indeed flowing south and, by using the measured velocities to adjust the reference level, a net transport to the south was computed for the region below about 1500 m. As Swallow and Worthington point out, such a transport can greatly change the total volume transport of the Gulf Stream by moving the reference level to mid-depths. Another important discovery during this cruise was that the velocity can increase with depth in the deep water and that water as deep as 2800 m could be moving at a velocity as high as 18 cm/sec.

Subsequent measurements made by Volkmann (in press) in the summer of 1959 in the area south of Woods Hole but north of the Gulf Stream showed a similar current (moving westward here) although it was much more widespread. In fact all the water below 1500 m across the 200 km section was moving in the same direction at 10 to 20 cm/sec and the transport was much larger than the transport reported by Swallow and Worthington off Charleston.

The following year separate measurements made by Barrett (in litt.) and by the author showed that all the deep water offshore was moving in the opposite direction, toward the east at velocities comparable to the westward velocities of the previous year. These opposing water motions occurred in the same area in the same water-mass and in the presence of similar slopes of the isotherms.

A long series of pinger measurements in the vicinity of Bermuda has just been completed by J. C. Swallow (in litt.). These also show an extreme variability of current speed and direction on a time scale of as little as two weeks and over a separation of as little as 20 km. While analysis is far from complete, it seems probable that a statistical analysis will show the significant features of the deep-water offshore circulation. The measured velocities were seldom less than 3 cm/sec and ranged up to 16–18 cm/sec and occasionally even higher.[1]

It would appear from these measurements that the deep circulation consists of a wide spectrum of motions containing some components with velocities at least an order of magnitude higher than the mean velocities. The significance of a single measurement in relation to the general circulation is now difficult to assess, although shorter series can be useful in specific problems and can, as they have in the Atlantic, serve as pilot studies. For the more general studies of oceanic circulation, it would seem that long-term continuous measurements are required.

## References

Knauss, J. A., 1960. Measurements of the Cromwell Current. *Deep-Sea Res.*, **6**, 265–286.

Knauss, J. A. and R. O. Pepin, 1959. Measurements of the Pacific Equatorial Countercurrent. *Nature*, **183**, 380.

Stommel, H., 1958. The abyssal circulation. *Deep-Sea Res.*, **5**, 80–82.

Stommel, H. and A. B. Arons, 1960. On the abyssal circulation of the world ocean—I. Stationary planetary flow patterns on a sphere. *Deep-Sea Res.*, **6**, 140–154.

Stommel, H. and A. B. Arons, 1960a. On the abyssal circulation of the world ocean—II. An idealized model of the circulation pattern and amplitude in oceanic basins. *Deep-Sea Res.*, **6**, 217–233.

Swallow, J. C., 1955. A neutral-buoyancy float for measuring deep currents. *Deep-Sea Res.*, **3**, 74–81.

[1] One 4000-m pinger travelled 60 km in 40 hours—a velocity of 42 cm/sec.

Swallow, J. C., 1957. Some further deep current measurements using neutrally-buoyant floats. *Deep-Sea Res.*, **4**, 93–104.

Swallow, J. C. and B. V. Hamon, 1960. Some measurements of deep currents in the eastern North Atlantic. *Deep-Sea Res.*, **6**, 155–168.

Swallow, J. C. and L. V. Worthington, 1961. An observation of a deep counter current in the Western North Atlantic, *Deep-Sea Res.*, **8**, 1–19.

Volkmann, G. H., in press. Deep current observations in the Western North Atlantic.

Wüst, G., 1935. Die Stratosphäre. *Wiss. Ergebn. Deut. Atlant. Exped. 'Meteor', 1925–1927*, Bd. 6, 1 Teil, 288 pp.

[Contribution No. 1316 from the Woods Hole Oceanographic Institution.]

# 14. DROGUES AND NEUTRAL-BUOYANT FLOATS

## J. A. KNAUSS

Velocity measurements in the open ocean have usually been made by one of two methods. Either the flow of water past a fixed point is measured by some device such as a propeller-driven current meter, or else the distance a water "parcel" moves in a given time is observed. The usual method of making this latter observation is to place something (such as a drogue or float) in the water and to follow it. For some applications this method of measuring velocity has two very distinct advantages over the current-meter method. The first is that the mean velocity can be measured very accurately. The error in the velocity observation is determined by the uncertainties, $\Delta_1$ and $\Delta_2$, of the positions of the drogue at the times of launching and recovery, respectively;

$$v = (d \pm \Delta_1 \pm \Delta_2)/t, \tag{1}$$

where $v$, $d$, and $t$ are velocity, distance and time respectively. Since navigation is a very real problem in the open ocean, $\Delta_1$ and $\Delta_2$ are usually not negligible;

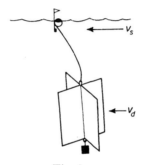

Fig. 1.

however, they are usually independent of both $d$ and $t$ and consequently the error in mean velocity can be made very small by following the drogue for a long enough period of time. The second advantage of this method is that the drogue is a "self-integrating" device. By following a drogue over several tidal cycles, it is possible to eliminate the effect of the tidal currents on the mean velocity. Similarly, the effect of "small-scale turbulence" superimposed on the mean flow is eliminated, by measuring velocity with a drogue. The main disadvantage of this method is that it takes a long time to make a single observation.[1]

A well designed drogue system is one in which the drogue moves at nearly the same velocity as the water at that depth. For a drogue designed to measure surface currents this is usually not a serious problem. It is a very real problem

---

[1] There is no reason why a pressure-indicating device cannot be attached to the drogue to measure its depth; however, so far as I am aware, this has never been done, and the lack of accurate information on the depth of the drogue must be considered a disadvantage.

[*MS received January, 1961*]    303

in the case of a drogue designed to measure subsurface currents as the following calculation indicates (Fig. 1). The drag force equation is

$$F = \tfrac{1}{2}C\rho Av^2, \qquad (2)$$

where $C$ is a non-dimensional drag coefficient whose approximate value is unity, $\rho$ is the density of the water, $A$ the cross-sectional area, and $v$ the velocity. Since there must be a balance of horizontal forces, the drag on the drogue must be balanced by the drag on the surface float and other appendages:

$$\tfrac{1}{2}C_s\rho A_s(v_s - v)^2 = \tfrac{1}{2}C_d\rho A_d(v - v_d)^2,$$

where the subscripts $s$ and $d$ refer to the surface float and drogue respectively, and $v$ is the velocity of the drogue system. Thus

$$v - v_d = (v_s - v)\left(\frac{C_sA_s}{C_dA_d}\right)^{\!\tfrac{1}{2}}. \qquad (3)$$

The criterion for a well designed drogue is that $v - v_d$ is made as small as possible, and this can only be done by making $(C_dA_d)/(C_sA_s)$ as large as possible; and, since this is a square-law relationship, a factor of a hundred in the ratio of cross-sectional areas results only in a factor of ten in the ratio of velocity differences. At best, drogues can give only approximate values of subsurface currents, although, with careful design and care in applying corrections, it is believed that subsurface currents can be measured to an accuracy of 5–10% of the surface current.

Drogues were used to measure subsurface currents in the Gulf Stream nearly a hundred years ago (Mitchell, 1867, 1868). Mitchell was also apparently the first man to anchor a buoy in deep water for the purpose of measuring currents. "In this last cruise we were able to anchor buoys and thus measure the current with accuracy. This can be done anywhere, and our charts ought to contain this kind of data. Our way of working was this: we first anchored a buoy, then having reeled the log-line within a free float, fastened the end of this line to the buoy, and as it reeled out, counted seconds. By this maneuver we avoided securing the boat to the buoy, which might have caused the anchor to drag" (Mitchell, 1868).

Drogues were used in 1871 to measure the subsurface outflow across the Straits of Gibraltar by W. B. Carpenter and Capt. G. S. Nares, the latter of whom served as captain on the first half of the *Challenger* Expedition (Carpenter, 1872). Eighty years later drogues were used to discover the Pacific Equatorial Undercurrent (the Cromwell Current) (Cromwell, Montgomery and Stroup, 1954). More recently drogues made from aviators' parachutes have been used successfully (Jennings and Schwartzlose, 1960).

The ideal drogue is one in which there are no surface appendages, and, therefore, there will be nothing to keep the drogue from moving at exactly the speed of the water in which it rests. This ideal was achieved in 1955 with the first successful use of neutral-buoyant subsurface floats by Swallow (1955) (see Chapter 13).

# References

Carpenter, W. B., 1872. Report on scientific researches carried on during the months of August, September and October, 1871, in H.M. surveying ship *Shearwater*. *Proc. Roy. Soc. London*, **20**, 535–644.

Cromwell, T., R. B. Montgomery and E. D. Stroup, 1954. Equatorial Undercurrent in Pacific Ocean revealed by new methods. *Science*, **119**, 648–699.

Jennings, F. D. and R. A. Schwartzlose, 1960. Measurements of the California Current in March, 1958. *Deep-Sea Res.*, **7**, 42–47.

Mitchell, H., 1867. Soundings in the Gulf Stream between Key West, Florida and Havana, Cuba. Report of the Superintendent of the United States Coast Survey, Appendix 15, 176–179.

Mitchell, H., 1868. Notes on Gulf Stream observations. Report of the Superintendent of the United States Coast Survey, Appendix 11, 166–167.

Swallow, J. C., 1955. A neutral-buoyant float for measuring deep currents. *Deep-Sea Res.*, **3**, 74–81.

# 15. ESTUARIES

## W. M. CAMERON and D. W. PRITCHARD

## 1. Definition of Estuaries

An estuary is a semi-enclosed coastal body of water having a free connection with the open sea and within which the sea-water is measurably diluted with fresh water deriving from land drainage.

Traditionally the term "estuary" has been applied to the lower reaches of a river into which sea-water intrudes and mixes with the fresh water draining seaward from the land. The term has been extended to include bays, inlets, gulfs and sounds into which several rivers empty and in which the mixing of fresh and salt water occurs.

Attempts have been made to enlarge the definition even further to include those zones along an open coast in which the salinity is significantly lower than in the open ocean. This extension allows the term "estuary" to include extensive regions off large rivers where the influence of the fresh-water drainage from the land can be clearly recognized. However, such a broad definition does not lead to a practical limitation of the term and we prefer our more restricted definition as expressed in the opening paragraph.

We shall also refrain from including as estuaries those embayments into which there is a negligible river discharge or in which evaporation exceeds the amount of fresh water draining into the system. Thus we exclude from consideration salt-water lagoons where the salinity exceeds that of the adjacent sea; we shall avoid the terms "inverse" or "negative" estuary as a suitable designation for stratified bodies of water in which the haline circulation is opposite in sense to that normally prevailing in estuaries. Thus, although we recognize that the layered inflow and outflow through the Straits of Gibraltar are controlled by factors analogous to those obtaining in typical estuaries, we do not consider anything to be gained by labeling the Mediterranean Sea a "negative estuary".

On the other hand, we have included in this chapter a discussion of the circulation patterns which occur in tributary embayments to estuaries, which in regard to the increment of fresh water added within the embayment would not be classified by our definition as estuaries, but in which the circulation is definitely related to the estuarine character of the adjacent water body.

It will be evident that many of the concepts developed here for estuaries will apply, with certain appropriate modifications, to non-estuarine coastal embayments and even to segments of the coast and of the open ocean.

## 2. General Considerations

The essential characteristics of an estuary result from the passage of river water through the system to the open ocean. It is in the estuary that the fresh water from the land meets the salt water from the sea, where mixing of salt

[MS received July, 1960]     306

and fresh water occurs, and where dynamic conditions are established which bring about the eventual discharge of the river water to the sea.

An important feature of estuaries is that although marked variations in the fresh-water inflow do occur, the system tends to be buffered against equivalent variations in the accumulation of fresh water in the estuary. Thus the changes in the total salt concentration are not in direct proportion to changes in the rate of fresh-water inflow. While an increase in river discharge will lead to a transient downstream displacement of the surface isohalines and some freshening at all depths, there is a counter response resulting in an increased inflow of salt water from the sea in the bottom layers, with a marked increase in vertical stratification.

Thus we are led to the concept of a quasi-steady dynamic balance tending to prevail in the system. On the one hand there is the continual discharge of river water into the estuary; at the same time, the salt water presses inward from the sea. These two influences interact, and the characteristics of the estuary reflect the balance of forces associated with each.

Early studies of estuaries have tended to emphasize the distribution of properties within an estuary. This emphasis reflected both the interests and specialties of the observers. Biological studies attempted to correlate the distribution of sessile organisms with the average distribution of salinity, of dissolved oxygen or of nutrient salts. The sampling of these scalar properties was a traditional and practical approach, while the measurement of velocities received much less attention, owing in part to technical difficulties.

The more recent appreciation of estuaries as dynamic systems has been associated with increased emphasis on the movement of waters within those systems. Studies of the circulation in estuaries and of the degree and extent of turbulent mixing have received appropriate attention. It is from this viewpoint that we propose to discuss estuarine oceanography.

## 3. Estuarine Circulation Patterns

Pritchard (1955) has described the various circulation patterns that are observed in estuaries, ranging from those occurring under highly stratified conditions to the patterns present in well-mixed vertically homogeneous types. In his treatment he has emphasized the kinematic processes which control the distribution of salt in the estuary, indicating the conditions under which advection and turbulent diffusion are important.

In this discussion we shall follow Pritchard's development of the estuarine sequences, emphasizing present understanding of the dynamic processes that are in control of the circulations. In what immediately follows we shall avoid formal reference to the pertinent equations of motion, deferring this consideration until the essential concepts of estuarine dynamics have been presented.

*A. The Highly Stratified Estuary*

### a. The frictionless model

We shall first consider an idealized coastal-plain estuary consisting of an elongated embayment in the coastline, into the upper end of which a single river source of fresh water empties. There is free connection with the sea at the lower end.

We shall assume the absence of tidal movements and that no significant mixing or friction is present. Under these conditions the sea-water would extend into the embayment along the river bottom to a position where the river surface is approximately at sea-level. The river water would flow out over the salt-water layer unimpeded by any frictional drag along its lower boundary. After entering the estuary, the fresh water would tend to hug the right shore of the embayment (looking toward the sea, in the Northern Hemisphere) under the influence of the Coriolis force. The interface between the two layers would be horizontal in the direction of flow but inclined downward to the right perpendicular to the flow, providing a transverse pressure force in the fresh upper layer and level pressure surfaces in the salt-water layer. The slope of the upper surface of the fresh water would be that governed by its geostrophic flow.

The distribution of salt and fresh water in such an idealized estuary would thus be compatible with the simplified dynamic conditions we have imposed. The circulation would consist merely of a static salt-water body and a seaward-moving stream of fresh water whose volume transport would be equivalent to that of the river discharge.

But such a situation does not occur in natural estuaries. Some mixing of the two types of water takes place. It is due to the variation in mixing agencies and of their vigor that variations in estuarine circulations are established.

### b. The non-tidal model

We shall next consider an estuary in which no tidal movement occurs, but where frictional forces are no longer negligible. This situation can be visualized by assuming an increased rate of river flow such that interfacial waves between the fresh surface water and the underlying salt water become unstable and break. This is essentially a one-way process which leads to an entrapment of some of the salt water in the upper fresh-water layer. The salt content of the upper water is increased as it moves toward the sea and the seaward volume transport is increased by the amount of sea-water entrained. There must be a slow compensating flow of sea-water toward the river to maintain the continuity of volume within the estuary.

It is to be expected that owing to the entrapment of salt water from the underlying layer into the surface water, the upstream margin of the salt water would retreat seaward due to this process alone. However, the retreat is also associated with the disturbed force distribution which results from the process of entrapment. In effect, not only is there a flux of volume into the fresh-water layer but also a flux of momentum, which acts on the upper layer as a force directed upstream. If a steady state is to be maintained this force must be balanced by an equal pressure force directed downstream. Such a force is established by a redistribution of density in the vicinity of the interface. This is facilitated by a tilt of the interface and an associated slope of the upper surface of the fresh water seaward. As the flow of the river increases, so does the frictional drag at the interface, with a consequently greater tilt to both the upper surface and the interfacial boundary, and a retreat of the upstream margin of the salt water further seaward. Keulegan (1949) has described such a process in his study of stratified flows in flumes.

It should be noted that the movement of sea-water toward the land is assumed to be relatively slow with a negligible frictional drag along the bottom of the estuary. Under this assumption the vertical gradient of the slopes of the pressure surfaces within the salt-water layer would be extremely small. Very little variation of density distribution within the salt-water wedge would occur.

In his original discussion of such a system, Pritchard (1955) pointed out that the two dominant processes in maintaining the salt balance are the horizontal and vertical advections of salt. We have emphasized the influence of the flux of momentum and the consequent compensating adjustment of the density field. It is clear that the maintenance of a steady state both of motion and of salt demands a nice adjustment of both these processes. However, it would seem that the character of the salt-wedge estuary is determined largely by the type of salt flux that prevails; that is, flux resulting primarily from advection across the interface boundary.

The salt-wedge estuary occurs when the ratio of river flow to tidal flow is relatively large, and when the ratio of width to depth is relatively small. The mouth of the Mississippi River shows the essential features of such an estuary.

## B. The Moderately Stratified Estuary

The next type of estuary occurs when the prime source of mixing is no longer the velocity shear along the interface, but rather the turbulence resulting from tidal motion.

If we assume a tide of moderate amplitude impressed on the estuary described above, the tidal velocities will give rise to random movements of water throughout the whole water column. Instead of the mixing being primarily due to an upward advection of salt water across the interface, the turbulent eddies tend not only to mix the salt water upward but also to mix fresh water from the upper layer down into the salt water below. This results in the content of salt increasing toward the sea in both the surface layers and the deeper layers. However, a vertical salinity gradient remains, with the deeper layers having a higher salt concentration than the surface layers.

This type of mixing adds a greater volume of salt water to the upper layer than in the case of a salt-wedge estuary. Consequently, flow in the upper layer is relatively much larger than the corresponding flow in the salt-wedge estuary, and compensating flow in the deep water is also greater.

The broader excursions of the turbulent elements resulting from the more vigorous activity cause a more extensive flux not only of salt but also of momentum. Instead of being localized at or near the interface, the eddy stresses will tend to extend throughout the layers from the bottom to the surface. A more complicated distribution of pressure gradients must be established if the estuary is to maintain a steady state from tidal cycle to tidal cycle. Relatively horizontal pressure surfaces in the deeper water were associated with the salt-wedge estuary. Here, in contrast, pressure surfaces in the deeper layer must slope downward toward the head to overcome the frictional effect of turbulent motion on the compensating headward flow.

In the upper layer, pressure surfaces must slope downward toward the sea. This variation with depth in the inclination of the pressure surfaces is brought about by the type of mass distribution resulting from the more intensive mixing.

It is commonly considered that the increased circulation that occurs in estuaries of moderate tidal mixing is powered by the increased potential energy of the system which results from the vertical mixing of fresh and salt water. What we are suggesting here is rather that the mixing process of the tide leads to the establishment and maintenance of stronger horizontal gradients of density within the estuary than would occur without tidal mixing. These horizontal gradients of density in turn permit horizontal pressure gradients of magnitude and distribution sufficient to maintain the relatively higher velocities in the presence of increased eddy frictional force. In other words, it is tidal mixing that promotes and permits the required distribution of potential energy within the estuary.

In the preceding paragraphs we have referred to the mean slope of the pressure surfaces over several tidal cycles. Superimposed on these mean slopes there will occur oscillating variations in slope associated with the flood and ebb of the tide. It is generally assumed that the variations in slope due to the tide are symmetrically distributed about the mean and that they are suppressed in the averaging process. To date it has also been assumed that the time variation in the tilt of the pressure surfaces is independent of depth and reflects the variation in slope of the water surface of the estuary. This assumption has permitted a concentration of attention upon the vertical gradient of slopes within the estuarine water and its association with the dynamics of the "mean" or "net" motion.

Recently, the averaging process has received more critical attention. There has been an increased appreciation of the nature of the non-linear inertia terms in the equation of tidal motion. Inasmuch as they involve products of the velocities, they do not average out over a tidal cycle. Thus, Pritchard (1956) has examined the effect of the non-linear term resulting from a variation in tidal excursion with distance along the estuary. Stewart (1957) has called attention to the effect of the curvature of the tidal channel in giving rise to a cross-channel inertial force of significance in the lateral equation of mean motion.

If these effects of the tidal oscillations on the mean pressure distribution in the estuary are independent of depth, they will be equally significant from the top to the bottom and will be detected only when the absolute slopes of the pressure surfaces are examined. However, if there is a large variation in tidal oscillations with depth, the mean effect of the non-linear terms might be expected also to vary with depth. On the average, therefore, a varying fraction of the mean pressure gradients might be required to balance these non-linear effects. Thus it is possible to conceive of a portion at least of the spatial variation of salinity within the estuary being governed by the inertial elements of the tidal motion, in addition to the mixing resulting from the motion of the tide itself.

Preliminary examination of the possible magnitude of the inertial effect described above suggests that it is probably of secondary importance compared to the effect of the flux of momentum associated with turbulence.

In our consideration of the moderately mixed estuary, we have emphasized the forces that act mainly in the direction of flow. As in the case of the salt-wedge estuary, the effect of Coriolis force will be evident laterally at right angles to the main direction of flow. There will be a tendency for a thicker and fresher layer of surface water to be present to the right of an observer looking seaward down the estuary. Pritchard (1956) has shown that the distribution of salinity across the James River estuary indicates that the Coriolis force resulting from the net horizontal motion is balanced primarily by the lateral pressure force. A similar balance in a deep inlet was reported by Cameron (1951).

It has been inferred that the horizontal pressure gradients in the longitudinal direction which are established by a sloping water surface and modified by the internal distribution of salinity are largely balanced by the frictional forces associated with tidal turbulence. Reference should be made to the magnitude of the forces required to accelerate the surface water seaward and decelerate the opposite flow of the deeper water. Pritchard (1956) has evaluated the field accelerations at a typical location in the James River estuary and shows them to be relatively small in comparison with the frictional terms. It would be expected, however, that the field acceleration terms would increase in importance as the salinity of the surface water more closely approached that of the sea-water. Stommel and Farmer (1953) have shown, for example, that the interrelation of the inertial term and the degree of salinity stratification in certain estuaries sets an upper limit on the volume of mixed water discharged past a control. They suggest that St. John Harbor (New Brunswick) represents such a condition.

The Chesapeake Bay and its tributary estuaries exhibit the characteristics of the moderately stratified estuary. Bousfield (1955) has reported the estuary of the Miramichi River to be of a similar type. The Savannah, Charleston and Delaware estuaries might also be classified as moderately stratified.

In estuaries of this kind the distribution of salinity is kinematically governed by both horizontal and vertical advection of salt and by the non-advective vertical flux.

## C. The Vertically Homogeneous Estuary

When mixing due to tidal motion becomes so vigorous that vertical homogeneity results, the dynamic and kinematic processes are associated with the horizontal variation in the distribution of salinity. The degree of lateral homogeneity has an important bearing on the type of circulation which occurs.

### a. With lateral variation

Estuaries which are vertically homogeneous may show not only a variation in salinity in a longitudinal direction, but also a variation in salinity across the axis. In this type of estuary the salinity on the right of an observer looking seaward is lower than on his left. The main stream of seaward-flowing water is concentrated to the right of center; the compensating flow of salt water, instead of being found below the outflowing current, is concentrated on the left side of the estuary. The existence of a cyclonic circulation within the estuary becomes more evident.

The absence of vertical gradients of salinity suggests that the salt distribution in such an estuary is not governed by vertical processes, but rather by lateral and longitudinal transfers by advective or non-advective means. On the other hand, it is recognized that vertical mixing of water elements is responsible for the homogeneity in a water column. The apparent inconsistency of these two factors arises from an as yet unsatisfactory representation or description of the non-advective flux process. More precise and sophisticated monitoring and interpretation of the short-term variability of salinity in the vertical are desirable before a satisfactory understanding of the role of vertical mixing processes is achieved.

In contrast to the apparently monotonous vertical profiles of salinity, there exist significant variations in the horizontal pressure gradient with depth. Vertical shears of velocity occur and important vertical fluxes of momentum are present. The magnitudes of these fluxes of momentum must be balanced by an appropriate pressure-gradient field, which is established and maintained by the free-surface slope and the horizontal variation in salinity. This again underlines the intimate relationship that must exist between the salinity distribution and the dynamic influence of the eddy frictional forces deriving from the intense tidal mixing.

### b. With lateral homogeneity

In certain estuaries, generally those with a relatively small width-to-depth ratio, the cyclonic circulation described above is not observed. The movement of water is essentially symmetrical about the main axis of the estuary. Variations in velocity are associated with the flood and ebb of the tide, and when the velocities are averaged over several tidal cycles the net flow is apparently seaward at all depths. This advective process tends to drive salt out of the estuary. The salt balance is maintained by a compensating non-advective longitudinal flux of salt from the sea toward the head.

Again, in contrast to the lack of vertical and lateral variation in salinity, there exists a significant variation in the vertical gradient of the horizontal pressure field. The longitudinal variation in salinity provides for a decreasing seaward slope of the pressure surfaces with increasing depth. We are unable at this stage to suggest what opposite force similarly varies with depth. It would appear to be in some way associated with a vertical variation in the asymmetry of the tidal flood and ebb. A better understanding of this aspect of estuarine dynamics may follow from more precise tidal velocity measurements and a more critical theoretical examination of the non-linear term in the tidal equation.

It is, in fact, quite possible that the vertically homogeneous estuary does not exist. Our observational methods may not be sufficiently sophisticated to show the slight degree of vertical stratification which might, on the average, exist in such systems. Only a small vertical stratification would be required to remove some of the anomalous factors mentioned above which are associated with this class of estuary.

### D. The Fjord Type of Estuary

The foregoing discussion has been concerned with estuarine types associated with relatively shallow basins. The long and deep inlets of the Canadian Pacific coast present conditions which facilitate the study of certain aspects of estuarine circulation, but which, on the other hand, make observation and study more difficult.

Pickard's (1956) comprehensive examination of the geomorphology of British Columbia inlets has demonstrated the wide variation in depths, widths and lengths, but confirms that the important difference between these bodies of water and the estuaries of the Atlantic coast lies in the greater relative depth of the former.

In many inlets the fresh-water discharge, generally concentrated at the head, is sufficient to establish and maintain a shallow layer of brackish water overlying a deep reservoir of denser salt water. The horizontal variation of salinity in the deep water is relatively small when compared to the horizontal gradient in the upper layer. Tully (1949) has indicated that the essential characteristic of the circulation in such an inlet is the entrainment of sea-water from below into the less saline seaward-flowing layer above.

Cameron (1951a) was able to demonstrate that the distribution of salinity in such an inlet is sufficient to provide for a seaward-accelerating flow of water in the surface layer in the presence of reasonable values of eddy viscosity associated with the simple vertical velocity profile he assumed. He also suggested that the field accelerations necessary to maintain a steady state demand an increasing proportion of the total pressure-gradient force toward the mouth of the inlet, leading to an upper limit of the permissible seaward velocity, analogous to the criterion established by Stommel and Farmer (1953) in their simpler two-layered model.

From a consideration of the heat balance in such an inlet, Pickard and Trites (1957) confirmed that the admixture of sea-water to the upper layer was from a relatively shallow depth of approximately seven meters.

Extensive observations of velocity profiles in Knight Inlet, British Columbia, by Pickard and Rodgers (1959) have shown a greater complexity of movement than Tully and Cameron inferred. They demonstrated that not only were there occasions when more than two opposing streams could be detected, but that extensive non-tidal movements occurred in the deeper waters of the inlet.

## 4. Some Variations on the Estuarine Sequence

The estuarine types described above can be considered to represent discrete band widths drawn from a continuous spectrum, or sequence, of possible estuarine types. Specific actual estuaries may well exhibit intermediate characteristics. In addition, there are variations, or offshoots, from this main sequence which deserve some consideration.

Bar-built estuaries are formed by the development of an offshore bar, or low-lying elongated island chain, along a coastline of low relief. There is usually a very restricted inlet between the sound and the sea. Where a river flows into the sound thus formed, an estuary is established which differs from the typical drowned-river-valley estuary as a result of this restricted communication. While quite strong tidal currents may occur at the inlet, tidal currents and the rise and fall of the astronomical tide are normally very small within the sound. The relatively great width and small depth of such systems permit wind-induced currents and wind tides to provide the major mechanism for movement and mixing of these estuarine waters. No adequate dynamic description of this type of estuary has yet been produced, and, in fact, this type of estuary has probably received the least systematic attention.

Another variation in estuarine circulation patterns occurs in certain tributary embayments to large estuarine systems. Thus, along the Chesapeake Bay there are several drowned valleys adjacent to and connected with the main estuary and into which very little fresh water is discharged. A circulation pattern develops in these tributary embayments which is definitely related to the estuarine character of the adjacent water body.

An example of such a tributary embayment is Baltimore Harbor, which is on the drowned valley of the Patapsco River, a small tributary to the Chesapeake Bay. So little fresh water is contributed to the harbor by the Patapsco River, however, that it is ineffective in contributing to the density distribution. The prime source of water in the tributary embayment is the adjacent Chesapeake Bay. In the bay the dynamic processes associated with fresh-water inflow at the head and salt-water inflow at the mouth continually maintain a vertical salinity (and hence density) gradient. Within the tributary embayment, however, these same processes do not occur, and vertical mixing tends to reduce the degree of vertical stratification. Thus, though the average salinity in the vertical within the harbor is very nearly the same as that found in the adjacent Chesapeake Bay, the surface waters of the harbor are saltier than the surface waters of the bay, while the bottom waters in the harbor are fresher than water at similar depths in the bay. An example of the vertical distribution of salinity in Baltimore Harbor and in the adjacent Chesapeake Bay is given in Fig. 1.

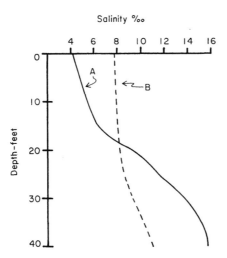

Fig. 1. Typical salinity *vs.* depth curves for (A) the Chesapeake Bay at the mouth of Baltimore Harbor and (B) near the head of Baltimore Harbor.

As a consequence, a horizontal density gradient exists between the waters of the bay and those of the harbor, and this gradient varies markedly with depth. In the surface layers an accelerative force is directed from the fresher waters in the bay toward the denser, more saline, surface waters of the harbor, and because of the constraints of the side boundaries, an inflow of the surface layers along this pressure gradient into the harbor occurs. On the other hand, the denser, more saline, deep waters of the bay also flow into the harbor since the waters on the harbor bottom are less saline than the bay waters at the same depth, just as sea-water flows into the estuary proper along the bottom. Outflow from the harbor to the bay occurs at mid-depth.

This three-layered flow pattern has the appearance of two simple estuarine flow patterns, one on top of the other. In the upper pattern the Chesapeake Bay appears as the source of fresh water and the harbor as the source of salt water, or as the "ocean". In the lower half of the water column, the roles of the harbor and the bay are reversed. Fig. 2 shows a typical longitudinal section of the salinity distribution and flow pattern for this system.

This three-layered type of flow pattern was first described by Pritchard and Carpenter (1960) and Carpenter (1960). In the case of Baltimore Harbor it

constitutes the major mechanism for exchange or renewal of the waters, domi-
nating the processes of tidal exchange and wind-induced circulation.

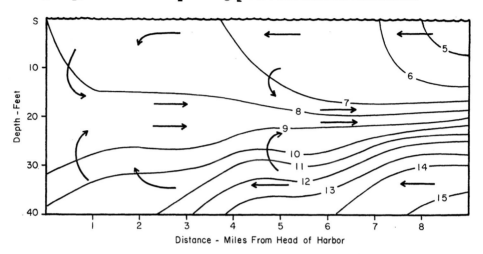

Fig. 2. Typical longitudinal section of the salinity distribution in Baltimore Harbor.
The Chesapeake Bay at the mouth of the harbor is at the right end of the figure.
The arrows show the net flow pattern.

## 5. Flushing in Estuaries

The circulation in estuaries has important implications in the field of pollution study and control. For any estuary in which this problem is significant it is
important to be able to calculate the extent to which a pollutant may accumulate in the estuary before a "steady-state" concentration is reached. Furthermore, the variation in concentration of the pollutant with location in the
system must be predictable.

The factors which bear on this important problem are associated with the
type of circulation that prevails in the estuary under study. In the salt-wedge
estuary, it is clear that a pollutant which is lighter than the water of the salt
wedge will be retained in the surface layer and pass out to sea at approximately the velocity of the surface flow. On the other hand, in a moderately
stratified estuary a portion of such a pollutant would find its way into the
bottom layer because of the vertical mixing which carries fresh water downward into that layer. It would then take part in the upriver flow of the layer, to
be distributed more broadly than in the first case considered. The prediction
of pollution, therefore, depends heavily on the ability to understand and
evaluate the essential processes, both advective and non-advective, which are
responsible for the distribution of properties in the estuary.

It was suggested by Ketchum (1951) that with certain fundamental assumptions (the most significant being that complete mixing occurs within volume
segments governed by the average excursion of a particle of water on the flood

tide), the average distribution of fresh and salt water within an estuary could be calculated.

Ketchum's empirical theory was formalized by Arons and Stommel (1951) into a mixing length theory of tidal flushing in which the characteristic length was associated with the horizontal tidal excursion. Stommel (1953) later questioned the acceptability of associating the mixing length with the tidal excursion and suggested that the coefficient of turbulent diffusion should be computed from the observed distribution of salinity within the estuary under consideration.

Preddy (1954) pointed out that the dispersal of a volume of water due to tidal mixing might be represented by a distribution curve appropriate to each point in the estuary. He represented these curves by simplified asymmetrical distributions involving two constants which could be determined from the observed salinity distribution. In applying his method to the Thames Estuary, he demonstrated that the dispersion of a unit of water varies with position along the estuary and at each position is greater to seaward than headward. He verified his method by successfully predicting changes in the salinity distribution resulting from the changes in river flow.

Gameson, Hall and Preddy (1957) applied Preddy's method to estimating temperature changes in the Thames which would follow an alteration in the amount of heat discharged into the estuary.

Dorrestein (1960) has proposed a numerical approach to the dispersion of an introduced contaminant in an estuary, somewhat similar to Preddy's method. However, Dorrestein has devised a practical technique for computing numerical solutions to the difference equations for the non-stationary distribution of concentration by means of matrix calculus.

The foregoing methods of computation assume that the significant process which maintains a steady state of salinity in the presence of a net advective flow of fresh water from the river is eddy diffusion along the longitudinal salinity gradient from the sea. There remains the question whether, in such cases, the possibility of advective transport of salt should be entirely neglected. The studies of Inglis and Allen (1957) on the distribution of mud in the Thames Estuary indicate an observable net upstream flow along the bed of the river to a point where the average salinity is as low as 5‰. The fact that this upstream flow could not be reproduced in models operated with fresh water throughout confirms that it is not associated with the geometry of the estuary nor the dispersal effect of the tide. It is clear that, in studies such as those associated with silting, the dynamic effect of the salinity distribution is significant, and over-simplification of the mechanism of the estuary will conceal important factors in flushing.

## 6. Dynamics of Estuaries

In this section some of the principles of estuarine circulation described above will be expressed more formally in mathematical terms. For this purpose, we consider the estuary as a geometrically simple, elongated indenture in the coastline with a river entering the upper end. A rectilinear, right-hand coordinate system is located with the origin at the landward end of the estuary. The x-axis is directed horizontally along the estuary toward the sea, the y-axis horizontally across the estuary, and the z-axis vertically upward.

The x-, y- and z-components of the velocity vector $\mathbf{u}$ are designated by $u_x$, $u_y$ and $u_z$, respectively. Further, we designate the specific volume by $\alpha$, the hydrostatic pressure by $p$, and the x-, y- and z-components of the angular velocity of the earth by $\Omega_x$, $\Omega_y$ and $\Omega_z$. Neglecting the molecular viscous stresses (for reasons to be explained later), the longitudinal or x-component of the equation of motion is expressed by

$$\frac{\partial u_x}{\partial t} + u_x \frac{\partial u_x}{\partial x} + u_y \frac{\partial u_x}{\partial y} + u_z \frac{\partial u_x}{\partial z} = -\alpha \frac{\partial p}{\partial x} - 2\left[\Omega_y u_z - \Omega_z u_y\right]. \tag{1}$$

The corresponding lateral or y-component is given by

$$\frac{\partial u_y}{\partial t} + u_x \frac{\partial u_y}{\partial x} + u_y \frac{\partial u_y}{\partial y} + u_z \frac{\partial u_y}{\partial z} = -\alpha \frac{\partial p}{\partial y} - 2\left[\Omega_z u_x - \Omega_x u_z\right]. \tag{2}$$

These equations apply to the instantaneous velocity field. Our observational evidence is at present insufficient to deal with the instantaneous distribution of properties, and it is necessary to treat these equations in such a way that mean values of the various parameters may be employed.

For this purpose, the instantaneous velocity $\mathbf{u}$ is considered to be composed of three terms:

(a) A time mean velocity, $\bar{\mathbf{u}}$, obtained by averaging over one or more tidal cycles. This mean velocity may vary slowly over time intervals longer than the period of averaging.

(b) An oscillatory velocity, $\mathbf{U}$, which is here assumed to vary according to a simple harmonic function of the tidal period.

(c) A velocity deviation, $\mathbf{u}'$, which results from those turbulent fluctuations having a time scale smaller than the period employed in the averaging process.

The x-, y- and z-components of the instantaneous velocity are then given by

$$u_x = \bar{u}_x + U_x + u'_x \tag{3a}$$

$$u_y = \bar{u}_y + U_y + u'_y \tag{3b}$$

$$u_z = \bar{u}_z + U_z + u'_z. \tag{3c}$$

Here the tidal velocity terms are given by simple harmonic functions of the type

$$U_x = U_{ox} \cos \varphi_x. \tag{4}$$

The term $U_{ox}$ is the amplitude of the longitudinal component of the tidal current, and $\varphi_x$ represents the sum of an angular time argument of tidal period plus a phase angle. Similar expressions apply to the lateral and vertical components. In most estuaries the tide is of the reversing type, and the longitudinal velocity component $U_x$ dominates. The lateral component $U_y$ is primarily related to changes in width of the estuary, while $U_z$ results directly from the tidal rise and fall of the water surface.

We neglect the turbulent fluctuations in density in comparison to the corresponding fluctuation in velocity, and assume water to be incompressible. The equations of continuity for the instantaneous and for the mean velocities are then

$$\frac{\partial u_x}{\partial x}+\frac{\partial u_y}{\partial y}+\frac{\partial u_z}{\partial z}=0 \tag{5a}$$

$$\frac{\partial \bar{u}_x}{\partial x}+\frac{\partial \bar{u}_y}{\partial y}+\frac{\partial \bar{u}_z}{\partial z}=0. \tag{5b}$$

Equations (3), (4) and (5) are used in (1) and (2) in obtaining the time mean equations of motion. In this operation cross-products of the type $\langle \bar{u}_x u'_y \rangle$ (where $\langle\ \rangle$ indicates a time mean) occur which are taken to be equal to zero, since by definition $\langle \bar{u}_x u'_y \rangle = \bar{u}_x \langle u'_y \rangle$ and $\langle u'_y \rangle = 0$. Likewise, terms of the type $\langle U_x u'_y \rangle$ are set equal to zero since there is no reason to suspect a correlation between the oscillatory tidal motion and the turbulent velocity fluctuations, even though there may be a relationship between the root mean square of the turbulent fluctuations and the magnitude of the tidal current. Cross-products involving mean velocities and the tidal components are also zero.

Available observational evidence also indicates that certain terms which occur in the mean equations are quite small compared to other, dominant terms. The mean vertical velocity, $\bar{u}_z$, is of the order of $10^{-5}$ m/sec, as compared to the horizontal velocity component of the order of 1 m/sec. The Coriolis terms involving the vertical velocity component are then negligible compared to the Coriolis terms involving the horizontal velocity. In the geometrically simple, elongated estuary here considered, the terms involving the lateral component of the velocity are also negligible.

The mean longitudinal equation of motion then is given by

$$\frac{\partial \bar{u}_x}{\partial t}+\bar{u}_x\frac{\partial \bar{u}_x}{\partial x}+\bar{u}_z\frac{\partial \bar{u}_x}{\partial z}+\frac{\partial}{\partial x}\langle U_x U_x \rangle+\frac{\partial}{\partial z}\langle U_x U_z \rangle$$
$$=-\left\langle \alpha\frac{\partial p}{\partial x}\right\rangle-\frac{\partial}{\partial x}\langle u'_x u'_x \rangle-\frac{\partial}{\partial y}\langle u'_y u'_x \rangle-\frac{\partial}{\partial z}\langle u'_z u'_x \rangle \tag{6}$$

and the corresponding mean lateral equation is given by

$$0=-\left\langle \alpha\frac{\partial p}{\partial y}\right\rangle-f\bar{u}_x-\frac{\partial}{\partial x}\langle u'_x u'_y \rangle-\frac{\partial}{\partial y}\langle u'_y u'_y \rangle-\frac{\partial}{\partial z}\langle u'_z u'_y \rangle, \tag{7}$$

where $f$ replaces $2\Omega_z$.

The eddy-stress terms in these equations, involving products of the type $\langle u'_i u'_j \rangle$, are many orders of magnitude larger than the molecular viscous-stress terms. Hence, the original omission of these latter terms in equations (1) and (2) is justified.

Pritchard (1956) argued that of the three eddy-stress terms in each equation, only $\partial(\langle u'_z u'_x \rangle)/\partial z$ would be significant in equation (6) and only $\partial(\langle u'_z u'_y \rangle)/\partial z$ should be retained in (7). On the basis of observations from the James River estuary, he showed that there was approximately a lateral geostrophic balance; that is, the mean lateral pressure force $-\langle \alpha \, \partial p/\partial y \rangle$ was very nearly balanced by the term $f\bar{u}_x$. The eddy-stress term in (7) was of secondary importance. Stewart (1957), on the other hand, raised the question whether the eddy-stress term could be retained in equation (7) in view of the fact that the mean lateral velocity was assumed to be negligible, and hence there would be no velocity shear corresponding to this stress term. He showed that the geostrophic imbalance could be explained as a result of the curvature of the boundaries of the James River estuary.

The mean longitudinal equation of motion retains two terms involving the tidal components of the velocity. This bears out the statements made in the earlier descriptive paragraphs that a portion of the mean pressure force is associated with the non-linear terms involving the tidal velocities, even though these velocity components may be purely harmonic in character and average to zero over the tidal period. If the tidal wave in the estuary is a progressive wave, then $U_x$ and $U_z$ will be 90° out of phase, and the second of the two terms in equation (6) involving tidal velocity components will disappear. The remaining term becomes, using equation (4),

$$\frac{\partial}{\partial x} \langle U_x U_x \rangle = U_{ox} \frac{\partial U_{ox}}{\partial x}. \tag{8}$$

If the tidal wave is a standing wave, then $U_x$ and $U_z$ will be 180° out of phase. In this case the last two terms on the left side of (6) become, after employing the equation of continuity,

$$\frac{\partial}{\partial x} \langle U_x U_x \rangle + \frac{\partial}{\partial z} \langle U_x U_z \rangle = \frac{1}{2} \left[ U_{ox} \frac{\partial U_{ox}}{\partial x} - U_{oz} \frac{\partial U_{ox}}{\partial z} \right]. \tag{9}$$

In the case of a progressive tidal wave in the estuary, a part of the mean pressure gradient is then balanced by a term involving the longitudinal variation in tidal-current amplitude. In the case of a standing tidal wave, both the longitudinal and vertical variations in the tidal-current amplitude are involved.

There are some interesting implications which result from the presence of these non-linear terms in the mean longitudinal equation of motion. For example, consider the simple case of a coastal embayment with no net fresh-water inflow and uniform density distribution. The mean pressure force $-\langle \alpha \, \partial p/\partial x \rangle$ would then be given simply by $\langle g \, \partial \eta/\partial x \rangle$, where $g$ is the acceleration due to gravity and $\eta$ the elevation of the water surface. In the absence of any density gradient, we would expect the mean velocity components to vanish, and for steady-state conditions in an embayment with a progressive tidal wave, equation (6) would be given by

$$U_{ox} \frac{\partial U_{ox}}{\partial x} = \left\langle g \frac{\partial \eta}{\partial x} \right\rangle - \frac{\partial}{\partial z} \langle u'_x u'_z \rangle. \tag{10}$$

This relationship shows that even in the absence of any density gradients, there would occur a mean slope of the surface of the embayment related to the longitudinal variation in tidal-current amplitude.

In an estuary, the density distribution is not uniform. The distribution of temperature and salinity can be utilized to compute the density distribution which, when employed in the hydrostatic equation, allows the determination of the pressure gradient relative to some pressure surface—for example, water surface of the estuary. Left undetermined is the absolute mean slope of the water surface. However, if the term $\partial(\langle u'_z u'_x \rangle)/\partial z$ is the only eddy stress of importance in (6), this equation can be integrated with depth from the surface to the bottom of the estuary. On the basis of boundary layer theory, reasonable estimates of the boundary values for the eddy flux of momentum can be made. Such a procedure allows the determination of the mean absolute slope of the water surface, since this term would be constant in the vertical integration of equation (6). Pritchard (1956) used the observed distribution of velocity and density in the James River estuary in (6) to compute the absolute longitudinal slope of the pressure surfaces. Using (7) in a similar manner, he determined the lateral slope of the pressure surfaces. This work showed that, in a partially mixed estuary, the water surface has a mean slope downward toward the sea and also downward to the left of an observer facing the sea. At about mid-depth the pressure surface is level in both the longitudinal and lateral directions, but in the deeper layers the pressure surfaces slope in the opposite direction to the slope of the water surface.

The estuary of the River Thames, in England, is reported to be vertically homogeneous in density. There is observed in this estuary, however, a net motion directed from the sea landward on the bottom, and a net flow directed seaward in the upper layers. Abbott (1960) argues that the non-linear tidal current terms in the mean longitudinal equation of motion, coupled with bottom friction, would produce just the mean flow observed. His argument should be weighed against the failure by Inglis and Allen (1957) to produce experimentally upstream bottom flow in a fresh-water hydraulic model of the Thames.

In the descriptive paragraphs it was pointed out that in certain embayments, tributary to large estuarine systems, a three-layered net flow pattern is observed, with inflow from the adjacent estuary to the embayment at both the surface and bottom, and outflow at mid-depths. We would suspect that in such an embayment the pressure surfaces would slope downward toward the head of the embayment at both the surface and bottom, and in the opposite direction at mid-depth. Level pressure surfaces would then occur at two depths, one at the boundary between the surface layers and the mid-depth layers, and the second at the boundary between the mid-depth layers and the bottom layers. Use of the mean longitudinal equation of motion with the observed distribution of velocity and density for one such embayment, Baltimore Harbor, confirms this mean distribution of pressure.

## 7. Kinematic Description of the Distribution of Properties in an Estuary

The mathematical relationship expressing the conservation of a water-borne constituent can be developed from the equation of continuity. We designate the local instantaneous concentration of some conservative property of the estuary, such as salt concentration, by $s$, and obtain the following differential equation

$$\frac{\partial s}{\partial t} = -\frac{\partial u_x s}{\partial x} - \frac{\partial u_y s}{\partial y} - \frac{\partial u_z s}{\partial z}, \tag{11}$$

where the molecular diffusion terms are omitted. Again, it is not possible to treat this instantaneous relationship. As with the velocity components, we consider the instantaneous concentration $s$ to be composed of three terms: a time mean concentration over one or more tidal cycles; a term which varies sinusoidally with the tidal period; and a turbulent deviation term. Hence,

$$s = \bar{s} + S_U + s'. \tag{12}$$

The major cause of fluctuations of tidal period in the local salt concentration is the advection of the longitudinal salinity distribution with the longitudinal component of the tidal current. The velocity component $U_x$ and the salinity term $S_U$ would then be approximately 90° out of phase, and cross-products involving these two terms would disappear from the mean form of (12). Using arguments similar to those presented in the development of (6) and (7), we obtain for the time mean of (11) over a tidal period

$$\frac{\partial \bar{s}}{\partial t} = -\frac{\partial \bar{u}_x \bar{s}}{\partial x} - \frac{\partial \bar{u}_y \bar{s}}{\partial y} - \frac{\partial \bar{u}_z \bar{s}}{\partial z} - \frac{\partial}{\partial x} \langle u'_x s' \rangle - \frac{\partial}{\partial y} \langle u'_y s' \rangle - \frac{\partial}{\partial z} \langle u'_z s' \rangle. \tag{13}$$

For each of the estuarine types described in the earlier paragraphs, a different set of terms will dominate (13). Thus, at the interface in a salt-wedge estuary the upward breaking of the interfacial waves can be considered as a net vertical advection, and the salt balance would be maintained primarily by the longitudinal and vertical advective terms in the equation. In the upper, seaward-flowing layers of such an estuary the vertical eddy flux of salt would also be important. Pritchard (1954) showed that in a partially mixed estuary the terms related to the longitudinal advective flux and the vertical eddy flux are dominant, with the vertical advective flux making a significant but lesser contribution. The other terms in (13) appear to be negligible. In a vertically homogeneous estuary only the horizontal terms would be significant.

There is very little basis for deciding on the form of the eddy-flux terms. The usual approach is to treat turbulent diffusion as analogous to molecular diffusion, and make substitutions of the type

$$\langle u'_x s' \rangle = -K_x \frac{\partial \bar{s}}{\partial x}, \tag{14}$$

where $K_x$ is the longitudinal eddy diffusivity. There is no adequate theoretical basis for determining the spatial and temporal variations in the eddy diffusivity. The usual approach is to determine empirically the eddy coefficients for a given environmental situation, using the observed distributions of some conservative property, and to use these coefficients in predicting the distribution for other conditions. Thus the one-dimensional form of (13), which applies to a sectionally homogeneous estuary, has been solved numerically for the temporal and spatial distributions of an introduced contaminant, using the longitudinal diffusivity computed from the steady-state distribution of salinity. A similar treatment has been used to predict the distribution of salinity in a sectionally homogeneous estuary under varying conditions of fresh-water inflow.

Another approach to the mathematical description of the distribution of salinity or of an introduced contaminant in an estuary involves the segmentation of the estuary into sections within each of which the concentration is assumed to be constant but different from that found in adjacent segments. A difference equation is developed involving so-called "exchange coefficients", which express the fractional rate of exchange of water between adjacent segments. These exchange coefficients are determined from observed distributions of salinity, and the relationship utilized for predicting purposes.

These engineering approaches are of considerable practical value. However, they remain completely unsatisfactory from the standpoint of actually explaining the mechanisms controlling the distribution of properties in an estuary. An adequate understanding of the eddy-flux terms is required. The observational data necessary as a basis for such understanding necessitate observational techniques different from those now in common use.

One observational approach to the problem is the direct measurement of turbulent fluctuations in velocity and salinity. Some progress has been made on the instrumentation for such observations. A second approach involves direct observation of the diffusion of an introduced tracer substance. Significant advances in detection methods and observational techniques for such direct measurements of diffusion in estuarine waters are reported by Pritchard and Carpenter (1960).

## References

Abbott, M. R., 1960. Boundary layer effects in estuaries. *J. Mar. Res.*, **18**, 83–100.

Arons, A. B. and H. Stommel, 1951. A mixing length theory of tidal flushing. *Trans. Amer. Geophys. Un.*, **32**, 419–421.

Bousfield, E. L., 1955. Some physical features of the Miramichi estuary. *J. Fish. Res. Bd. Canada*, **12**, 342–361.

Cameron, W. M., 1951. On the dynamics of inlet circulations. Doctoral dissertation, Scripps Institution of Oceanography, University of California.

Cameron, W. M., 1951a. On the transverse forces in a British Columbia inlet. *Trans. Roy. Soc. Canada*, **45**, 1–8, Ser. 3, Sect. 5.

Carpenter, J. H., 1960. The Chesapeake Bay Institute study of the Baltimore Harbor. *Proc. 33rd Ann. Conf. Md.-Del. Water and Sewage Assoc.*, 62–78.

Dorrestein, R., 1960. A method of computing the spreading of matter in the water of an estuary. *Disposal of Radioactive Wastes*, vol. 2, Intern. Atomic Energy Agency, 163–166.

Gameson, A. L. H., H. Hall and W. S. Preddy, 1957. Effects of heated discharges on the temperature of the Thames Estuary. *The Engineer*, **204**, 816–819; 850–852; 893–896.

Inglis, C. C. and F. H. Allen, 1957. The regimen of the Thames Estuary as affected by currents, salinity and river flow. *Proc. Inst. Civ. Eng.*, **7**, 827–878.

Ketchum, B. H., 1951. The exchanges of fresh and salt waters in tidal estuaries. *J. Mar. Res.*, **10**, 18–38.

Keulegan, G. H., 1949. Interfacial instability and mixing in stratified flows. *J. Res. Nat. Bur. Standards*, **43**, 487–500.

Pickard, G. L., 1956. Physical features of British Columbia inlets. *Trans. Roy. Soc. Canada*, **50**, 47–48, Ser. 3.

Pickard, G. L. and R. W. Trites, 1957. Fresh water transport determination from the heat budget with applications to British Columbia inlets. *J. Fish. Res. Bd., Canada*, **16**, 605–616.

Pickard, G. L. and K. Rodgers, 1959. Current measurements in Knight Inlet, British Columbia. *J. Fish. Res. Bd., Canada*, **16**, 635–678.

Preddy, W. S., 1954. The mixing and movement of water in the estuary of the Thames. *J. Mar. Biol. Assoc. U.K.*, **33**, 645–662.

Pritchard, D. W., 1954. A study of the salt balance in a coastal plain estuary. *J. Mar. Res.* **13**, 133–144.

Pritchard, D. W., 1955. Estuarine circulation patterns. *Proc. Amer. Soc. Civ. Eng.*, **81**, 717/1–717/11.

Pritchard, D. W., 1956. The dynamic structure of a coastal plain estuary. *J. Mar. Res.* **15**, 33–42.

Pritchard, D. W. and J. H. Carpenter, 1960. Measurements of turbulent diffusion in estuarine and inshore waters. *Bull. Intern. Assoc. Sci. Hydrol.* No. 20, 37–50.

Stewart, R. W., 1957. A note on the dynamic balance for estuarine circulation. *J. Mar. Res.*, **16**, 34–39.

Stommel, H., 1953. Computation of pollution in a vertically mixed estuary. *Sewage and Industrial Wastes*, **25**, 1065–1071.

Stommel, H. and H. G. Farmer, 1953. Control of salinity in an estuary by a transition. *J. Mar. Res.*, **12**, 13–20.

Tully, J. P., 1949. Oceanography and prediction of pulpmill pollution in Alberni Inlet. *Bull. Fish. Res. Bd., Canada*, **83**, 169.

# 16. APPLICATIONS OF THE GYROPENDULUM

## W. S. von Arx

### 1. Introduction

From a ship at sea the confusion of horizontal accelerations prevents men and most instruments from being able to detect in any precise way the direction of local gravity. The bob of an ordinary pendulum tends to lag behind the motion of its point of support as it is accelerated. Only under conditions of rest will the center of mass of a short pendulum lie exactly and continuously on a vertical line beneath its point of support.

But if the earth is approximated by a non-rotating sphere and it is considered that the pivot point of a simple pendulum is moving along a great circle at some slow, constant speed, it is apparent, once the initial oscillations have been damped out, that the line connecting the pivot point with the center of mass of the pendulum can be extended to intersect the geocenter at all times. As seen from inertial space this requires the rest point of the pendulum to execute one rotation around the pivot for each revolution of the pivot around the circumference of the great circle.

It was shown by M. Schuler in 1923 that if, on a non-rotating spherical earth, a physical pendulum is allowed to align itself with local gravity, the direction of the line joining the center of mass and the point of support will always coincide with the direction toward the earth's center, regardless of horizontal accelerations, if its period of oscillation is 84 min. A physical pendulum with a period of 84 min is equivalent to a simple pendulum having a length equal to the mean radius of the earth. With the bob, in effect, at the geocenter, the point of support can be moved in any arbitrary way over the curve of a non-rotating spherical earth with the results that Schuler described.

A difficulty with Schuler's suggestion is that any practical physical pendulum with a period near 84 min has its center of mass separated from the point of support by the same order of distance as the unit-cell dimensions separating the atoms of the material of which the pendulum is composed.

### 2. The Gyropendulum

Schuler's ideal system can still be realized if a gyrating mass is substituted for the inert mass of the pendulum bob. A gyropendulum with its rotating member spinning about an axis passing through the pivot point transforms the inert mode of pendulum oscillation in a plane into a conical motion about the vertical. The precessional period of a gyropendulum can be made very long in comparison with its period as an inert physical pendulum.

If $r$ is the radius of the gyration of the rotor, $S$ the angular speed of the rotor, $L$ the distance from the pivot to the center of mass of the gyropendulum and $g$ the acceleration due to gravity, the period of oscillation, $T$, of the system is

$$T = 2\pi \frac{(r^2 S/L)}{g}.$$

[*MS received June, 1960*]

The Schuler-tuning condition is satisfied for a given station when $T = 2\pi\sqrt{(R/g)}$, where $R$ is the local radius of the earth.

Schuler tuning of a reference vertical (Wrigley, 1941, 1950) becomes of interest when the motion of a ship generates geocentric angles which approximate the order of accuracy required of the vertical reference. Since one minute of arc is related by definition to the length of the nautical mile, it is readily apparent that a second of arc approximates the length of a ship. Even hove-to, a ship may easily move a distance comparable with its own length in the few minutes needed to make an observation.

A Schuler-tuned gyropendulum erects to vertical through a succession of 84-min oscillations. When these oscillations are larger in amplitude than the desired accuracy of vertical indication, damping must be applied. But since, in existing systems, the damping function does not satisfy the Schuler-tuning condition (the dash-pot of the equivalent simple pendulum is not at the center of the earth) such a system responds to changes in the motion of the supporting vehicle. For this reason it is probably desirable that the system be undamped when measurements are in progress and to average observations through the period of at least one conical oscillation.

A further qualification of Schuler-tuned systems is that they tend to compensate for geographic motion of the supporting vehicle in a geocentric mode. To provide a continuous indication of the local-gravity vertical, it is necessary to re-tune the system to correspond to the local radius of the earth according to some idealized model of the earth's figure, and, for this reason, to know the geographic position of the pendulum. For the oceanographic purposes to be discussed, it seems better to employ a system that yields gravity vertical more directly.

To this end it has been shown that if a physical system has a period of conical oscillation which is very great compared with the period of roll and pitch of a ship, but less than 84 min, it will still tend to average through horizontal disturbances and gradually, with isotropic damping, tend to seek the gravity vertical at sea. Motion of the pivot point will misalign the spin axis of the gyropendulum with the direction of local gravity, thereby producing a gravitational torque on the pendulum which will cause it to precess. If the translation is steady in inertial space and angular equilibrium has been attained, it is possible for a gyropendulum to maintain a fixed attitude relative to both gravity and the path of motion of its pivot similar to that shown in the right or left half of Fig. 1. This behavior of a gyropendulum causes its length to sweep out a flat cone having an axis normal to the plane of its orbit. Schuler (1923) has pointed out that two gyropendula spinning in opposite directions, but having otherwise identical properties, could be used to generate two cones sharing the same base but with their apices oppositely directed along the central normal to their circular orbit, as shown in Fig. 1. Were the angle between the two translating gyropendula to be continuously bisected, local gravity vertical would be indicated so long as the rate of translation was slow compared with the tangential velocity of the earth at the latitude of the system.

However promising this suggestion seems, a physical system has yet to be developed for practical use.

One that has been developed employs a different principle; that of slaving to a north-seeking gyrocompass rotor a second horizontal rotor having its spin axis oriented east–west. When suitably damped by torques controlled by signals from electrolytic levels or linear horizontal accelerometers this combination of rotors will precess to seek both gravity vertical and the geographic north point and thus to define the local meridian plane. Meridian gyro systems tend to drift in azimuth and to a far lesser extent depart from gravity vertical when exposed to sustained horizontal motions or accelerations of the supporting vehicle. Still, by these means the direction of gravity can be sensed at sea with an accuracy of about $10^{-4}$ rad.

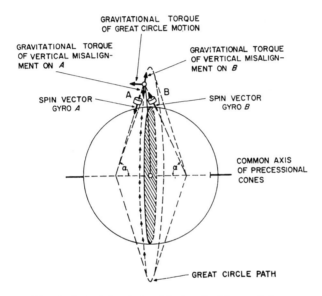

Fig. 1. Principle of the Schuler double pendulum.

Having a means by which the direction of gravity can be determined both closely and reliably from surface ships a number of oceanographic applications come within range of useful consideration:

A. Recently tried at sea is the possibility that celestial navigation can be practiced at any time that clear skies exist. Given the local vertical and north point to define the local meridian plane, Greenwich Mean Time and the celestial co-ordinates of a body, it is possible to fix the momentary latitude and longitude of a moving ship *from a single observation*, and to track its change of position with time by taking serial observations on a *single celestial body*.

B. The same navigational apparatus also suggests uses for a gravity vertical as part of a system for direct measurements of the shape of the earth at sea.

C. It is conceivable that gyro verticals may one day be used to determine the steric and dynamic departures of the ocean surface from horizontal and thus to permit absolute measurements of the horizontal gradients of pressure.

### A. Celestial Navigation

For celestial navigation an indication of local vertical to an accuracy of one minute of arc, or better, makes it unnecessary to employ the horizon as a fiducial mark. By the same token it is unnecessary to transform the celestial coordinates of a body to the horizon system, or to make separate determinations of latitude and longitude if the plane of the local meridian can be known continuously from shipboard (Fig. 2). Except for observations in very high latitudes, this can be done both day and night with the help of a meridian gyrocompass.

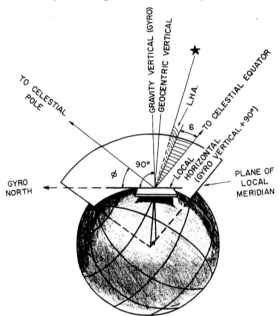

Fig. 2. Given gravity vertical and the local meridian by gyro, the astronomical triangle can be solved from a single sight on a single celestial object at a known time.

Recent experiments with a Sperry Mark 19 Mod 3 meridian gyrocompass mounted on the Research Vessel *Chain* of the Woods Hole Oceanographic Institution show that it is within the range of present technical achievement to maintain a vertical reference on shipboard that can be relied upon to a small fraction of a minute of arc and north reference to a small fraction of a degree (Fig. 3). Ideally, to reduce the confusion of horizontal accelerations caused by waves and steering, the meridian gyrocompass should be mounted near the metacenter but preferably between the turning center and the intersection of the pitch and roll axes of the ship as shown in Fig. 4. The point from which celestial observations are made would, in consequence, have to be remote from the master gyropendulum. In recent practice, it seemed simplest to accept the effects of horizontal accelerations and to avoid the problem of correcting angular misalignment between the master gyro and a remote observing station (which involve the vibrational, shearing and torsional modes of motion of the decks on which the master gyro and slave systems are mounted) by locating

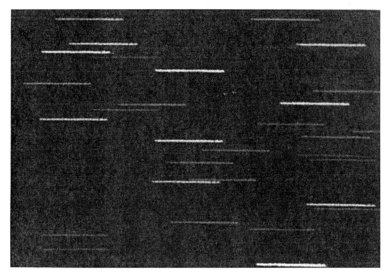

Fig. 3. Star trails across the zenith recorded in latitude 00° 08′ South with a camera fixed rigidly to the gyro head. The trails, 75 minutes (arc) in length, show by their width the 1-minute (arc) error produced by backlash in the pitch and roll servomotor gearing, and by their overall straightness the higher quality of average vertical that is maintained.

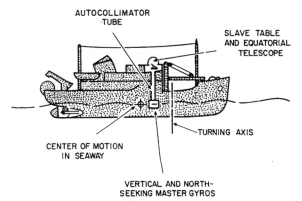

Fig. 4. Arrangement of master gyropendulum and observing station required to minimize the effects of horizontal accelerations.

the master compass above the metacenter, and to take observations directly from the compass head.

With the system mounted at a level on the ship that provides a clear view of the sky, preferably unobstructed by stack gases, it is possible to adjust the polar axis and set the hour angle and declination axis of an equatorially mounted telescope well enough to bring the image of a bright celestial object into the field of view in daylight or darkness.

The equatorially mounted optical train used in these experiments is shown in Fig. 5. It consisted of a Kern DKM-1 precision theodolite with its azimuth

Fig. 5. Small theodolite mounted on the telescope tube of a larger theodolite, fixed to the meridian gyro head. This arrangement permitted the upper theodolite to move as an equatorial mounted telescope and its circles to read Right Ascension and Declination, while the upper circle of the lower instrument gave the altitude of the celestial pole or ship's latitude. The camera clamp mounted on the hour circle box of the upper theodolite was used in making Figs. 9 and 10.

circle set in accurate alignment with the former position of the objective lens of a Kern DKM-2 theodolite. In this impromptu arrangement there are four accurately graduated circles: (1) *the azimuth circle of the lower theodolite* to calibrate adjustment of the lower trunions (these should lie parallel to the level east–west line maintained by the meridian gyro; squaring-on is accomplished by means of a long-radius ball-and-socket mounted directly on the servo-driven meridian gyrocompass housing), (2) *the altitude circle of the lower theodolite* to give the altitude of the polar axis or latitude, (3) *the azimuth circle of the upper theodolite* to serve as an hour circle, and, finally, (4) *the altitude circle of the upper theodolite* to serve as a declination circle. The following observational procedure proved effective:

(i) set the declination of the celestial body on circle 4,
(ii) set the approximate ship's latitude on circle 2,
(iii) set the approximate L.H.A. of the body on circle 3.

Then, having sighted the body, make the fine adjustments of the altitude of the polar axis required to bring the object to the equatorial cross-hair, lock the polar axis, and, having set the hour angle cross-hair slightly west of the position of the object in the field, note the G.M.T. of its crossing the given L.H.A. The latitude of the sight position is given directly upon reading circle 2. The longitude is easily reckoned from a knowledge of the G.H.A. of the body and its observed L.H.A.

The precision and reliability of the assembly is such that once the lower theodolite has been squared on the gyro head (by taking sights on the horizon in daylight) and the azimuth of the lower trunions adjusted (by taking sights on celestial objects low in the eastern and western skies), it is possible to determine the ship's position to an accuracy of $\pm 1$ nautical mile (Figs. 6, 7 and 8) during the course of the ensuing 36 h.[1]

To keep small the effects of atmospheric refraction and hunting of the gyro-compass north point it is desirable to restrict sights to objects near the zenith.[2] This is easily accomplished at night, when the usual navigation stars are accessible along with many other celestial objects, but less so by day. During mid-day the sun is the obvious target, but in the early morning or late afternoon it is preferable to choose a bright star or planet higher in the sky. Favoring the visibility of such objects during the early and late hours of daylight is the fact that the brightness of the sky tends to fall off rapidly with increasing angular distances from the solar disc. Under favorable conditions it is possible to

---

[1] This accuracy was achieved in 1960–61. During 1962 improvements in gyro performance and in observing technique have resulted in a decrease in the average error of position fixing to about $\pm 0.5$ nautical mile, provided the ship has been underway at constant speed and on a relatively straight course for two hours or more. In recent experiments, the errors in fixing position have been assessed through a comparison of the charted range and bearing to points on nearby islands with the range and bearing determined by ship's radar.

[2] When a suitable object is not to be found near the zenith in daylight, the polar axis can be elevated to the zenith and the altitudes and azimuths of two or more objects measured (without reference to the horizon) to determine lines of position in the conventional way. Since both altitude and azimuth can be preset to lie within the field of view of the telescope, the problem of finding celestial objects in full daylight is eased.

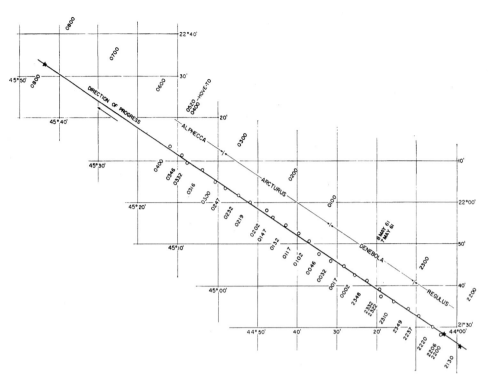

Fig. 6. Consecutive position fixes by *Geon* ○ compared with track established by conventional celestial navigation ★ aboard R.V. *Chain* (2200, 7 May, 1961–0400, 8 May, 1961). Course 305° true; speed 13 knots.

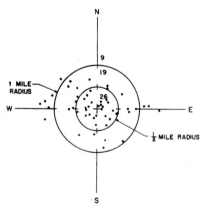

Fig. 7. Scatter among 54 *Geon* fixes taken from R.V. *Chain* lying at dockside in Woods Hole.

observe in full daylight objects generally brighter than magnitude zero with astronomical telescopes having apertures as small as 30 mm and to reach first magnitude objects with apertures of about 50 mm.

The brighter stars and planets (listed in Table I) can often be seen with telescopic aid in daylight because the light of a point source or star can be concentrated optically by the ratio of the area of the entrance and exit pupils of the telescope (when exit pupil equals the pupilary diameter of the eye), while the brightness of an extended source, such as the background sky, is not increased by optical means. Optical systems can, however, reduce the intensity of background illumination to values considerably below the naked-eye level.

Successful observation of stars in daylight involves physiological as well as physical effects. Since the threshold of sensitivity of the human eye decreases as the intensity of total stimulation rises, there are some advantages to be

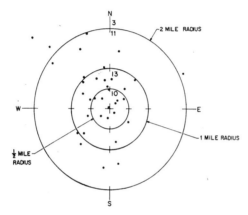

Fig. 8. Difference between 37 *Geon* and *Loran* fixes taken simultaneously from R.V. *Chain* at sea.

realized by placing the telescope in shaded observing conditions. It is also found that elimination of the polarized component of scattered light can heighten the apparent contrast of star images observed 90° or so from the sun. Polaroid filters dim the stars nearly a full magnitude but under favorable conditions, as when the air is free of large particles, they may enhance contrast by reducing the background brightness of the sky an even greater amount.

The apparent sharpness and contrast of a foveal image can be enhanced by stimulating the surrounding portions of the retina—the phenomenon of spatial induction. But if the extra-foveal illumination is too bright the contrast of the foveal image is reduced. Thus a variable level of extra-foveal stimulation could prove useful, or a very wide field eyepiece could be used so that the sky itself might provide appropriate levels of extra-foveal stimulation through an angle some 30 to 35° off axis.

The principle of the navigational method has been shown to be sound but there are numerous improvements possible in the optical and mechanical

TABLE I[a]

Celestial Objects Brighter than Magnitude Zero

| | | |
|---|---|---|
| Sun | − 26.7 | |
| Moon | − 12.55 | At mean opposition |
| Venus | − 4.01 | Max. at elongation |
| Mars | − 1.85 | At mean opposition |
| Jupiter | − 2.23 | At mean opposition |
| Saturn | − 0.18 | Max. rings open at opposition |
| Sirius | − 1.58 | |
| Canopus | − 0.86 | |

Objects Brighter than First Magnitude

| | | |
|---|---|---|
| α Centauri | 0.06 | |
| Vega | 0.14 | |
| Mercury | 0.16 | Max. at elongation |
| Capella | 0.21 | |
| Arcturus | 0.24 | |
| Rigel | 0.34 | |
| Procyon | 0.48 | |
| Archenar | 0.60 | |
| β Centauri | 0.86 | |
| Altair | 0.89 | |
| Betelgeuse | 0.92 | |

[a] Data from Russell, Dugan and Stewart, *Astronomy*. Ginn and Company, Boston, 1926.

system presently mounted on the Mark 19 meridian gyro. For example, it was found that the observing skill needed to manipulate the adjusting screws while the eye is being held at the eyepiece of a telescope, which does not share in the pitching, rolling and yawing motion of the rest of the ship, is not inconsiderable. The selection of objects near the zenith makes it necessary to use an elbow eyepiece almost exclusively. A system in which the line of sight is taken along the polar axis would be simpler both optically and mechanically, and, with catadioptric optics, permit both the weight of the system and the excursions of the eyepiece to be greatly reduced.

As a by-product of the navigational experiments the meridian gyro apparatus also made it possible to take 30-min exposures of star fields with small cameras mounted on the hour circle box of the upper theodolite, the latter serving as an equatorial mounting. The diurnal motion of the earth was cancelled by manual operation of the hour-circle slow-motion screw and occasional corrections of the altitude of the polar axis were effected by adjustments of the latitude slow-motion screw. The accompanying photographs of the star clouds in the Crux-Argo (Fig. 9), the Sagittarius–Scorpius regions (Fig. 10) and mosaic of the Milky Way (Fig. 11) were obtained by these methods. The moon, sun and brighter planets have also been photographed on a scale of 1 cm/deg.

The applications of the meridian gyro to other problems in astronomical and meterological photography and theodolite measurements from shipboard are numerous and may include direct photography of solar eclipses as well as their associated flash and coronal spectra, records of meteor and satellite trajectories,

Fig. 9. Milky Way in the Crux-Argo region photographed on 19 March, 1961, from R.V *Chain* while under way at 13 knots in lat. 00° 10′ South and long. 18° 44′ West. During the 30 min exposure (0130 to 0200 G.M.T.) with a Zeiss Biotar of 50 mm focal length, working at f/2 on Tri-X film, compensations for roll, pitch and yaw were provided by the Sperry Mark 19 meridian gyrocompass. Corrections for the ship's motion in latitude and longitude as well as for the earth's rotation were provided by hand guiding on Spica. North is at the top. The Southern Cross and "coal sack" appear to the left of center.

and whole-sky time-lapse auroral or cloud photographs, to name a few; all with the option of placing the observing station wherever geographic and climatic conditions are most favorable. It is worth noting that in the vicinity of the magnetic equator the absence of aurora and air glow is marked, and, being far from the smoke and scattered light of populated regions on land, the night sky at sea is so black that faint celestial objects are often unexpectedly easy to observe.

Fig. 10. Star clouds and obscuring nebulosity of the galactic center in the Sagittarius–Scorpius region of the Milky Way photographed on 23 April, 1961, from R.V. *Chain* in lat. 01° 22′ North, long. 17° 22′ West during the period 0430 to 0500 G.M.T. with the same lens and equipment used for Fig. 9. The guide star was Rasalhague. North is at the top.

Fig. 11. A mosaic of the Milky Way from Deneb in Northern Cygnus to Sirius in Canis Major made to demonstrate that astronomical photography is practicable at sea on a routine basis. This mosaic is composed of 16 exposures, each of 15 min duration, made from R. V. *Chain* in latitudes 12° to 14° north on the 65° west meridian during the nights of 4th and 5th May, 1962, using a Schneider Xenon lens of 50-mm focal length working at $f/2$ on Tri-X film. The principal objects are from left to right: Deneb, Vega, Altair, Scutum Sobieski, Sagittarius, Scorpio (Antares), $\alpha$ Centauri, the southern coal sack, southern Cross, Argo Navis, Canis Major (Sirius). The Argo region is faint because it remained low on the southern horizon.

## B.  Figure of the Earth

The Gyro Erected Optical Navigation system (or *Geon* as the navigational device has been called) provides fixes of position which are indistinguishable in principle from those obtained by sextant, transit, zenith tube or other standard astronomical methods making reference to the local direction of gravity. Owing to deflections of the vertical, it is to be expected that the course of given meridians and latitude circles over the face of the earth must be somewhat irregular as well as being arranged on an ellipsoidal format. On the other hand, the proposed system of satellite navigation (*Transit*) will tend to take reference to the geocenter as the altitude of the satellites is increased and, therefore, to describe geographic position in terms of a system of meridians and latitude circles which have more regular geometrical configurations. *Transit* navigation, as now proposed, promises to provide geocentric position to an accuracy of about 0.1 nautical mile; or 6 seconds of arc. If *Geon* can be made to match this precision it would be possible from a comparison of *Geon* and *Transit* positions to measure directly the more conspicuous configurations of the geoid. For example, the rotational flattening of the earth produces a discrepancy of 11 minutes of arc ($0.3 \times 10^{-2}$ rad) between the direction of the geocenter and gravity at latitude 45°. Deflections of the vertical associated with higher order harmonics of the earth's figure and local effects of inhomogeneities in the crustal composition in the order of $10^{-4}$ rad could be sensed and measured at sea if these two systems were made equally precise.

## C.  Regional Slopes of the Sea Surface

Because of rotation, the earth's figure is nearly ellipsoidal and the field of plumb-line verticals is roughly described in terms of the resultants of the centrifugal force of rotation and geocentric gravitation at each point on the earth. While this yields an approximation of the field of gravity verticals, it is distinguished from it by its high degree of physical idealization.

There are local anomalies in the density distribution within the substance of the earth which cause the true field of verticals to depart from ellipsoidal symmetry. Gravity verticals tend to converge toward local excesses of mass and to diverge in the vicinity of local mass deficiencies. Were the solid earth to be entirely covered by motionless water of uniform density the surface would be horizontal (at right angles to local gravity), have uniform geopotential and represent the actual figure of the earth—the geoid.

For oceanographic considerations we would wish to measure the regional departures of the sea surface from horizontal. Such departures presumably arise from organized motion within the fluid produced by wind stress or horizontal gradients of water density.

Fluid dynamical reasoning suggests that the regional slopes of the sea surface associated with either the baroclinic or barotropic modes of geostrophic motion become important when they reach the order of $10^{-6}$ rad or about 0.2 second of arc. Only the strongest surface currents, such as the Gulf Stream or

Kuroshio, would be expected to develop transverse slopes in the order of one part in $10^{-5}$ (2 seconds of arc). Technically impossible as the measurement of such small slopes are at the moment from ships, their importance to oceanography (and marine geodesy) is so great and the rate of technical advancement in gyroscopic systems is so rapid that a few experiments have been made in anticipation of the day when sensing of vertical to the required accuracy may become feasible.

Since the real fluid covering the earth consists of two components, air and water, it is to be expected that motion in either component will be accompanied by a surface-pressure anomaly and a deformation of the interface. Because of the relatively great depth of water in the major ocean basins, it is possible for the gravity wave developed by an anomaly of atmospheric pressure to keep step with the relatively slower motions of atmospheric high- and low-pressure centers that cross the ocean surface. It can be shown that the velocities of deep water needed to satisfy continuity, as the ocean surfaces elevated or depressed by atmospheric pressure change, are so small that the readjustments of the pressure field are quasi-hydrostatic within the volume affected. Only when a very low pressure system, such as a hurricane, or possibly a sharp front passes over water of coastal depths at speeds approaching or exceeding that of a shallow-water gravity wave will non-hydrostatic effects appear. In deeper

## TABLE II

Magnitudes of Regional Sea-Surface Slopes Produced by Synoptic Influences

| | | |
|---|---|---|
| Equilibrium tidal slopes in mid-ocean | | $10^{-10}$ |
| Wind set-up in open ocean (trades) | | $10^{-7}$ |
| Sea-surface slopes accompanying baroclinic or barotropic geostrophic flows (middle latitudes) | 1 cm/sec | $10^{-7}$ |
| | 10 cm/sec | $10^{-6}$ |
| | 100 cm/sec | $10^{-5}$ |
| Shelf tides; Tsunamis in deep water | | $10^{-5}$ |
| Quasi-hydrostatic barometric loading | | |
|    Margins of anticyclones | | $10^{-5}$ |
|    Cores of extra-tropical cyclones | | $10^{-5}$ |
|    Hurricanes (near eye) | | $10^{-4}$ to $10^{-3}$ |
| Wind set-up by coastal gales | | $10^{-4}$ |
| Tidal slopes in embayments and canals | | $10^{-4}$ |
| Bores, capillary and ordinary gravity waves | | (not regional) |

water the changing loads of atmospheric pressure on the sea surface are very soon (within 12 pendulum hours) compensated by changes in the surface elevation (about 1 cm change of water elevation per mb of atmospheric pressure change) which cancel the horizontal gradients of pressure in the remainder of the water column. Since the slopes of the sea surface associated with barometric loading can be assessed and eliminated and those associated with the astronomical tide are small (Table II), it seems probable that slopes related to seasonal or longer period influences may be observed in relatively pure form. These arise, it is thought, from the regional average stress of wind on the sea surface or the climatological factors influencing the distribution of water densities within the body of the ocean.

Except near coasts, where set-up may occur, the stress of the mean wind should be accompanied by an equilibrium slope transverse to the average wind direction. Slopes associated with horizontal gradients of water density should parallel these gradients when motional equilibrium has been attained. The time required to establish an equilibrium circulation and associated sea-surface slope may range from pendulum days to many years depending upon the latitude and on whether the response of the ocean is of a barotropic or baroclinic nature (Veronis, 1956; Veronis and Stommel, 1956).

As can be seen from Table II, the departures of the mean sea surface from the horizontal are so small that in good weather it has been useful to consider the average horizon as the boundary of a level surface on the earth. But the circle of the horizon does not lie at 90° to the zenith, being in excess of this due

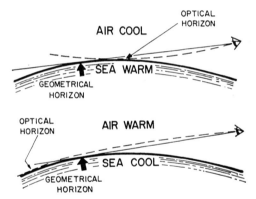

Fig. 12. Effects of thermal lapse rate on the range to the visible horizon from a given height of eye. (After Minnaert, 1954.)

to dip or occasionally less than one right angle because of the effects of meteorological refraction (Brocks, 1950). The surface joining the circle of the horizon with the observer's eye usually tends to be a very flat cone, apex up (or occasionally down), that is apparently buckled by horizontal gradients in the lapse rate of temperature in the air and sea, actually buckled by waves, tilted by regional gradients of atmospheric pressure, and the steric or dynamic gradients of surface elevation associated with fluid motion.

The horizontal distance to the optical horizon varies with the air–sea temperature contrast. When the air over the sea surface is colder than the water the visible horizon tends to move closer to the observer than the locus of tangent rays, and be pushed beyond the geometrical horizon when the air is warm and the sea is cool, as suggested in Fig. 12. Altogether these effects tend to make optical measurements of the zenith angle of the horizon rather uncertain and, possibly, of dubious value for the measurement of regional slopes, except when the air–sea temperature contrast is almost nil.

As a means for studying the changing behavior of the horizon, a dipmeter was built, as shown in Fig. 13a, b, to make optical measurements of the zenith angle of the horizon from land-based stations. Basically this device consists of a good-quality 5-in. refracting telescope fitted with a reticle eyepiece against which the altitude of the visible horizon is measured. The telescope is leveled by first blocking the view of the horizon with a shutter beyond the pentaprism and then illuminating the reticle from the eye end so as to produce a collimated beam of light. The collimated beam enters a pentaprism and is reflected back upon itself from a quiet mercury surface housed in the cylindrical compartment between the tripod legs. When properly focused, both the reticle and its image can be clearly seen at the same time and made to coincide through suitable rotations of the alignment screws.

In principle, the reticle and its image define a line normal to the mercury surface. If the reticle and its image are made to coincide before each observation of the horizon on a new azimuth, the apparent departure of the horizon from 90° zenith angle can be measured directly.

But the pentaprism deviation differs slightly from 90°. This error introduces a small systematic angular elevation or depression of the optical axis of the telescope from horizontal. Owing to dip and the prism error, the regional inclination of the circle of the horizon with respect to local vertical can be obtained only by plotting the apparent zenith angle of the horizon as a function of azimuth relative to the center of a polar diagram. If the horizon figure lies some 90° from the local vertical, the apparent zenith will be uniformly separated from the center of the polar diagram on all azimuths and a circle of best fit will have its center at the origin, as shown in Fig. 14. In the presence of a regional slope, however, the center of the circle of best fit will depart from the center of the polar diagram by an amount, $\tau$, proportional to the angle between the local vertical and the normal to the horizon plane. The actual patterns of points obtained from any station may not be circular owing to real or apparent buckling of the horizon, as, for example, when the azimuth of view swings across the backs of waves and into their troughs. The assumptions made in this treatment of the observations are that the hydrostatic effects of barometric loading can be measured and eliminated, and that the direction of gravity within the circle of the horizon is conically symmetrical around the point from which the observations are made. It is also assumed that the effects of atmospheric refraction are uniform on all azimuths. This last is, perhaps, the weakest point of the method except for the fact that, at sea, the lapse rate of temperature through the sea surface into the lower layers of the air can be remarkably uniform over large areas.

In good weather, when the temperature difference between the air and the sea is not greater than 1°C and the winds are so light as to produce only small ripples (when the sea is glassy smooth the horizon is invisible), the horizon can

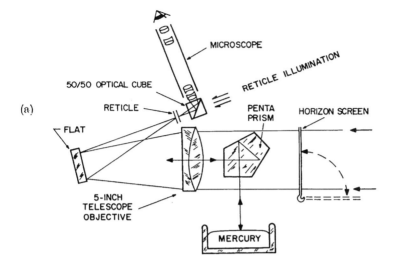

(a)

(b)

Fig. 13. (a) Arrangement of parts in the dipmeter shown in Fig. 13b. (b) The 5-in. dipmeter for land stations, shown diagrammatically in Fig. 13a, with the horizon shutter
open and the mercury dish exposed. The resolving power of the optical train is about
1 second of arc.

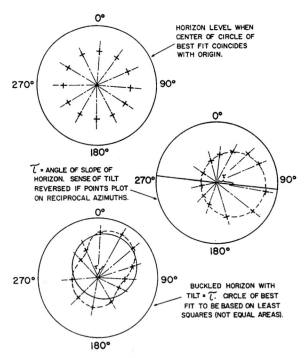

Fig. 14. Dipmeter observations on several azimuths can be composed on polar diagrams to reveal the altitude of the horizon circle with respect to the zenith. Three cases are shown.

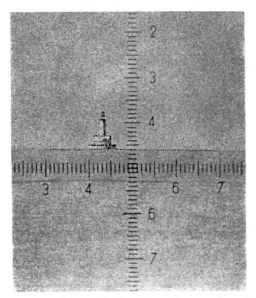

Fig. 15. Horizon at a range of 6 nautical miles under nearly calm conditions. Reticle interval is 60 seconds of arc per division.

be exquisitely sharp (Fig. 15). Slow changes in the mean atmospheric refractive index along the line of sight cause the horizon line to rise and fall each 5 min or so through an amplitude of some 3 to 5 seconds of arc. An average position can probably be obtained to an accuracy approaching one second of arc from 100 observations scattered through a total of some 20 min time.

Under less favorable conditions, as when the "seeing" is bad, the image of the horizon appears to heave violently under even moderate optical magnification (Fig. 16). Mirage and loom effects also contribute important disturbances

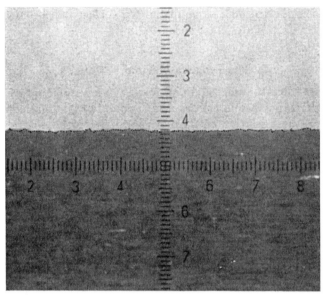

Fig. 16. Horizon at a range of 15 nautical miles under conditions of poor seeing. Reticle interval is 60 seconds of arc per division; sea state is force 2.

amounting to several minutes of arc. One of the most interesting of these refractive effects is the pronounced vertical extension of objects on the horizon when cold air moves out over warm water in the fall of the year. Not only is the horizon closer but the amplitudes of low waves are greatly exaggerated. This effect can be so extreme as to lead to alarming overestimates of the sea state or, indeed, for wave crests to appear square shouldered or even as lenticles hanging in the air. In any of these conditions the zenith angle of the horizon cannot be measured with any certainty, except possibly from stations having a height of eye in excess of 10 m. From such elevations the zone where light passes through the steepest lapse rate of air temperature near the sea surface is far enough removed from the observer to make its angular disturbances small. While this provides an advantage, too much height of eye may permit the haze on long light paths to make the horizon quite indefinite.

From preliminary inspection of the image quality under a variety of temperature conditions, it would seem that observations having sufficient coherence to deal with slopes as small as $10^{-5}$ must be made under light winds and at times when the air–sea temperature difference is less than 1°C. Unfortunately the principal regions of the ocean where air–sea temperature contrasts tend to be this small are also regions where currents, and hence sea-surface slopes, also tend to be minimal. In the vicinity of strong but warm currents, such as the Gulf Stream, air–sea temperature differences can be quite large. At best, then, one may expect optical methods to be useful in the vicinity of strong currents only under unusual circumstances. But given these, and a suitably precise reference vertical, it seems to lie within the range of future possibility that measurements of the regional slope of the sea surface can be made by optical means where the slopes are as large as $10^{-5}$.

### 3. The Problem of Time Variations

Were any method to succeed in providing measurements of the regional slope of an isobaric surface in the sea as a function of long periods of time, and thus yield the time variation of the horizontal pressure field in the ocean, it would be possible to deal much more completely than it is now practicable with the question of the variability of the ocean circulation around the climatological mean. Because of the very difficulty of this problem, we have not only tended to become preoccupied with those simplified forms of the equations of frictionless (geostrophic) motion which neglect accelerations but have further restricted their uses to the components of the horizontal pressure field associated with the internal distribution of density. This limitation is the result of the practical assumption, made in the method of dynamic sections, that the sea surface departs from level only insofar as *steric* anomalies permit it to do so. Yet it is well known that an entirely homogeneous ocean can develop sea-surface slopes, possess horizontal pressure gradients and develop accompanying barotropic circulations which accommodate themselves to environmental changes much more quickly than any of the baroclinic processes (Veronis and Stommel, 1956). That the barotropic circulation, in steady state, can resemble the observed ocean circulation very closely has already been shown in experiments with rotating models (von Arx, 1957). These experiments have also revealed the conspicuous nature of the variability that may be expected in an ocean responding barotropically to seasonal changes in the wind field.

Since the slope of the sea surface is related to both barotropic and baroclinic influences, direct measurements of this regional property could help to describe the ocean circulation not only in absolute terms and at all depths but with reference to volumes of water large enough to have significance in the general circulation.

### References

Brocks, K., 1950. Die Lichtstrahlkrümmung in Bodennähe. *Deut. Hydrog. Z.*, **3**, 241–248.
Brocks, K., 1951. Refraktionsmessungen über See. *Z. Vermess.*, Hft. 3.

Minnaert, M., 1954. *The Nature of Light and Color in the Open Air*. Dover Publications Inc., New York, 362 pp.

Schuler, M., 1923. Die Störung von Pendel-und-Kreisel Apparaten durch die Beschleunigung der Fahrzenges. *Phys. Z.*, **24**, 344–350.

Veronis, G., 1956. Partition of energy between geostrophic and non-geostrophic ocean motions. *Deep-Sea Res.*, **3**, 157–177.

Veronis, G. and H. Stommel, 1956. The action of a variable wind stress on a stratified ocean. *J. Mar. Res.*, **15**, 43–75.

Wrigley, W., 1941. An investigation of methods available for indicating the direction of the vertical from moving bases. ScD. Thesis, Dept. of Phys., Mass. Inst. Technology, Cambridge, Mass.

Wrigley, W., 1950. Schuler tuning characteristics in navigational instruments. *Navigation*, **2**, 282–290.

von Arx, W. S., 1957. An experimental approach to problems in physical oceanography. In *Physics and Chemistry of the Earth*, **2**, 1–29. Pergamon Press, London.

# IV. BIOLOGICAL OCEANOGRAPHY

## 17. GEOGRAPHIC VARIATIONS IN PRODUCTIVITY

### J. H. RYTHER

### 1. Introduction

The production of organic matter on land is determined by numerous environmental factors which limit photosynthesis or plant growth. The more important of these include the nature of the physical substratum, temperature, moisture, carbon dioxide, chemical nutrients and illumination. Several of these factors may also limit the growth of marine plants in restricted coastal and estuarine situations, but in the open sea it is only the latter two, nutrients and light, which are significant. It is the variability of these factors and of the hydrographic features which control their availability that are responsible for the geographic variations in marine productivity. It is, therefore, pertinent to discuss in some detail and without reference to specific locations, the manner in which and the degree to which light and nutrients control and limit the rate of primary production in the sea. With that information as a background, the range and magnitude of the process in representative parts of the ocean will be considered. While these regions will be selected to illustrate the various controlling mechanisms which will be discussed, the selection is limited by the availability of data. This chapter, then, will consist primarily of a discussion of illustrated principles. No attempt will be made to describe the productivity of the oceans as a whole or even to summarize all of the existing data on the subject.

### 2. Incident Radiation

One of the most obvious variable factors influencing plant production is the amount of solar energy reaching the ocean's surface. Changing weather patterns may cause extreme short-term variations in incident radiation of an order of magnitude or more which may appreciably influence day-to-day rates of primary production (e.g. Sorokin and Koblentz-Mishke, 1958; Currie, 1958; McAllister et al., 1959). However, these effects would presumably average out over longer periods of time to values relating to the mean radiation for the period. Radiation data, averaged over periods of a month or longer, are more or less constant and predictable for a given season and latitude, and such data have been compiled by Kimball (1928), Kennedy (1949), Budyko (1955) and others.

On the basis of an empirically derived relationship between light intensity and the photosynthesis of some marine planktonic algae, the relative

[MS received June, 1960]     347

photosynthesis of phytoplankton beneath a square meter of sea surface may be expressed as a function of total daily incident radiation (Ryther, 1956). By means of this relationship and the average radiation data of Kimball (*loc. cit.*) the effects of seasonal and latitudinal changes in radiation on potential plant growth may be demonstrated (Fig. 1). This picture is, of course, unrealistic,

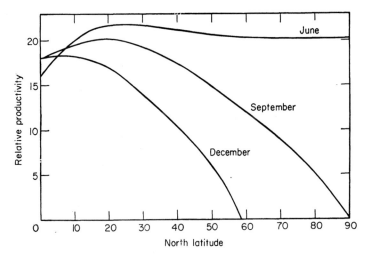

Fig. 1. Relative photosynthesis as a function of mean radiation at different latitudes and seasons.

for the effects of radiation must be considered in relation to other factors. It does illustrate the seasonal and geographical variability of the radiation factor *per se*, which is obviously critical at high latitudes and relatively unimportant in temperate and tropical seas.

### 3. Transparency

Of more immediate significance to organic production than the radiation falling on the sea surface is the amount of light available to the plants themselves. This is, of course, a function of the incident radiation, but is also influenced by the transparency of the water, the concentration of phytoplankton and the presence of other particulate or dissolved substances which absorb light.

Plant growth may occur whenever photosynthesis exceeds respiration; the light intensity at which the two processes are equal is the compensation intensity. This is somewhat variable between species and even between differently adapted members of the same species, but is of the general order of 100 ft-candles.

The compensation depth in the sea is the depth at which the compensation intensity occurs, and the water layer above this, where plant growth may occur, is known as the euphotic zone. For a given compensation intensity, the compensation depth is a function of the incident radiation and the transparency of the water. It is a common misconception that the compensation depth is constant for a given transparency. Obviously it changes throughout the day and has no meaning at night. Hence it is convenient to define the compensation depth more liberally as the maximum depth at which plant growth may occur under clear skies and with the sun overhead. Thus qualified, this parameter is more or less constant for a given body of water and hence is a useful concept for comparative purposes. Since the maximum intensity of visible sunlight at noon is of the order of 10,000 ft-candles and the compensation intensity about 100 ft-candles, it is adequate for most purposes to consider the compensation depth as that depth to which 1% of the incident visible radiation penetrates.

Sunlight in the ocean is attenuated with depth at an exponential rate which may be expressed as the attenuation or extinction coefficient ($k$), the natural logarithm of the fraction of incident light penetrating to a given depth. Conventionally, $k$ is expressed as light extinction per meter of depth. There have been many local studies of light penetration, at various times and places, and extensive surveys of large parts of the earth's oceans by Jerlov (1951) and Steemann Nielsen and Jensen (1957).

If we disregard the selective absorption of light of different wavelengths and consider the total visible spectrum, we may say that for clear ocean water $k$ equals 0.04–0.05, representing a compensation depth of 100–120 m. These values are typical of tropical and semi-tropical seas (the Sargasso, western Pacific, Mediterranean and Caribbean Seas) and appear to be fairly constant throughout the year.

The temperate and northern parts of the oceans tend to be more turbid and more seasonally variable, with values for $k$ typically ranging between 0.10 and 0.20 and corresponding compensation depths of 50–25 m. This is largely due to the greater abundance of phytoplankton, other organisms and their decomposition products in these waters. Although the annual rates of primary production in temperate and tropical seas may not differ appreciably, as will be discussed below, much larger populations of plants and animals develop in the former regions. These populations and the detritus associated with them are an important factor in reducing the transparency of the water.

In coastal and inshore regions, transparency is too variable to be assigned an average value, but it is generally more turbid than offshore waters, with extinction coefficients as high as 1.0 (compensation depth 5 m) not uncommon. This is again due partially to higher densities of plankton and organic detritus, but may also result from suspended bottom sediments, particularly in turbulent shallow waters. Furthermore, there is associated with coastal waters a dissolved organic "yellow substance", described and studied by Kalle (1938), which may contribute significantly to the absorption of daylight.

The highest possible rates of organic production occur in situations approaching a hypothetical optimum where all incident radiation is absorbed by plants. This possibility is precluded in the clear blue open sea where most of the radiation is absorbed by the water itself. Since the extinction coefficient of visible light in pure sea-water is about 0.035, a coefficient of 0.040 implies that some 90% of the light is absorbed by the water. High transparency, then, is symptomatic of, but not in itself a cause of, low productivity. It merely reflects a paucity of light-absorbing organisms, but does not explain why they are scarce.

As a corollary to the above statement we may say that organic production beneath a unit of surface area is inversely proportional to the depth of the euphotic zone. This was well demonstrated by Steemann Nielsen and Jensen (1957), whose illustration is reproduced here (Fig. 2), and by Berge (1958), but

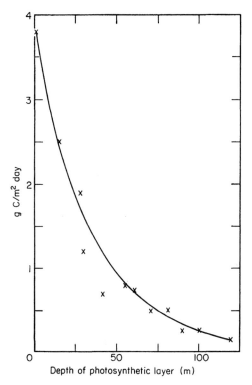

Fig. 2. The relation between productivity and the depth of the euphotic zone. (After Steemann Nielsen and Jensen, 1957.)

it is true only to the extent that increased turbidity is due to increased concentrations of photosynthetic organisms. Such is not necessarily true, as discussed above.

Riley (1956) has determined empirically the relationship between transparency and phytoplankton density (expressed as chlorophyll) in regions where he assumed light absorption by material other than water and plants to be negligible. His expression is

$$k = 0.04 + 0.0088C + 0.054C^{\frac{2}{3}},$$

where $C$ is chlorophyll in micrograms per liter and the extinction coefficient of pure sea-water is taken as 0.04. By comparing coefficients calculated by this means with those observed in Long Island Sound, Riley concluded that two-thirds of the light extinction of that region is caused by non-living particulate matter in suspension. The same method indicated that no more than half the turbidity of continental-shelf waters off the eastern United States is attributable to living phytoplankton (Ryther, unpublished data).

Looking at organic production as the utilization of solar radiation then, we see the contrast between the clear tropical seas on the one hand, where the organisms are so scarce that the water absorbs most of the energy, and the more turbid northern seas and coastal regions on the other, where the organisms must compete with their own excreta, exuvia and remains and with terrigenous and sedimentary materials for the available light. Though the causes are quite different, the end result in both situations is the utilization of an almost equally small fraction of the light energy which penetrates the sea surface. This is one reason why organic production, when considered over long periods of time, does not appear to differ greatly between temperate and tropical seas and between oceanic, coastal and estuarine environments—a hypothesis, first advanced by Riley (1941), which will be explored below.

## 4. Nutrients

The availability of nutrients is the other environmental factor which critically limits organic production in the ocean as a whole. This, however, must be considered in relation to the previously discussed factor, illumination, for it is only the nutrients available in the upper, euphotic layers which are pertinent to the discussion.

A brief comparison of the situation in the ocean with that on land will serve to illustrate the relative infertility of the former. Nitrogen will be used as an example in this discussion and it will be assumed that other nutrients are present relative to nitrogen in amounts roughly proportional to their con-centrations in plant tissue, an assumption which has some validity at least for the ocean (e.g. Redfield, 1934). Rich fertile soil contains some 5% organic humus and as much as 0.5% nitrogen. This and the accompanying nutrients in a cubic meter of rich soil, together with atmospherically-supplied carbon, hydrogen and oxygen, can support a crop of some 50 kg of dry organic matter, an amount equivalent to more than 200 tons per acre of soil 3 ft deep.

Under optimal conditions plants are capable of converting the solar energy falling on a square meter of surface to an organic yield of the order of 10 g/day in excess of their own metabolic requirements (Ryther, 1959), an annual produc-tion of several kilograms per square meter. If terrestrial plants can sink their roots into 3 ft of rich soil, they have access, then, to enough nutrients to grow at their maximum potential rate for periods of several to many years. Thus forests, representing the accumulation of decades of organic production, are possible.

There are no comparable forests of plankton in the sea. The richest ocean water, exclusive of local polluted areas, contains about 60 μg atoms/l. or 0.00005% nitrogen, four orders of magnitude less than fertile land. A cubic meter of this sea-water could support a crop of no more than about 5 g of dry organic matter.

As discussed above, the maximum depth of the euphotic zone is about 100 m. This surface layer is normally poor in nutrients relative to the deeper waters and seldom contains more than 10–20 μg atoms/l. of available nitrogen (e.g. Riley, 1951). A 100-m euphotic zone containing nitrogen at a concentration of, say, 15 μg atoms/l. represents a reservoir of 21 g of nitrogen in a 1-m² column. However, as phytoplankton grows in this water column, the organisms themselves absorb more light, the euphotic zone becomes progressively shortened, and the reservoir of nutrients available to the plants is correspondingly diminished in size. As the plant population grows, the organisms not only consume nutrients but also shut themselves off from their supply. Table I illustrates this

<h2 style="text-align:center">TABLE I</h2>

Relationships between Chlorophyll, Standing Crop of Organic Matter, Transparency, Organic Production and Nitrogen Availability and Requirement under Conditions Stipulated in Text

| 1 | 2 | 3 | 4 | 5 | 6 | 7 | 8 |
|---|---|---|---|---|---|---|---|
| Chl. a, g/l. | Standing crop, mg/m³ dry wt. | Extinc. coeff., $k$ | Depth euphotic zone, m | Calc. org. prod., g dry wt./m²/day | N initially available in euphotic zone, mg | N required to produce existing population, mg | N requirement, mg/day |
| 0 | 0 | 0.04 | 120 | 0 | 25,300 | 0 | 0 |
| 0.1 | 10 | 0.07 | 66 | 0.20 | 13,800 | 66 | 20 |
| 0.5 | 50 | 0.13 | 36 | 0.50 | 7600 | 180 | 50 |
| 1.0 | 100 | 0.19 | 24 | 0.75 | 5000 | 240 | 75 |
| 2.0 | 200 | 0.29 | 15 | 1.00 | 3100 | 300 | 100 |
| 5.0 | 500 | 0.50 | 10 | 1.40 | 2100 | 500 | 140 |
| 10.0 | 1000 | 0.79 | 6 | 1.80 | 1260 | 600 | 180 |
| 20.0 | 2000 | 1.30 | 3.5 | 2.20 | 735 | 700 | 220 |

by showing the relationship between the standing crop of phytoplankton, the rate of organic production, water transparency and the daily requirement and availability of nitrogen in a hypothetical situation where the water initially, with no phytoplankton present, contained 15 μg atoms/l. of nitrogen. The table is based upon the relationship between chlorophyll and transparency proposed by Riley and discussed above. It also assumes that dry phytoplankton contains 1% chlorophyll and 10% nitrogen. Production is calculated from chlorophyll and light penetration for an incident radiation of 400 g cals/cm²/day from the equations of Ryther and Yentsch (1957).

According to Table I, by the time a phytoplankton crop of 2 g/m³ has developed, the euphotic zone is limited to 3.5 m and the nitrogen in this water is exhausted. The same situation is depicted in Fig. 3, showing the maximum standing crop which can develop at each depth before the light becomes extinguished or the nitrogen consumed. It is obvious that only at the very surface can dense populations develop. Even if all the organisms produced were able to persist in the water column after the light had been shut out from

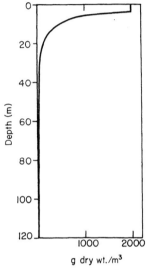

Fig. 3. The population of phytoplankton, expressed as g dry wt./m³, which can develop at different depths within a water column under conditions stipulated in the text.

above, the maximum amount of organic matter which could accrue beneath a square meter of ocean surface (obtained by integrating the curve in Fig. 3) is no more than about 25 g—less than one-thousandth the potential standing crop on land.

In terms of the population of plants which may exist at any time, then, the sea is a desert compared to moderately productive land. Furthermore, although the rate of phytoplankton production in the sea may momentarily approach or equal the highest observed rates on land (Ryther, 1959), these high rates may be maintained *in a given body of water* for no more than a matter of days in the sea in contrast to years on land. High levels of marine production may be maintained only by the constant replacement of the water in which the organisms have grown with a new supply of nutrient-rich water. In other words, high production may persist only in terms of a geographical location through which there flows a continuous supply of new, nutrient-rich water.

Upwelling of rich water from intermediate depths to the surface does occur more or less continuously in certain restricted areas. Over most of the ocean, however, no mechanism exists for such regular and persistent vertical transport; the only process contributing to surface enrichment is vertical turbulence. This varies seasonally and geographically and its variability determines, by and large, the variability of the ocean's fertility. To some extent, then, productivity may be correlated directly with the degree of vertical mixing. Here again, however, light and nutrients must be considered together, for it is the effect of turbulence upon these two factors which controls plant production. For example, vertical mixing not only brings nutrients to the surface, but also carries plant cells downward. When the mixed layer exceeds the depth of the euphotic zone by several-fold, the organisms may spend a majority of their time in darkness, unable to grow. The concept of a "critical depth" of the wind-mixed layer, the depth below which respiration exceeds photosynthesis for the whole population within the mixed layer, has been employed by Gran and Braarud (1935), Riley (1942) and Sverdrup (1953), and a quantitative treatment of the subject by Riley (1957) will be discussed later in some detail.

Thus in providing the phytoplankton with one essential requirement, that is nutrients, turbulence deprives the organisms of their other basic need, namely illumination. For this reason neither vertical stability nor strong turbulence, in themselves, contribute to high productivity. Favorable conditions for growth stem rather from an alternation of these conditions, turbulence enriching the surface layers followed by stability allowing the phytoplankton to remain in the euphotic zone long enough to utilize the nutrients thus provided. Over a large part of the ocean the alternation from vertical instability to stability is a seasonal phenomenon, and it is this periodicity which causes the well known seasonal cycles of phytoplankton maxima and minima. In the following discussion we shall see how variations in the degree of turbulence and/or stability, as well as incident radiation, water transparency and depth, determine geographic variations in the general level, annual rate and seasonal cycles of primary production.

### 5. Latitudinal Variations in the Stability of Surface Water

Before discussing the productivity of specific parts of the ocean, we may examine briefly, and in the light of the preceding discussion, the variations with latitude of the thermal structure of the surface layers. We have already described the latitudinal variability of incident radiation (Fig. 1) but pointed out that this factor *per se* could not be assessed without simultaneous consideration of nutrients. Since vertical turbulence affects, in opposite ways, the availability of both light and nutrients, it is clear that the large difference in thermal structure between tropical and polar surface waters must play an important role in determining the relative fertility of the two regions. This effect is heightened by the accompanying changes in incident radiation.

Fig. 4 shows the seasonal cycle of temperature in the upper 300 m at latitudes of 0°, 20°, 40° and 60° in the North Atlantic Ocean. The figure represents a rather subjective interpretation of bathythermograph records on file at the

Woods Hole Oceanographic Institution. As nearly as possible the data are taken from one-degree squares of latitude and longitude, but records from the closest position available have been interpolated liberally. Data for the three southern latitudes represent mid-oceanic positions with corresponding west longitudes of approximately 35°, 38° and 31° from south to north respectively. The data for 60°N originated from the Norwegian Fisheries Investigations and represents temperatures off the west coast of Norway.

Such data as these do not, of course, provide a reliable index of vertical mixing, surface enrichment and other related phenomena. They do give some indication of the relative stability of the surface waters on a seasonal basis at the four different latitudes. It is clear, for example, that the euphotic zone is thermally stratified throughout the year at both 0° and 20°. At the equator,

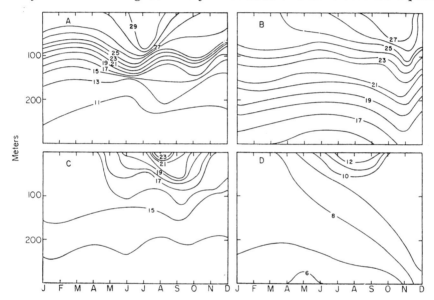

Fig. 4. The thermal structure of the upper 300 m throughout the year at the following approximate locations: (A) 0°N, 35°W; (B) 20°N, 38°W; (C) 40°N, 31°W; and (D) 61°N, 03°W.

particularly, the thermocline is strongly developed just at the compensation depth. At the temperate latitude of 40° the upper 100 m shows pronounced stratification only during the summer though a thermal gradient exists for over half the year. From November to April, however, the surface waters are isothermal and potentially wind-mixed to depths below 100 m. Off the Norwegian coast, at 60°, vernal warming is far less pronounced; surface temperatures never exceed 12°C and the surface layers are only weakly stratified in mid-summer. In winter, these waters are isothermal to depths greater than 200 m.

Although the degree to which the euphotic layer is enriched from below by wind-mixing and eddy diffusivity cannot be estimated quantitatively, it is clear that the process is minimal and seasonally invariable at the equator, maximal at 60° and most pronounced in the temperate and northern seas in winter. On the basis of nutrient availability alone, one would predict a low and nearly constant rate of organic production in the tropics, low summer but relatively high winter rates at 40° and appreciably greater production throughout the year at 60°.

Turning to light as a limiting factor, we must keep in mind three aspects of this factor: (1) incident radiation, (2) the transparency of the water and (3) the depth of the wind-mixed layer. In the tropical latitudes, radiation is relatively high and constant throughout the year, the water is exceptionally clear and thermal stability prevents the plants from being mixed out of the euphotic zone. At 40°, however, and more strikingly at 60°, radiation is appreciably lower in winter, the waters tend to be more turbid and mixing may carry the plants to depths 5–10 times as great as the euphotic zone. To the extent that isothermal water is indicative of such mixing, plant growth is clearly inhibited during the winter at 40°, and for half the year or more at 60°.

By now it is apparent that the two factors which control plankton growth in the sea, though seldom limiting at the same time, are also seldom available simultaneously in amounts capable of sustaining high production. When and where, in the four situations we have discussed, are both light and nutrients present in non-limiting quantities? In tropical seas probably never except in those peculiar regions to be discussed later where hydrodynamic features bring rich water to the surface; in temperate and northern seas, only for that brief period in spring when surface waters, enriched by winter mixing, become thermally stratified. Possibly again in the fall, when the thermocline begins to break down and periods of mixing alternate with periods of stability, conditions may again become favorable for moderately high production. The timing of the spring and fall maxima varies from year to year, depending upon weather conditions, but generally occur respectively later and earlier at progressively higher latitudes and may merge together as a single summer flowering in boreal and austral regions.

The low level of production which persists in most of the tropics and in temperate waters during the summer is maintained by the slow upward diffusion of nutrients through the thermocline and, probably more important, by the complete cycling of these materials within the euphotic zone itself. There is at present no suitable way of evaluating either of these processes. It seems probable, however, that the bacteriological remineralization of organic matter would be enhanced by the relatively high temperatures which prevail in the tropics and in summer temperate conditions.

There are, of course, many exceptions to the general picture presented above. One in particular deserves mention here. A column of water of the same density is unstable, but instability is a non-dynamic characteristic which requires a force such as wind stress to produce vertical turbulence. There are probably many interludes of fine, calm winter weather when an unstable water-mass is not appreciably mixed. These conditions are eminently favorable for phyto-plankton growth and may result in brief intermittent bursts of plant growth throughout the winter in temperate seas. Riley (1957) discusses such occurrences in the North Atlantic and one particular example has been described by Ryther and Hulbert (1960). Conversely, violent storms may temporarily destroy the thermocline in tropical and temperate summer seas, enriching the surface layers sufficiently to stimulate production briefly. The importance of this factor is perhaps questionable in tropical regions where the waters are highly stratified. Francis and Stommel (1953) noted that gale force winds were capable of deepening the mixed layer by no more than 20–30 ft in the Gulf of Mexico. However, Steele (1956) considers wind action an important stimulus to summer production in the less stable waters of the North Sea.

## 6. Productivity of the Major Oceanic Regions

The following section will consist of a brief résumé of selected studies of primary production in the principal oceanic regions of the world, i.e. in tropical, semi-tropical, temperate and sub-polar, arctic and antarctic. Excluded from this discussion will be areas of special interest, such as upwelling, convergences and divergences, which will be treated separately in the following section. It will be the purpose of this review to illustrate the principles and substantiate the hypotheses presented in the preceding discussions. The realization of the objectives is made difficult by the fact that data are widely scattered geo-graphically, usually limited to single observations for a given location, and often may not be comparable because of the use of different methods. Although no attempt will be made to include all of the pertinent literature on the subject, some of the major oceanic regions are not represented by any productivity studies as such and must be evaluated by indirect evidence seasoned with speculation. Productivity data reported here will be limited to those expressed on an areal basis as grams of carbon fixed beneath a square meter of sea surface per day and, for comparative purposes, will be largely restricted to those obtained by the [14]C technique or to methods which are believed to give comparable results.

### A. Tropical Seas

During the *Galathea* expedition, Steemann Nielsen and Jensen (1957) measured productivity in the tropical Atlantic, Pacific and Indian Oceans. Jitts and Rotschi (1958), on expedition "Equipac", made measurements in the Coral Sea and between the New Hebrides and Gilbert Islands. Holmes *et al.* (1957) conducted studies in the tropical Eastern Pacific (expedition "Eastropic"). Ryther and Menzel (1960) have measured production between Bermuda and the West Indies on two oceanographic cruises in March, 1959, and January–February, 1960. In all of these studies variations of the [14]C method of Steemann Nielsen (1952) were employed.

In general, though with a few notable exceptions, production rates in tropical open-ocean waters were found to be almost uniformly and consistently low, usually ranging between 0.05 and 0.15 g carbon/m²/day. Their consistency, regardless of the time of year the measurements were made, make it appear that the lack of seasonal coverage is not a serious omission. There is no need, therefore, to consider the majority of these data in any further detail.

One of the above-mentioned exceptions concerns the productivity of the Pacific equatorial region which Holmes *et al.*, Jitts and Rotschi, and Steemann Nielsen and Jensen all found to be appreciably higher (*ca.* 0.50 g carbon/m²/day) than that of waters to the north and south of the equator. The equatorial part

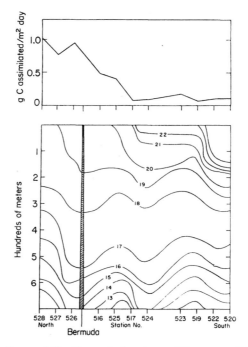

Fig. 5. Primary production and the thermal structure of the upper 700 m along a section between 34°N, 67°W and 25°N, 64°W in March, 1959. (After Ryther and Menzel, 1960.)

of the Indian Ocean was also shown by Steemann Nielsen and Jensen to have significantly higher production rates (0.20–0.25 g carbon/m²/day) than tropical waters in general.

Two *in situ* ¹⁴C measurements in the Mediterranean two miles south of Monaco gave values of 0.03 and 0.04 g carbon/m²/day in July and October respectively (Brouardel and Rink, 1956). Three measurements by Steemann Nielsen and Jensen and three by the author (unpublished) in the Caribbean Sea (10°–20°N lat.) gave values between 0.10 and 0.20 g carbon/m²/day. The *Galathea* section from the West Indies to Europe, which crossed the Atlantic between 20° and 40° in mid-summer, provided the lowest productivity values encountered on the entire expedition.

The results from the cruise by R.V. *Crawford* from 35° to 25°N latitude in March, 1959, are interesting since they show the transition from semi-tropical to tropical conditions during the peak of the spring flowering at the northern end of the section. Fig. 5, from Ryther and Menzel (1960), illustrates how high productivity (*ca.* 1.00 g carbon/m²/day) is associated with isothermal wind-mixed surface water in the Bermuda region, and how low values (*ca.* 0.1 g carbon/m²/day), typical of the tropics, were encountered to the south as soon as thermal stratification of the euphotic zone became apparent. A similar section in January–February, 1960, from 30°N, 53°W southwest to the West Indies (11°N) confirmed the earlier results. Since these studies were made in mid-winter, when cooling and mixing are presumably most pronounced, it seems doubtful that higher production rates normally occur in that part of the ocean. Unfortunately no measurements have been made in the equatorial Atlantic, much of which is affected by African and South American coastal influences.

### B. The Sargasso Sea

Productivity values for the Sargasso Sea obtained by the *Galathea* section in June were the lowest encountered on that expedition (*ca.* 0.05) and were at variance with earlier estimates for the same area (Riley *et al.*, 1949). The reason for this became apparent when the seasonal cycle of organic production for this area was considered. From 1957 to 1960 measurements were made of primary production and related physical, chemical and biological features of the Sargasso Sea a short distance from Bermuda but in typical oceanic conditions. Observations were made at two-week intervals for a period of three years, representing the only attempt of its kind to study intensively the seasonal cycle and annual variability of primary production in the open sea. Observations for the first half of the investigation have been described by Menzel and Ryther (1960).

That part of the Sargasso Sea (32°N) may be considered semi-tropical, having characteristics intermediate between temperate and truly tropical situations. In addition there are certain unique hydrographic features which play an important role in controlling the productivity of the area, most important of which is the existence of a permanent thermocline beginning at 400–500 m. A typical summer thermocline develops in the upper 100 m, below which, and above the permanent thermocline, there lies a large body of water at a temperature of approximately 18°C. In winter the surface-water temperature may fall to 18°C, in which case the summer thermocline is completely destroyed and the surface layers become isothermal and mixed to depths in excess of 300 m. The 18° water is relatively poor in nutrients (0.1–0.3 µg atom/l. $PO_4^{3-}$-P, 1–3 µg atoms/l. $NO_3^-$-N) but richer than the euphotic zone where these nutrients are almost undetectable in summer. Thus mixing of the 18°

water to the surface provides a definite though limited stimulus to plant production. At the same time, because nutrient concentrations are not great enough to permit the development of dense plant populations and because incident radiation is relatively high in winter, the euphotic zone remains deep. It is an interesting fact that, although chlorophyll values vary seasonally by an order of magnitude (0.1–1.0 µg/l.), the transparency remains almost unchanged, presumably maintained by a balance between living and non-living material. Thus the phytoplankton is not mixed out of the euphotic zone in winter for long enough periods to prevent its growth, and production is maintained at relatively high levels throughout the winter. In spring, at the onset of thermal stratification, there is a population maximum, but it is more the climax of a gradual build-up than the sudden outburst of the spring flowering typical of more temperate regions.

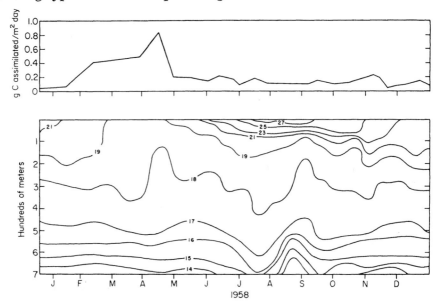

Fig. 6. Primary production and the thermal structure of the upper 700 m in the Sargasso Sea off Bermuda. (After Menzel and Ryther, 1960.)

The seasonal cycle of primary production and the vertical temperature profile of the upper 700 m for 1958 (Fig. 6) reveal that production increased with the breakdown of the summer thermocline, built up during the winter and early spring to a maximum of 0.89 g carbon/m²/day in April, and then quickly fell off to low summer rates of 0.1–0.2 g carbon/m²/day, as soon as thermal stability returned to the euphotic zone. Annual production for the period was 72 g carbon/m², perhaps twice the value which could be expected in the tropics.

The winter of 1957–1958 was severe enough to destroy completely the summer thermocline by March (Fig. 6). The winters of 1958–1959 and 1959–1960 were milder, surface temperatures did not fall below 19°–20° and a slight thermal gradient persisted throughout the year. This had a pronounced effect upon the rates of organic production, which never attained values in these two years as high as in 1958. Although analyses of data were not completed at the time of this writing, it is clear that climatic variability causes considerable year-to-year fluctuations in the seasonal pattern and the total annual primary production of this region.

## C. Temperate and Sub-polar Waters

As is true of productivity studies elsewhere in the world, most of those conducted in temperate and sub-polar seas have been made as a part of general oceanographic surveys of specific areas during restricted and often brief periods of the year. The lack of data on a seasonal or annual basis need not, of course, detract from the values of such investigations, most of which were carried out for purposes quite outside the objectives of this discussion. However, the seasonal variability of primary production in this part of the ocean is so great that regional comparisons, beyond the scope of the immediate investigation, are meaningless without knowledge of the cycle of production throughout at least one and, preferably, several years. Therefore, two investigations of north temperate areas which were concerned with the seasonal cycle of primary production will be emphasized, one in the Atlantic Ocean off the northeast coast of the United States, the other in the North Sea. Other studies of the same oceanic region may then be considered briefly in the light of the seasonal studies.

During the period September, 1956–May, 1958, primary production was measured as part of a broad investigation of the plankton ecology of the Continental Shelf waters off New York by the Woods Hole Oceanographic Institution. This discussion will be restricted to these measurements made at three stations located outside the 1000-fathom contour which are believed to be representative of oceanic conditions. Most of the productivity estimates for the area were made from chlorophyll and radiation data by the method of Ryther and Yentsch (1957). Due, however, to the problematical relationship of these data to those obtained by the $^{14}C$ method, only the relatively few values obtained by the latter technique will be reported here. Measurements by both methods during the investigation have been reported by Ryther and Yentsch (1958). Although the area was visited at rather scattered intervals throughout the year, care was taken to select those periods when the greatest contrast in biological activity could be anticipated, and it is believed that a reasonably accurate picture of the seasonal cycle of primary production resulted from these efforts (Fig. 7).

For the seventeen in situ $^{14}C$ measurements, the mean rate of production is 0.56 g carbon/m²/day, the standard deviation, 0.57, indicating the high degree of variability in the data and the need for measurements throughout the year. The weighted annual mean is 0.33 g carbon/m²/day, equivalent to a total annual production of 120 g carbon/m², twice the estimate for the Sargasso Sea and four times the annual production of the tropics. Noteworthy are the low winter and high summer rates of production relative to the Sargasso Sea. Apparently the combined effects of vertical turbulence, greater turbidity and reduced radiation are sufficient at this latitude (*ca.* 40°N) to inhibit winter production markedly. On the other hand, the higher summer rates may result from regeneration within the euphotic zone of nutrients from a considerably larger spring flowering than occurs in the Sargasso Sea. The higher production during the spring flowering (maximum observed value of 1.93 g carbon/m²/day), in turn, resulted from the enrichment of the surface layers through winter mixing with water containing nitrate and phosphate at concentrations approaching 10.0 and 1.0 μg atoms/l. respectively. These levels are nearly an order of magnitude greater than ever are encountered in the surface water of

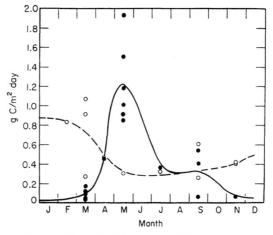

Fig. 7. Primary production as determined by *in situ* $^{14}$C measurements at offshore stations (solid line, filled circles) and shallow stations (broken line, open circles) off New York.

the Sargasso Sea, where sub-surface concentrations down to the level of mixing are low to begin with, and where the nutrients are continually assimilated as quickly as they become available throughout the winter.

Productivity studies in the North Sea were made by Steele (1956, 1958), first between the years 1951 and 1953 at a single location, the Fladen Ground, and later between 1954 and 1956 over the entire northern part of the sea. The method used was a refinement of the technique employed earlier by Kreps and Verjbinskaya (1932), Cooper (1938) and Riley *et al.* (1949) which involves the indirect estimation of plant growth from the rate of removal of phosphate from the water. Steele's treatment considered the vertical transport of phosphate within the water column but not its rate of regeneration within the euphotic zone. For that reason conservative estimates might be expected, but the author found close agreement between values obtained by that and the $^{14}$C method (Steele, 1957). Therefore the North Sea data are assumed to be comparable to others reported here.

The cycles of production on the Fladen Ground are shown in Fig. 8 and Table II. Note that no winter values (November–March) are given. There is the tacit assumption that winter production is negligible, since the total for the rest of the year is referred to as "annual production". However, the lack of winter data appears to have been due mainly to the occurrence of horizontal advection, a complication which renders Steele's method impractical.

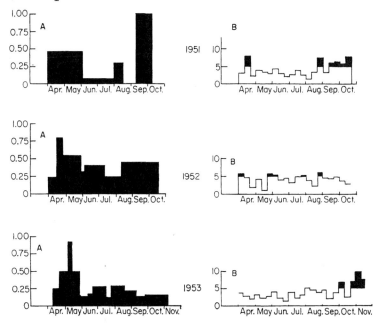

Fig. 8. Primary production in g carbon/m²/day (A) and wind strength (m/sec) (B) on the Fladen Ground. (After Steele, 1956.)

## TABLE II

Primary Production on the Fladen Ground (g carbon/m²). (After Steele, 1956.)

| Year | Spring | Summer | Autumn | Total |
|------|--------|--------|--------|-------|
| 1951 | 28.0 | 8.9 | 28.0 | 64.9 |
| 1952 | 30.4 | 22.4 | 29.5 | 82.3 |
| 1953 | 26.1 | 17.6 | 13.7 | 57.4 |

The advantage of estimating production from phosphate utilization, on the other hand, lies in the fact that the process is thereby integrated over relatively long periods of time (the intervals between sampling). This eliminates errors inherent in attempting to describe annual cycles on the basis of a series of scattered, short-term photosynthesis measurements which may be highly variable from day to day at a given place (see, for example, Fig. 7; also Currie, 1958).

For the reasons given above, Steele's data are less variable and the seasonal range is of smaller magnitude than appears to be the case with the studies based upon $^{14}C$ measurements. Nevertheless, a well-defined seasonal cycle is obvious, with spring and fall maxima and a summer minimum. Production during the spring (April 1–June 15) was relatively constant for the three years of the study, which is consistent with the hypothesis that the spring flowering follows the establishment of the thermocline and is limited by the amount of nutrients brought into the euphotic zone by winter mixing. During the summer and fall periods, however, production was quite variable from year to year, a fact which Steele has attributed largely to the variability of winds (Fig. 8) and their effect upon the vertical transport of nutrients through the thermocline in summer and the breakdown of the thermocline in autumn.

The annual production at the Fladen Ground ranged from 57.4 to 82.3 g carbon/m². Steele's subsequent estimates for 27 locations in the northern North Sea (Steele, 1958) ranged from 45 to 110. Thus production in this enclosed, rather shallow, boreal environment appears to be somewhat lower than that estimated for the open Atlantic 20° of latitude further south. It is noteworthy that the difference is due primarily to the presumed absence of plant production during North Sea winters.

Other productivity studies in temperate and northern waters over shorter periods of time deserve mention here because they illustrate local geographic variations in production, and also because they emphasize the wide range and extreme variability of the process in this general oceanic region. The measurements by Saijo and Ichimura (1959) in the northwest Pacific south of Japan during summer (May–September) show the contrast between the productivity of the warm Kuroshio Current (0.05–0.10 g carbon/m²/day) and the cold, coastal, more northerly Oyashio Current (0.25–0.50 g carbon/m²/day). Studies of the same general area during the spring (Sorokin and Koblentz-Mishke, 1958) gave equally low values for the Kuroshio, suggesting the absence of a seasonal cycle of production typical of tropical water. Their measurements further north, on the other hand, where the Oyashio is mixed with waters from the Sea of Okhatsk and from the general sub-arctic water-mass, were extremely variable, ranging from 0.006 to 5.1 g carbon/m²/day. The higher values were usually associated with "cold spots", presumably indicative of local vertical turbulence.

The boreal North Atlantic and the waters surrounding southern Greenland, Iceland and the Faeroe Islands were investigated for several years during the summer months by the Danish Institute for Fisheries and Marine Research (Steemann Nielsen, 1958; Hansen, 1959). Productivity measurements of this area (1954–1957) revealed a high degree of variability related to the complexity of different water-masses and current systems. Summer rates of production typically ranged from 0.1 to more than 2.0 g carbon/m²/day.

In general, low production was observed in Davis Strait, the East Greenland Polar Current, the Irminger Current and the North Atlantic Ocean west of the

Faeroes and south of Iceland. One of the factors contributing to the low production of the Polar Current is presumed to be the low salinity of the surface waters, which, combined with weak thermal stability, effectively prevents intermixing of the euphotic layer with nutrient-rich deep water (Steemann Nielsen, 1958). In the Irminger Sea, south of Greenland, and in the Labrador Sea to the west, winter cooling may produce vertical convection currents extending from the surface to the bottom (Sverdrup et al., 1946). Such intensive mixing must enrich the surface waters with high concentrations of nutrients. Similar locations exist north of Iceland and southwest of Spitzbergen. The productivity of these areas has not been investigated but high spring and summer rates could be expected.

Investigations in the northern Norwegian Sea in June (Berge, 1958) disclosed an area of high productivity in the central part of the sea south of Spitzbergen extending in a north–south direction for some 500 miles. This zone of high production is clearly identifiable with the tongue of warm, highly saline North Atlantic water which intrudes into the Norwegian Sea. The rate of production within the North Atlantic water was as high as 2.4 g carbon/$m^2$/day as compared with values of 0.40–0.60 in the fresher Arctic waters to the west and east. The associated hydrographic features of the area were not discussed.

## D. The Arctic

Quantitative phytoplankton investigations at Scorsby Sound, East Greenland (Digby, 1953), and at Allen Bay, Cornwallis Island, North West Territories, in 1956 by S. Apollonio (unpublished manuscript) suggest that primary production at these latitudes (70°–75°N) may attain relatively high levels during the brief period of open water. Although no productivity measurements were made, chlorophyll concentrations averaged more than 1.0 µg/l. and were found in excess of 5.0 µg/l. at Allen Bay, values comparable to those observed during the spring flowering of temperate coastal waters.

In the Polar Basin itself, however, phytoplankton production appears to be extremely low. Biological studies of this region were initiated with the Norwegian North Polar Expedition, 1883–1896 (Gran, 1904), and were continued by Soviet scientists beginning in 1937 at the drifting ice-floe station, "North Pole I" (Shirshov, 1944). All indications pointed to low or negligible plant production under the permanent ice pack. This was substantiated by subsequent investigations, including a few [14]C productivity measurements, during the International Geophysical Year 1957–1958, at the U.S. drifting stations "Alpha" (T. S. English, unpublished manuscript) and "Bravo" (Apollonio, 1959).

According to these authors, the floes of the permanent ice pack average 2–4 m thick and are covered with wind-packed snow for ten months of the year. During July and August the snow gradually melts and ponds of melt-water appear over 25–35% of the area. Open leads also develop at this time, but probably never represent more than about 1% of the ice field. Plant growth occurs in the open leads, but a more important stimulus to the production of the region as a whole is believed to be the light which penetrates the ice beneath the surface pools of melt-water, which may act as lenses. Three *in situ* [14]C measurements at Station "Alpha" in 1957 within an open lead revealed no detectable photosynthesis on July 1 and values of 0.006 and 0.005 g carbon/m$^2$/day in late July and early August. At Station "Bravo" five *in situ* [14]C measurements gave values ranging from undetectable levels in June to a maximum rate of 0.024 g carbon/m$^2$/day in July, averaging 0.01 g carbon/m$^2$/day for the period June 20–August 20. It was estimated by both Apollonio and English that the annual production of organic matter in the Polar Sea is something less than 1 g carbon/m$^2$.

## E. The Antarctic

No published measurements of primary production in the Southern Ocean were available prior to the preparation of this chapter, an unfortunate discrepancy for purposes of regional comparison. However, more than a decade of intensive exploration on the part of the *Discovery* Committee of Great Britain has made this one of the best-known oceans of the world, particularly with respect to its biology, and has left little doubt that it is the world's most fertile large oceanic region. The following brief discussion is based on publications concerning the region's physical oceanography by Deacon (1933), its nutrient chemistry by Clowes (1938) and Ruud (1930), and its phytoplankton by Hart (1934, 1942) and Hasle (1956).

The explanation for the high fertility of the Southern Ocean lies in its physical circulation. Surface water around the entire Antarctic continent spreads northward to about 50° where, at the Antarctic Convergence, it sinks below the sub-Antarctic surface water. This cold, fresh layer then continues to spread to the north beneath the surface until, at about 40°, intensive wind action causes it to become mixed with the warmer, more saline waters above and below it. Here, at the sub-tropical convergence, the mixed water sinks and moves northward as the Antarctic Intermediate Water. There is also sinking of cold water in locations near the Antarctic Continent, forming Antarctic Bottom Water which also moves to the north.

The loss of water from these two sources is compensated by the southerly flow of water from intermediate depths (2000–3000 m) which rises to the surface near the edge of the continent and thence spreads northward as the above-mentioned surface drift. Water from such great depths is, of course, extremely rich in nutrients. Where it reaches the surface, concentrations of nitrogen and phosphorus are probably higher than in any other unpolluted surface waters of the world. As this water moves away from the continent during summer, plant growth reduces the concentrations of dissolved nutrients appreciably, but even at the Antarctic Convergence the levels of these substances are still high enough for production probably never to be seriously nutrient-limited in most parts of the region.

The upwelled Antarctic surface water is fresher, warmer and hence less dense than the water 50–100 m beneath it ($\sigma t = 26.8$–$26.9$ as against $27.4$). Thus the euphotic layer, in addition to its high fertility, is very stable. The high concentrations of phytoplankton found in these waters are limited to depths of 25–50 m. North of the Antarctic Convergence, on the other hand, the waters are far less stable and the phytoplankton is an order of magnitude less dense and is dispersed over a depth of 100–150 m (Hasle, 1956).

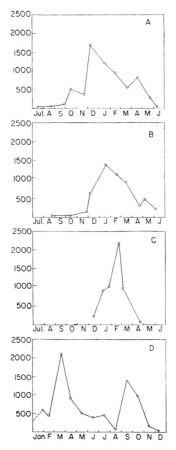

Fig. 9. Plant pigments (Harvey units/m³) in the northern (A), intermediate (B) and southern (C) regions of the Antarctic Ocean and in the English Channel (D). (After Hart, 1942.)

The generalized picture presented above is, of course, greatly oversimplified. Topographic features of the continent and sea-floor influence the vertical circulation, which, in turn, results in considerable geographical variability in the degree of surface enrichment, the details of which cannot be treated here.

Quantitative studies of the plant pigments of the region by Hart (1942) give the best indication of its relative productivity (Figs. 9 and 10). His results are

summarized on a seasonal basis for three arbitrarily defined subdivisions of latitude between the Antarctic Convergence and the continent. The data, expressed
as Harvey pigment units, were obtained from plankton-net hauls, which normally give much lower values than are obtained by modern ultra-filtration
techniques. Thus Hart's results cannot be compared with more recent studies of
plankton-pigment distribution. However, he has shown the comparison between
his data and those obtained by comparable methods in the English Channel,
indicating that the quantities of phytoplankton which develop in both regions

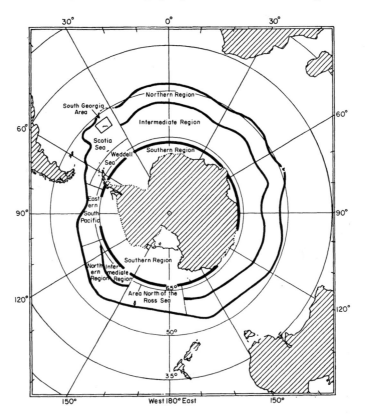

Fig. 10. The Antarctic Ocean showing regions referred to in Fig. 9. (After Hart, 1942.)

are nearly the same. Note that plant growth in the whole Antarctic is restricted
to a single midsummer flowering, in contrast to the English Channel, and is
probably typical of productivity at high latitudes where radiation is the primary
controlling factor (page 356).

As a rough approximation, based on Cooper's (1938) estimates for the
English Channel, an annual production of the order of 100 g carbon/m² seems
reasonable for the Antarctic Ocean as a whole. Certain locations of exceptionally
high fertility, such as the South Georgia area and the Scotia Sea, probably
exceed this figure by several-fold.

## 7. Hydrodynamic Features Which Influence Productivity

Throughout the world's oceans there are specific areas whose productivity is influenced by local hydrodynamic features obscuring the generalized regional picture presented above. The physical mechanisms involved in these situations are varied and often complex, but the end result in every case is the vertical transport of nutrient-rich water up into the euphotic zone by means other than turbulence or diffusion. To this extent, and for the purposes of this discussion, these processes may be referred to collectively as "upwelling", a term which may be employed more restrictively elsewhere. Some areas of upwelling are small and local; others are extensive and may influence large parts of the ocean. We have, in fact, already discussed, without so designating the process, upwelling around the Antarctic Continent as the factor responsible for the high fertility of that entire oceanic region.

Graham (1941) first presented biological evidence of upwelling in the equatorial Pacific. As stated earlier, subsequent investigations of the region have clearly shown increased production as the equator is approached (Doty et al., 1956; Steemann Nielsen and Jensen, 1957; Austin, 1957; King et al., 1957; Holmes et al., 1957; Jitts and Rotschi, 1958). The hydrodynamic features of the upwelling have been investigated by Cromwell (1951, 1953, 1954) and will not be discussed here in detail. Briefly, the effects of Coriolis's force north and south of the equator on the westerly drift of the Equatorial Current results in the poleward divergence of the surface layers, which are replaced with water upwelling from depths within the thermocline (150–200 m). The biological effects of this upwelling are detectable across the entire Pacific, as revealed by expedition "Eastropic" (King et al., 1957), and are evidenced by the abundance of zooplankton and pelagic fish as well as by increased rates of primary production (Sette, 1955).

The upwelling of water from sub-surface and intermediate depths on a smaller scale may occur in many parts of the oceans, and it will not be attempted here to record every documented case of the phenomenon. The process is most conspicuous along the western coasts of continents where prevailing offshore winds carry surface waters seaward, drawing colder, nutrient-rich water from moderate depths to the surface. The effects appear to be heightened where there are strong coastal currents. Familiar examples of coastal upwelling are those along the coasts of Southwest Africa (Défant, 1936), Peru (Gunther, 1936) and California (Sverdrup and Fleming, 1941). Currie (1953) has pointed out that the vertical circulation associated with the upwelling in the Benguela Current is somewhat more complicated than the "one-sided divergence" pictured by Défant. Rather than a continuous line of divergence along the coast, there appear to be formed a series of eddies of varying size and strength, somewhat similar to the situation along the Peruvian coast as described by Gunther. Thus high productivity associated with coastal upwelling is irregular and may be highly localized.

On the other hand, coastal upwelling may influence the fertility of large

13—S. II

oceanic areas, as illustrated by the distribution of total phosphorus in the North
Atlantic trade wind belt (Fig. 11, unpublished data of B. H. Ketchum, Woods
Hole Oceanographic Institution). High concentrations of this element obviously
associated with upwelling off the northwest coast of Africa spread well into the
mid-Atlantic. Note also indications of an upwelling of lesser magnitude of
unknown origin off the South American coast.

Upwelling may also be discontinuous with time if the driving wind forces are
intermittent. A weakening or migration of the trade-wind belt may have
pronounced effects upon the hydrography of West African and South American
coastal waters. The resulting cessation of upwelling and warming of the surface
waters, *in situ* or through incursions of warm coastal currents, causes notorious
mass mortalities of fishes and other sea life which are discussed elsewhere.

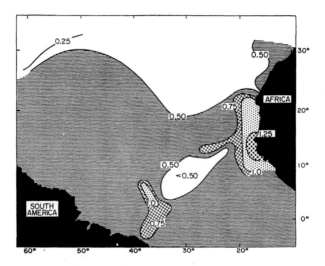

Fig. 11. Mean concentration of total phosphorus (μg atoms/l.) in the upper 50 m of the
    equatorial Atlantic. (Unpublished data of B. H. Ketchum.)

Upwelling has also been reported for much of the Indian coastal waters (e.g.
LaFond, 1954 and 1958; George, 1953; Jayaraman and Gogate, 1957; Banse,
1959) as well as off the East African coast of Somali (Sverdrup *et al.*, 1946). In
these cases, it is the offshore monsoon winds which produce the divergences
and upwelling. The seasonal periodicity of the southwest and northeast monsoons cause alternating periods of high and low fertility at different places
along the coast depending upon their geographical orientation.

Indications of high productivity, in terms of the abundance of marine life,
in regions of upwelling are too numerous to review here. Relatively few productivity measurements have been made, but the existing data (Steemann
Nielsen and Jensen, 1957; Currie, 1958; Holmes *et al.*, 1957) consistently reveal
high rates (*ca.* 0.50–1.00 g carbon/m²/day). Measurements in the vicinity of the
Benguela Current included the highest values observed during the *Galathea*

Expedition, the maximum production (2.4 g carbon/m²/day) occurring in Walvis Bay, Angola.

The vertical circulation of the equatorial Pacific, discussed above, becomes obscured at the eastern end of the ocean by coastal topographic and bathymetric features, the influence of the Peru Current, and other complicating factors. Upwelling is somewhat less common, but over large areas the thermocline occurs at relatively shallow depths of 25–50 m. Since this lies within the euphotic layer, plant production may draw upon nutrients in the relatively rich thermocline waters and hence the productivity of the area is relatively high. One such location, off Costa Rica (9°N, 90°W), is of particular interest since the thermocline depth is extremely shallow and may frequently "outcrop" during the winter months. The productivity of this "dome" area in November was 0.41–0.80 g carbon/m²/day (Holmes et al., 1957). The physical basis of this and similar areas of "ridging" was investigated by Cromwell (in press). While the hydrodynamic features were not completely elucidated, it was apparent that the thermocline depth of the Costa Rican dome is related to the strength of the trade winds and cyclonic shear over the dome.

Another physical process which increases the fertility of the sea surface is the vertical turbulence which occurs at current fronts and, particularly, between two opposing current systems. Sette (1955) has presented evidence of high fish production in the North Pacific (34°N) in the region which he refers to as the shear zone between the North Equatorial Current and the North Pacific Drift. A temperature profile across this zone of mixing revealed numerous isolated spots where isotherms rise from thermocline depths to the surface, suggesting eddy formation in the shear zone. As with most vertical transport mechanisms in the sea, the physical explanation of this situation is not entirely clear.

Finally, rich waters from moderate depths may be carried into the euphotic zone by turbulence or mass uplift where currents or drifting water-masses cross submarine ridges or banks, or when the horizontal flow is interrupted by outcropping islands. Again Sette (1955) has related the presence of a small but productive fishery near Hawaii to large scale, semi-permanent eddies, both convergent and divergent, downstream from the Hawaiian Islands.

In the boreal North Atlantic, which has already been discussed in some detail, the presence of land-masses, submarine ridges and numerous current systems and water-masses cause upwellings by all the afore-mentioned processes. According to Dr. Unnsteinn Stefánsson (in litt.), the productivity of the waters around Iceland is influenced by (1) vertical turbulence to the southeast at the boundary between the Atlantic and Arctic waters, (2) vertical turbulence to the northeast at the polar front and (3) upwelling off the southwest coast of Iceland and at the shelf along its west coast. Steemann Nielsen (1958) describes productive areas around Greenland as (1) at the front between the North Atlantic Current and the Irminger Sea, (2) at the edge of the East Greenland Current, (3) at the front between the Labrador Current and the ocean water in the center of the Davis Strait, and (4) in an area of upwelling off the coast of West Greenland (62°–67°N). Hansen (1959) found high rates of production over the Faeroe–Iceland Ridge (0.65–2.70 g carbon/m²/day), between the East Greenland Polar Current and the Irminger Current (0.55 g carbon/m²/day) and over the Reykjanes Ridge (0.53–1.30 g carbon/m²/day). By contrast, the productivity of the surrounding waters (in summer) generally ranged between about 0.15 and 0.25 g carbon/m²/day. Thus a significant part of the entire region's fertility would appear to stem from numerous local areas of upwelling and hydrodynamic mixing.

## 8. Plankton Production in Coastal and Inshore Waters

When the depth of the wind-mixed layer exceeds that of the euphotic zone beyond a certain point (Sverdrup's "critical depth"), the phytoplankton does not receive sufficient light to grow. The situation obtains in temperate and sub-polar seas during most of the winter, as we have seen. In such cases, it is usually the onset of thermal stability, together with increasing radiation, which provides conditions suitable for the initiation of the spring flowering.

However, mixing may be confined within the critical depth by a factor other than thermal stratification, the depth of the ocean bottom. In shallow water, the plants may receive enough light to grow throughout the year. In somewhat deeper water, the low radiation in mid-winter may prevent growth, but increasing sunlight in the early spring may be sufficient to stimulate plant production even though the waters are still mixed to the bottom.

Gran (1932) noted that phytoplankton blooms first appeared in the shallow coastal waters of Norway (60°N) in early March, a week or two later at the edge of the continental shelf and several weeks later still further offshore. Hart (1942) made similar observations in the Antarctic.

Riley (1957) has expressed the critical depth concept in quantitative terms as the average light intensity ($\bar{I}$) within the mixed layer.

$$\bar{I} = \frac{I_0(1 - e^{-kz})}{kz},$$

where $I_0$ is the incident radiation and $k$ the extinction coefficient. Table III from Riley shows the mean radiation in the water column for varying depths, times of year and transparency, as calculated from the above expression. He concludes that a mean radiation of 0.03 g cal/cm²/min is critical for plant production. Thus reference to Table III will show the depths of water, or mixing, where initation of the spring flowering could be expected at different times of year.

In the shallow coastal waters off New York and New Jersey (50 m or less) we have observed dense phytoplankton populations and high rates of production throughout the winter (Fig. 7). The relationships between productivity, season and water depth is more clearly shown in Fig. 12 from Ryther and Yentsch (1958). The data here are mean values for separate groups of shallow, intermediate and deep stations. They were obtained from chlorophyll and light

TABLE III

Mean Amount of Light (g cal/cm²/min) in a Water Column with Values of
Depth ($z$) and Incident Radiation ($I_0$) and Extinction Coefficient ($k$) as Indicated.
(After Riley, 1957.)

| Depth | Dec. | Jan. | Feb. | Mar. | Apr. | May |
|---|---|---|---|---|---|---|
| $k = 0.15$ | | | | | | |
| 0 ($= I_0$) | 0.086 | 0.094 | 0.138 | 0.212 | 0.272 | 0.306 |
| 10 | 0.045 | 0.049 | 0.072 | 0.110 | 0.140 | 0.158 |
| 20 | 0.027 | 0.030 | 0.044 | 0.067 | 0.086 | 0.097 |
| 30 | 0.019 | 0.021 | 0.030 | 0.047 | 0.060 | 0.067 |
| 40 | 0.014 | 0.016 | 0.023 | 0.036 | 0.045 | 0.051 |
| 50 | | 0.013 | 0.018 | 0.028 | 0.036 | 0.041 |
| 60 | | | 0.015 | 0.024 | 0.030 | 0.034 |
| 80 | | | | 0.018 | 0.023 | 0.025 |
| 100 | | | | | 0.018 | 0.021 |
| $k = 0.075$ | | | | | | |
| 0 | 0.148 | 0.117 | 0.124 | 0.154 | 0.230 | 0.279 |
| 40 | 0.047 | 0.037 | 0.039 | 0.049 | 0.073 | 0.088 |
| 50 | 0.039 | 0.031 | 0.032 | 0.040 | 0.060 | 0.073 |
| 60 | 0.033 | 0.026 | 0.027 | 0.034 | 0.051 | 0.061 |
| 80 | 0.025 | 0.019 | 0.021 | 0.026 | 0.038 | 0.047 |
| 100 | 0.019 | 0.015 | 0.016 | 0.020 | 0.030 | 0.036 |

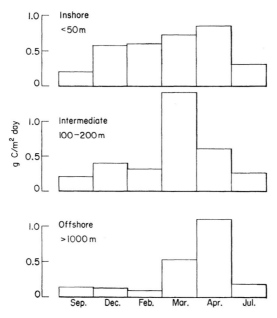

Fig. 12. Primary production in shallow, intermediate and deep stations of the continental
shelf off New York. (After Ryther and Yentsch, 1958.)

measurements (Ryther and Yentsch, 1957) and hence are not comparable to values obtained by the [14]C technique, but they clearly show the relation between depth and the timing of the spring flowering. A rough estimate of the annual production of the shallow coastal stations, based on the [14]C data (Fig. 7), is 165 g carbon/m² as compared to 120 g carbon/m² for the deep stations off the edge of the shelf. Thus it would appear that annual production varies only slightly between coastal and offshore environments, at least in temperate regions, but that the timing of the annual cycle is decidedly related to water depth.

However, in addition to the effect of depth on the amount of light available to the phytoplankton, the productivity of shallow coastal and inshore waters is influenced by nutrients and suspended material of terrigenous and sedimentary origin, and by the circulatory patterns common to many embayments and estuaries. Redfield *et al.* (Chapter 2) have described the situation in positive estuaries, where surface outflow, sub-surface inflow and the sinking of particulate

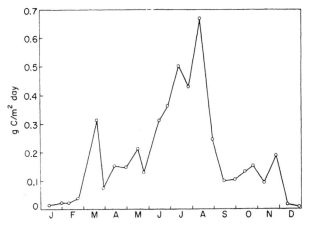

Fig. 13. Primary production in the Great Belt off Denmark, 1955. (After Steemann Nielsen, 1958.)

organic matter combine to provide a nutrient trap. By this mechanism, the fertility of estuaries may build up to levels far exceeding those of the contiguous ocean waters.

The seasonal cycle of production in shallow temperate regions characteristically differs from that of offshore waters of the same latitude. Rather than a bimodal cycle with a summer minimum, high production normally persists for the entire summer in shallow waters. This has been observed in Long Island Sound (Riley, 1956a), in Danish coastal waters (Steemann Nielsen, 1951, 1958; see also Fig. 13) and by the present author in numerous embayments of the Woods Hole region. Persistence of a summer thermocline is rare in these environments, and nutrients available from the bottom and from land drainage apparently support production throughout the summer months.

Steemann Nielsen (1958) and Riley (*in litt.*) have both suggested that the mid-summer peak in production may be related to high water temperatures and to the role of this factor in the bacterial regeneration of nutrients, particularly from the bottom.

According to Steemann Nielsen and Jensen (1957): "It was found during the [*Galathea*] expedition that the rate of organic production is high practically anywhere in the tropics in shallow water." The productivity of ten shallow stations in the Indo-Malayan region ranged from 0.24–1.08, averaging 0.61 g carbon/m²/day, almost an order of magnitude greater than offshore waters of the same region.

It is less evident that the average rate of production of shallow, temperate regions is appreciably higher than that of the deep ocean. Annual net production (roughly comparable to the $^{14}$C values reported elsewhere) for Long Island Sound was estimated at about 180 g carbon/m². Values for Danish inshore waters averaged about 75 g carbon/m². These estimates are not appreciably different from those reported earlier for offshore waters of comparable latitudes. The reasons for this are not entirely clear. One undoubtedly important factor is the preponderance in shallow environments of light absorbing, non-living particulate matter which may compete seriously with the phytoplankton for the available light (see page 350). Another factor is simply the limitation of depth. If production is considered on a volume basis, as carbon fixation per cubic meter, there is no question of its far greater magnitude in shallow regions. Values of 10–100 mg carbon/m³/day are common for the latter, whereas a range of 1–10 is typical of oceanic water. This advantage, however, is offset by a disadvantage of equal magnitude. Plant growth in the open sea may occur over depths of 50–100 m, while in estuaries and embayments the depth of the euphotic zone, if not the bottom itself, seldom exceeds 5–10 m. Production integrated over the entire water column is thus comparable for the two situations. A rather dramatic example of this compensatory effect is the comparison between production in a shallow, extremely rich, sewage oxidation pond and the Sargasso Sea during the spring flowering (Fig. 14, from Ryther, 1960). In this case, a difference of two orders of magnitude between rates of production per unit volume is countered by an equal but opposite difference in euphotic zone depths. Production per square meter differed by less than 20%.

The question of which of the two regions is the more fertile or the more productive is one of definition. Although equal areas may produce equal amounts of organic matter, the advantage to herbivores and succeeding trophic levels of having their food concentrated in a small volume is perhaps a matter of somewhat greater significance to the animals in question.

## 9. Production of Benthic Plant Communities

Discussion of the productivity of coastal waters would not be complete without consideration of the contribution made by the benthic plants, both the algae and phanerogams. Studies of their rates of production are disappointingly few, but those which are reported leave little doubt that the benthic community plays an important, if not dominant, role in the organic production of the coastal region.

The California coast maintains some of the densest stands of vegetation known. Crops of the large kelps may amount to several kilograms per square meter. Compare this with the estimated maximum potential standing crop of 25 g carbon/m² of phytoplankton (page 353). Measurements of the rate of production of these kelps give staggering values, as high as 33 g carbon/m²/day (Blinks, 1955). In two months, a crop of 4400 g/m² of *Laminaria* was produced, a figure comparable to over 1000 g carbon/m²/month, ten times the annual production of phytoplankton in temperate waters.

Cold, rich water upwells along the California coast, as we have discussed. What of benthic production in less fertile regions? According to Odum and

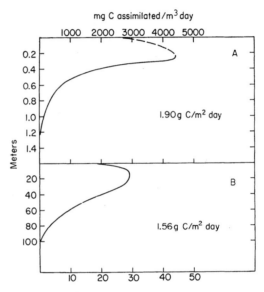

Fig. 14. Primary production in a sewage oxidation pond (A) and in the Sargasso Sea off Bermuda (B). (After Ryther, 1960.)

Odum (1955) some 12 g carbon/m² are produced daily by the algal members of a coral reef in the tropical Pacific. Comparable values were obtained for a community of marine turtle grass (*Thalassia*) in Florida (Odum, 1957). These values, unlike those reported elsewhere in this chapter, represent gross production (photosynthesis uncorrected for the respiratory requirements of the plants themselves). In most natural environments, net plant production is equivalent to 50–75% of gross production (Ryther, 1959). Thus, even the coral reef could be expected to produce a net 2000 g of plant carbon/m²/year, a synthesis which incidently is consumed as quickly as it is produced by the animal members of the coral reef community. These figures, then, give some idea of the relatively huge production rates of the benthic flora. How can they be maintained even in impoverished tropical seas? The answer lies in the great advantage which benthic plants have in the problems of obtaining nutrients. Unlike the phytoplankton, they are not confined to a given parcel of water which they can leave only through the force of gravity or feeble swimming. Low though the concentrations of nutrients may be in the water which bathes them, its continual renewal by tides and currents provides the sessile plants with an inexhaustable supply of these essential materials.

During the summer of 1957, a brief and rather preliminary study was made of the relative importance of benthic and planktonic algae in the primary production of a small estuarine pond near Woods Hole (Robert Barth, unpublished manuscript). At the time of the study, the pond contained a dense bloom of dinoflagellates and a rather inconspicuous quantity of green algae (*Ulva* and *Enteromorpha*) around its edges. Yet productivity measurements indicated that the phytoplankton contributed no more than about 1% of the daily organic production of the pond.

Vinogradov (1953) has estimated that the benthic algae cover a minimum area of $10^{11}$ m$^2$, about 0.1% of the ocean's surface, and that their mean standing crop is some $10^{15}$ g. If we may assume, conservatively, that these plants are annuals, in other words that their standing crop is roughly equivalent to their annual production, the latter would amount to at least one-tenth of the most recent estimates of the ocean's phytoplankton production (Steemann Nielsen and Jensen, 1957; Ryther, 1960). More extensive and precise quantitative studies of the benthic algae may well reveal that they play a considerably greater role in the relative productivity of the sea, perhaps equally as important as the phytoplankton. This interesting, if tenuous, idea must remain a subject for present speculation, a challenge for future investigators. It is obvious, however, that estimates of primary production of coastal and inshore regions which fail to take the benthic flora into account must be quite unrealistic.

## References

Apollonio, S., 1957. Plankton productivity studies in Allen Bay, Cornwallis Island, N.W.T., 1956. (Unpublished manuscript.)

Apollonio, S., 1959. Hydrobiological measurements on IGY drifting station Bravo. *Nat. Acad. Sci., I.G.Y. Bull.*, **27**, 16–19.

Austin, T. S., 1957. Summary, oceanographic and fisheries data, Marquesas Islands Area, August–September, 1956 (*Equipac*). *Spec. Sci. Rep., U.S. Fish. Wild. Serv., Fish.*, **217**, 1–186.

Banse, K., 1959. On upwelling and bottom-trawling off the southwest coast of India. *J. Mar. Biol. Assoc. India*, **1**, 33–49.

Berge, G., 1958. The primary production in the Norwegian Sea in June 1954, measured by an adapted $^{14}$C technique. *Rapp. Cons. Explor. Mer*, **144**, 85–91.

Blinks, L. R., 1955. Photosynthesis and productivity of littoral marine algae. *J. Mar. Res.*, **14**, 363–373.

Brouardel, J. and E. Rink, 1956. Determination de la production de matière organique en Mediterranée à l'aide du $^{14}$C. *C. R. Acad. Sci. Paris*, **243**, 1797–1799.

Budyko, M. I., 1955. *Atlas teplovago Balansa*. Leningrad, 41 pp.

Clowes, A. J., 1938. Phosphate and silicate in the southern ocean. *Discovery Reps.*, **19**, 1–120.

Cooper, L. H. N., 1938. Phosphate in the English Channel, 1933–1938, with a comparison with earlier years, 1916 and 1923–32. *J. Mar. Biol. Assoc. U.K.*, **23**, 181–195.

Cromwell, T., 1951. Mid-Pacific oceanography. January through March, 1950. *Spec. Sci. Rep., U.S. Fish. Wild. Serv., Fish.*, **54**, 1–9.

Cromwell, T., 1953. Circulation in a meridional plane in the central equatorial Pacific. *J. Mar. Res.*, **12**, 196–213.

Cromwell, T., 1954. Mid-Pacific oceanography II. Transequatorial waters, June–August, 1950, January–March, 1951. *Spec. Sci. Rep., U.S. Fish. Wild. Serv., Fish.*, **130**, 1–13.

Cromwell, T. (in press). Thermocline topography, horizontal currents and "ridging" in the Eastern Tropical Pacific. *Proc. 9th Pacific Sci. Cong. Bangkok, 1957*.

Currie, R., 1953. Upwelling in the Benguela Current. *Nature*, **171**, 497.

Currie, R., 1958. Some observations on organic production in the North-East Atlantic. *Rapp. Cons. Explor. Mer*, **144**, 96–102.

Deacon, G. E. R., 1933. A general account of the hydrology of the South Atlantic Ocean. *Discovery Reps.*, **7**, 173–238.

Défant, A., 1936. Das Kaltwasserauftriebsgebiet vor der Küste Südwestafrikas. *Landerkdl. Forsch. Festchr. N. Krebs*, 52–66.

Digby, P. S. B., 1953. Plankton production in Scorsby Sound, East Greenland. *J. Anim. Ecol.*, **22**, 289–322.

Doty, M. S., M. Oguri and R. Pyle, 1956. Aspects of oceanic productivity in the eastern tropical Pacific. *Proc. Hawaii Acad. Sci. 31st Ann. Meeting, 1955–56*.

English, T. S., 1959. Primary production in the central North Polar Sea, Drifting Station Alpha, 1957–58. Unpublished manuscript, paper presented at 1st Intern. Oceanog. Cong., New York, 1959.

Francis, J. R. D. and H. Stommel, 1953. How much does a gale mix the surface layers of the ocean? *Q. J. Roy. Met. Soc.*, **79**, 534–535.

George, P. C., 1953. The marine plankton of the coastal waters of Calicut with observations on the hydrological conditions. *J. Zool. Soc. India*, **5**, 76–107.

Graham, H. W., 1941. Plankton production in relation to character of water in the open Pacific. *J. Mar. Res.*, **4**, 189–197.

Gran, E. H., 1904. Diatomaceae from the ice-floes and plankton of the Arctic Ocean. In: Nansen, F. Norwegian North Polar Exped., 1893–1896. *Sci. Res.*, **4** (11).

Gran, H. H., 1932. Phytoplankton. Methods and Problems. *J. Cons. Explor. Mer*, **7**, 343–358.

Gran, H. H. and T. Braarud, 1935. A quantitative study of the phytoplankton in the Bay of Fundy and Gulf of Maine. *J. Biol. Bd. Canada*, **1**, 279–467.

Gunther, E. R., 1936. A report on oceanographical investigations in the Peru Coastal Current. *Discovery Reps.*, **13**, 107–276.

Hansen, V. K., 1959. Danish investigations on the primary production and the distribution of Chlorophyll *a* at the surface of the North Atlantic during summer. Unpublished manuscript; paper presented at Intern. Counc. Explor. Mer, Spec. I.G.Y. meeting, Copenhagen, 1959.

Hart, T. J., 1934. On the phytoplankton of the south-west Atlantic and the Bellingshausen Sea, 1929–31. *Discovery Reps.*, **8**, 1–268.

Hart, T. J., 1942. Phytoplankton periodicity in Antarctic surface waters. *Discovery Reps.*, **21**, 261–356.

Hasle, G. R., 1956. Phytoplankton and hydrography of the Pacific part of the Antarctic Ocean. *Nature*, **177**, 616–617.

Holmes, R. W., M. B. Schaefer and B. M. Shimada, 1957. Primary production, chlorophyll and zooplankton volumes in the tropical Eastern Pacific Ocean. *Bull. Inter-Amer. Trop. Tuna Comm.*, **4**, 129–156.

Jayaraman, R. and S. S. Gogate, 1957. Salinity and temperature variations in the surface waters of the Arabian Sea off the Bombay and Saurashtra Coasts. *Proc. Indian Acad. Sci.*, **45B**, 151–164.

Jerlov, N. G., 1951. Optical studies of ocean waters. *Rep. Swed. Deep-Sea Exped., 1947–1948*, **3** (Physics and Chemistry), 1–59.

Jitts, H. R. and H. Rotschi, 1958. Mesure par la méthode au $^{14}$C de l'activité photosynthétique de quelques masses d'eau du Pacific Sud-Ouest en relation avec l'étude de la fertilité de ces mêmes eaux. *Radioisotopes in Scientific Research* (Proc. 1st UNESCO Intern. Cong.), **4**, 607–632.

Kalle, K., 1938. Zum Problem der Meereswasserfarbe. *Ann. Hydrog. Mar. Met.*, **66**, 1–13.

Kennedy, R. E., 1949. Computation of daily insolation energy. *Bull. Amer. Met. Soc.*, **30**, 208–213.

Kimball, H. H., 1928. Amount of solar radiation that reaches the surface of the earth on the land and on the sea and the methods by which it is measured. *Monthly Weather Rev.*, **56**, 393–398.

King, J. E., T. S. Austin and M. S. Doty, 1957. Preliminary report on expedition *Eastropic*. *Spec. Sci. Rep., U.S. Fish. Wild. Serv., Fish.*, **201**, 1–155.

Kreps, E. and N. Verjbinskaya, 1932. The consumption of nutrient salts in the Barents Sea. *J. Cons. Explor. Mer*, **7**, 25–46.

LaFond, E. C., 1954. On upwelling and sinking off the East Coast of India. *Andhra Univ. Series* No. 49, 117–121.

LaFond, E. C., 1958. On the circulation of the surface layers off the East Coast of India. *Andhra Univ. Series* No. 62, 1–21.

McAllister, C. D., T. R. Parsons and J. D. H. Strickland, 1959. Primary productivity and fertility observations at ocean weather station "P". C.G.S. St. Catharines Cruise P-59-3, July 7–August 24, 1959. Circular 1959–20, Fish. Res. Bd., Pac. Oceanog. Group, Nanaimo, B. C.

Menzel, D. W. and J. H. Ryther, 1960. The annual cycle of primary production in the Sargasso Sea off Bermuda. *Deep-Sea Res.*, **6**, 351–367.

Odum, H. T., 1957. Primary production measurements in eleven Florida springs and a marine turtle grass community. *Limnol. Oceanog.*, **2**, 85–97.

Odum, H. T. and E. P. Odum, 1955. Trophic structure and productivity at a windward coral reef community on Eniwetok Atoll. *Ecol. Monog.*, **25**, 291–320.

Redfield, A. C., 1934. On the proportions of organic derivatives in sea water and their relation to the composition of plankton. *James Johnstone Memorial Volume*. The University Press, Liverpool.

Riley, G. A., 1941. Plankton studies III. Long Island Sound. *Bull. Bingham Oceanog. Coll.*, **7**, 1–93.

Riley, G. A., 1942. The relationship of vertical turbulence and spring diatom flowering. *J. Mar. Res.*, **5**, 67–87.

Riley, G. A., 1951. Oxygen, phosphate, and nitrate in the Atlantic Ocean. *Bull. Bingham Oceanog. Coll.*, **13**, 1–126.

Riley, G. A., 1956. Oceanography of Long Island Sound, 1952–1954. IX. Production and utilization of organic matter. *Bull. Bingham Oceanog. Coll.*, **15**, 324–344.

Riley, G. A., 1956a. Oceanography of Long Island Sound, 1952–1954. II. Physical oceanography. *Bull. Bingham Oceanog. Coll.*, **15**, 15–46.

Riley, G. A., 1957. Phytoplankton of the North Central Sargasso Sea, 1950–1952. *Limnol. Oceanog.*, **2**, 252–270.

Riley, G. A., H. Stommel and D. F. Bumpus, 1949. Quantitative ecology of the Western North Atlantic. *Bull. Bingham Oceanog. Coll.*, **12**, 1–169.

Ryther, J. H., 1956. Photosynthesis in the ocean as a function of light intensity. *Limnol. Oceanog.*, **1**, 61–70.

Ryther, J. H., 1959. Potential productivity of the sea. *Science*, **130**, 602–608.

Ryther, J. H., 1960. Organic production by plankton algae and its environmental control. In *The Ecology of the Algae*. Spec. Pub. 2, Pymatuning Laboratory of Field Biology, the University of Pittsburgh, pp. 72–83.

Ryther, J. H. and E. M. Hulburt. 1960. On winter mixing and the vertical distribution of phytoplankton. *Limnol. Oceanog.*, **5** (3), 337–338

Ryther, J. H. and D. W. Menzel, 1960. The seasonal and geographical range of primary production in the Western Sargasso Sea. *Deep-Sea Res.*, **6**, 235–238.

Ryther, J. H. and C. S. Yentsch, 1957. The estimation of phytoplankton production in the ocean from chlorophyll and light data. *Limnol. Oceanog.*, **2**, 281–286.

Ryther, J. H. and C. S. Yentsch, 1958. Primary production of continental shelf waters off New York. *Limnol. Oceanog.*, **3**, 327–335.

Ruud, J. T., 1930. Nitrates and phosphates in the southern Seas. *J. Cons. Explor. Mer*, **5**, 347–360.

Saijo, Y. and S. Ichimura, 1959. Primary production in the Northwestern Pacific Ocean. Unpublished manuscript; paper presented at 1st Intern. Oceanog. Cong., New York, 1959.

Sette, O. E., 1955. Consideration of midocean fish production as related to oceanic circulatory systems. *J. Mar. Res.*, **14**, 398–414.

Shirshov, P. P., 1944. Scientific results of the drift of Station *North Pole*. *Akad. Nauk S.S.S.R. Obschee Sobranie*, February, 1944, 110–140.

Sorokin, Y. I. and O. I. Koblentz-Mishke, 1958. Primary production in the sea of Japan and the Pacific Ocean near Japan, Spring, 1957. *Doklady Akad. Nauk S.S.S.R.*, **112**, 1018–1020.

Steele, J. H., 1956. Plant production on the Fladen Ground. *J. Mar. Biol. Assoc. U.K.*, **35**, 1–33.

Steele, J. H., 1957. A comparison of plant production estimates using $^{14}$C and phosphate data. *J. Mar. Biol. Assoc. U.K.*, **36**, 233–241.

Steele, J. H., 1958. Plant production in the Northern North Sea. Scottish Home Dept., *Marine Research 1958*, No. 7, 1–36.

Steemann Nielsen, E., 1951. The marine vegetation of the Isefjord—a study on ecology and production. *Medd. Kumm. Dan. Fisk. Havunders. Ser. Plankton*, **5**, 1–114.

Steemann Nielsen, E., 1952. The use of radioactive carbon ($^{14}$C) for measuring organic production in the sea. *J. Cons. Explor. Mer*, **18**, 117–140.

Steemann Nielsen, E., 1958. A survey of recent Danish measurements of the organic productivity in the sea. *Rapp. Cons. Explor. Mer*, **144**, 92–95.

Steemann Nielsen, E. and E. Aabye Jensen, 1957. Primary oceanic production. The autotrophic production of organic matter in the oceans. *Galathea Rep.*, **1**, 49–136.

Sverdrup, H. U., 1953. On conditions for the vernal blooming of phytoplankton. *J. Cons. Explor. Mer*, **18**, 287–295.

Sverdrup, H. U. and R. H. Fleming, 1941. The waters off the coast of Southern California, March to July, 1937. *Bull. Scripps Inst. Oceanog. Univ. Calif.*, **4**, 261–378.

Sverdrup, H. U., M. W. Johnson and R. H. Fleming, 1946. *The Oceans: their physics, chemistry and general biology*. Pentice-Hall, New York. 1087 pp.

Vinogradov, A. P., 1953. The elementary chemical composition of marine organisms. *Mem. Sears Found. Mar. Res.*, No. 2. New Haven.

# 18. BIOLOGICAL SPECIES, WATER-MASSES AND CURRENTS

M. W. JOHNSON and E. BRINTON

## 1. Introduction

Oceanographers have long been aware of the existence in the oceans of areas that can be more or less clearly defined as "water-masses". These areas are recognized because of individual characteristics, particularly with respect to the temperature–salinity relationship as established by $T$–$S$ curves for sub-surface water. For fullest discussion the reader is referred to *The Oceans* (Sverdrup, Johnson and Fleming, 1942).

It is also recognized by marine biologists that of the vast array of marine organisms many are sensitive to small changes in environmental conditions during all or part of their life history. Salinity and temperature are outstanding factors that are most clearly demonstrated as environmental variables having critical ranges for the lives of many species. In some species, notably the oceanic, the range of tolerance for fluctuations in these factors is relatively small. Such species are known as either stenohaline or stenothermic or both, and the limitations may be most pronounced with respect to reproduction, or during the larval stage. Usually the adult vegetative stage is the most tolerant. Ranges of tolerance may be within different salinity or temperature limits. Thus, an animal may be stenohaline to either high or low salinities. Although the coastal animals are usually considered the more tolerant to environmental changes, still many of these have characteristic limits for reproduction, as shown by investigations carried out by Runnström (1927) on thermal relations. There are also other factors at play that are less well understood which influence endemic distribution, such as non-conservative chemical elements, inter-specific competition and duration of maximum or minimum temperatures.

It is to be expected, then, that bodies of water with distinctive characteristics, such as water-masses, will produce environments for distinctive faunas. This is, indeed, what has been found to be the case in many instances where detailed studies have been made. The preference of certain species for North Sea waters and of others for the adjacent oceanic waters provide classical examples (Russell, 1939). Thus, a biological means is provided to aid in the identification and delineation of water-masses or to trace the source, direction and extent of water currents, and the isolations and pathways they may have provided in speciation of the organisms. It is increasingly realized that the main faunistic regions of the high-oceanic pelagic fauna (certain features of which were heralded by Steuer, 1933) are more strongly characterized taxonomically than they were formerly believed to be (cf. Ekman, 1953, p. 319).

Before discussing representative investigations showing a relationship of species distribution with specific water-masses or currents, it will be useful first to consider some generalizations as to the types of organisms that can best be used as "indicators", and to consider some of their limitations.

Although the geographic distribution of benthic or bottom-living animals is doubtless often dependent directly upon the type of water and the nature of

[*MS received July, 1960*] 381

prevailing currents, these animals are less used as indicators than the plankton. Therefore, we shall confine our remarks to the latter category. It should be mentioned that certain phytoplankton organisms have also been used as indicators, especially the photosynthetic dinoflagellates. However, the short individual life of unicellular autotrophic organisms and their spontaneous reproductive response to local environmental changes makes their use more complicated, especially as regards indicators of water currents.

## A. Zooplankton Indicators

In considering the application of organisms in this category, it is necessary to distinguish first between the "permanent" plankton (holoplankton) and the "temporary" plankton (meroplankton), for each division has its own uses and limitations. The permanent plankton consists of organisms that are planktonic throughout their entire life, and includes especially such abundant forms as the chaetognaths, pteropods, euphausiids and most of the copepods. The relatively long life of many of these forms contributes to their value, particularly as indicators of water movements. In addition to precise identification of the species, a knowledge of the life history and information on the breeding season are of great importance in order to ascertain the area in which the animals find optimal or tolerable conditions for reproduction. This is the "home area", which is either a slowly moving water-mass or a section of a continuing, or more or less well-developed, current system. This system may be one in which there is a semi-closed circulation, with some incoming flow balanced by an outward flow at some other point or depth. Thus, depending upon the tolerance of the organisms being swept into the system but not endemic to it, or of endemic forms swept out with outward flow, zooplankton can serve as an indicator of the exchange of water (Bigelow, 1926). Usually these expatriates, if carried far out of their endemic breeding area, are found only as adults or submature animals, whereas in the "home areas" all stages of development may be found during the reproductive season.

Although no analyses have been made to determine the relative distance expatriates may be carried out of their endemic reproductive area before succumbing to changing conditions, it is reasonable to believe that the border of the endemic area may lie nearer the extreme range of distribution when the transport outward is into gradually increasing temperatures than it would be under cooling conditions; for it is well known that, in general, the optimum temperature lies closer to the lethal high temperature than to the lethal low temperature. The cooler temperatures tend to retard the vital processes but, if not extreme, lead to an extension of the animal's life.

The temporary plankton is made up of floating larval stages of the benthic or nectonic (swimming) animals. Most benthic and nectonic animals pass through a drifting planktonic stage. The duration of this stage does, however, differ greatly in different species, and only larvae of relatively long-floating existence can be useful as indicators of water movements.

Not much study has been made of the tolerance of the temporary plankton to admixture of other waters, or to changes in temperature, and so forth, experienced during dispersal from the hatching or nursery area. But it should be pointed out here that there is some reason to believe that the larval forms may succumb more readily to changes than do the expatriates of the permanent plankton, although some larvae apparently do survive as drifting organisms for periods longer than considered normal for the species, when conditions favorable to metamorphosis to the benthic stage are not encountered. However, larval stages generally cannot be expected to be found, except as mere stragglers, at very great distances from the distributional range of the adult. The limits of the adult range may be controlled by the tolerance of the planktonic larvae, rather than by that of the benthic stage. However, the floating larvae of some near-shore shallow-water or intertidal adults can be traced to distances ranging up to 150 or more miles offshore over great depths, giving evidence of offshore flow of water.

To illustrate how these biological constituents may be useful as indicators in oceanographic, zoogeographic and ecological studies, the remainder of our discussion will be concerned chiefly with a review of a number of recent surveys, mainly in the North and South Pacific and adjacent seas, that have aided our concepts and understanding of the dynamics of the oceans and of their inhabitants.

The literature contains many reports, largely from the Atlantic, showing the application of this kind of plankton study. Some of these have been reviewed by Hardy (1956) and extended most recently by Tebble (1960). For recent work by Japanese and Russian investigators in the West Pacific see especially Bogorov (1955), Brodskii (1957) and Tokioka (1959).

It is not always possible to distinguish sharply between the application of biological indicators toward identification of water-masses *per se* as opposed to application as indicators of water currents. Often these are one and the same. However, in studies of the terminal extensions of water-masses or of the more local movements of water intruding from or into separate areas, the application becomes quite distinct. For purpose of discussion, we shall treat the surveys under distinct headings although much overlapping is of necessity implied.

These discussions will not be deeply concerned with the precise physiological or ecological actions of temperature, salinity, etc. that function to confine an organism to a particular water-mass. This is a field that also needs much research, but we shall mainly be occupied in presenting some aspects that contribute to the descriptive phase of oceanography; these aspects are currently experiencing an upsurge in the less explored Pacific and adjacent waters.

## 2. Water-Masses and Biological Species

The studies of distribution and systematics are interdependent in the development of biogeographical concepts. Oceanic populations found to be small in range sometimes prove to be ecotypically or genotypically distinct from other populations of the same morphological type. The apparently broad character of other ranges has prompted study leading also to the recognition of complexes of regional species or subspecies. For example, *Salpa fusiformis* Cuvier, long recognized as cosmopolitan and variable in form, is now seen to consist of four species, two tropical–subtropical, one living south of the subtropical convergence, and one limited to the Pacific sector of the Antarctic (Foxton, 1961). Such forms thereby become meaningful in the framework of oceanic ecology. The discussion in this section will be concerned mainly with the euphausiacian Crustacea and is based on earlier work by Brinton (1962a).

Many of the zooplankton species became known from collections made by exploratory expeditions in the late 19th and early 20th centuries. Classification of plankton at the level of the subspecies or geographical race, requiring exhaustive systematic and ecological data, is a recent consequence of sampling by comprehensive oceanwide surveys and of morphometric comparisons of specimens from different regions. Investigations carried out by various nations in the Atlantic and Antarctic Oceans, and by North American, Japanese and Russian surveys in the Pacific, 1950–1960, provided material from nearly the full extent of the distributional ranges of many species. These surveys have employed improved sampling techniques making possible better quantitative estimates.

The water-mass boundary is recognized as a zone of discontinuity frequently associated with sharp horizontal gradients in temperature. The distributions of subspecific forms (subspecies, races, ecophenotypes), as well as species, may correspond with physical zones. In other instances, two or more closely related forms live in adjoining parts of one well defined water-mass, forming an allopatric complex. Nevertheless, each distribution derives its identity from a system of circulation. The biological evolution of passively drifting organisms is believed to have proceeded according to the availability of habitats that can conserve a stock.

Furthermore, the vertical range of certain planktonic species may correspond with the thickness of the physical water-mass and its contained currents. For example, in temperate and tropical latitudes where the euphausiid and sergestid (decapod) crustaceans perform diurnal vertical migrations between the surface and 300–800 m of depth, the water-masses are identified according to the temperature–salinity characteristic between depths of 150 and 1000 or more meters. The migrations and water-masses are mainly in the mesopelagic zone, a stratum between about 150 to about 700 m. Above this lies the epipelagic or photic zone and below it the bathypelagic and abyssopelagic zones. Bruun (1957) discusses the temperature relations for this zonation and Hedgpeth (1957) has reviewed certain works pertinent to the latitudinal aspects of the epipelagic.

The geographical ranges of species inhabiting the epipelagic zone appear in many cases to be limited by near-surface isotherms, while boundaries of mesopelagic species can be shown to conform with isotherms lying at some greater depth within the vertical range of the species. Inasmuch as many species have an extensive diurnal range that carries them through a vertical temperature gradient of 10°–12°C, it is not yet possible through the study of distribution to establish that a boundary isotherm can limit a horizontal range of such species.

### A. The General Applicability of Zooplankton–Water-Mass Relationships

In each of the several systematic groups thus far studied in the zooplankton of the Pacific (based mainly on samples collected by the California Cooperative Oceanic Fisheries Investigations, The Pacific Oceanic Fisheries Investigations and the Scripps Institution of Oceanography) certain species are recognized as having geographical distributions that are in close agreement with the major water-mass provinces. Pelagic Foraminifera (Bradshaw, 1959), Chaetognatha (Bieri, 1959; Alvariño, 1962), pteropod molluscs (McGowan, MS) and euphausiid crustaceans (Brinton, 1962b) each include species limited to subarctic, subtropical and tropical zones. These three ecological zones are designated *subarctic*, *central* and *equatorial* in this discussion, following the clear role played by water-masses in the zonation of the high seas. Studies of the distribution of polychaetous annelids (Dales, 1957; Tebble, 1962) and certain pelagic tunicates (Berner, 1960) and copepods (Johnson, 1938) in northern and northeastern Pacific waters are in general agreement with the recurring patterns.

A fourth Pacific faunal zone lies in temperate waters at the northern limit of the central region. In the western and mid-Pacific this is a zone of transition between central and subarctic faunas, and occupies water of the North Pacific Drift, 38°–45°N. Toward its eastern limit the zone diverges: a northern branch enters the eastern Gulf of Alaska while a southern branch occupies the California Current, terminating off Baja California. The California Current and the deep and coastal countercurrents which aid in maintaining plankton of the Current lie mainly within this zoogeographic province. However, the characteristic species of the zone are abundant only to the north of about 30°N. A "transition region", distinguished by Sverdrup, Johnson and Fleming (1942) on the basis of physical properties, lies in the California Current where subarctic, central and equatorial waters converge. Certain widely distributed species maintain their highest concentrations in this *transition zone*. Other species are endemic to it.

An analogous zone of transition lies in the Southern Hemisphere, 35°–45°S, reaching northward in offshore waters of the Peru Current. This belt includes the region of the Subtropical Convergence and the northern part of the West Wind Drift. It is the habitat of endemic species and of certain bisubtropical species (i.e. species occurring on each side of the tropics but not within them) found also in the transition zone of the North Pacific.

The quantitative aspects of many of the distributions included in this discussion are based on samples obtained above a depth of 300 m. These generally contained epipelagic populations, including the developmental stages of many species having adults concentrated in the stratum below 200–300 m.

In the case of the strongly migrating euphausiids, sub-adults and larvae of most species are present above 200–300 m during the day-time, while adults are found in the upper layers at night. Samples obtained below 300 m, including those from multiple hauls using opening-closing nets, are used to plot generalized profiles of vertical distribution. Such hauls serve also to establish which of the species distributions can be reliably mapped on the basis of the standard collections—that is, whether presence or absence at a locality can be determined from collections made above 300 m.

### B. The Pacific Subarctic Species

The Pacific subarctic sustains a high biomass, although its geographical extent is small (six million square miles) compared with that of the North Pacific central (10 million square miles) and Pacific equatorial zones (15 million square miles). In the northwestern Pacific, the subarctic zone is characterized by the copepod species *Calanus tonsus*, *C. cristatus*, *Eucalanus bungii bungii*, the amphipod *Parathemisto japonica* and the euphausiid *Euphausia pacifica* (Brodskii, 1955; Bogorov and Vinogradov, 1955; Beklemishev and Burkov, 1958). *Eucalanus bungii bungii*, the northern variety of this species, is also characteristic of the eastern portion of this zone and the northern portion of the California Current (Johnson, 1938 *et seq.*), as is *Calanus cristatus*.

The distribution of the euphausiid *Thysanoessa longipes* is representative of the subarctic zone (Fig. 1a). Banner (1949) showed that a large form of *T. longipes* bears spines on the posterior abdominal segments. A smaller form has keels on those segments. The "spined" form is more northern than the "unspined" form. Both occur off Hokkaido, Japan, though their densities diminish sharply at the convergence where the cold water of the Oyashio submerges beneath the Kuroshio system. The "spined" form is predominant in the Gulf of Alaska while only the "unspined" form is found in the California Current, southward to 39°–40°N.

Water-mass envelopes for the two forms of *T. longipes* (Fig. 2) confirm the subarctic nature of the water in which both live. Water inhabited by the "spined" form is more nearly uniform with respect to the *T–S* property than that inhabited by the more southern "unspined" form. The environment of the latter, while remaining nearly subarctic in character below 300 m, is strongly influenced above that depth by admixture of the North Pacific Drift. (The layer of maximum density of *T. longipes* during both day and night is 0–280 m.) Populations of "unspined" *T. longipes* are intermingled with the transition zone species *T. gregaria* (Fig. 4) in the region 41°–45°N.

An area of abundance is evident in the distribution pattern of each of the forms of *T. longipes*. The areas are separated from each other in mid-ocean by a belt of lower density having an east–west axis near 50°N, but converge off northern Japan at the western limit of both ranges. The two forms are regarded as ecophenotypes. It is not known whether they are genetically separate. The morphological distinctions, not evident in the young, appear to be brought

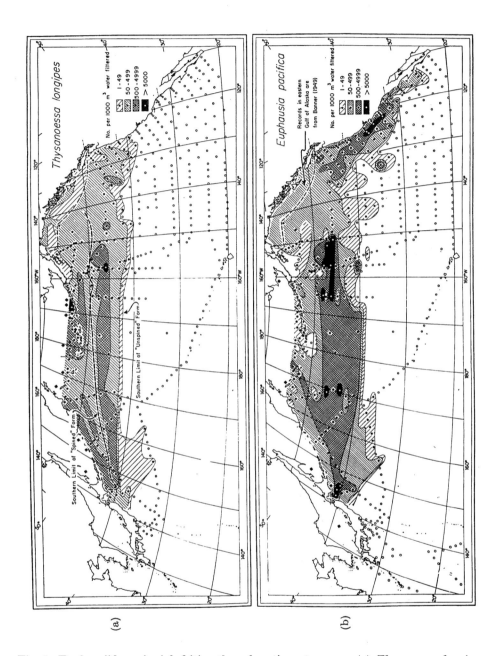

Fig. 1. Euphausiid species inhabiting the subarctic water-mass. (a) *Thysanoessa longipes* and (b) *Euphausia pacifica*. (Based mainly on sampling by the Scripps Institution and the California Cooperative Oceanic Fisheries Investigations between the surface and 270–300 m.)

about by differences between the two optimal environments: the temperature at 200 m is less than 4°C in the range of the "spined" form, but may be as high as 7°C for the "unspined" form.

Other northern species have wider environmental tolerances, permitting them to range southward in the cool coastal waters off California. This group includes *Euphausia pacifica* (Fig. 1b), the copepods *Calanus cristatus*, *Eucalanus bungii bungii*, and the pteropod *Limacina helicina*. Some of these species persist until subarctic and upwelled components of the California Current become thoroughly mixed with equatorial water toward the south, and with central water in the offshore region. McGowan (MS) has shown that a northern

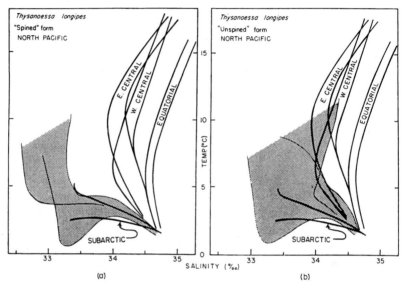

Fig. 2. The shaded part represents temperature–salinity characteristics below about 150 m in the environment of *Thysanoessa longipes*. These observations were made during the "Transpacific" expedition (1953) and the "Norpac" cruise (1955). Certain of the curves that limit the *T–S* envelope for the species are shown as fine lines. The more northern "spined" form (a) is seen to be more typically subarctic than the "unspined" form (b). (The water-mass envelopes are from Sverdrup, Johnson and Fleming, 1942.)

high-spired form of *L. helicina* may be distinguished from a low-spired form occupying the transition zone and the California Current. He regards these as geographical races or ecotypes. The limits of the distribution of *Euphausia pacifica* off Baja California fall where subarctic and equatorial waters are mixed in nearly equal proportions, according to the temperature–salinity characteristics of that region (Tibby, 1943). Seasonal variability in the southward extent of the distribution of *E. pacifica* in the California Current is small (Brinton, 1960). In general, this species is absent from Baja California waters during the winter and recolonizes that region at the onset of spring upwelling.

Distributions like that of *E. pacifica* are included in the subarctic group because the centers of distribution are (1) subarctic water near the northern boundary of the North Pacific Drift, and (2) upwelled subarctic water of the California Current.

A relationship between the distribution of the planktonic worm *Poeobius meseres* and the subarctic water-mass was found by McGowan (1960). Though

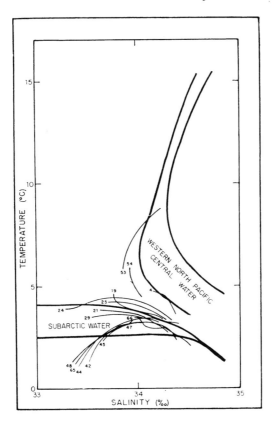

Fig. 3. Temperature–salinity curves from the stations where *Poeobius meseres* was caught during the "Transpacific" expedition, compared with subarctic and North Pacific central *T–S* envelopes. Station numbers are indicated on the fine lines. Only that segment of the curve within the depth range of *Poeobius* is plotted. (After McGowan, 1960.)

widely distributed in the 150–300 m layer in the subarctic region, this species is found at greater depths in the California Current and in the terminal part of the Peru Current between 6°S and the Galapagos Islands (0°). Unlike *Euphausia pacifica*, *Poeobius* submerges toward the southern limits of its range. The tropical records were of specimens believed to be sterile expatriates: it was implied that their occurrence near the equator might be due to a relationship between eastern tropical intermediate water and subarctic water.

McGowan showed that, strictly speaking, the part of a $T$–$S$ curve which represents a precise depth of capture should be used to identify the water-mass habitat. This procedure is valid where the population maximum is encountered below the mixed layer. Divided hauls carried out at 150-m intervals by the "Transpacific" expedition in 1953 made it possible to show a stratum-of-capture. In the case of *Poeobius*, the segments of $T$–$S$ curves plotted for these layers generally fell within the envelope of the subarctic water-mass (Fig. 3).

Surface plankton has been related to the surface $T$–$S$ property (Bary, 1959). Distributions included in the present discussion are based on oblique open-net hauls. These may sample (1) only the upper limits of an extensive vertical range, or (2) a stratum thicker than the local vertical range of the species. In these cases it can be useful to identify the water-mass habitat either by means of the full $T$–$S$ curve at the locality or by the part relating to the usual vertical range of the species, if this is known.

Recurrent species groups (in the sense of Fager, 1957) were distinguished among certain North Pacific chaetognaths, euphausiids and pteropods (Fager and McGowan, in press) on the basis of the "Transpacific" data of Alvariño, Bieri, Brinton and McGowan. A subarctic group concorded best with shape-of-$T$–$S$-curve (50–1000 m), supporting the concept of the water-mass habitat.

### C. The Transition Zone Species

The region of transition between subarctic and central populations in the mid-Pacific and between subarctic and equatorial populations in the California Current is regarded as a biogeographical zone because it harbors (1) endemic species, (2) bisubtropical species, limited in the North Pacific to this belt, and (3) maximum densities or regions of dominance of certain species having broader total ranges. A zone between the subantarctic and central regions in the Southern Hemisphere is distinguished in the same way.

In the western Pacific, where the cold subarctic region is in contact with the warm Kuroshio and the Kuroshio Extension, a narrow transition zone was recognized in which the copepod *Calanus pacificus* is particularly abundant (Bogorov and Vinogradov, 1955; Brodskii, 1955). In mid-ocean the transition zone was identified by Hida (1957) as an area of mixed fauna and variable biomass, usually characterized by large numbers of the pteropod *Limacina inflata*. A chaetognath identified by Hida as *Sagitta lyra* was abundant in the transition zone. This form, now distinguished from the more widespread *S. lyra* as *S. scrippsae* (Alvariño, 1962), appears to be endemic to the transition zone.

The euphausiid *Nematoscelis difficilis* lives in this narrow transition belt. A relict population of this species is also maintained in the northern part of the Gulf of California. *N. megalops*, a sibling species close to *N. difficilis*, occupies the transition zone of the Southern Hemisphere, occurring in the Indian and

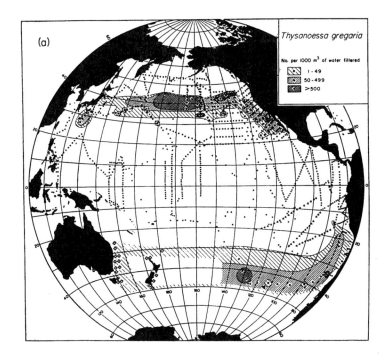

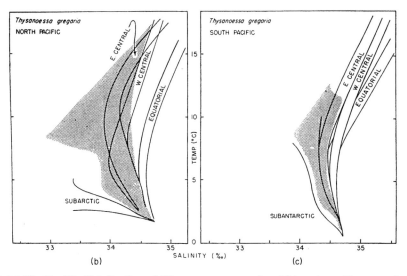

Fig. 4. (a) The Pacific distribution of *Thysanoessa gregaria* within a transition zone between central and subarctic (or subantarctic) water-masses. *T–S* characteristics of water below about 150 m for the North Pacific habitat (b) and the South Pacific habitat (c) of *T. gregaria* are compared with the *T–S* envelopes of the principal Pacific water-masses. South Pacific *T–S* curves for the *T. gregaria* habitat are all from the eastern half of the ocean.

Atlantic Oceans as well. *N. megalops*, rather than *N. difficilis*, occupies the transitional belt in the North Atlantic.

The bisubtropical euphausiid *Thysanoessa gregaria* (Fig. 4) lives in the transition zones of both hemispheres in the Pacific. In mid-ocean its low-latitude boundary is near the 30° parallel in both hemispheres. In the eastern boundary currents the range reaches toward the tropics, terminating in equatorial water. This is near 15°S in the Peru Current and 18°–20°N off Baja California.

The temperature–salinity envelopes for *T. gregaria* (Fig. 4b, c) show that both the northern and southern habitats are intermediate in character between equatorial and subarctic (or subantarctic) waters. In both cases there is overlapping with the central water-masses. In the South Pacific the *T–S* habitat of this species also extends southward into subantarctic water and northward into modified equatorial water. Distributions of the southern transition zone species are closely related to the east–west belt of the Subtropical Convergence, which curves northward as the South American continent is approached and becomes poorly defined in offshore waters of the Peru Current. Northward-reaching tongues in the distributions of *T. gregaria* and *N. megalops* are maintained, in part, by a northward-flowing subantarctic current component and a deeper remnant of southward-flowing equatorial water. *T–S* curves indicate the similarity of the equatorial water to both the South Pacific central water and the subantarctic water.

The term "transition zone" is particularly descriptive of the North Pacific habitat of *T. gregaria* where the *T–S* envelope of the species falls in part between the envelopes of typical subarctic and central water and in part within the central envelopes. In the South Pacific it almost completely overlaps both envelopes. It is to be noted that the *T–S* habitat of *T. gregaria* in the North Pacific differs somewhat from that in the South Pacific. The differences lie mainly in the salinity. The temperatures in the two zones are nearly the same. Apparently a broader range of salinity is found in the northern zone within the range of temperature tolerance. It will be seen that the disjunct habitats of central species, when described in terms of the relationship of temperature to salinity, differ from each other in exactly the same way as those of the transition zone. Salinity *per se* probably has little direct effect in controlling the geographical distributions of oceanic species. Rather, the essential features of the oceanic habitat are (1) continuity in the system of circulation (water-mass) implying a permanent or semi-permanent place of origin of the water, (2) a range of temperature to which the species can adapt, (3) an adequate food supply, and (4) an area sufficiently large to enable restocking despite loss by dispersal.

In an alternative explanation there may be genetic differences, not yet discernible, between the two populations of *T. gregaria*. Such differences are already suggested in the *Nematoscelis difficilis–N. megalops* pair. Geographically separate populations of a species will, in time, diverge genetically as a consequence of selective pressures that differ between the two environments.

## D. The Pacific Central Species

Plankton species of the central waters may be grouped according to the extent to which their ranges conform with the distribution of the most typical central water.

The euphausiids *Nematoscelis atlantica, Euphausia brevis* (Fig. 5) and *E. hemigibba* are most numerous in the warm, barren mid-parts of the North Pacific gyrals. *N. atlantica* and *E. brevis* are bisubtropical, occurring also in the South Pacific central zone, while *E. hemigibba* is replaced in the Southern Hemisphere by *E. gibba. T–S* envelopes for these species, as shown for *E. brevis* (Fig. 5b, c), are in agreement with the *T–S* characteristics of the central water-masses.

The distributions of *E. brevis* and *N. atlantica*, unlike those of many other central species, do not extend into the region south and east of Japan, where central water is affected by the warm Kuroshio system and the submerging subarctic water.

A second central group may be represented by *Euphausia mutica, Sagitta pseudoserratodentata*, and the foraminiferan *Globorotalia truncatulinoides*. These species are found in low concentration toward the central parts of the oceanic gyrals, while areas of abundance are associated with the margins of the ranges. These margins are places where the central environment impinges on more fertile waters of the subarctic region, the eastern boundary currents, and the equatorial water-mass.

Ranges of the species of this second central group frequently enter near-shore waters in the southern part of the California Current at 20°–34°N. This is a region south of Pt. Conception, California, where the climate of the northern epipelagic region is modified by the effect of the subtropical latitude and by admixture of southern and offshore water.

The species of a third central group have North and South Pacific zones of distribution, joined in the western Pacific but separated in the eastern Pacific. The eastern region from which these species are excluded lies in the equatorial water-mass, characterized by low sub-surface values for temperature and dissolved oxygen. The chaetognath *Sagitta californica* and the euphausiid *Stylocheiron abbreviatum* are examples of this group that may be showing either incipient bisubtropicality or a coalescence of northern and southern zones of distribution.

The composite range of the species-pair *Thysanopoda aequalis–T. subaequalis* (Boden and Brinton, 1957) falls in the third group. In the North Pacific, *T. subaequalis* is found in the western part of the ocean while *T. aequalis* lives in the eastern part. This is the only known case in which the eastern and western central water-masses of the North Pacific support distinct populations of closely related zooplankton species. (The eastern and western North Pacific central water-masses were distinguished by Sverdrup, Johnson and Fleming (1942), but are not now generally believed to be distinct gyrals.) In the South Pacific the ranges of the two forms overlap; *T. aequalis* lives in cooler water than *T. subaequalis* and is the more widespread form.

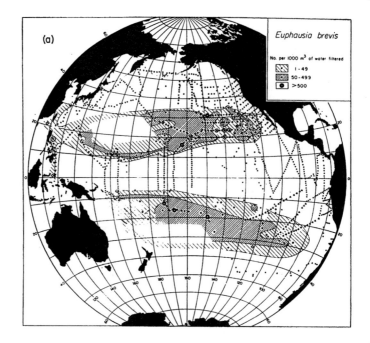

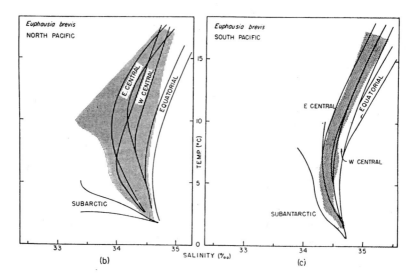

Fig. 5. The Pacific distribution of *Euphausia brevis* in the central water-masses. *T–S* characteristics below about 150 m for the North Pacific habitat of this species (b) and the South Pacific habitat (c) are compared with the *T–S* envelopes of the principal Pacific water-masses. South Pacific *T–S* curves for the habitat of *E. brevis* are all from the eastern half of the ocean.

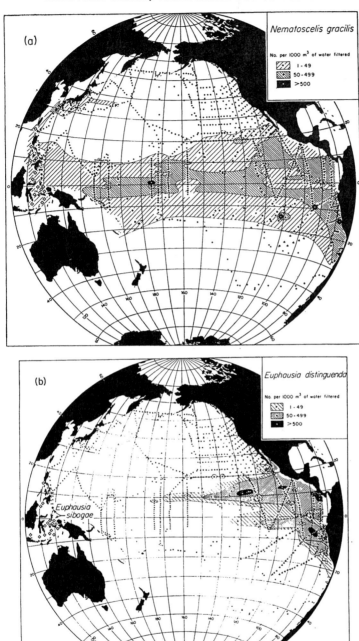

Fig. 6.  Geographical distribution of equatorial euphausiid species, *Nematoscelis gracilis* (a)
and *Euphausia distinguenda* (b). The only other localities for these two species are
in the equatorial water-mass of the Indian Ocean. Crosses in (a) indicate non-
quantitative records for *N. gracilis*, and in (b) all known localities for *E. sibogae*, a
species related to *E. distinguenda*.

## E. The Equatorial Species

The euphausiids *Euphausia distinguenda* and *Nematoscelis gracilis* provide examples of two patterns of distribution (Fig. 6a, b), both related to the region of the Pacific equatorial water-mass. These extend into the Peru and California Currents at the eastern boundary of the ocean. *Euphausia distinguenda* is carried westward by the North and South Equatorial Currents, until the range becomes attenuated in mid-ocean. The influence of the Equatorial Counter-current, $2°–8°N$, is evidently a disadvantage to this species. (The distributions of other eastern equatorial species, *E. eximia* and *E. lamelligera*, are completely split along the east–west axis of the countercurrent, according to sampling carried out in 1952 and 1955.) *E. sibogae*, very similar to *E. distinguenda*, is found in the warmer waters of the Indo-Australian Archipelago.

The equatorial water-mass is narrowed in the western Pacific. The range of *Nematoscelis gracilis* narrows in the same way, though this species is carried northward to Japan by the warm Kuroshio.

Four Pacific equatorial euphausiids, including the two just discussed, are found also in equatorial waters of the Indian Ocean but not in the Atlantic. This is in conspicuous agreement with the distribution of equatorial water-masses, recognized in the Pacific and Indian Oceans, but not in the Atlantic.

Each of the planktonic groups thus far studied with respect to distribution contains tropical species which occupy both the equatorial region and the central regions of the western Pacific. The limits of these distributions generally follow the $24°–26°C$ summer surface isotherm. This assemblage was called the *equatorial-west-central fauna* by Bieri (1959) and Bradshaw (1959), and is made up of numerous epipelagic species including *Sagitta robusta*, *Euphausia tenera*, the foraminiferan *Pulleniatina obiquiloculata* and the pteropod *Limacina inflata*.

## F. Species of the Antarctic Ocean

The niches in the Antarctic Ocean appear to have evolved in relation to four boundaries: the Antarctic continent, the edge of the pack ice, the Antarctic Convergence and the Subtropical Convergence. The meridional component of Antarctic circulation, together with sinking at the Antarctic and Subtropical Convergences, partially separates each of the epipelagic zones from its neighbors. Mackintosh (1934) noted that, here, the limits of distribution of species may be controlled by vertical migrations between surface water which has a northward-moving component and southward-moving deep water.

John (1936) showed that *Euphausia crystallorophias* is associated with the coast of the Antarctic Continent, *E. superba* with the zone between the continent and the Antarctic Convergence, and *E. frigida* with the region between the ice edge and the Antarctic Convergence. These may be regarded as species of the antarctic water-mass. Three species of polychaetes, *Rhynchonerella bongraini*, *Vanadis antarctica* and *Tomopteris carpenteri*, are believed to be endemic to waters south of the Antarctic convergence (Tebble, 1960). *Euphausia vallentini*,

*E. longirostris* and *E. lucens* live in subantarctic water bounded by the Antarctic and Subtropical Convergences. This habitat is comparable in both latitude and temperature to the subarctic habitat of *E. pacifica*, a species related to the *E. lucens–E. vallentini* line of the Antarctic Ocean. A thorough study of *E. triacantha* (Baker, 1959) demonstrated a close relationship between maximum density of this circumpolar species and the position of the Subantarctic Convergence. Smaller numbers were found on either side of the convergence, within the belt of the West Wind Drift. David (1955) described separate races of *Sagitta gazellae*: a "large southern" form is antarctic, while a "small northern" form is subantarctic.

## G.  Distribution and Speciation

Reference was made in the preceding section to forms that may or may not be sufficiently distinct genetically to be regarded as species. The importance of colonization of neighboring waters as compared with absolute geographical isolation in the differentiation of zooplankton species is an open question.

It may be noted, however, that there are several instances in which forms of uncertain taxonomic rank live on opposite sides of convergences or other oceanographic boundaries, while there are few instances in which closely related, *well defined species* are found in such adjoining regions of the high seas. The forms of *Sagitta gazellae*, *Thysanoessa longipes*, and *Limacina helicina* and the *Vanadis antarctica–V. longissima* and *Thysanoessa aequalis–T. subaequalis* "species pairs" are instances in which the extent of reproductive isolation between closely related forms is not clear. These, and the intergrading forms of the euphausiid *Stylocheiron affine*, discussed in later paragraphs, may be sympatric complexes in which morphological differences between populations are caused by environmental differences across the geographical range of a single genotype. Oceanic forms clearly recognized to be sibling species, as in the cases mentioned above of *Nematoscelis difficilis–N. megalops* and *Euphausia distinguenda–E. sibogae*, are often more widely separated geographically.

In view of the uncertain influence of water-mass boundaries in evolutionary processes, only the three north–south continental land-masses that partition the seas into the Atlantic, Pacific and Indian Oceans can be recognized as absolute barriers between pelagic populations of low latitudes. Certain warm-water species (e.g. *E. brevis*, Fig. 5a) are today split up by these barriers into isolated populations living in the separate oceans.

Similarly, the antitropical pattern of distribution, in which parts of a range are separated by a tropical or subtropical belt, can establish absolute isolation of northern and southern population elements.

The paired roles of continental and tropical barriers in the differentiation of euphausiid species are considered by Brinton (1962b). In that discussion, isotherms that agree with the limits of Recent distributions are used to extrapolate the limits of hypothetical distributions that may have existed during past epochs when the oceans were warmer or cooler.

The positions of ancient water-mass habitats are not mapped as easily. The persistence of belts of oceanic productivity, as measured in the sediments by Arrhenius (1955), suggests that some of the major currents have retained their positions since the close of the Tertiary. If the zones of the water-masses have similarly persisted, it is unlikely that plankton distributions have migrated in strict harmony with postulated latitudinal migrations of isotherms. When ocean-wide warming or cooling compelled a species to occupy a new geographical range, still tolerable with respect to temperature, salinity and food, this new environment had also to include a sufficiently closed current system to maintain the population. Nevertheless, temperature, reflecting obvious responses of the ocean to climatic change, was necessarily used in the euphausiid study.

Of the species with disjunct ranges, those with the widest latitudinal range-split occupy the two transition zones (zones of sinking isohaline and isothermal surfaces, Fig. 7). Given sufficient time, the separated populations of *Thysanoessa gregaria* would be expected to undergo independent selective adaptations. The species pair *Nematoscelis difficilis–N. megalops* is an example of antitropical evolutionary divergence already in an early stage. Compared with *N. difficilis–N. megalops*, *Thysanoessa gregaria* ranges somewhat farther into warm waters, by 1°C, interpreting 100-m isotherms as limiting. The northern and southern populations of *T. gregaria* are morphologically indistinguishable. They may have exchanged genetic material across the tropics more recently than the precursor of *Nematoscelis difficilis–N. megalops*.

Reconstructed Pleistocene distributions for *Thysanoessa gregaria* (Fig. 9) are based on $2\frac{1}{2}$°C warming, or cooling, at a depth of 200 m. Hubbs (1952) suggested that if the surface near-shore waters of the eastern Pacific were 8°C cooler in the winter than during present winters and 3°C cooler in the summer, continuity and exchange between fish faunas now antitropically separated would have been possible. The 7°C and 11°C isotherms at 200 m are associated with the limits of the Recent range of *T. gregaria*. The temperature change required to cool the warm waters west of Central America enough to allow equatorial transgression by *T. gregaria* would have made the California Current extremely cold. The bisubtropical transition-zone fauna would then have moved far offshore into central waters, so that it would no longer have access to prevailing currents entering the eastern equatorial basin. The northern and southern populations might, therefore, have coalesced far offshore in the zone of 0°–12°N (Fig. 9c). With 2°–3°C of cooling, the coastal fauna in the region 20°–30°N would have consisted of subarctic species (*Euphausia pacifica*, *Thysanoessa longipes*, *Tessarabrachion oculatus*). These occur only in the northern hemisphere. It is probable that they have never transgressed the tropics. Similarly, the Peru Current species *Euphausia mucronata* and all subantarctic euphausiids appear to have never crossed the equator.

Thus, during moderately cool epochs, either the transition-zone populations would have coalesced at the equator, or, if the species had previously been present in only one hemisphere, access to a new habitat in the opposite hemisphere would have been provided. During warm epochs the belts of distribution would have moved toward the poles, establishing isolated pockets of distribution in the northern oceans (there, forms would differentiate if they were to remain isolated sufficiently long), while allowing circum-global exchange of genetic material to take place south of the continents.

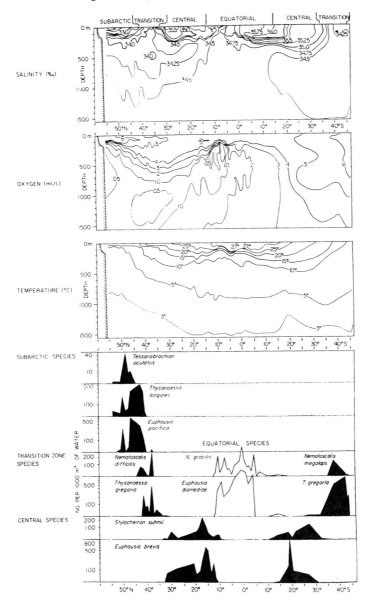

Fig. 7. Abundance of certain euphausiids, belonging to four faunas, from a north–south mid-Pacific track (Fig. 8) compared with salinity, oxygen and temperature profiles from the same stations.

Interglacial warming of the oceans is of equal importance to oceanic cooling in developing or maintaining antitropical distribution. The sub-surface equatorial region is characterized by a belt of cool water (Fig. 7) which extends across the ocean. While species of the transition zone required cooling of the eastern Pacific to accomplish equatorial transgressions, the central species presumably crossed the tropics on the western side of the ocean during epochs of general oceanic warming. Only then would temperatures in the sub-surface equatorial belt become tolerable. *E. mutica* and *Thysanopoda subaequalis*

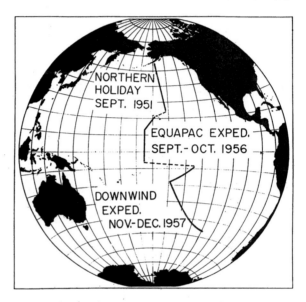

Fig. 8. Composite track of portions of three expeditions on which the physical profiles and species concentrations shown in Fig. 7 are based.

transgress the equator in the western Pacific today. During epochs of oceanic cooling, the central species would be withdrawn into the warm mid-parts of the central gyrals.

It is significant that the bisubtropical euphausiid species, including most of the *Euphausia* species occupying the central water-masses of the oceans, are also panoceanic, occurring in the Atlantic and Indian Oceans as well as the Pacific. In contrast, subarctic euphausiid species (excepting the two most northern species, *Thysanoessa inermis* and *T. raschii*, which appear to have passed between the Atlantic and Pacific Oceans by way of the Arctic Ocean) and the most tropical species, e.g. *Euphausia americana* of the Atlantic and *Nematoscelis gracilis* of the Indo-Pacific equatorial water-mass, do not occur in both the Atlantic and Indian-Pacific Oceans, but are endemic to one or the other. Burkenroad (1936) adduced evidence from penaeid distributions that only shallow water connected the Atlantic with the Pacific at Panama during the Cenozoic. It appears probable that oceanic populations have not crossed these connections since the Cretaceous. All antarctic and subantarctic species of euphausiids are confined to the Antarctic Ocean and are circumpolar.

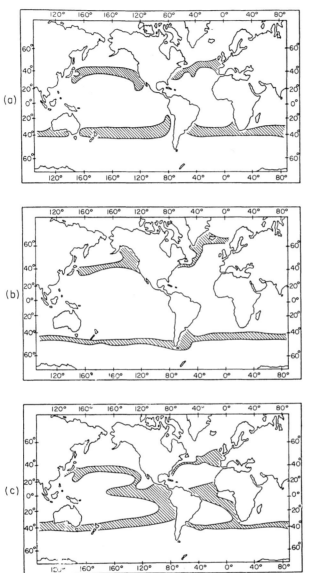

Fig. 9. (a) Present range of the euphausiid *Thysanoessa gregaria* compared with hypothetical circumstances of oceanwide warming (b) and cooling (c), both by $2\frac{1}{2}$°C. The 7°C and 11°C isotherms at 200 m depth, associated with present limits of distribution of *T. gregaria*, are considered limiting for the purpose of the extrapolation. With warming (b), the 7°C isotherm (at the high-latitude boundary) moves to position of present $4\frac{1}{2}$°C isotherm, and the 11°C isotherm (at the low-latitude boundary) moves to position of present $8\frac{1}{2}$°C isotherm. With cooling (c) the limiting 7°C and 11°C isotherms move to positions of present $9\frac{1}{2}$°C and $13\frac{1}{2}$°C isotherms, respectively.

Groups of four related species occur in each of the three major euphausiid genera that inhabit the upper layers—*Euphausia* (2 groups), *Nematoscelis* (1 group), *Stylocheiron* (1 group). These may be explained by postulating long-term isolation of populations, split up from the once-widespread distribution of each parent species, into the four major subtropical oceans—the Atlantic, North Pacific, South Pacific and Indian Oceans. Each quartet of species contains three central and one equatorial Pacific–Indian Ocean species. For example, *Euphausia brevis*, *E. mutica* and *E. recurva* are central, while *E. diomediae* is equatorial. The central species have, in most cases, re-established themselves in the other central regions already occupied by sibling species when these other regions became accessible during Pleistocene periods of latitudinal oscillations of isotherms. The equatorial species remained bound to the equatorial water-masses.

Three closely related mesopelagic species occur in each of three euphausiid genera: the "*Thysanopoda orientalis* group", the "*Stylocheiron maximum* group" and the three species comprising the genus *Nematobrachion*. All occupy depths to 1000 m, mainly in the zone 45°N–45°S. Differentiation from the three parental forms may have followed availability of the three isolated mesopelagic habitats—the low-mid-latitudes of the Atlantic, Pacific and Indian Oceans.

Furthermore, the ten endemic antarctic and subantarctic species of *Euphausia* and *Thysanoessa* may have differentiated within the southern zones of the Atlantic, Pacific and Indian Oceans when circumpolar stocks were split up by the southern continents during cool epochs in the cycles of change in ocean climate.

In this way, latitudinal thermal fluctuations occurring simultaneously within a series of intercommunicating oceans could have led to the rise of groups of euphausiid species. Invasion of equatorial waters and of passageways around the southern tips of continents can account for the present sympatric occurrence of species that initially differentiated as allopatric forms. Earlier transgressions may have brought about the present sympatric occurrence of related genera. Successive climatic revolutions might, therefore, be accompanied by an increasing rate of species formation were it not for the probability that few environmental changes have been of sufficient magnitude or geographical extent to allow the differentiation of species in the partitioned stocks. In addition, successive epochs may not always be enough separated in time to permit allopatric differentiation before coalescence again occurs, either around the tips of continents or across the tropics. Of course, environments may be so altered by climatic change that those species unable to adapt to the changing conditions become extinct.

Less may be inferred about the possible effectiveness of *oceanographic* barriers in isolating planktonic populations from one another. The findings of geographical races or subspecies suggests the possibility that morphological differentiation may arise in segments of the overall population of a species, in each of a series of communicating oceanographic regions. For example, several Pacific variants of the euphausiid *Stylocheiron affine* are distinguished on the basis of morphometric characteristics (ratios of measurements of the eye and abdomen) (Brinton, 1962b). They appear to be adapted to separate geographical areas and make up a *Rassenkreis* along the anticyclonic pathway of the circulation of the subtropical and tropical North Pacific. The biogeographical sequence includes a "California Current" form, an "Eastern Equatorial" form, a "Western Equatorial" form, and a "Central" form. None are regarded as species because each appears to intergrade with neighboring upstream or downstream forms. Thus, they are sub-species, in the sense that the bulk of the population of each form is morphologically uniform. However, the intergrades have been found only in small parts of the overlapping ranges. The "California Current" and "Central" forms overlap, but are regarded as terminal elements of the cline because they do not intergrade with each other.

In another complex, *Euphausia nana*, occupying the Tsushima water-mass of the East China Sea, is more nearly isolated from *E. pacifica*, its large sub-arctic counterpart (Brinton, 1962a). *E. pacifica*, ranging eastward from the Japan Sea to the California Current (Fig. 1b), differs conspicuously from *E. nana* in that it is about three times larger at maturity. Morphological differences are small. Though these differences might have ecotypic bases, the absence of intergrading specimens has led to the assignment of specific status to both forms. The partial barriers of the Tsushima Strait and the Kuroshio may be insufficient to have brought about genetic isolation of the two populations; however, it appears possible that unequal maturation rates associated with the different water-mass habitats may have reproductively isolated the large northern adults from the small southern ones. On the other hand, eggs of the large form, after drifting into the southern habitat, might develop into small adults, thereby sustaining a uniform *E. pacifica–nana* gene pool.

It is not possible at present to provide a satisfactory solution of these complex relationships in terms of oceanographic factors. Environmental gradients exert continuing, however variable, pressure on the adaptive capacity of all plankton populations. The populations are thereby constantly subject to both physiological and genetic adaptations, but the implications of these to species formation are clear only when gene pools that are not in communication with each other have been created.

## 3. Water Currents and Biological Species

Water-masses are conceived of as being sufficiently large in volume to maintain rather steady and uniform conditions throughout circumscribed geographic areas. When these masses are maintained as such over large areas or are confined as semi-closed systems of circulation, a characteristic assemblage of planktonic species may develop. However, these waters are constantly being formed anew by admixtures of water from other systems, each with their own identity. These blended waters, although maintaining their physical-chemical identity for the main body, are changed at their margins and extremities by admixture of yet other waters, or by physical changes resulting from seasonal changes or from gradual flow into higher or lower latitudes where the effects of temperature or evaporation alter the distinguishing features.

It is often of interest to follow the course of these moving, changing masses, which have now lost part or all of their chemical-physical identities, and exist as modified extensions in the form of currents, tongues or wedges, into other geographic positions. Assuming that the planktonic fauna characterizing the water-mass is well known, it is here that further application of an "indicator species" can serve to corroborate physical-chemical studies, or even provide a clue not otherwise obtainable as to the water's origin or circulation.

Many planktonic species, although dependent on certain waters for reproduction and development, may be swept from these endemic areas and survive for some time as expatriates. Some reproduction may take place at first, but eventually only adults and submature specimens will be found and these, too, finally succumb to changing environmental conditions.

## A. Transport from the Bering Sea into the Arctic Ocean

Beginning in 1934 and continuing in the next two decades, considerable oceanographic study, both physical and biological, has been concerned with determining the extent and character of flow of North Pacific and Bering Sea water and its floating life through the Bering Strait into the Arctic Ocean. Parts of these studies will provide some idea of the application of plankton species in following the movement of water-masses, and, conversely, of the kind of dispersal pressure that planktonic species must be subject to under the influence of a current system sweeping them relentlessly downstream into more and more unfavorable living conditions. In this instance, the contrast between the endemic area and the area involuntarily invaded is probably so great that it militates against the evolution and establishment of epiplanktonic species or forms adapted to the more severe environment of the Arctic. However, it has become increasingly evident that some planktonic species hitherto considered common to widely separate areas, such as the North Pacific and the Arctic and boreal waters of the Atlantic, may be distinct species or varieties that have evolved in the Arctic. Outstanding among these is *Calanus glacialis*, established by Jaschnov (1955) as a new species. Although relatively larger than *Calanus finmarchicus* Gunnerus, it was previously identified with that species. Also, Brodskii (1950, 1957) has named a surprising number of new copepod species, some of which are closely allied to common Atlantic or arctic forms. Taxonomically, this affects the present analysis only in that *Metridia lucens* and *Calanus tonsus* as previously recognized by authors for the Pacific form are probably Brodskii's *Metridia pacifica* and Marakawa's *Calanus plumchrus*, both of which are North Pacific and Bering Sea species.

Parts of the arctic areas that we shall be here concerned with were widely surveyed during the short arctic summers by the U.S. Coast Guard cutter *Chelan*, the U.S.S. *Nereus* and H.M.S. *Cedarwood* in the Bering Sea and parts of the Chukchi Sea in 1934, 1947 and 1949 respectively, and by the U.S. ice-breaker *Burton Island* in the Chukchi and Beaufort Seas in 1950 and 1951. Thus, a seasonal picture of the plankton was obtained which shows its relative stability with respect to occurrence and distribution of certain species for separate years. The study will suffice to illustrate the kind of integrated information that can be gleaned by piecing together oceanographic observations that are applicable to conditions in new or little known areas. Additional details of the plankton analysis made on these surveys is given by Johnson (1953, 1956).

In the eastern portion of the Bering Sea and the southeastern Chukchi Sea, there is a characteristic assemblage of holoplanktonic species of copepods which, while not always abundant, are found at many scattered sampling stations over much of the area. Most of these species are also common along the coast of North America for varying distances south of the Aleutian Islands in subarctic water. To the north in the Chukchi Sea and along the Alaskan coast into the Beaufort Sea, the frequency of catches of these species falls off, and eventually the species disappear in the northern reaches of the Chukchi Sea and in all but the near-shore stations in the Beaufort Sea (Figs. 10 and 11). The abrupt disappearance of *Acartia clausi, Calanus tonsus, Epilabidocera amphitrites, Eucalanus bungii bungii, Eurytemora herdmani, Metridia lucens* and *Tortanus discaudatus* in the region of Point Barrow, Alaska, is interpreted as resulting from an eastward flow along the coast to this area, where it is met by a clockwise flow from the Beaufort Sea as it swings to the northwest in this region. It is obvious that these copepods found off Point Barrow have their main connection with the major population to the southwest in the Chukchi and Bering Seas, where, as a group, they occurred at about 80% of the stations.

Probably the most striking example of this kind in these waters is shown by the distribution of *Acartia longiremis*, a more abundant species of wide distribution in coastal waters of the arctic and subarctic regions (Fig. 11). This is generally considered a coastal species, but with a wide tolerance for offshore conditions. It is abundant in the Bering Sea and southeastern Chukchi Sea, where in 1934 and 1947 it was found at 75% and 85% respectively of the oceanographic stations. From here it may be carried in considerable numbers far into the Arctic Ocean to 74° 17′N, 168° 56′W, where, in 1951, a single station yielded 1040 adult or submature specimens in a 0–100-m haul with a $\frac{1}{2}$-m net. The species was absent from stations north of 76°N, and also totally absent from the many offshore stations in the Beaufort Sea, where in the summer the ice-free water would appear to be as ecologically suitable as the northern Chukchi Sea. This absence in the main part of the Beaufort Sea may be explained by assuming that, except at the near-shore stations, the water entering the Beaufort Sea flows down from high arctic regions as the eastern portion of a large clockwise gyral, the southern part of which is shown in Fig. 10. That such a gyral does exist is also evidenced by physical studies (Worthington, 1953). The offshore portions of the Beaufort Sea affected by such a current could not be expected to harbor any *Acartia longiremis*, which, as shown in

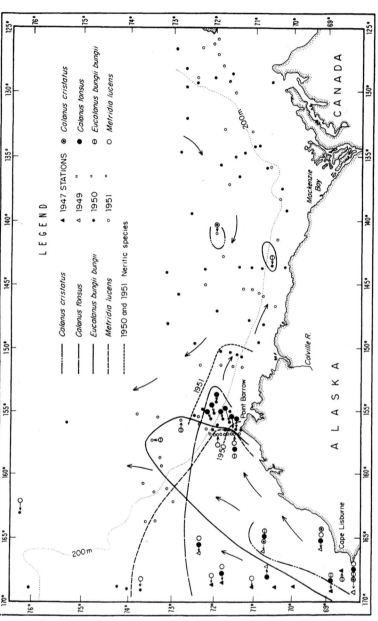

Fig. 10. The extent of Bering Sea and southern Chukchi Sea influence in the arctic region off northern Alaska as indicated by locality records of certain planktonic copepods for the summers of various years. Data are based on vertical hauls with 40 cm (1947, 1949) and 50 cm (1950–51) nets towed from 100–0 m or from bottom to surface at shallower depths. The lines enclose the approximate usual extent of penetration of the species. Note isolated instances.

Shown also are the 1950 and 1951 observed extent of the more strongly neritic species *Acartia clausi*, *Centropages abdominalis* (syn *mcmurrichi*), *Epilabidocera amphitrites*, *Tortanus discaudatus*. The long arrows show general direction of flow based on species distribution.

Fig. 11, was eliminated by either high arctic or oceanic conditions in the northern portion of the Chukchi Sea.

It should be further pointed out that the occurrence of *Metridia lucens* at 73° 42′N, 169° 01′W and 76° 22′N, 163° 16′W in the Chukchi Sea can also be taken to indicate an influx of Bering Sea water into the Arctic Ocean, at least to that latitude. A comparison with Fig. 12 shows this to be farther than reported for penetration of Bering Sea water based on earlier physical-chemical studies (Barnes and Thompson, 1938; LaFond and Pritchard, 1952; Saur,

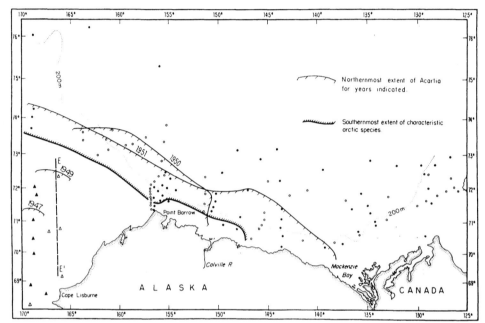

Fig. 11. The extent of observed arctic penetration of *Acartia longiremis* for 1947, 1949, 1950 and 1951. The species occurred at 73%, 100%, 100% and 97% respectively for each year within the distributional areas outlined by light lines. The southernmost observed extent of the two characteristically arctic species, *Calanus hyperboreus* and *Metridia longa*, is shown by the heavy line. The former occurred at 96% of the stations above the line in 1950 and 82% in 1951; and the latter at 83% in 1950 and 82% in 1951. All stations south of the line were negative.

Solid triangles, 1947 stations; open triangles, 1949 stations; open circles, 1950 stations; solid circles, 1951 stations.

Water structure along line E–E′ is shown in Fig. 12.

Tully and LaFond, 1954) made concurrently with plankton collecting. More recent studies (Coachman and Barnes, 1961) do, indeed, indicate the presence in the Arctic Basin of a stratum of water between 50 and 130 m with temperatures as much as 0.5 to 0.8°C warmer than the water immediately above or below it. It diminishes from the Chukchi Sea toward the Pole and toward Ellsmere Island. It is probable that certain hardy expatriates of *Metridia lucens*, found near 74°N in the Chukchi Sea, *Calanus cristatus*, near 72°N in the Beaufort Sea, and the two species discussed in the next paragraph were transported in such a water stratum.

A recent analysis of zooplankton collected in 1957–58 during the drift of the ice floe "Drift Station Alpha" in latitudes 81° 14′N to 85° 16′N shows additional expatriate species from the North Pacific Ocean–Bering Sea populations (Johnson, MS). Individual adults of *Eucalanus bungii bungii* Giesbrecht were taken at 83° 41′N, 155° 13′W and at 85° 6′N, 167° 40′W. This same species was also reported by Brodskii and Nikitin (1955) at 77° 5′N and 80° 51′N. Another species, *Mimocalanus distinctocephalus* Brodskii, occurred in the ice floe collection taken at 83° 4′N and is probably an expatriate there. Brodskii (1957) found it at 50% of the oceanic stations from 500- to 1000-m depths in the Bering Sea.

Under conditions of cooling associated with northward flow of Pacific and Bering Sea water, the survival of contained southern species far beyond their endemic area would be enhanced by the drop in temperature. That this might occur is supported by the experiments carried out by Clarke and Bonnet (1939) in which the survival of *Calanus finmarchicus* was definitely better at 3°C than at temperatures of 6° and 9°C.

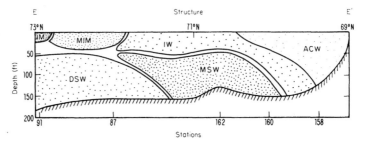

Fig. 12. Water-mass structure from Cape Lisbourne to the ice pack, based on temperature and salinity relations. See Fig. 11 for position of section E–E′.

ACW, Alaskan Coastal Water; MSW, Modified Shelf Water; MIM, Modified Ice Melt; IM, Ice Melt; DSW, Deep Shelf Water. This DSW is not considered continuous with the Bering Sea as are the others. (After Saur, Tully and LaFond, 1954.)

The northward and eastward extensions of the U.S.S. *Burton Island* collections in the Chukchi and Beaufort Seas make it also possible to consider here, in contrast, the distribution of two characteristically arctic species of copepods —*Calanus hyperboreus* and *Metridia longa*. The southern limit of distribution observed for these species shows that a good deal of overlapping of areas may occur (Fig. 11), but the main population of arctic species is held back in the Chukchi Sea by the flow from the Bering Sea. A broad transition zone seems to exist in the northern Chukchi Sea, if one may judge by the 1947 and 1951–1952 collections. The position and extent of this zone doubtless varies somewhat from year to year. In the Beaufort Sea area the truly arctic zone is more sharply defined east of Point Barrow where there is, however, a clear overlapping of the arctic and the more southern species, suggesting considerable mixing of waters in the coastal part of this area. The line of demarcation between these two faunas may result in part from the width of the Continental Shelf, which is narrow in the Beaufort Sea and relatively broad in the Chukchi Sea as shown by the 200-m depth contour in Fig. 11. This would tend to influence the relative position of the neritic and of the more offshore species such as *Calanus hyperboreus* and *Metridia longa* which occur in deep water but are not characteristically bathypelagic species and were caught regularly in tows from 100 to 0 m depth, and also in a tow from 700 to 430 m depth. Although the geographic limits shown here for these two arctic species appear to be valid for the boundary of their *continuous* distribution, it must be mentioned that both species have been reported far south of this range. Anraku (1954) lumps *M. longa* with *M. lucens* from the Aleutian area; Campbell (1929) reports *M. longa* from the Vancouver Island region; and Wilson (1942) lists *M. longa* from both the North and South Pacific and *Calanus hyperboreus* from the North Pacific. Some of these identifications appear to be in error. Further work on deep-water plankton is needed to determine the extent of subarctic penetration by these and other cold water forms into the deep cold water of the Pacific.

The truly abyssal species found in the Arctic Basin have their main connection with the Atlantic Ocean, since the shallow depth of the Bering Strait permits transport of only the surface species and such bathypelagic forms as also frequent the upper layers.

Parts of the area here covered by the plankton surveys have also been studied by Russian biologists. Their reports substantiate the conclusion that Bering Sea plankton is carried northward through the Bering Strait (Stepanova, 1937; Brodskii, 1957).

## B. The Meroplankton

We turn now to illustrate briefly a different aspect of plankton dispersal with emphasis on the floating larval stages of bottom-living invertebrates.

The dispersal of meroplankton has been of interest to marine biological studies mainly because of its importance in the ecology, recruitment and general life cycle of the animals involved. But the extent and pattern of dispersal of these forms can also contribute to the study of water currents, especially on a more restricted scale, although the most widely known and remarkable study ever made on the drift of marine plankton is that of Johannes Schmidt (1925), dealing with the larval drift of the European eel. This eel, after hatching, is carried during its two-to-three-year larval life some 3000 miles by currents across the Atlantic Ocean. The planktonic larval life of most invertebrates is, however, usually much shorter, and their sensitivity to hydrographic changes probably greater. Hence, they are not so likely to be carried great distances away from the spawning area of adult distribution. The usefulness of these larvae as indicators lies in the fact that the spawning area of the adult is often known, and the duration of the larval stage may also be known within limits. Some information is, therefore, provided that can be of aid in considering the source, direction, time lapse or complexity of currents. During the growth of larvae there is often a succession of well defined stages (or sizes) which, when well known, are further aids in considering the source and time of larval entrance into the waters as floating organisms. The information on water currents that can be gleaned from a study of the distribution of these larvae cannot, as a rule, be expected to indicate a direct line of flow, and thus to provide a clear-cut answer as to the source and time involved. More often, the occurrence of the larvae gives only partial answers and poses problems that need clarification through more detailed oceanographic investigations. They often bring to light oversimplifications that are likely to be introduced in dynamic computations of flow. Thus, more questions about water currents may often be asked than are answered; but this is an important contribution, for it is widely recognized that in oceanographic research questions must first be asked of the sea before answers are obtained.

The implications of a study of this kind are well illustrated by a seven-year investigation of the occurrence of the floating larvae of the California spiny lobster (*Panulirus interruptus*). In this survey it was found that it requires about $7\frac{3}{4}$ months to complete the planktonic larval life in which there are a succession of 11 larval stages (Johnson, 1960). During this relatively long, hazardous time, the larvae are at the mercy of prevailing currents. A characteristic picture of the geographic distribution of various larval stages is given in summary form in Fig. 13.

In the light of this long larval period, it is remarkable that the lobster fishery on the California coast remains as stable as it does. Precisely what are the details of circulation that enable retention of sufficient numbers of larvae to bring about restocking in the area? *A priori*, it appears that the system of eddies and countercurrents that retard outflow from the area is far more complicated than usually pictured on the basis of computations giving dynamic topography. Although major eddies are usually clearly detected by this method, the direction and rate of surface flow suggest that it would be unlikely that any considerable packet of water and its contained larvae would remain for seven to eight months within the distributional area of the adult lobster on the Southern California and Baja California coasts. Only a few drift bottles set adrift 40 miles from the coast to test the currents have been recovered, and the inshore current is said to flow southward from March through October (Reid, Roden and Wyllie, 1958). Nevertheless, the biological data, although evidencing a dispersal from the shore where the larvae are hatched, do show that there must be a fair retention of larvae, and in addition to this retention within the area, a mixture of larval stages does at times occur in a single sample. This would indicate a great deal of mixing within the water by eddy diffusion.

In the interpretation of the meaning of the distribution of these larvae (and, indeed, also of the holoplankton) in terms of water movements, it is important to have in mind at all times that here one is dealing with living creatures that, unlike inanimate drift bottles, may exercise some degree of modification of their distribution. This could be accomplished mainly by swimming from one depth level to another during vertical diurnal migrations, or even through seasonal migrations of this type. During these shifts of level, which occur under the directive stimulus of light, even weakly swimming animals may conceivably spend a good deal of time alternately in currents flowing in different directions, or at different speeds. In this way a retardation or prevention of

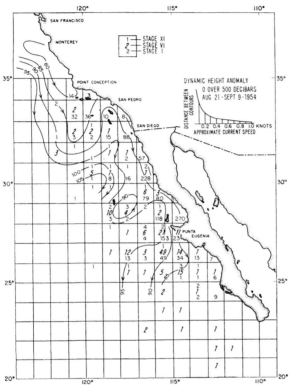

Fig. 13. Summary of the distribution of the Stage I, Stage VI and Stage XI (last) phyllosoma larvae of the spiny lobster *Panulirus interruptus* based on approximately monthly cruises, 1949–1955. Superimposed on the grid is the surface-current pattern for Aug. 21–Sept. 9, 1954. There is considerable variation in currents from month to month but the general flow is to the south. August and September are months with many early and intermediate larvae.

wholesale outwash is effected. That something of this nature occurs in many parts of the sea where a continuous yet drifting population is found is suggested by many studies of both the meroplankton and the holoplankton.

The mystery of how specific planktonic populations can be maintained as such within rather restricted geographical areas is not easy to solve. The answer will most likely be found to include some aspects of water movements correlated with the behavior and life history of the animals. Semi-closed circulation with horizontal flow, such as exists in the Gulf of Main (Bigelow, 1926; Redfield, 1941) and, on a smaller scale, in coral atoll lagoons, where a semi-closed system is set up in a vertical rather than horizontal plane (Johnson, 1954), provide the more simple and obvious conditions for interpreting the interrelations of species within their ever-moving aquatic environment.

# References

Alvariño, A., 1962. Two new Pacific chaetognaths; their distribution and relationship to allied species. *Bull. Scripps Inst. Oceanog. Univ. Calif., Tech. Ser.*, **8**, 1–50.

Anraku, M., 1954. Gymnoplea Copepoda collected in Aleutian waters in 1953. *Bull. Fac. Fish., Hokkaido Univ.*, **5**, 123–136.

Arrhenius, G., 1955. Sediment cores from the East Pacific. *Rep. Swed. Deep-Sea Exped.*, **5** (1), 1–227.

Baker, A. de C., 1959. Distribution and life history of *Euphausia triacantha* Holt and Tattersall. *Discovery Reps.*, **29**, 309–340.

Banner, A. H., 1949. A taxonomic study of the Mysidacea and Euphausiacea (Crustacea) of the North Pacific, Pt. III. Euphausiacea. *Trans. Roy. Soc. Canadian Inst.*, **28** (58), 2–49.

Barnes, C. A. and T. G. Thompson, 1938. Physical and chemical investigations in Bering Sea and portions of the North Pacific. *Univ. Wash. Pub. Oceanog.*, **3**, 35–39.

Bary, B. M., 1959. Species of zooplankton as a means of identifying different surface waters and demonstrating their movements and mixing. *Pacific Sci.*, **13**, 14–54.

Beklemishev, K. V. and V. A. Burkov, 1958. The connection between plankton distribution and the distribution of water masses in the frontal zone of the northwestern part of the Pacific Ocean. Investigations of the Kurile-Kamchatka Trench and the Northwest Part of the Pacific Ocean, Pt. IV. *Trudy Inst. Okeanol. Acad. Sci., U.S.S.R.*, **27**, 55–65. (In Russian.)

Berner, L. D., 1960. Unusual features in the distribution of pelagic tunicates in 1957 and 1958. *Calif. Coop. Oceanog. Fish. Invest. Reps.*, **7**, 133–136.

Bieri, R., 1959. The distribution of the planktonic Chaetognatha in the Pacific and their relationship to the water masses. *Limnol. Oceanog.*, **4**, 1–28.

Bigelow, H. B., 1926. Plankton of the offshore waters of the Gulf of Maine. *U.S. Bur. Fish.*, **40** (for 1924), pt. 2, 509 pp.

Boden, B. P. and E. Brinton, 1957. The euphausiid crustaceans *Thysanopoda aequalis* Hansen and *Thysanopoda subaequalis* Boden, their taxonomy and distribution in the Pacific. *Limnol. Oceanog.*, **2**, 337–341.

Bogorov, V. G., 1955. Regularities of plankton distribution in North-west Pacific. *Proc. UNESCO Symp. Phys. Oceanog. Tokyo.*, 260–276.

Bogorov, V. G. and M. E. Vinogradov, 1955. Some essential features of zooplankton distribution in the northwestern Pacific Ocean. *Trudy Inst. Okeanol. Acad. Sci., U.S.S.R.*, **18**, 113–123. (In Russian.)

Bradshaw, J. S., 1959. Ecology of living planktonic Foraminifera in the North and Equatorial Pacific Ocean. *Contrib. Cushman Found. Foram. Res.*, **10**, 25–64.

Brinton, E., 1960. Changes in the distribution of the euphausiid crustaceans in the region of the California Current. *Calif. Coop. Oceanog. Fish. Invest. Reps.*, **7**, 137–146.

Brinton, E., 1962. Two new euphausiids, *Euphausia nana* and *Stylocheiron robustum*, from the Pacific. *Crustaceana*, **4**, 167–179.

Brinton, E., 1962a. The distribution of Pacific euphausiids. *Bull. Scripps Inst. Oceanog. Univ. Calif., Tech. Ser.*, **8**, 51–270.

Brodskii, K. A., 1950. Copepoda, Calanoida, of the far-eastern waters of the U.S.S.R. and the Polar Basin. *Zool. Inst., Acad. Sci., U.S.S.R., Moscow*, 441 pp. (In Russian.)

Brodskii, K. A., 1955. Plankton of the northwestern part of the Kuroshio and the waters of the Pacific Ocean adjacent to the Kurile Islands. *Trudy Inst. Okeanol. Acad. Sci., U.S.S.R.*, **18**, 124–133. (In Russian.)

Brodskii, K. A., 1957. The Copepoda (Calanoida) fauna and zoogeographic division into districts in the northern part of the Pacific Ocean and of the adjacent waters. *Zool. Inst., Acad. Sci., U.S.S.R., Moscow*, 222 pp. (In Russian.)

Bruun, A. F., 1957. Deep sea and abyssal depths. Chap. 22 in Vol. 1, *Treatise on Marine Ecology and Paleoecology* (Ed. by J. W. Hedgpeth), Geol. Soc. Mem. 67.

Burkenroad, M. D., 1936. The Aristaeinae, Solenocerinae and pelagic Penaeinae of the Bingham Oceanographic Collection. Materials for a revision of the oceanic Penaeidae. *Bull. Bingham Oceanog. Coll.*, **5** (Art. 2), 1–152.

Campbell, M. H., 1929. Some free-swimming copepods of the Vancouver Island region. *Trans. Roy. Soc. Canada*, Third Ser., **23**, 303–332.

Clarke, G. L. and D. D. Bonnet, 1939. The influence of temperature on the survival, growth, and respiration of *Calanus finmarchicus*. *Biol. Bull.*, **76**, 371–383.

Coachman, L. K. and C. A. Barnes, 1961. The contribution of Bering Sea water to the Arctic Ocean. *Arctic*, **14**, 147–161.

Dales, R. P., 1957. Pelagic polychaetes of the Pacific Ocean. *Bull. Scripps Inst. Oceanog. Univ. Calif.*, **7**, 99–168.

David, P., 1955. The distribution of *Sagitta gazellae* Ritter-Zahony. *Discovery Reps.*, **27**, 235–278.

Ekman, S., 1953. *Zoogeography of the Sea*. Sidgwick and Jackson Ltd., London, 417 pp.

Fager, E. W., 1957. Determination and analysis of recurrent groups. *Ecology*, **38**, 586–595.

Fager, E. W. and J. A. McGowan, in press. Recurrent groups of zooplankton species in the North Pacific. *Science*.

Foxton, P., 1961. *Salpa fusiformis* Cuvier and related species. *Discovery Reps.*, 32, 1–32.

Hardy, A. C., 1956. *The Open Sea: The world of plankton*. Collins, London, 335 pp.

Hedgpeth, J. W., 1957. Marine biogeography. Chap. 13 in Vol. 1, *Treatise on Marine Ecology and Paleoecology* (Ed. by J. W. Hedgpeth), Geol. Soc. Mem. 67.

Hida, T. S., 1957. Chaetognaths and pteropods as biological indicators in the North Pacific. *U.S. Fish Wild. Serv. Spec. Sci. Rep.* No. 215, 13 pp.

Hubbs, C. H., 1952. Antitropical distribution of fishes and other organisms. Symposium on Problems of Bipolarity and of Pan-Temperate Faunas. *Proc. 7th Pacific Sci. Cong.* **3**, 324–330.

Jaschnov, V. A., 1955. Morphology, distribution and systematics of *Calanus finmarchicus s. l., Zool. J. Acad. Sci. U.S.S.R.*, **34**, 1201–1223.

John, D. D., 1936. The southern species of the genus *Euphausia*. *Discovery Reps.*, **14**, 193–324.

Johnson, M. W., 1938. Concerning the copepod *Eucalanus elongatus* Dana and its varieties in the northeast Pacific. *Bull. Scripps Inst. Oceanog. Univ. Calif., Tech. Ser.*, **4**, 165–180.

Johnson, M. W., 1953. Studies on plankton of the Bering and Chukchi seas and adjacent areas. *Proc. 7th Pacific Sci. Cong.*, **4**, 480–500.

Johnson, M. W., 1954. Plankton of the North Marshall Islands, Bikini and nearby atolls, Part 2, Oceanography (Biological). *U.S. Geol. Surv. Prof. Paper*, 260-F, 301–314.

Johnson, M. W., 1956. The plankton of the Beaufort and Chukchi Sea areas of the Arctic and its relation to the hydrography. *Arctic Inst. North Amer. Tech. Paper* No. 1, 32 pp.

Johnson, M. W., 1960. Production and distribution of larvae of the spiny lobster, *Panulirus interruptus* (Randall), with records on *P. gracilis* Streets. *Bull. Scripps Inst. Oceanog. Univ. Calif.*, **7**, 413–462.

LaFond, E. C. and D. W. Pritchard, 1952. Physical oceanographic investigations in the Eastern Bering and Chukchi Seas during the summer of 1947. *J. Mar. Res.*, **11**, 69–86.

Mackintosh, N. A., 1934. Distribution of the macroplankton in the Atlantic Sector of the Antarctic. *Discovery Reps.*, **9**, 65–160.

McGowan, J., 1960. The relationship of the distribution of the planktonic worm, *Poeobius meseres* Heath, to the water masses of the North Pacific. *Deep-Sea Res.*, **6**, 125–139.

McGowan, J., MS. The systematics, distribution and abundance of the Euthecosomata of the North Pacific. Doctoral Dissertation, in Library, Univ. Calif., La Jolla, Calif.

Redfield, A. C., 1941. The effect of circulation of water on the distribution of the calanoid community in the Gulf of Maine. *Biol. Bull.*, **80**, 86–110.

Reid, J. L., G. I. Roden and J. G. Wyllie, 1958. Studies of the California Current system. *Calif. Coop. Oceanic Fish. Invest., Prog. Rep.* 1 July, 1956, to 1 January, 1958, 27–56.

Runnström, S., 1927. Über die Thermopathie der Fortpflanzung und Entwicklung mariner Tiere in Beziehung zu ihrer geographischen Verbreitung. *Bergens Mus. Aarb., Naturv. rekke.* no. 2.

Russell, F. S., 1939. Hydrographical and biological conditions in the North Sea as indicated by plankton organisms. *J. Cons. Explor. Mer*, **14**, 171–192.

Saur, J. F. T., J. P. Tully and E. C. LaFond, 1954. Oceanographic cruise to the Bering and Chukchi Seas, Summer 1949, Part IV. Physical Oceanographic Studies: Vol. 1, Descriptive Report, U.S. Navy Electronics Lab. Rep., 416, 31 pp.

Schmidt, J., 1925. The breeding places of the eel. *Ann. Rep. Smithsonian Inst.*, 1924, 279–316.

Stepanova, V., 1937. Biologische Zeichnungen der Stromungen im Norlichen Teil des Bering und im Sudlichen Teil des Tschuktschen Meeres. *Pub. Hydrolog. Inst. Leningrad.*, **5**, 175–216.

Steuer, A., 1933. Zur planmässigen Erforschung der geographischen Verbreitung des Haliplanktons, besonders der Copepoden. *Zoogeographica*, **1**, 269–302.

Sverdrup, H. U., M. W. Johnson and R. H. Fleming, 1942. *The Oceans; their physics chemistry and general biology.* Prentice-Hall, New York, 1087 pp.

Tebble, N., 1960. Distribution of pelagic polychaetes in the South Atlantic Ocean. *Discovery Reps.*, **30**, 161–300.

Tebble, N., 1962. The distribution of pelagic polychaetes across the North Pacific Ocean. *Bull. British Mus. (Nat. Hist.), Zool.*, **7**, 371–492.

Tibby, R. B., 1943. Oceanographic results from the "E. W. Scripps" Cruise VIII, May 10 to July 10, 1939. *Rec. Obs. Scripps Inst. Oceanog.*, **1**, 67–80.

Tokioka, T., 1959. Observations on the taxonomy and distribution of chaetognaths of the North Pacific. *Pub. Seto Mar. Biol. Lab.*, **7**, 350–456.

Wilson, C. B., 1942. The copepods of the plankton gathered during the last Cruise of the *Carnegie*. Sci. results of Cruise VII of the *Carnegie* during 1928–1929 under the Command of Captain J. P. Ault. *Biology*, **1**, 1–237.

Worthington, L. V., 1953. Oceanographic results of Project Skijump I and Skijump II in the Polar Sea, 1951–1952. *Trans. Amer. Geophys. Un.*, **34**, 543–551.

# 19. COMMUNITIES OF ORGANISMS

E. W. FAGER

## 1. Introduction

The species composition and general qualitative characteristics of level-bottom benthic communities have recently been thoroughly reviewed (Thorson, 1957). There are also recent, less detailed, reviews of the communities of rocky intertidal regions (Doty, 1957) and of sand beaches (Hedgpeth, 1957), and an excellent introduction to the concepts of community and ecosystem (Hedgpeth, 1957a). Each of these has an extensive bibliography from which can be obtained references giving the details of particular communities. Very little seems to have been done with the community ecology of plankton or abyssal benthic organisms. In view of the preceding, no attempt has been made to present a complete review of the published work on marine communities in this chapter. Instead, the emphasis is on the identification, description and analysis of communities. Most of the more recent development and use of formal ways of identifying and describing communities has been done by plant ecologists and students of soil fauna. Similarly, the analysis of communities in terms of energy relations has been most completely worked out in freshwater habitats. Thus many of the ideas and methods discussed will be from these fields. While it is true that one cannot assume that everything done in freshwater or terrestrial ecology is directly applicable to the marine environment, general ecological concepts, such as food webs, pyramid of numbers, trophic levels, etc., do apply in the latter, no matter where they may have been developed. There may seem to some to be too much space devoted to the problem of the identification and description of communities. The writer believes that one cannot adequately study something until it can be fairly objectively identified and described. In this view, the structural aspects of a community are as important as, but no more important than, the functional aspects; both are necessary before an even approximately complete understanding of the community is possible. As is evident in the section on community theory (page 433), the separation between structure and function is purely formal and convenient; in fact, they are completely interwoven in the community.

It is probably as well to state at the start that the writer believes there are such things as communities in the sense of recurrent organized systems of organisms with similar structure in terms of species presence and abundances, and that it is possible and meaningful to identify and study them, always with the fact firmly in mind that they are open systems. The opposite viewpoint is that there are only randomly assembled collections of organisms whose ecological tolerances allow them to exist in a particular environment, that each collection is an individual point on a continuum and that any grouping of them is, at best, artificial. Evidence supporting this view is presented later.

Precise and complete repetition of species composition and of relative abundances of all species present probably never occurs and, therefore, individual assemblages will appear more or less dissimilar when compared in

[*MS received August, 1960*]    415

detail. However, anyone who has looked at the plants and animals living in the intertidal region or at samples of plankton or benthic organisms is aware of the existence of similar, although not identical, recurrent groups of species. The dissimilarities tend to be emphasized when one looks at the rarer species; the similarities often appear more important when one considers the more common species. It is true that no species will be represented in a sample unless its ecological tolerance range covers the conditions in the environment sampled, but among the factors determining these conditions are the presence and activities of the other organisms. As Elton and Miller (1954) have pointed out, organisms do not exist in nature as isolated single-species populations which have to cope only with the physical environment but rather are influenced by and in turn influence all other organisms with which they come in contact, directly or indirectly. Four of the biotic relationships which are considered important factors in the limitation of numbers of a species are between different species: interspecific competition, herbivore–plant relations, predator–prey relations, parasite–host relations; and only one, intraspecific competition, might be studied out of the community context. The relative importance of the physical and biological aspects of the environment changes from place to place, the biological generally becoming more important as the physical become less variable, but neither is ever entirely ineffective. When studying assemblages in nature it is necessary, then, to conceive of a complex network of action–reaction involving both physical and biological factors, and defined by a set of co-ordinates in time and space. It will seldom, if ever, happen that such a network will have sharp and impassable limits but it will often be found that there are broad areas of considerable internal similarity bordered by relatively narrower regions of rapid change, these in turn grading into other areas with internal consistency. No one of the internally similar areas will be exactly like another but if each must be treated as an unique individual, generalization will be difficult. One needs, therefore, to decide how often and how precisely an assemblage must be repeated in space before it can be considered a general enough phenomenon to merit careful study in the hope of discovering some widely applicable principles. This decision ultimately rests on experience and intuition, but there are some objective ways of identifying, describing and comparing assemblages which will help. These are discussed in the section on community structure (page 424).

Variation in time must also be considered. Should each seasonal phase be considered as a separate community or should they be thought of as aspects of one community? The plant ecologists take the latter view; e.g. a meadow community is accepted although the spring flora may be quite different from the summer flora and that from the autumn and winter floras. This seems a reasonable point of view for there is continuous change throughout the year and any subdivision would seem to be unnecessarily arbitrary. The same question arises in connection with succession. In the normal course of events, cleared or partially cleared patches will be formed in intertidal and benthic communities and may be, by grazing, predation, exhaustion of a particular micronutrient, etc., in plankton communities. Watt (1947) has discussed this problem in connection with a number of plant communities and has concluded that in many cases a community is best considered as a sequence of inter-related processes which lead to a certain assemblage; this then is broken down to nearly the initial conditions and the sequence starts again. At any one point in time, all phases are found at some point in the space occupied by the com-munity. Something like this seems to happen in *Mytilus* beds on exposed rocks; e.g. a buildup from bare rock to a crop of *Mytilus* with its associated organisms which is too heavy to withstand winter storms; practically bare rock exposed by the breaking off of sections of the bed and a sequence of stages leading again to the *Mytilus*. Although the different phases might be separated for convenience of study, it seems more reasonable to consider them all as parts of the *Mytilus* community.

A multi-species assemblage has certain properties which are not shared by the individual species populations, in the same sense that the mortality rate or birth rate or age structure of a population are properties not shared by individual organisms in the population. There is a structure in terms of numbers and kinds of organisms in each of the trophic levels represented; there is an overall pattern of flow of energy and matter through the assemblage; there is a multiplicity of possible pathways for the flow of energy and matter and this multiplicity confers stability on the assemblage (MacArthur, 1955); there is often an ordered succession of species abundances in either time or space or both; there are effects upon the environment which are the result of several species working in sequence. From a theoretical point of view, there seems, therefore, good reason to study assemblages or communities and the species interactions occurring within them.

From a practical point of view, the study of communities is exceedingly complex and, with the present state of our knowledge, a complete study is probably impossible. A major difficulty is taxonomy. No one person has much hope of becoming a competent taxonomist in all the groups which will be represented in even a moderate-sized community, certainly ranging from bacteria, fungi and algae to small invertebrates and, perhaps, to higher plants and large vertebrates. There are a number of approaches to the solution of this difficulty. One is to cut off arbitrarily at some size level(s) and study just a part of the community (this is what the exponents of the study of single species do in an extreme way); another is to guess at the functions of the organisms present and talk in terms of energy flow and the biomass of producers, herbi-vores, carnivores and decomposers (this tends to obscure the biological nature of the system in a cloud of thermodynamics); another is to take identifications to large classificatory groupings and then assign numbers or letters to what seem to be separate entities within these groupings (this is much like suggesting that a modern organic chemist should be satisfied to report that he analyzed the leaves of a certain plant and found three aromatic compounds and two heterocyclic compounds, none of which he has identified but all of which he thinks are important). The first approach can include a respectable part of the community and can be made relatively sound taxonomically with the assistance of specialists in the groups encountered. The thermodynamic approach involves the danger that guesses at functions may seem obvious and yet be quite incorrect. For example, Sanders (1960) has demonstrated that a species of the gastropod genus *Turbonilla*, all members of which were formerly believed to be highly specific ectoparasites, is a deposit feeder and that species of the polychaete genera *Nephythys* and *Ninoe*, which would ordinarily be considered carnivorous, are also deposit feeders. There is no question that the determination of the rates and pathways of energy flow is a basic and necessary part of any complete study of a community, but it is just a part and as such largely ignores behavioral, physiological and populational differences between species in the same trophic level. As Bray (1958) says: "If the natural history approach has sometimes failed to quantify its methods, it is also possible that the mechanist approach has occasionally forgotten the problem." In the present distressing state of the taxonomy of many groups, particularly among the invertebrates, the third approach is attractive. It is, however, an almost universal experience that such a procedure results in failure to separate morphologically similar species which may have quite different functions.

In addition to the biological and philosophical difficulties discussed above, there are the technical ones associated with the necessity of using several different sampling methods, of summarizing and understanding great masses of data, of long-term repetitive work in a limited area, of defining the area, of accounting for import into and export from the area, etc. Despite these difficulties, the complexity of nature is a fact and an attempt should, therefore, be made to study it.

## 2. Definition of Community

No formal definition of community has yet been given. The writer feels that this should be an operational definition such as: a group of species which are often found living together. As pointed out earlier, the decision on what is sufficiently "often" must ultimately be a subjective one, though objective procedures can help and will promote consistency. The decision as to whether species are "living together" or not depends upon an estimate or determination of the mobility, activity and sphere of influence of individuals and of the interspersion of the populations. As most marine studies, except possibly in the intertidal zone, will depend upon samples take over areas or volumes which are unseen, some care is necessary in interpreting co-occurrence in a sample as indicating that the organisms were "living together". This definition puts no limits on the size of communities: any assemblage from the algae and associated invertebrates on a spider crab carapace to the plankton in the upper 50 m over broad areas in the North Pacific Ocean may qualify. There may be some value in identifying as major communities those which are essentially independent of all external sources of energy except the sun (Park, 1949) but these are only one end of a continuum in terms of the relative importance of internal and external sources of energy. In many cases, a community will be associated with a particular physical habitat, but if emphasis is to be on the action–reaction aspect of relations between organisms it seems sounder practice to base the definition on the species group rather than the habitat type.

### 3. Identification of Communities

Given a series of samples in which co-occurrence is expected to have biological meaning, how can the species groups be identified? The usual method is to calculate a matrix of indices of relationship between all possible pairs of species and then rearrange this matrix so that species showing high positive values of the index are grouped together (Kontkanen, 1957). Product–moment correlation coefficients may be used as indices but, because organisms are nearly always clumped, some transformation of the raw abundance values will be needed in order to make their distribution sufficiently close to normal. With the usual sets of samples there is the additional problem raised by the fact that, in order to avoid the use of too many zero values, the correlation coefficients of different pairs of species will often be based on different numbers of observations. As is well known, the significance of a given value of coefficient changes with a change in the number of observations upon which it is based. Probably the best way of making such values comparable is to change them to unit normal variates by way of Fisher's $z$ transformation (Fisher, 1950, pp. 197–204). The organisms can then be grouped either by putting together all species which have positive values over a certain limit (see later discussion of procedure for use with dichotomies) or by cutting the matrix in columns and moving these around until adjacent columns "match" more or less well in terms of the sign and magnitude of values in each row. There will, however, seldom be perfect "matches" and, as no rules are available for deciding when they are good enough, there will often be many species whose association with one group or another will have to be based on purely subjective judgment. In the hands of a worker with much experience with the organisms involved, this method is rapid and fairly reliable. Its major drawbacks are the necessity for experience and the practical impossibility of defining exactly the basis of decisions involved in the assignment of species to groups.

If one is in the fortunate position of having a matrix in which all correlation coefficients are based on the same number of samples, there are a number of grouping procedures which start with the pair of species having the highest correlation and build up groups from this by adding the species with highest average correlation with the initial pair, then adding a fourth to the trio, etc. (Olson and Miller, 1958; Holzinger and Harman, 1941). Rules are given for deciding on the limits of groups, based on the relative magnitudes of the average intra- and extra-group correlations of the last-added species. These procedures have the disadvantage that the buildup of groups is slow.

It has also been suggested that tetrachoric correlation coefficients may be used because, although less efficient, they are easier to calculate and do not require normalization of the data. Based on these Tryon (1955) has published a grouping procedure which involves matching matrix columns by eye. The matching is then tested by determining how well correlation within and between clusters reproduces the original correlations between individual variables.

One result of all methods based on correlation coefficients is that they separate species whose abundances are negatively correlated or uncorrelated even though these species may be very often or always found together. Such separations are reasonable in the case of the materials for which the methods were developed—anatomical, sociological or psychological measurements— but, for the purpose of identifying assemblages, it seems better to the writer to use methods which put together species which are frequently part of each other's environment and then to look at abundance relations within these groups. This results if presence and absence are used as the basis of the index.

For this purpose, no doubt the simplest indices are based on those proposed by Jaccard (1912) and Sørensen (1948) for measuring the similarity between floras. Both have a theoretical range from 0 to 1 and measure the proportion of total occurrences of two species which are co-occurrences: Jaccard's, $c/(a+b-c)$; Sørensen's, $2c/(a+b)$; where $a$ is the number of occurrences of species A, $b$ is the same for species B and $c$ is the number of joint occurrences of the two species. Both indices have several serious limitations: if $a$ and $b$ are not equal, the maximum possible value is less than 1; even if $a$ and $b$ are equal, the maximum observable value of the index will not be 1 unless both species are very abundant or the samples are very large because species of moderate to low abundance will often be absent from small samples by chance; neither takes account of the number of occurrences, a value of 0.5 based on ten occurrences of each species is not distinguished from one based on several hundred occurrences of each species although the latter is certainly more meaningful; as long as $c$ and the sum of $a$ and $b$ remain constant, the values of the indices do not change with changes in the relative sizes of $a$ and $b$, yet the probability of co-occurrence of randomly distributed organisms (probably also of non-randomly distributed organisms) decreases as the frequencies $a$ and $b$ depart from equality; there is at present no way of calculating an expected value for either index. The preceding list of limitations suggests that, despite their simplicity, these two indices are apt to be misleading.

Indices based on a $2 \times 2$ contingency table and the associated $\chi^2$ value avoid many of the difficulties outlined above (Cole, 1949). They have, however, certain properties which make them unsuitable as a basis for grouping: if two species always occur together and also are found in most of the samples taken, they will be unassociated by this procedure although they should certainly be grouped together on the basis of being a frequent part of each other's environment; conversely, if two species seldom occur together but are sufficiently rare, they will be found to be associated. In order to overcome these drawbacks, Fager (1957) proposed a modified version of the $2 \times 2$ table. Correspondence with W. H. Kruskal has made it evident that the parent distribution of the modified table is not the simple hypergeometric and, therefore, significance tests based upon this distribution are not precise. It is suggested that the $\chi^2$ index be replaced by one based on the geometric mean of the proportions of joint occurrences, corrected for the number of occurrences on which the index is based: $c/\sqrt{ab} - 1/2\sqrt{b}$, where $a$ is the number of occurrences of species A, $b$ is the same for species B, $c$ is the number of joint occurrences and species are assigned to the letters so that $a$ is less than or equal to $b$. The index may be used as a basis for grouping if an arbitrary value is required in order to consider two species associated. A requirement of 0.50 has been found to give bio-logically reasonable groups when applied to certain zooplankton from the North Pacific, to invertebrates in decaying oak wood, etc. If this value is required, $a$ can never be less than $b/4$ so that one is grouping only species which are of about the same frequency of occurrence.[1] This seems an advantage if one is looking for groups of species all members of which are frequently enough parts of each other's environment that one may suspect the existence of interactions; it may be a disadvantage if one is trying to associate rare species with more common ones although the grouping procedure will generally show the former as "associates" of the group containing the more common species with which they are linked.

Williams and Lambert (1959, 1960) have suggested a grouping procedure based on $\chi^2$ values calculated for the $2 \times 2$ contingency tables for all possible pairs of species in the samples, excluding those which were found in all samples or in less than 2% of the samples. Values of $\chi^2$ for individual species are summed and the first division of samples is made on the basis of the presence or absence of the species which has the greatest sum. These two parts are then divided on the basis of the presence or absence of the species with the next largest sum of $\chi^2$ values, etc. In their study, this led to assemblages of plant species which were reasonable in terms of soil types, burning regime, etc. The same sort of approach has been used by Goodall (1953). (See earlier discussion of $\chi^2$ indices for possible disadvantages.)

The writer has outlined a grouping procedure which is applicable to any set of data which can be put into dichotomous form; in the present case one would use the dichotomy, associated/not associated, based on an arbitrary significance level for $\chi^2$ or a correlation coefficient or on an arbitrarily chosen value of an

---

[1] The values of $c$ for all appropriate pairs of species can be easily gotten if the occurrences of each species are recorded by punching out in the margin of a card or cards positions corresponding to the samples in which they were found. When such cards for two species are held together, the number of coincident punch-outs is the value of $c$. If the cards are put in order of frequency of occurrence of the species, the card for the least frequent species may be compared with those for all species up to those which occurred more than four times as often; its card can then be laid aside and that of the next most frequent species can be compared; etc. This systematic procedure makes comparisons for as many as 100 species go rather rapidly even when done by hand and could be adapted for mechanical sorting machines. The cards are also useful when it comes to picking out samples which are representative of the groups. This procedure, the calculation of the index and the grouping process have been programmed for the CDC 1604 computer. The program will handle up to 150 species.

index such as the modified geometric mean suggested above. Given such a
"score" for all pairs of species in the samples, one can write down a set of
rules which will lead to the largest possible groups within which all species are
"associated" and will decide between possible groups of the same size which
have members in common. The rules and mechanics of isolating the groups are
given in Fager (1957) and will not be repeated here. It may be pointed out that
although the description appears complicated when read, the mechanics are
quite straightforward and easily worked out on a set of samples. The procedure has the advantages of being definable and, therefore, repeatable by any
worker using the same samples, of requiring a minimum of subjective judgment
and of separating out the largest group or groups first and thus considerably
reducing the work as compared with procedures which build up groups from
pairs of species.

A quite different approach is advocated by those ecologists who feel that
every assemblage of organisms is an individual point on a continuum and that
attempts to group assemblages into community types lead to artificial classifica-
tions (Gleason, 1939). Instead of trying to identify recurrent groups with more
or less constant species composition they work toward an ordination of sets of
samples, each set taken from what is thought to be a homogeneous area (stand).
Curtis and his co-workers (Curtis and McIntosh, 1951; Brown and Curtis, 1952;
Bray and Curtis, 1957) have been particularly active in the application of this
viewpoint to forest ecology in the northern United States. In some of their
earlier work they were able to order the different stands by using importance
values for the four dominant tree species based on the sum of % relative
density, % relative dominance and % relative frequency in the samples taken
within each stand. There was a good deal of overlap in the plots of importance
values of the four species versus the stands ordered according to moisture
conditions, but the higher importance values for two species tended to be
associated with dry conditions, for another species they were associated with
less dry conditions and for the fourth with the most moist locations sampled.
In the 1957 paper they calculated scores for stands based on frequency of
occurrence of shrubs and herbs in a number of quadrats within the stand,
density of the major tree species within the stand and a measure of size of
the same tree species within the stand. These were all converted to percentages
relative to the highest value for a species in the set of 59 stands and then the
scores of all the species present in each stand were converted to percentages of
the total score for that stand. An index of similarity was then calculated for
each pair of stands: index of similarity $= 200w/(a+b)$, where $w$ is the sum of the
lesser of the two score values for each species common to the two stands, $a$ is
the total score of one stand, $b$ that of the other stand; as $a$ and $b$ are both 100
in their case, the index reduces to $w$. The theoretical range of the index is 0 to
100 but the maximum value found between replicate sets of samples from
the same stand was 80. The two stands with least similarity were then placed
at the ends of a line ($x$ axis) and each of the other stands was ordered upon this
line in terms of its "distance", measured as 80 minus the appropriate value of

$w$ from the terminal stands. Two stands which were near the center of the $x$ axis and close together on this axis but had the maximum "distance" between them in terms of 80 minus $w$ were then used as the basis of a $y$ axis placed at right angles to the $x$ axis. This was repeated for a $z$ axis at right angles to the other two axes. The stands were thus ordered in three dimensions. Various physical and chemical properties of the stands and their soils were then tested for correlation with the ordering along each axis. The $x$ axis seemed to represent the change from drier, more open forest to mesic, closed-canopy forest; the $y$ axis was related to drainage and soil aeration; the $z$ axis was related to the influence of recent disturbances such as fires, grazing, etc. This same sort of approach could be applied to planktonic or benthic communities if multiple samples were taken at each station and might help, particularly in regions where physical and chemical characteristics change over relatively short distances, to identify the properties of the habitat which most affect each species. A major objection to the method is that two samples with the same fauna and the same numbers of individuals for all species but one may have a very low score if that one species is rare in one sample and overwhelmingly abundant in the other. It also tends to put great weight on the abundant species and minimizes the effects of qualitative differences involving less common species.

Williams (1947) has suggested a method of comparing floras which avoids some of these difficulties. It assumes that the numbers of individuals per species follow a logarithmic series; i.e. if $n$ is the number of species with one individual and $x$ is a constant less than unity, $n$, $nx/2$, $nx^2/3$, $\cdots$ represent the numbers of species with 1, 2, 3, $\cdots$ individuals. Fisher (Fisher, Corbet and Williams, 1943) has shown that when this series holds, as it seems to in some cases, the ratio $n/x$ is a constant for all random samples from the same population of species. This ratio, called an index of diversity, is denoted by $\alpha$. It follows from the properties of the logarithmic series that the number of species expected in a sample is given approximately by $\alpha \log_e (N/\alpha)$, where $N$ is the number of individuals in the sample. On the assumption that two samples are random samples from the same population, one can calculate the expected number of species in common and compare this with the observed number (Greig-Smith, 1957, pp. 122–126). The ratio observed/expected could be used as an index of similarity upon which to base ordination. This may be a more reliable method but, as Greig-Smith points out, some of the basic assumptions are unproven; e.g. one of the consequences of the logarithmic series is that rare species should be the most numerous, whereas the experience of field botanists is that moderately common species are more numerous than either rare or very common species.

The preceding review covers most of the methods which have been used to identify assemblages of organisms. It has been noted that most of them have more or less serious limitations. Some take account of abundances, others use only presence and absence. The investigator must, therefore, choose the sort which emphasizes the group properties in which he is most interested and then try to remain aware of its limitations, particularly if the groups found agree too well with his preconceptions. One fact which cannot be too strongly emphasized is that all groups based on samples, however formed, depend strongly on the sampling techniques used, the sample size chosen, the areas over which sampling was done, the season or time of day during which sampling was carried out, the distribution patterns of the organisms being sampled, the mobility and sensitivity of the organisms, etc. They are in this sense "artifacts" of the sampling program. This does not deny the existence of recurrent groups of species in nature, but does require that very careful consideration be given to the relation between the sampling and the organisms before the groups based on samples are equated with natural groups. No doubt the fact that results may be relatively easily checked in the case of terrestrial plant communities has been one of the main reasons leading to the advanced development of community studies in plant ecology.

## 4. Community Structure

After a community has been identified, it can be described in terms of various distributional statistics of the component species, in terms of the behavior and physiology of the species or in terms of trophic levels and energy flow. These are, of course, not really separable but they may be so considered as an aid to understanding the community. The following discussion will assume, contrary to fact, that, by the use of an appropriate grouping procedure, the near-surface plankton communities of the North Pacific have been identified as to species composition and their positions in the region have been determined. A start has been made on this and it is hoped that the suggestions made here may provide an impetus to work in this region and elsewhere. The first requirement for further analysis of the communities would be a large number of samples with the appropriate characteristics (size of net, depth and volume sampled, etc.) taken throughout the year and, preferably, at randomly determined positions within the area of each community. On the basis of these one could start to describe in some detail the structure of the communities in terms of the distribution patterns of each species.

Perhaps the first statistics to be looked at would be the related ones of frequency and abundance. The first gives some idea of the spread of a species throughout a community. Species with high frequency and high fidelity (see below) will be good indicators of the presence or absence of a particular community in an unknown area. Abundance may be thought of as either the mean number of individuals per unit volume, the mean number of individuals per unit volume in the samples in which the species was present or the number of individuals relative to those of other species within each sample. The clearest picture of overall abundance is probably given by a combination of frequency, the second abundance measure suggested, and a statement of whether the organisms are aggregated, randomly dispersed or evenly spaced. A convenient index number expressing the extent of aggregation is the constant $k$ of the negative binomial distribution (Bliss and Fisher, 1953). This has a low value when there is much aggregation and a high value when there is little; it is infinity for random distributions. If the species are aggregated, as many are, the median plus quartile values may be a more realistic abundance measure than the usual mean plus standard deviation. Either aggregation or evenness immediately raise problems of cause which may involve the behavior of the organisms, their life history (particularly their method of reproduction) and discontinuities in the physical properties of the habitat. As Hutchinson (1953) has pointed out, one or all of these may be the basis for an observed aggregated distribution.

The abundance of a species relative to the abundance of all other species which are found with it provides more information about the structure of a community than does absolute abundance. It is general experience that within the different size ranges a few species provide the bulk of the individuals in each community, and usually in each sample, while the other species are represented by relatively few individuals (e.g., Sanders, 1960). This may be expressed for each species in terms of an average percentage of the total individuals in the samples but a more informative measure is the following: in each sample the species are ordered in terms of the number of individuals; individuals are then summed in order starting with the species with the greatest number of individuals, and those species which have been included in the summation when it reaches the value of half the total individuals present are considered numerical dominants in the sample. The frequency of this dominance is an important measure of the species' position in the community. It is probably most meaningful if the summation is done using animals which fall within a moderate range of sizes. A supplementary way of looking at dominance is suggested in Fager (1957). The species are ranked in terms of abundance within each sample. The ranks for each species are summed over all samples and the set of sums is tested for concordance (Kendall, 1955, pp. 94–102). If there is significant concordance, it may be concluded that the dominance relations between species tend to be constant over the set of samples and the overall dominance position of a species is given by the rank of its sum of ranks.

A number of different parent distributions have been suggested to account for the observed distributions of individuals per species in communities. Of these, the most extensively examined are the logarithmic series (Fisher, Corbet and Williams, 1943), the truncated log-normal distribution (Preston, 1948) and the distribution of relative lengths marked out by the random placement of $n-1$ points on a stick of unit length (MacArthur, 1957). As noted earlier, the logarithmic series predicts that rare species will be the most numerous while experience suggests that moderately common species are in fact the most numerous. The latter is predicted by the log-normal distribution, and Williams (1933) has accepted the closer relation to observation of Preston's formulation. As Goodall (1952) points out, however, the difference between expectations for the two distributions is seldom sufficiently great to make an objective decision between them. MacArthur's distribution is, at least formally, based on more biological assumptions than the preceding two. The lengths marked out on the stick are supposed to represent the "sizes" of non-overlapping, contiguous niches in the community and, therefore the numbers of individuals of the different species occupying them. Niche is, of course, used in its original sense, i.e. what an animal does in the community. MacArthur reported satisfactory agreement between theory and observation in the case of bird species from relatively small, and therefore presumably homogeneous, areas. When bird censuses from large areas were considered, the common species were more common and the rare species were rarer than predicted. Kohn (1959) has reported extraordinary fit of observation with expectation in the case of species of *Conus* in Hawaii. MacArthur has suggested that fit or non-fit with prediction might be used as a measure of the relative homogeneity of the area sampled. Hairston (1959), working with soil fauna from samples carefully selected for homogeneity, found, however, that the common species were more common and the rare species were rarer than predicted. He suggests that this is an expression of the organization of the community as opposed to the random structure assumed in MacArthur's formulation. If this suggestion is correct, one should find increasing divergence between observation and prediction as one adds together more and more samples from a homogeneous community and thus more completely shows up the organization, and decreasing divergence if one combines samples from quite different communities thus reducing the order. This is the case in the examples he presents. Again, one is faced with the question of whether a community is an organized system or simply a random collection of species which are able to adjust to the prevailing conditions. Hairston's evidence agrees best with the assumption of at least some organization.

The concept of diversity of a community is directly related to the structure in terms of relative abundances. A number of indices of diversity have been proposed but most of them, like the constant $\alpha$ in the Fisher and Williams distribution, are of doubtful reliability because of the unproven assumptions underlying the parent distributions. Two measures have been proposed which are independent of the form of the parent distribution. Margalef (1958) has suggested the use of the natural logarithm of the number of permutations of $N$ objects of which $N_a$ are alike, $N_b$ are alike, etc., multiplied by a constant which converts the value into the number of "bits" of information contained in the community. Information in the community in this form may be considered as the number of questions with yes/no answers which would have to be asked to define the community completely. As the formula for permutations contains factorials, the computation of the index becomes formidable when $N$ is even a moderate size. Margalef, therefore, suggests the use of a simpler index due to Gleason (1922) which is based on the assumption that the number of species in a sample is a linear function of the logarithm of the number of individuals. Hairston (1959) has shown that Margalef's justification for this substitution, the stability of the index with increasing sample size when tested on the frequencies of letters in a Spanish text, is invalid because the letters tend to be randomly or evenly dispersed whereas organisms are most often aggregated. The index would be expected to be stable when applied to samples of increasing size from even or random distributions but quite unstable when applied to such samples from aggregated distributions. The other index of diversity which is distribution-free is that suggested by Simpson (1949). It is the probability that two individuals selected independently and at random from a sample will be of the same species: $\sum n_x(n_x-1)/N(N-1)$, where $n_x$ is the number of individuals of species $x$, $N$ is the total number of individuals in the sample, and the sum is over all species present. If all individuals are of one species, the index takes its maximum value, 1; if there are $S$ species all with the same number of individuals and the total number of individuals, $N$, is much greater than $S$, the index takes a minimum value, approximately $1/S$. A limitation of this index, and even more of the information-based one suggested by Margalef, is that their values are determined primarily by numbers of individuals of the abundant species and little affected by considerable differences in abundances of the rarer species. In view of the preceding discussion, it is probably fair to say that indices of diversity are likely to be misleading though they may in particular situations give interesting results (Margalef, 1958).

Three other statistics which the plant ecologists (Braun-Blanquet, 1932) have found useful in their descriptions of the position of a species in a community are fidelity, vitality and periodicity. Fidelity is a measure of the restriction of a species to the community. This may range from species which are found only in the community to those which are found elsewhere but are most abundant in the community and, finally, to those which are about equally abundant in a number of communities. As mentioned previously, species with high frequency and high fidelity are excellent indicators of the community. In addition, those with high fidelity are most likely to have specific physiological and behavioral characteristics related to the conditions in the area occupied by the community. They are, therefore, probably the best choices for study of the biological adjustments necessary for successful coping with the specific conditions. Vitality is a measure of the completeness of the life cycle in a community. Thus, there may be some species which are often present but in poor physiological condition, others which are present as healthy adults but do not breed and others which carry out their complete life cycles within the community. Periodicity measures changes of abundance with time, usually in terms of seasons. This has a clear-cut meaning for intertidal and benthic communities where one may have seasonal changes in both the animals and plants associated with a definite location and these changes occur in more or less the same manner every year. In the case of plankton communities, one will seldom, if ever, be dealing with the same piece of water at any one geographic location and yet the annual recurrence of phytoplankton blooms in the spring and in the late summer or fall in north temperate regions, involving essentially the same species in the same sequence, suggests that the concept of periodicity within a community is relevant here too (Allen, 1945; Lillick, 1940).

Table I is an example of the sort of information which it is suggested should be obtained for each species as a basis for understanding the structure of the community.

## TABLE I

Statistics Associated with Three Hypothetical Species of Small Planktonic Crustacea Found in Samples Taken within the Area Occupied by a Particular Plankton Community

|              | Species A                                  | Species B                    | Species C                                      |
| ------------ | ------------------------------------------ | ---------------------------- | ---------------------------------------------- |
| Frequency    | 104/110                                    | 17/110                       | 76/110                                         |
| Abundance    | $1-40-700/m^3$                             | $1-1-2/m^3$                  | $1-3-10/m^3$                                   |
| Dispersion   | Strongly aggregated $k=0.2$                | Random                       | Aggregated $k=1.2$                             |
| Dominance    | 72/110                                      | 0/110                        | 1/110                                          |
| Overall rank | 2nd ex 15                                   | —                            | 12th ex 15                                     |
| Fidelity     | Exclusive                                   | Accidental                   | Equally abundant in several other communities  |
| Vitality     | Breeds                                      | Dying                        | Adults only                                    |
| Periodicity  | Rare Nov.–Feb., abundant rest of year      | Only present in Mar.–May     | No seasonal changes                            |

Species A: An important and typical member of the community.
Species B: Most probably a stray from another community, perhaps distant.
Species C: A member of some importance in the community but not typical.

Additional information can be obtained from the numbers of individuals of different species within the samples characteristic of the assemblage. In order to determine whether any particular sample or group of samples is better than the others, as indicated by higher numbers of individuals of the species in the community, the set of samples is ranked in terms of each species separately and the ranks are summed for the different samples. These sums are then tested for concordance (Kendall, 1955, pp. 94–102). Significant concordance indicates that, despite species differences, the community in general agrees on what is a good or bad habitat. One can, therefore, pick out the "best" and "worst" samples and examine their characteristics in the hope of getting an idea of conditions favorable and unfavorable to the community as a whole. Pairs of species can also be examined for correlation between abundances, always remembering that multiple comparisons within the same set of data considerably change the significance levels (e.g. the probability of getting at least one comparison significant at the 0.05 level when the ten possible pairs of comparisons between five species are made is about 0.40 not 0.05). Despite this limitation, a highly significant correlation, either positive or negative, probably indicates something worth investigating further in terms of possible biological interaction (cf. Fager, 1957, for examples of the procedures in this paragraph).

## 5. Community Function

After a community has been identified and its structure in terms of individuals and species has been described in quantitative terms, including variations in time and space, the pathways and amounts of energy flowing through the community can be related to the structural features. A good description of the function of a community is provided by a paraphrase of a statement due to Bertalanffy (1950): communities are not in being, they are happening; they are the expression of a perpetual stream of matter and energy which passes through the community and at the same time constitutes it.

Although others had previously investigated the food relations of organisms in communities, it was Lotka (1925, reprint 1956) who first treated this in a formal manner. He pointed out that "aggregates of living organisms are, in their physical relations, energy transformers" and that what we have to consider is "essentially the evolution of a system of energy transformers" and "the progressive redistribution of the matter of the system among these transformers". This brings up a point which, as Macfadyen (1948) has shown, was not always clearly apprehended in earlier publications, i.e. that the matter is recycled but the energy is transformed, eventually to heat, and finally lost to the system. The transformers may work in parallel, as for example a number of different herbivores feeding on one plant, or be coupled in a series, as for example a food chain involving plant–herbivore–carnivore. Each of these transformers in addition to using some of the available energy for the performance of work acts as an accumulator, storing up energy which can be used later by the individual or by the next step in the food chain. Lindeman (1942) applied these ideas to a specific example, Cedar Bog Lake, using values from the literature for estimation of the amounts of energy used in respiration, growth, etc. within each trophic level. Some of his calculations were in error and his choice of values from the literature does not always seem to have been wise, but his paper is still of primary importance because it showed that thinking about communities in these terms is possible and leads to interesting conclusions concerning the system and its parts. His definition of productivity, following Hutchinson's (1953) suggestion, as the rate of contribution of energy from the next lower trophic level is, however, confusing and not in accord with the usual meaning of the word. It would seem better, as Odum (1956) suggests, to consider net rate of increase in energy content, including losses due to predation and death and decomposition, as productivity. Energy content is preferred to biomass because of differences in composition. At the producer level it is relevant to consider gross productivity as well as net, but only the latter seems to have a clear meaning in higher levels. Whichever method is used, the annual productivity decreases as one progresses to higher trophic levels. This is another way of saying that larger animals spend a greater percentage of their income on respiration than smaller ones do and that the higher trophic levels are usually composed of larger and less numerous animals; the "Eltonian" pyramid.

Lindeman's definition of efficiency is also rather special. It is the ratio of the productivity (his definition) of the level being considered to that of the next lower level. On this basis, efficiencies increase as you go to higher trophic levels. A more reasonable definition would seem to be input minus respiration, divided by input, $(I - R)/I$. This is nearly the same as the definition given by Patten (1959). This would give a measure of the efficiency of a trophic level in transforming input into accumulated energy. In this sense, efficiency decreases with increase in trophic level for the same reason that productivity decreases. The producer level is a special case. If one considers the energy available in sunlight as the input, the efficiency is very low; if one considers the gross photosynthetic production as the input, the efficiency is probably about equal to that of the primary consumers—50 to 75% (Ryther, Chapter 17). Harvey (1950), using a definition like the preceding but with input equated to respiration plus growth, calculates an efficiency of 70% for primary consumers (herbivorous copepods) and about 5 to 10% for carnivores. A decision between the two definitions would turn on the question of whether materials which were ingested but which the organism had no chance of digesting, such as chitin, should be included in the input and how losses due to predation or death and decomposition should be handled. In either case, the amount of energy available as food decreases as one goes to higher levels and, though this may be partly compensated for by increased predatory ability, eventually becomes too small to be exploited by larger animals. As a result, ascending food chains seldom have as many as five links. Other than limiting the number of links in a food chain or as an aid in deciding whether to harvest fish, zooplankton or phytoplankton, there is some question of what efficiency means in terms of the community as a whole. Most communities are not rapidly filling up their habitats with stored energy in the form of organic materials and, therefore, in terms of the above definition, they must be quite inefficient. As Lotka (1925) says "the great world engine or energy transformer composed of a multitude of subsidiary units . . . seems, in a way, a singularly futile engine, which, with a seriousness strangely out of keeping with the absurdity of the performance, carefully and thoroughly churns up all the energy gathered from the source."

In broad terms, it is now clear what proportion of the chemical energy originally fixed by photosynthesis reaches each trophic level and how each level portions it out among various activities. It would be more accurate to have everything in terms of free energy instead of total energy content but this would probably not materially change the picture. What is not available is a detailed analysis of the mechanisms involved in a particular community, e.g. the physiological and behavioral properties of species A which cause it to get only 10% of the energy available to a certain level while species B gets 60%, etc. When one gets down to the problem of trying to set up an accurate balance sheet for even a simple community, all sorts of troubles arise which are glossed over if one looks only at large units (trophic levels). For example, Teal (1957), in his study of a temperate cold spring containing a very limited community, used the spring as a thermostat when measuring the respiration of the animals. He thus avoided uncertainties in correcting respiration for temperature, but, as the respiratory quotient was not known, used an average oxycalorific coefficient to convert to calories and assumed a linear relation between weight and respiration whereas most studies have suggested a non-linear relation (Bertalanffy, 1957). He obtained an indirect measure of mortality in the field by measuring growth under constant conditions in the laboratory and assuming a reasonable, but unproven, relation between population weight, growth rate and respiratory rate. Using this estimated mortality rate, he corrected for losses occurring between sampling periods. In some cases the caloric content of a species was calculated from an analysis of a more or less closely related species taken from the literature. He took multiple random samples so that he could calculate confidence limits for the mean number of each species but only presented these fully for one species, the fourth in importance in terms of energy flow. Because the animals were aggregated, 90% confidence limits for the mean often covered a range of one-half to twice the mean value. Therefore, energy flow based on mean values of assimilation per individual and respiration per individual could easily be in error by these factors. This makes the perfect agreement between credit and debit columns in his community energy-balance chart a bit difficult to believe unless the figure for deposition, 28% of the debit column, was added to achieve the balance.

None of the preceding should be taken as a condemnation of Teal's work. It is presented only to outline some of the problems involved and the numerous assumptions and simplifications, made without rigorous justification, required in striking an energy balance in even a relatively simple community. In terms of replicate determinations and attention to sampling theory, population changes, feeding habits, etc., his study of community productivity is one of the most complete and careful yet done and should be looked at by everyone contemplating such a study. By comparison, some marine community productivity studies (Harvey, 1950; Odum and Odum, 1955), although admittedly done on much more complex communities, leave a great deal to be desired.

Teal's results emphasize the necessity for work continuing over at least a year; despite the fact that the spring he studied was almost constant chemically and thermally, the species which used most of the energy available during the summer was unimportant during the winter and over the year was only third in terms of energy flow.

Even though phytoplankton productivity is discussed in other chapters in this volume (Steemann Nielsen, Chapter 7, page 129; Ryther, Chapter 17, page 347), because of its primary importance as a source of energy in marine communities it seems appropriate to mention briefly some of the limitations of the methods now used for its estimation (see the review by Steele, 1959, for documentation). A combination of chlorophyll, light intensity at the surface and the extinction coefficient has been used to estimate photosynthesis which is then corrected for respiration by assuming a ratio of photosynthesis to respiration of 10/1. It has, however, been shown that the ratio can vary from 20/1 to 1.4/1, depending on the physiological state of the phytoplankton and environmental conditions. The light–dark bottle method of productivity measurement has two disadvantages: it may require very long periods during which it is uncertain what may happen to the phytoplankton in the unnaturally restricted space and in the presence of solid surfaces; it measures only gross production, and net production must be estimated by some assumption concerning respiration of the phytoplankton. The use of $^{14}C$ has many advantages —speed, direct measurement of photosynthesis, etc.—but it gives an estimate somewhere between gross and net production. The value of the correction which should be used to get gross production may vary depending on the physiological state of the phytoplankton. It will be noted that a major problem in the estimation of net production is the correction for respiration of the phytoplankton; this cannot be easily separated from that of the zooplankton and bacteria in the sample. As mentioned earlier, Ryther has estimated that net production is usually in the range 50–75% of gross production, but this is a rather large spread if one wishes to strike any sort of energy balance and the estimate is based on certain assumptions which may not be valid under all conditions.

Another problem which arises when measurements are infrequent is that the productivity measurement may be only a somewhat sophisticated measure of the standing crop at the moment of measurement. For it to have a meaning as production over a period of time, it is necessary to assume that, during the period, environmental conditions do not change in such a way as to affect photosynthesis and that the standing crop does not change in numbers, species composition, or any other manner that will affect the rate of photosynthesis. Such assumptions may be difficult to justify unless measurements are made with considerable frequency (see Steemann Nielsen, Chapter 7, page 129). Attempts have been made to take into account certain of the changes by setting up more or less complicated equations for the rate of change of plant material in a vertical water column under unit surface area (Riley, Stommel and Bumpus, 1949; Cushing, 1959; Steele, 1959). These involve estimates of photosynthetic rates, grazing rates by zooplankton, sinking rates of the phytoplankton, mixing, changes in nutrients (usually phosphorus), etc. As Steele points out, however, there is an element of circularity in some of the calculations and a rather serious doubt as to whether certain of the estimates are relevant to conditions in the ocean. For example, laboratory determinations of the filtering rate of *Calanus*, upon which the estimate of grazing rate has been based, range from 1 to 240 ml per animal per day while *Calanus* respiration rates combined with phytoplankton concentrations in the sea require rates of 1 to 3 liters per animal per day. There is some evidence that part of the discrepancy may be due to the smallness of the volumes of water used in the laboratory experiments.

Cushing (1955) has used a method which overcomes some of these problems, although it can, unfortunately, be used only on diatoms and thus is inappropriate unless diatoms make up most of the phytoplankton crop. It also requires identification of plankton patches and repeated sampling of the same patches. Because of the mechanism of division in diatoms, there is a progressive decrease in mean cell width in a diatom population which is growing and this decrease can be related to the number of divisions which have occurred between samplings. Given the initial standing crop, the number of divisions and the average volume of plant material per diatom, one can estimate the theoretical net production. Given the observed standing crop of diatoms at some time after the initial measurement and an estimate of natural (non-grazing) mortality (Cushing uses the number of intact, empty frustules to get this), a comparison can be made with the theoretical net production and a grazing rate for herbivores can be estimated from the difference if it is assumed that the change in the diatom population with time is a function only of the division rate, the grazing rate and the natural mortality rate. When applied to a major, and perhaps rather simple, food chain, diatom–*Calanus*–herring, off the northeast coast of England, it gave reasonable results. This may be a special situation but it seems that the method might be more widely applied.

## 6. Community Theory

In an earlier section the methods used for describing the structure of communities were discussed. This structure is concerned with two concepts: order and complexity (Bray, 1958). The order of a system may be considered as directly related to its content of negative entropy or to its information content or to the amount and kind of energy within and flowing through the system, with particular emphasis on the availability of the energy within it. Complexity is a measure of the number of parameters needed to define a system fully in time and space. These two concepts are not necessarily related although increasing complexity often goes with higher order. The difference is perhaps most clearly shown by an example presented by Patten (1959): a single cell of *Bacterium pycnoticus* has been estimated to contain about $10^{13}$ "bits" of information compared to about $2 \times 10^{28}$ "bits" for a man; $2 \times 10^{15}$ cells of *B. pycnoticus*, therefore, contain as much information as a man but certainly are not as complex.

As a community develops from a situation where there are no organisms present in a volume of water or on an area of substratum to a condition of more or less complete filling up of the space or area by organisms, the order of the community increases. It may be regarded as acquiring negentropy from the environment, storing it as coded information in the organisms and then converting it to enthalpy and entropy as required. At first the negative change in entropy made possible by import of complex organic substances far overbalances the increase of entropy associated with irreversible processes such as respiration. If plants are part of the community, import may become less and less important as they take over the role of suppliers of negative entropy through photosynthetic processes. Even after photosynthetic production has

15—s. II.

reached its maximum, a further increase in order can occur by an increase in the numbers and kinds of organisms which utilize various waste products that still contain available energy. The community tends toward a steady state where the entropy change for the system as a whole is a minimum: a situation where the rate of decrease in entropy, resulting from the import and internal production of complex organic materials, approximately balances the rate of entropy increase resulting from irreversible processes.

Concurrent with an increase in order, there develops an increase in complexity. This is usually described in terms of numbers and kinds of organisms present, their distribution and their interactions. Bertalanffy (1950) has shown that, in a steady-state open system, the ratio of concentrations (numbers/unit measure) of different components (organisms) depends only on the system constants (kind and intensity of interactions). Therefore, as long as the constants do not change, the composition remains the same. He has shown that such a system is self-regulatory in the sense that it reacts against a disturbance of the steady state in a manner which offsets the effects of the disturbance. In the case of an individual organism, this reaction may overshoot the mark and lead to a series of physiologic oscillations around the steady-state position. In the case of a community, as Bray (1958) points out, the movement is one-sided for a community cannot overshoot; what is seen is the steady state and various less-ordered stages leading up to it.

Patten (1959) has developed a general theory of ecosystem dynamics—with appropriate modifications it can be applied to a smaller community—on the basis of modern information theory. In this context he has formally defined efficiency and the conditions for succession, steady state and senescence of communities. He concludes that Odum and Pinkerton's (1955) theory that open systems in the steady state tend to operate at low efficiencies which are optimum for maximum output of power is theoretically incorrect and is not supported by the evidence which has been cited. He re-emphasizes the perhaps obvious but worth repeating fact that the standing-crop biomass is not directly related to either gross or net productivity and is only "a relatively static side issue to the main stream of energy flow". As Patten admits, one of the major problems preventing the use of information theory in the study of community dynamics is that of measurement. It is, therefore, necessary to continue using energy flow. However, as entropy units can be directly converted to "bits", energy flow can be converted to information if the assumption is made that the rate of entropy increase is closely approximated by the rate of energy release in respiration. He applies this relation to four studies of the energy flow in freshwater habitats. This is an interesting mental exercise but does not seem to provide much more understanding of the situation than the original papers.

In considering biological phenomena, it is evident that the community stands far up in a hierarchy based on organization. The general theory of organization suggests that there is always an interplay of action–reaction between the higher and lower levels of organization, the higher ones depending for their existence upon the continuance of the lower and the lower being modified or directed by the higher. The formal separation which we make between an individual of a species, the population of the species and a community containing this and other species, is useful as each of these levels has characteristics of its own, but no level can be understood or analyzed completely without reference to at least those nearest it in the hierarchy. Thus, for example, studies of the behavior of individuals and of the birth and death rates of single species populations are a necessary part of a complete community analysis, and the community provides the context within which the results of such studies can be understood in their relation to the existence of the species in nature.

# References

Allen, W. E., 1945. Seasonal occurrence of marine plankton diatoms off Southern California in 1938. *Bull. Scripps Inst. Oceanog. Univ. Calif.*, **5**, 293–334.

Bertalanffy, L. von, 1950. Theory of open systems in biology. *Science*, **111**, 23–29.

Bertalanffy, L. von, 1957. Quantitative laws in metabolism and growth. *Q. Rev. Biol.*, **32**, 217–231.

Bliss, C. I. and R. A. Fisher, 1953. Fitting the negative binomial distribution to biological data. *Biometrics*, **9**, 176–200.

Braun-Blanquet, J., 1932. *Plant Sociology. The study of plant communities.* (Translated by G. D. Fuller and H. S. Conard.) McGraw-Hill, New York, 439 pp.

Bray, J. R., 1958. Notes toward an ecologic theory. *Ecology*, **39**, 770–776.

Bray, J. R. and J. T. Curtis, 1957. An ordination of the upland forest communities of southern Wisconsin. *Ecol. Monog.*, **27**, 325–349.

Brown, R. T. and J. T. Curtis, 1952. The upland conifer-hardwood forests of northern Wisconsin. *Ecol. Monog.*, **22**, 217–234.

Cole, L. C., 1949. The measurement of interspecific association. *Ecology*, **30**, 411–424.

Curtis, J. T. and R. P. McIntosh, 1951. An upland forest continuum in the prairie-forest border region of Wisconsin. *Ecology*, **32**, 476–496.

Cushing, D. H., 1955. Production and a pelagic fishery. *Fish. Invest., Lond.*, (2), **18**, No. 7.

Cushing, D. H., 1959. On the nature of production in the sea. *Fish. Invest., Lond.*, (2), **22**, No. 6.

Doty, M. S., 1957. Rocky intertidal surfaces. Pp. 535–585 in Treatise on Marine Ecology and Paleoecology, Vol. I, Ecology. *Geol. Soc. Amer. Mem.* 67 (J. W. Hedgpeth, ed.).

Elton, C. S. and R. S. Miller, 1954. The ecological survey of animal communities: with a practical system of classifying habitats by structural characters. *J. Ecol.*, **42**, 460–496.

Fager, E. W., 1957. Determination and analysis of recurrent groups. *Ecology*, **38**, 586–595.

Fisher, R. A., 1950. *Statistical Methods for Research Workers* (11th ed.). Hafner, New York, 354 pp.

Fisher, R. A., A. S. Corbet and C. B. Williams, 1943. The relation between the number of species and the number of individuals in a random sample of an animal population. *J. Anim. Ecol.*, **12**, 42–58.

Gleason, H. A., 1922. On the relation between species and area. *Ecology*, **3**, 158–162.

Gleason, H. A., 1939. The individualistic concept of the plant association. *Amer. Midl. Naturalist*, **21**, 92–110.

Goodall, D. W., 1952. Quantitative aspects of plant distribution. *Biol. Revs.*, **27**, 194–245.

Goodall, D. W., 1953. Objective methods for the classification of vegetation. I. The use of positive interspecific correlation. *Austral. J. Bot.*, **1**, 39–63.

Greig-Smith, P., 1957. *Quantitative Plant Ecology.* Academic Press, New York, 198 pp.

Hairston, N. G., 1959. Species abundance and community organization. *Ecology*, **40**, 404–416.

Harvey, H. W., 1950. On the production of living matter in the sea off Plymouth. *J. Mar. Biol. Assoc. U.K.*, **29**, 97–137.

Hedgpeth, J. W., 1957. Sandy beaches. Pp. 587–608 in Treatise on Marine Ecology and Paleoecology. Vol. I, Ecology. *Geol. Soc. Amer. Mem.* 67 (J. W. Hedgpeth, ed.).

Hedgpeth, J. W., 1957a. Concepts of marine ecology. *Ibid.*, 29–52.

Holzinger, K. J. and H. H. Harman, 1941. *Factor Analysis: A synthesis of factorial methods.* Univ. Chicago Press, 417 pp.

Hutchinson, G. E., 1953. The concept of pattern in ecology. *Proc. Acad. Nat. Sci., Philadelphia*, **105**, 1–12.

Jaccard, P., 1912. The distribution of the flora in the alpine zone. *New Phytol.*, **11**, 37–50.

Kendall, M. G., 1955. *Rank Correlation Methods* (2nd ed.). Charles Griffin, London, 196 pp.

Kohn, A. J., 1959. The ecology of *Conus* in Hawaii. *Ecol. Monog.*, **29**, 47–90.

Kontkanen, P., 1957. Delimitation of communities in research on biocoenotics. *Cold Spring Harbor Symp. on Quant. Biol.*, **22**, 373–378.

Lillick, L. C., 1940. Phytoplankton and planktonic protozoa of the offshore waters of the Gulf of Maine. Part II Qualitative composition of the planktonic flora. *Trans. Amer. Phil. Soc.*, **31**, 193–237.

Lindeman, R. L., 1942. The trophic-dynamic aspect of ecology. *Ecology*, **23**, 399–418.

Lotka, A. J., 1925. *Elements of Physical (Mathematical) Biology*. Williams and Wilkins, Baltimore (Dover Pub., N.Y.), 465 pp.

MacArthur, R. H., 1955. Fluctuations of animal populations and a measure of community stability. *Ecology*, **36**, 533–536.

MacArthur, R. H., 1957. On the relative abundance of bird species. *Proc. Nat. Acad. Sci.*, **43**, 293–295.

Macfadyen, A., 1948. The meaning of productivity in biological systems. *J. Anim. Ecol.*, **17**, 75–80.

Margalef, R., 1958. Temporal succession and spatial heterogeneity in phytoplankton. Pp. 323–349 in *Perspectives in Marine Biology*. Univ. California Press. (A. A. Buzzati-Traverso, ed.).

Odum, H. T., 1956. Efficiencies, size of organisms and community structure. *Ecology*, **37**, 592–597.

Odum, H. T. and E. P. Odum, 1955. Trophic structure and productivity of a windward coral reef community on Eniwetok Atoll. *Ecol. Monog.*, **25**, 291–320.

Odum, H. T. and R. C. Pinkerton, 1955. Times speed regulator, the optimum efficiency for maximum power output in physical and biological systems. *Amer. Sci.*, **43**, 331–343.

Olson, E. C. and R. L. Miller, 1958. *Morphological Integration*. Univ. Chicago Press, 317 pp.

Park, O., 1949. The community, introduction. Pp. 436–441 in *Principles of Animal Ecology*. Saunders, Philadelphia.

Patten, B. C., 1959. An introduction to the cybernetics of the ecosystem: the trophic dynamic aspect. *Ecology*, **40**, 221–231.

Preston, F. W., 1948. The commonness and rarity of species. *Ecology*, **29**, 254–283.

Riley, G. A., H. Stommel and D. F. Bumpus, 1949. Quantitative ecology of the plankton of the western North Atlantic. *Bull. Bingham Oceanog. Coll.*, **12**, 1–169.

Sanders, H. L., 1960. Benthic studies in Buzzards Bay. III The structure of the soft-bottom community. *Limnol. Oceanog.*, **5**, 138–153.

Simpson, E. H., 1949. Measurement of diversity. *Nature*, **163**, 688.

Sørensen, T., 1948. A method of stabilizing groups of equivalent amplitude in plant sociology based on the similarity of species content and its application to analyses of the vegetation on the Danish commons. *Biol. Skr.*, **5**, 1–34.

Steele, J. H., 1959. The quantitative ecology of marine phytoplankton. *Biol. Revs.*, **34**, 129–158.

Teal, J. M., 1957. Community metabolism in a temperate cold spring. *Ecol. Monog.*, **27**, 283–302.

Thorson, G., 1957. Bottom communities (sublittoral or shallow shelf). Pp. 461–534 in Treatise on Marine Ecology and Paleoecology, Vol. I Ecology. *Geol. Soc. Amer. Mem.* 67 (J. W. Hedgpeth, ed.).

Tryon, R. C., 1955. Identification of social areas by cluster analysis: A general method with an application to the San Francisco Bay area. *Univ. Calif. Pub. Psychol.*, **8**, 1–100.

Watt, A. S., 1947. Pattern and process in the plant community. *J. Ecol.*, **35**, 1–22.

Williams, C. B., 1947. The logarithmic series and the comparison of island floras. *Proc. Linnean Soc. Lond.*, **158**, 104–108.

Williams, C. B., 1953. The relative abundance of different species in a wild animal population. *J. Anim. Ecol.*, **22**, 14–31.

Williams, W. T. and J. M. Lambert, 1959. Multivariate methods in plant ecology. I. Association analysis in plant communities. *J. Ecol.*, **47**, 83–102.

Williams, W. T. and J. M. Lambert, 1960. Multivariate methods in plant ecology. II. The use of an electronic digital computer for association-analysis. *J. Ecol.*, **48**, 689–710.

# 20. THEORY OF FOOD-CHAIN RELATIONS IN THE OCEAN

## G. A. Riley

## 1. Introduction

Descriptive oceanography reveals a variety and complexity of species relationships and environmental influences that defy quantitative analysis except by recourse to simplifying procedures of one sort or another. The obvious solution to many of these problems is laboratory experimentation under simplified and controlled conditions. Such work leads to a better quantitative understanding of ecological relations than mere observation of nature; however, it is desirable to take the further step of recombining experimental results into a quantitative synthesis of the interaction of processes in nature. This is the only way we can demonstrate to our satisfaction that the experiments are realistic. In order to effect such a synthesis, the experiments must be combined within a logical framework which duplicates, as closely as our knowledge permits, the relationships of organisms and environmental factors that we find in nature.

All of the work that has been done on the theory of food-chain relations has stemmed from the prey–predator relationship postulated by Volterra (1928). Defining $N_1$ as the number of prey and $N_2$ as the number of predators, Volterra's equation for the rate of change of prey with respect to time was

$$\frac{dN_1}{dt} = N_1(b_1 - k_1 N_2), \tag{1}$$

where $b_1$ is the growth coefficient of the prey, and $k_1$ is a predation coefficient having the sense that the predator "hunts" a constant amount of space in unit time and consumes all of the prey therein. The equation for predators is

$$\frac{dN_2}{dt} = N_2(k_2 N_1 - d_2), \tag{2}$$

where $k_2$ is a coefficient of food assimilation which in the first approximation may be synonymous with $k_1$ or may be some constant fraction of it. The coefficient of death, $d_2$, is also a constant.

According to these equations, an increase in prey leads to an increase in the number of predators, which in turn reduce the net rate of growth of the prey until the latter begins to decline. Continuation of the sequence leads to the well known prey–predator oscillation.

In this ideally simple form neither the equations nor their integrated results are biologically realistic. Smith (1952) has discussed some of the defects. However, the equations provide a logical framework, and changes that make them more realistic do not alter the basic structure. The constants are not likely to remain the same through the whole cycle, but this can be corrected on the basis of known facts about the organisms in question. There is likely to be a time lag between feeding and reproduction, which will vary considerably in different

[*MS received August, 1960*]     438

animals. Wangersky and Cunningham (1957) have examined this problem. With variations in the time lag in relation to certain other parameters, they were able to demonstrate a variety of results including damped oscillations, monotonic approach to equilibrium and an expanding oscillation which probably ultimately approaches a limit cycle. True Volterra oscillations were possible with a finite time lag, but only within a narrow range.

Bartlett (1957) studied the effects of stochastic fluctuations on several kinds of population models with results that cast doubt on the accuracy of deterministic solutions. In the case of the prey–predator equations, his stochastic model showed three distinct peaks of prey and predators, then a sudden change to a smaller and poorly defined cycle, and finally extinction of the predators.

Ecological models present more problems than the classical population models. In all that thus far have been devised, the prey and predators have been plankton populations containing an assemblage of species. An attempt must be made to frame physiological coefficients in average terms. The fact that there are species differences in behavior and a variable species composition inevitably increases the margin of error. Furthermore, ecological models generally are devised to include effects of variations in physical and chemical properties as well as biological relationships. This is clearly a more difficult problem than the laboratory models which postulate a constant environment. The effects of stochastic fluctuations have not been thoroughly investigated in ecological models, but they must be accepted as a possible cause of error.

For these reasons an attack on ecological problems on a purely theoretical basis would inspire little confidence. It is necessary to apply the equations to a particular situation in nature and compare theoretical results with field observations. In some cases it will be seen that the comparison is quite accurate, particularly when random errors are minimized by using average values for environmental factors and plankton which have been obtained at a series of closely spaced stations. Useful conclusions have already been obtained by means of the theoretical approach, crude as it is in its present state, and it seems to be a promising tool for future development.

## 2. Mathematical Models of Plankton Populations

Harvey, Cooper, LeBour and Russell (1935) described the spring diatom flowering in the English Channel and proposed that its termination was primarily due to grazing by the zooplankton population, which increased during, and as a result of, the flowering. Fleming (1939) postulated a relationship based on equation (1), taking the form

$$\frac{dP}{dt} = P[a - (b + ct)], \tag{3}$$

where $P$ is the total phytoplankton population underlying a unit area of sea surface, and $a$ is a constant growth coefficient during the flowering period; $b$ is the initial grazing coefficient; the zooplankton is postulated to increase linearly, so that grazing increases at a constant rate, $c$, per unit of time $t$. Using values for the coefficients that had been estimated by Harvey *et al.*, Fleming integrated equation (3) and obtained a bell-shaped, symmetrical curve that approximated observed conditions in nature.

Riley (1941) obtained a series of observations on Georges Bank, including measurements of phytoplankton, zooplankton, certain environmental characteristics of the water and experimental determinations of phytoplankton photosynthesis. These later (Riley, 1946) provided the basis for a theoretical evaluation, again based on equation (1). A general equation was stated,

$$\frac{dP}{dt} = P(P_h - R - G) \tag{4}$$

in which $P$ again is the total phytoplankton population per unit area of sea surface, $P_h$ is a photosynthetic coefficient, $R$ is the coefficient of phytoplankton respiration, and $G$ is a grazing coefficient. $P_h - R$ is analogous to $b_1$ in equation (1) and to $a$ in (3), except that these terms are now to be regarded as ecological variables rather than constants.

Photosynthetic experiments in winter and early spring indicated a more or less linear relationship between photosynthesis and incident radiation. Later in the season there was a reduction in the rate that was correlated with phosphate depletion. The experiments were carried out on surface water, but it was reasonable to suppose that the depth of the euphotic zone, as indicated by transparency measurements, would affect the mean photosynthetic rate of the total population. This mean rate was roughly computed by integrating the illumination from the surface to a depth where the light intensity had a value of 0.0015 g cal $cm^{-2}$ $min^{-1}$ and dividing by the depth of the layer. In winter the depth of the mixed layer exceeded that of the euphotic zone, so that part of the population was mixed downward to a depth where photosynthesis was negligible. It was assumed in such cases that the photosynthetic coefficient was reduced in proportion to the ratio of the depths of the two layers. In short, photosynthesis was postulated to depend upon incident radiation, transparency of the water, depth of the mixed layer, and phosphate. The equation was

$$P_h = (pI_0/kz_1)(1 - e^{-kz_1}NV), \tag{5}$$

where $p$ is an experimentally derived photosynthetic coefficient, $I_0$ is incident radiation, $k$ is the extinction coefficient of visible light, and $z_1$ is the arbitrarily defined depth of the euphotic zone. $N$ is a nutrient factor, here designated as phosphate in µg atoms P/l., and

$$N = \frac{\text{µg atoms P/l.}}{0.55}, \quad \text{when } P < 0.55. \tag{6}$$

$V$ is a correction for vertical turbulence given by

$$V = z_1/z_2, \quad \text{when } z_1 < z_2,$$

in which $z_2$, the depth of the mixed layer, is defined as the maximum depth at which the density is no more than $0.02\sigma_t$ units more than the surface value.

Phytoplankton respiration was assessed statistically from experimental data. The respiratory coefficient $R$ increased with increasing temperature. The data were not accurate enough to assess the form of the curve reliably, but it was assumed to be a logarithmic relationship. The equation was

$$R = 0.0175e^{0.069T}, \tag{7}$$

where $T$ is the temperature in °C. The grazing rate was postulated to be

$$G = gH, \tag{8}$$

where $g$ is a grazing coefficient and $H$ is the observed herbivore zooplankton population.

Equations (5), (7) and (8) restate equation (4) in terms of the six environmental factors that have been listed. Average numerical values were derived from Georges Bank data and were plotted on a seasonal basis. An approximate integration was then obtained over successive short intervals of time, assuming for each variable a constant average value during that time. In the time interval 0 to $t$,

$$\ln P_t - \ln P_0 = \overline{P_h} - \overline{R} - \overline{G}, \tag{9}$$

and by a series of such integrations a relative curve of seasonal change was obtained. This was arbitrarily transformed into absolute terms by statistically determining the best fit of the curve for all of the cruise averages for phytoplankton. Results are shown in Fig. 1a.

Subsequently other seasonal cycles were examined (Riley, 1947; Riley and Von Arx, 1949). The first of these was based on observations in southern New England waters off Woods Hole by Lillick (1937), Clarke and Zinn (1937) and Clarke (1938). The second was an analysis of Kokubo's (1940) study of Husan Harbor, Korea. Subsidiary data obtained by the Imperial Fisheries Experiment Station of Korea were also utilized. The theoretical equations were similar to those described above except for minor alterations that were introduced in an attempt to improve the accuracy, such as the use of a photosynthetic temperature coefficient and a nitrate depletion factor in the Husan analysis. Results are shown in Figs. 1b and c.

Although the agreement between observed and calculated values was by no means perfect, the models succeeded in depicting recognizably observed regional differences in seasonal cycles and, in the case of Husan Harbor, differences from one year to the next. In the latter case, the large autumn flowering of 1932 was due to a complex of favorable factors including light, nutrients and a sparse zooplankton crop, whereas in 1933 all three factors were less advantageous. Indeed, these analyses indicated that few major events in

the seasonal cycle were controlled by a single factor. For example, the onset of
the spring flowering was controlled by light and stability, and its termination
was brought on by grazing, nutrient depletion and reduction in transparency.
The conclusion of Harvey *et al.* (1935), which has been reiterated by Cushing
(1958), that flowerings are controlled by grazing, may be true in British waters
but is not applicable to the New England area. There the relative importance
of grazing varies from one part of the area to another, suggesting that this is
not a subject for broad generalization.

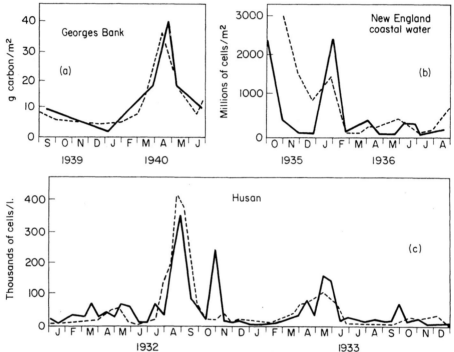

Fig. 1. Comparison of observed seasonal cycles of phytoplankton (solid lines) with
theoretical cycles (dotted lines) computed according to equations (4) to (9).

The Volterra equation for predators was applied, with some modifications,
to the Georges Bank zooplankton by Riley (1947a). The equation for the
herbivore population, $H$, was

$$\frac{dH}{dt} = H(A - R - C - D), \tag{10}$$

where $A$ is a coefficient of assimilation, $R$ is the respiratory coefficient, $C$ is
predation by carnivores, and $D$ is natural death. $R$ was determined from
laboratory experiments, but other coefficients contained arbitrary elements.
$A$ was postulated to equal $gP$, using the same numerical value for $g$ that was
established in the phytoplankton analysis (equation 8). However, it was neces-

sary to assume a maximum limit for assimilation, which was set at 8% of the animals' weight per day, otherwise the rate of increase would have been too rapid during the spring flowering. The limitation was qualitatively realistic although empirically derived, because it is well known that herbivores at such times eat more than they can digest. Carnivorous predation was attributed to *Sagittae* and was proportional to their number, although they were not the sole predators. A predation coefficient and the death rate were determined statistically. The results, shown in Fig. 2, were moderately realistic, but the empirical elements made the analysis less acceptable than the phytoplankton treatment.

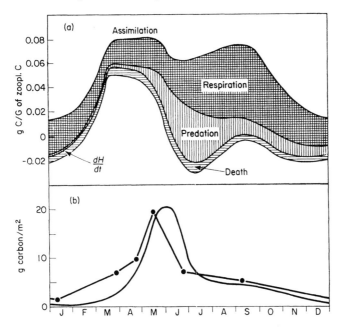

Fig. 2. Seasonal cycle of zooplankton on Georges Bank. (a) Summation of growth processes. Upper curve is the postulated seasonal cycle of the coefficient of assimilation. From it the coefficients of respiration, predation and natural death are successively subtracted. The remainder, the lowest curve, is the estimated rate of change of the zooplankton population. (b) Dots represent observed zooplankton population; the smooth curve is a theoretical seasonal cycle obtained by approximate integration of the rate curve in (a).

The equations that have been described transformed the Volterra theory into a promising ecological tool, but an important aspect of the theory was lost, namely the mutual dependence of equations (1) and (2). Fleming postulated an arbitrary rate of zooplankton increase, and the writer used observational data. Ecologists recognize the existence of this mutual dependence in nature. Zooplankton growth depends on the quantity of phytoplankton available, and conversely, grazing affects the phytoplankton concentration. Also, the animals and bacteria release inorganic nutrients that influence phytoplankton growth, which in turn reduces the nutrient concentration. The biological association as a whole—plants, animals and their component chemical elements—forms a vast and intricately related system. The short cuts that have been discussed accomplish the immediate purpose of describing seasonal changes in plankton but fail to elucidate the broader principles of ecological relations.

Ordinary ecological reasoning supplies the principles for a broader quantitative theory. Every environment possesses certain distinctive physical features —properties of geography and climate—that determine what kind of population can invade it and which continue to exercise a control of the population by influencing the rates of all life processes. These are independent factors; they exist whether there is a population or not, although the way they operate depends upon the physiological behavior of all organisms in the community. In addition, the population brings with it a mass of interdependent biological relationships. But a sound analysis must recognize that only the physical factors are basically causal, and the biological relations must be resolved by simultaneous solutions of a series of inter-related equations.

Riley, Stommel and Bumpus (1949) made preliminary tests of this type of analysis, using data from five areas in the western North Atlantic, ranging from temperate coastal waters to an oceanic region in the subtropics. An abbreviated food chain was postulated consisting of phytoplankton, herbivores, their predators and phosphate. Equations were written for each group and the necessary physiological coefficients were inserted. A review of ecological and physiological data provided a basis for developing relationships between these coefficients and their environment; however, lack of sufficient knowledge of some processes necessitated the use of arbitrary assumptions.

Five independent environmental factors were used in the analysis. They were incident radiation, temperature, vertical eddy diffusivity, transparency of the water, and the deep-water concentration of phosphate. Only the first three were fully justified as basic factors. The phosphate in deep water is admittedly affected by biological processes, but the concentration remains sensibly constant so that its effect on the surface layer through the agency of vertical turbulence is both fundamental and predictable. Transparency can hardly be regarded as a basic factor because it is affected by phytoplankton. The relationship could have been included in the analysis, but for the time being it seemed desirable to avoid this added complication. The problem was further simplified by assuming the existence of a steady state. In other words, the calculation determined the quantity of each ecological group that could exist as an equilibrium population with any particular array of primary environmental factors. The steady-state assumption required application to situations in which one could expect a reasonably stable population.

The mathematical treatment consisted of equations for phytoplankton and phosphate at a series of about twenty depths, extending from the surface to well below the euphotic zone. Herbivores and carnivores were treated simply as a mean concentration in the vertical column. Simultaneous solution was effected by the "relaxation" method of Southwell (1946), an arithmetic procedure of successive approximation.

The equation for phytoplankton, $p$, at any given depth $z$ is

$$\frac{dp}{dt} = p(p_h - r_p - gh) + \frac{\delta}{\delta z} \cdot \frac{A}{\rho} \cdot \frac{\delta p}{\delta z} - v\frac{\delta p}{\delta z}, \qquad (11)$$

where $A$ is the coefficient of vertical eddy diffusivity (calculated as eddy conductivity from seasonal temperature changes and assumed to be synonymous), $v$ is the sinking rate of phytoplankton, and $\rho$ is the density of the water. The concentration of phytoplankton, $p$, and other variables at particular depths are denoted by lower case symbols to distinguish them from the former usage in which capital letters indicated the total population per unit area of sea surface. The respiratory coefficient $r_p$ has an identifying subscript, as will the respiratory coefficients for other groups.

Solution by Southwell's method requires a steady-state, finite difference form of equation (11), which is

$$p_0(p_h - r_p - gh) + \frac{1}{z}\left(\frac{A_1}{\rho} \cdot \frac{p_1 - p_0}{z} - \frac{A_{-1}}{\rho} \cdot \frac{p_0 - p_{-1}}{z}\right) - \frac{v(p_1 - p_{-1})}{2z} = 0, \qquad (12)$$

where $p_0$ is the phytoplankton concentration at any given depth and $p_{-1}$ and $p_1$ are the concentrations at equidistant intervals $z$ above and below $p_0$. $A_{-1}$ and $A_1$ are average eddy coefficients for the depth intervals above and below the level of $p_0$ respectively.

There is a series of equations for a controlling nutrient, in this case phosphate, taking the form

$$\frac{1}{z}\left(\frac{A_1}{\rho} \cdot \frac{n_1 - n_0}{z} - \frac{A_{-1}}{\rho} \cdot \frac{n_0 - n_{-1}}{z}\right) + v[hr_h + cr_c - p_0(p_h - r_p)] = 0. \qquad (13)$$

Vertical exchanges are expressed in the same way as the diffusion term in equation (12). The biological rate of change depends upon phytoplankton productivity and excretion by herbivores, $h$, and carnivores, $c$. Excretion is assumed to be proportional to the product of the biomass of animals and their respiratory coefficients $r_h$ and $r_c$. Since the biological populations are expressed in terms of carbon, their inclusion in the nutrient equation requires a proportionality factor $v$.

Vertical migration of animals is postulated to permit an equal amount of time at each depth. This is equivalent to assuming that there exists over a period of time a statistically uniform distribution, so that a single equation giving the mean concentration is sufficient. In the steady state the quantity of herbivores can be derived from equation (12). Thus we have

$$h = \int_0^{z'} p(p_h - r_p)\, dz - vp_{z'} \bigg/ \int_0^{z'} gp\, dz, \qquad (14)$$

where $z'$ is the deepest level in the model, and $p_{z'}$ is the concentration of phytoplankton at that depth. The sense of the equation is that herbivores eat all of the phytoplankton except that which is lost by sinking. In a few cases it might also be desirable to include a term for loss by eddy flux, but in general, vertical diffusion and the plankton gradient are slight at $z'$, so that the error is negligible. It may also be noted in equation (14) that phytoplankton appears in both the numerator and denominator. Technically this does not cancel the effect of phytoplankton biomass on the zooplankton population, but in practice it very nearly does so. Thus the size of the equilibrium population of herbivores largely depends on the ratio of the production coefficient to the grazing coefficient. This is in no way contradictory to the sense of (2) and (10), in which the growth rate of predators was postulated to depend on the quantity of prey. In the one case an increase in herbivore biomass simply requires an abundance of food at the moment, while in the other case the maintenance of the animal population requires sufficient productivity to ensure renewal of the food supply.

The equation for the rate of change of herbivores may be restated for present purposes

$$\frac{dh}{dt} = h(gp - r_h - fc),\tag{15}$$

where $f$ is the feeding rate of the carnivores, $c$. Then in the steady state

$$c = (\overline{gp} - \overline{r_h})/\overline{f}.\tag{16}$$

Certain boundary conditions must be met in order to obtain a steady-state solution, but neither these nor the derivation of the physiological coefficients need be discussed in detail. Any further work of this type would require revision of the coefficients in order to make use of new physiological findings.

The equations can be solved simultaneously because they are all related either through physiological coefficients or through shared biomasses and chemical quantities at successive depth intervals. Solution of the equations leads to relative vertical distributions of phytoplankton and phosphate and relative quantities of animals. One absolute value has been postulated, namely phosphate at the lowest depth interval, and this serves to convert all quantities to absolute terms.

An example of the kind of results obtained is shown in Fig. 3. Most of the vertical curves were not highly realistic, but total populations were predicted with an average error of not more than 25%.

Steele (1956) made a theoretical study of plant production on the Fladen Ground. An approximate integration of equation (11) was obtained, since a steady-state solution was inadequate for his seasonal study. Regarding $g$ and $v$ as unknowns, he first applied the equations to two depth intervals and obtained a simultaneous solution of grazing and sinking rates. They were then used in an examination of the form of the vertical profiles. The equation was

$$\left(1 + \frac{g\Delta t\, h_2}{2}\right) \int_{140}^{z} p_2\, dz - \frac{\Delta t\, y p_2}{2} = P_r + \left(1 - \frac{g\,\Delta t\, h_1}{2}\right) \int_{140}^{z} p_1\, dz$$

$$+ \frac{\delta p/z}{\delta T/z} \int_{140}^{z} (T_2 - T_1)\, dz + \frac{\Delta t\, y}{2}\, (p_1 - [p_1 + p_2]_{140}^{z}),\tag{17}$$

where

$$\frac{\delta p}{\delta z} = \frac{1}{40} [p_1 + p_2]_{z+10}^{z-10}. \tag{18}$$

$P_r$ is production below $z$ meters during the time interval $\Delta t$; $p_1$, $p_2$ are the phytoplankton populations at the beginning and end of the interval; $\delta p/\delta z$, $\delta T/\delta z$ are mean gradients of population and temperature at depth $z$; $h_1$, $h_2$ are the concentrations of zooplankton.

The left-hand side of equation (17) contains only $p_2$. The right-hand side, in addition to $p_1$, also contains $p_2$ as is apparent in equation (18). However, the third term on the right is much smaller than the others, so that equation (17) can be used to predict $p_2$ by successive approximation.

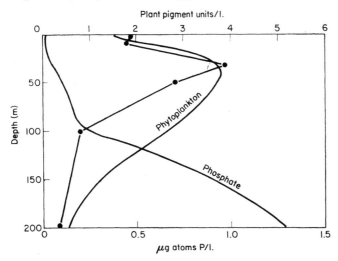

Fig. 3. Dots are mean values for phytoplankton (Harvey plant pigment units) at four stations in the slope water off southern New England in May, 1939 (Riley, Stommel and Bumpus, 1949). Smooth curves are theoretical vertical distributions of phytoplankton and phosphate calculated from equations (12) and (13).

The mean value for the upper 20 m was computed. This was used to determine the 30-m value, and so on to 50 m or more. Calculations were first made with the derived values for $g$ and $v$, but in some cases these were altered later in order to see if more realistic results could be obtained. In general there was good agreement between the model and observations.

Steele's method of calculating $g$ and $v$ is a useful biproduct of mathematical models. In both processes, laboratory experiments have given highly variable and sometimes conflicting results, and the sinking rate in particular cannot be predicted even to the nearest order of magnitude. But if most of the terms in the equation are known with some degree of confidence, one or two unknowns can be evaluated. In the present case there was evidence of a seasonal change in the sinking rate, which could hardly have been discovered in any other way.

Riley, Stommel and Bumpus (1949) used the same method to estimate the sinking rate and vertical eddy diffusivity simultaneously. Their estimates of the sinking rate were essentially in agreement with Steele's. Most of the values in both of these papers fell within a range of 1 to 5 m per day, which is well within the rather large spread of measured values.

Cushing (1958) discussed the difficulties of estimating grazing rates experimentally and recommended the use of a simple model in which grazing is derived from the difference between phytoplankton production and observed changes in the standing crop. This presupposes that the effects of vertical turbulence and sinking are negligible, but the method is a useful first approximation. The oversimplifications in his method were largely justified by successful application to the spring growth period in the North Sea (Cushing, 1959). The latter paper is particularly notable for a careful review of the various kinds of information that can be brought to bear on problems of diatom growth and zooplankton grazing and for the introduction of new methods for dealing with these problems.

### 3. Complex versus Simple Models

The main purposes of most of the models that have been described are to test physiological knowledge and ecological hypotheses; hence they must be as realistic as possible. However, useful purposes have been served by models that are deliberately oversimplified in order to facilitate computation, so that a wide range of variables can be examined without prohibitive labor. Fleming's (1939) model was of this sort, as was an analytical solution of the relative vertical distribution of phytoplankton described by Riley, Stommel and Bumpus (1949). The latter postulated a two-layered system in which the surface layer had a positive net production coefficient, $\omega_1$, that was uniform with respect to depth, and the lower layer had a negative coefficient, $\omega_2$, i.e.

$$\omega_1 = p_h - r - gh,$$
$$\omega_2 = r + gh.$$

Vertical mixing and the phytoplankton sinking rate were postulated to be uniform with respect to depth throughout both layers. The steady-state solution for phytoplankton in the surface layer, $p_1$, is

$$p_1 = E\, e^{az - b_2 l} (\beta \cos \beta_1 z + a \sin \beta_1 z) \tag{19}$$

and in the lower layer

$$p_2 = E\, e^{(a - b_2)z} (a \sin \beta_1 l + \beta_1 \cos \beta_1 l), \tag{20}$$

where $E$ is an integration constant, $z$ is depth, and $l$ is the depth of the boundary between the two layers. Further,

$$a = v/2A, \quad \beta_1 = \sqrt{\left(\frac{\omega_1}{A} - a^2\right)}, \quad b_2 = \sqrt{\left(a^2 - \frac{\omega_2}{A}\right)},$$

where $v$ and $A$ are the phytoplankton sinking rate and vertical eddy coefficient as before.

The analysis shows that there must be an exact balance of processes in order to maintain a steady state, which must satisfy the equation

$$\beta_1(a + b_2)/(\beta_1{}^2 - ab_2) = \tan \beta_1 l \qquad (21)$$

Furthermore, a steady state is not possible if $v^2 > 4A\omega_1$, and the depth of the

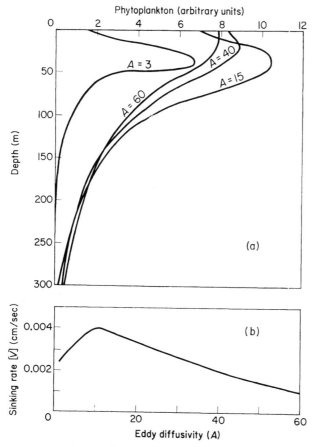

Fig. 4. (a) Vertical distribution of phytoplankton calculated from equations (19) and (20), using four different values for vertical eddy diffusivity. (b) Relationship between eddy diffusivity and phytoplankton sinking rate that is required for maintenance of a steady state according to equation (21).

productive layer must fall within the limits of $0 < l < \pi\beta_1$.

Fig. 4a is a family of curves computed from equations (19) and (20). $E$, $\omega_1$ and $\omega_2$ are the same throughout. $\omega_1$ is equivalent to a daily increase of 10% in the upper layer, and $\omega_2$ is a 7% decrease below the depth $l$, which is set at 50 m. Only $A$ and $v$ vary, and their relationship, which must satisfy equation (21), is shown in Fig. 4b. The family simulates vertical distributions that have often been observed in nature, where summer conditions correspond to low values of $A$ and winter distributions are simulated by strong mixing. The maximum population is attained with an intermediate mixing rate. With strong turbulence, plankton is rapidly removed from the lower part of the productive layer, so that the largest population occurs near the surface. When the diffusion coefficient is low, the surface water is depleted by sinking, and a maximum is attained in the lower part of the productive layer. Some investigators have postulated that this type of distribution is due to increasing density and decreasing sinking rate near the bottom of the euphotic zone. The model shows that this is not necessarily so. However, Steele and Yentsch (1960) have pointed out that the maximum sometimes occurs below the limit of the euphotic zone, and they are correct in stating that this requires a reduction in sinking rate if it is to be a stable, continuing situation.

Steele (1958) has used an idealized two-layered system to good effect in simple equations that may be used either for steady-state determinations or time sequences. In his model the thermocline has a fixed depth, and the layers above and below are postulated to be homogeneous with respect to all variables. A simple mixing coefficient is postulated, consisting of a daily exchange of a fraction of the water across the boundary between the two layers. Similarly, phytoplankton sinking is represented by a daily transfer of a constant fraction of the population from the upper layer to the lower one.

Phytoplankton production is assumed to be linearly proportional to phosphate when the latter is less than 0.4 μg atom P/l. and independent at higher concentrations. No other environmental variables were introduced into the production coefficient. The phosphate concentration is postulated to depend upon phytoplankton growth and is independent of zooplankton. The rather difficult problems of zooplankton growth and loss were by-passed by deriving suitable constants from observed seasonal changes on Fladen Ground. In a case where the depth of the surface layer is 40 m, and phosphate is less than 0.4 μg atoms P/l., the equations for this layer are

$$\frac{dp}{dt} = (0.75n - 0.11 - 0.024h - m)p, \tag{22}$$

$$\frac{dn}{dt} = -(0.58n - 0.027)p + m(0.70 - n), \tag{23}$$

$$\frac{dh}{dt} = 4p - 0.01h^2, \tag{24}$$

where $p$ is phytoplankton, $n$ phosphate, $h$ herbivores, and $m$ is the mixing coefficient. The constant in (22), $-0.11$, includes a term for phytoplankton respiration derived from (7) and also a term for loss by sinking based on an estimated sinking rate of 3 m/day. The first term on the right-hand side of (23) is derived from $-(p_h - r)vp$, as previously postulated in (13). The right-hand term assumes that phosphate in the lower layer is constant at 0.70 μg atom.

Equation (24) was derived empirically but has the sense that carnivores vary in proportion to the square of the herbivore population.

Steele's equations retain the framework of equations (12) to (14) and preserve the interdependence of different food-chain levels but in a more tractable form. They are solved numerically by stepwise integration.

A sample of the results is shown in Fig. 5. Here seasonal cycles of three food

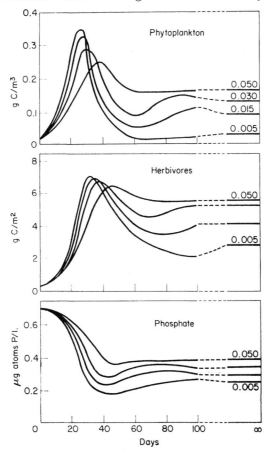

Fig. 5. Simulated seasonal cycle of phytoplankton, herbivore zooplankton and phosphate computed by means of equations (22) to (24), using four different values for the mixing coefficient $m$.

chain elements are calculated, using four different values for the mixing coefficient $m$, ranging from 5% to 0.5% interchange of water between the two layers per day. The well known situation of a diatom flowering accompanied by nutrient decrease and followed by zooplankton increase is faithfully depicted. According to Steele, the tendency to approach a summer steady state by a series of damped oscillations may be a mathematical artifact. Although phosphorus is strongly limiting in summer, the spring flowering is terminated by grazing before phosphate declines to the point that is postulated as limiting in the model. In this respect the model agrees with the conclusions of other investigators cited above as to the importance of the grazing effect in British waters. Steele also calculated other cycles, examining the effects of variations in sinking, grazing and photosynthetic rates.

In a later paper (Steele, 1961), equations (22) and (24) were slightly modified to analyze a situation in which light intensity varied from day to day in accordance with observed radiation at Aberdeen, Scotland, in May and June, 1957. In essence this is a simple stochastic treatment. Large random fluctuations from day to day during the early part of the period had only a slight effect on the phytoplankton population. Later there was a week of bright weather followed by a slightly longer period of low radiation. These systematic variations resulted in the growth and subsequent decay of a phytoplankton flowering of considerable size. Changes in phosphate and zooplankton were less pronounced.

Steele also presented a theoretical analysis of the effects of zooplankton patchiness on the distribution of phytoplankton and phosphate. His model demonstrated that zooplankton patches produced a significant reduction of phytoplankton if they remained in contact with the same plant population for about a week. Changes in phosphate were relatively small.

Steele's models of time series represent a simple and orderly series of events. But if pairs of values for phytoplankton and zooplankton or any other two biological variables are taken from the graph and plotted together, the correlation is distinctly non-linear, because of time lags and other non-linearities in the relationships that are involved. In the case of phytoplankton and zooplankton, the curve is roughly elliptical. Steele pointed out that with sufficient random scatter the elliptical form would not be obvious, and the relationship would simply appear to be a poor correlation. This was demonstrated in the analysis of the effect of stochastic fluctuations of zooplankton.

The importance of this concept lies in the fact that relationships observed in a time series are essentially analogous to horizontal variations found in nature. In certain cases this is clearly demonstrable, as when a spring flowering begins in shallow water and gradually spreads to deeper water, so that the time series can be equated with onshore–offshore distance. Similarly, in a tropical area of upwelling, time is synonymous with distance along the path of movement away from the area of upwelling. Commonly, horizontal variations are more randomized, but the existence of a variety of stages of development at successive sampling points tends to be supported by horizontal variations in the quantity of plankton, species composition and photosynthetic coefficients. Thus the analysis demonstrates that, with conventional sampling, apparent randomness and poor correlations between different biological variables can result from well defined processes. Steele further points out that lines or grids of stations perhaps are not well suited for the study of plankton problems and that detailed sampling of small and unusually homogeneous areas may be necessary in order to reveal the biological patterns of the growth process.

## 4. Prognosis for Mathematical Models

Simple models of the type developed by Steele are invaluable in improving our insight into plankton problems, and the results generally are sufficiently realistic to fulfill most of the functions of a quantitative analysis.

The more complicated models which attempt to be faithful to nature are limited by the state of our physiological knowledge. In recent years some interesting new techniques have been developed for measuring physiological processes in nature and in the laboratory. However, these are too new to be thoroughly trustworthy, and in some cases there are unresolved conflicts between new and old methods. The future is promising, but at the moment it must be admitted that little progress is being made in the development of complex models.

The study of phytoplankton production is a typical example of an unresolved physiological problem. Riley, Stommel and Bumpus (1949) developed an empirically realistic equation for surface photosynthesis in which the ecological variables were light, temperature and phosphate. The equation provided an adequate expression for some 215 experiments on surface populations in temperate and subtropical waters of the western North Atlantic. However, there was not enough information available to devise a realistic treatment of subsurface production. The models used the same equation for all depths, so that photosynthesis declined with depth in proportion to the decrease in light intensity. Admittedly this was a lame device, and the method needs to be overhauled.

Ryther and Yentsch (1957) proposed a simpler method involving only light, which was applied later (Ryther and Yentsch, 1958) with a fair degree of success to southern New England coastal waters. According to these authors, there appears to be an adjustment of the population to any particular light intensity such that the rate of production and the nutrient supply are in balance. Therefore nutrients do not need to be specifically represented in the formula. The validity of this concept appears to be supported by data in the paper cited, but it remains to be determined how generally applicable the method will be to the varying degrees of non-steady states that sometimes are found in nature. It is not applicable to Georges Bank data obtained by Riley (1941) except in experiments with an initial phosphate concentration of 0.4 µg atom P/l. or less. At the highest concentrations encountered, the average observed photosynthesis was twice the predicted value.

Ryther and Yentsch handled the problem of subsurface photosynthesis more realistically than Riley, Stommel and Bumpus, and the overall accuracy of their method may be superior. However, it is not suitable for mathematical models. The interplay between nutrients and phytoplankton that is found in (12) and (13) and in (22) and (23) is the only guarantee of stability in the system. If phytoplankton production is limited by the supply of nutrients but is independent of nutrient concentrations, as postulated by Ryther and Yentsch, there is no theoretical reason for the system to achieve a steady state. A maximum limit would be imposed on production by the rate of nutrient supply, but wide variation would be possible below this limit. Their theory, reduced to its simplest terms, is expressed by Volterra's equations for prey and predators, modified only to the extent that a maximum limit is postulated for the con centration of prey. At lower concentrations there would be prey–predator oscillations.

Moreover, if nutrient limitation is removed from (14), the zooplankton population will depend only on light and temperature. This leads to unacceptable results. For example, there would be little difference in zooplankton concentrations in upwelling and non-upwelling areas in the tropics, or between temperate and subtropical waters in summer.

If the [14]C method essentially measures net production, as claimed by Ryther (1956), these difficulties could be avoided by devising an ecological equation for [14]C uptake. It is too early to decide what factors would need to be used, but presumably light intensity would be involved and probably a nutrient factor. Holmes, Schaeffer and Shimada (1957) have clearly shown that high net production coefficients are associated with nutrient enrichment in the equatorial dome of the eastern Pacific. King, Austin and Doty (1957) obtained similar results in the equatorial belt of upwelling in the central and eastern Pacific. These authors kindly put their data in the hands of the writer for a comparison of [14]C uptake per unit of chlorophyll and the initial phosphate concentration in the experiments. There was a large amount of scatter, but the correlation coefficient of 0.32 was significant ($P < 0.001$). This is a promising result, especially since phosphate levels in the Pacific are so high that probably phosphate is merely serving as a crude index of variation of some other element that is more important as a limiting factor.

However, in contrast with these results, Ryther and Yentsch found no correlation between [14]C production coefficients and nutrients in New England waters, a result later confirmed by Steele and Baird (unpublished MS) in the North Sea. But the latter authors found that the chlorophyll content of the phytoplankton declined, relative to total carbon, when phosphate was deficient. Thus a production coefficient stated in terms of [14]C uptake per unit of phytoplankton carbon would be nutrient dependent. These results tend to invalidate the use of chlorophyll in production coefficients that are to be applied to model work unless there is thorough knowledge of chlorophyll:carbon ratios. And unfortunately there are many circumstances in which measurement of phytoplankton carbon is difficult or impossible because of associated detritus.

Some of the other physiological coefficients, as well as sinking rates, exist in a similar state of flux and uncertainty. Thus it is premature to attempt to make models more realistic until some of the immediate problems have been solved. However, it seems worthwhile to discuss briefly some of the elements that might be embodied in any future attempt.

Although the main concern thus far has been with one-dimensional, steady-state models, two-dimensional models are possible. These could be either a seasonal cycle in a given locus or a steady-state model of a vertical profile. More than two dimensions would be extremely awkward to handle.

Four of the five independent factors previously employed are acceptable for further use in analyses of open ocean conditions, namely light, temperature, vertical eddy diffusivity and the deep-water concentration of a limiting nutrient. Transparency should be treated in part as a biological variable. Riley (1956) found that a satisfactory equation for use in the open sea is

$$k = 0.04 + 0.0088C + 0.054C^{2/3}, \tag{25}$$

where $k$ is the extinction coefficient of visible light, and $C$ is μg of chlorophyll per liter. In shallow water additional terms would have to be added for extinction due to suspended silt and bottom sediment.

Some types of problems would require inclusion of horizontal diffusion, currents and, perhaps, upwelling as primary factors. Coastal waters present a special problem in that the nutrient supply is influenced by runoff and by transport along the bottom from offshore. This could be handled as a steady-state profile extending from shore to deep water. However, a seasonal cycle in a particular area would require some rather arbitrary coefficients of external enrichment.

## 5. Higher Elements of the Food Chain

All of the models thus far constructed have dealt almost exclusively with plankton. There are serious difficulties in attempting to extend the analysis to higher food-chain levels. The multiplicity of species with individual food habits does not lend itself to division into broad ecological groups. The alternative of dealing with individual species not only would be impossibly laborious but also would create some theoretical problems for which there are no clear solutions in sight. Above the plankton level, food chains tend to become intermeshed and often are more properly called food webs. A realistic treatment must permit organisms to feed at more than one level of the food chain and must make allowance for varying degrees of competition between species that have an overlapping diet. There must also be provision for detritus feeders.

The prey–predator equations are ideal for working with a simple, linear food chain but obviously are ill-suited for a food web. It would be necessary to set up a series of partial feeding coefficients corresponding to the various sources of supply. Competition for the same food supply would be an even more difficult problem. In the present state of our knowledge, any attempt to deal theoretically with these problems would probably be artificial and trivial.

Thus there seems to be little point in trying to create realistic models of higher food-chain elements until we possess better descriptive knowledge. However, deliberately oversimplified models might be instructive, since the problem of overall food-chain efficiency is not well understood. Riley (1956a) has noted that there is not a close relationship between plankton productivity and commercial fish yields in coastal and bank waters. Even in the clear correlations that were established by Sette (1955) for phosphate, zooplankton, and tuna in the tropical Pacific, the relationship appeared to be non-linear. Moderate regional variations in zooplankton were accompanied by larger variations in tuna.

Smith and Swingle (1939) demonstrated a similarly non-linear relation between particulate organic matter in fertilized fish ponds and production of blue gills. In this work the fish were introduced as fry and fingerlings, and the ponds were drained at the end of the test period to recapture the product of a season's growth. Thus the results were more accurate than is commonly the case in fisheries work.

This is a type of relationship that can be examined readily with the aid of simple mathematical models. Certain other non-conformities between basic productivity and fish production probably are the result of competition between fishes and commercially unimportant species. For reasons given above, such relations do not warrant theoretical investigation at the present time.

Aside from its complexity, the upper end of the food chain presents certain other problems. It is only in certain special cases that the true biomass of a fish population can be assessed accurately. If the model is to be compared with nature, a relationship generally must be established between theoretical biomass and catch per unit of effort. Also physiological studies and feeding experiments on fishes have been somewhat different from the zooplankton investigations, necessitating slight changes in model methods. Thus it seems desirable to develop a simple model of the upper end of a food chain, perhaps more as an investigation of methods than for any intrinsic merit.

A pelagic food chain in tropical and subtropical waters is postulated, consisting, as in earlier studies, of phytoplankton, herbivores and first-order carnivores; but in addition a second order of carnivores is proposed. The first order carnivores are in actual fact a mixed group consisting of small fishes, squids and Crustacea. The common collective term for this group is forage animals. More than one level of the food chain may be included, and probably the situation is more accurately characterized as a food web rather than a food chain. However, in the first approximation, the forage animals are regarded as a single level having the characteristics of small fish.

The second order carnivores are a hypothetical entity called potential tuna. This means that the model seeks to determine the quantity of tuna that could be supported in a given area on the food that is available. This is expected to be a maximal estimate because the tunas compete for food with other groups such as sharks and marlin. Also any variations not associated with the food supply of adult fish, such as migrations and year class fluctuations, are ignored. Thus the actual quantity of tuna in a given area at any one time might be more or less than the potential crop. Because of these simplifications the model has little immediate value, but on the other hand it deals with one of the weakest aspects of fisheries biology, namely an understanding of the supporting power of a given area and the potential yield.

The analysis presupposes that there is information on phytoplankton and zooplankton which has been obtained either by theoretical methods or field observations. Hence we shall be particularly concerned with feeding and growth efficiency of forage fishes and tuna, and how this information might be used to compute their biomass. The study will also show the kind of information that needs to be acquired in order to understand the dynamics of the fish population.

Although not much is known about the production of forage animals, it seems safe to assume that their growth processes qualitatively resemble those of small temperate-water fishes which have been studied, so that it is only the magnitude of the processes which is uncertain. Hence the general form of the relationship between food and growth will be derived from some of the experiments that have been reported. The work of Dawes (1930, 1931) is instructive. Small plaice in their third year (20–120 g) were fed varying amounts of food, from a maintenance ration, $M$, which was just enough to hold the weight approximately constant, to about five times this amount. A $2M$ ration was sufficient to double the size during one summer growth season. There was a further increase with maximal feeding but at reduced efficiency. The daily maintenance ration was of the order of 1–2.4% of the weight of the fish, the percentage decreasing with increasing size of the animals. Between $M$ and $2M$ the growth increment averaged about $M/3$.

These relations can be put into a simple equation provided the available food is relatively scarce, permitting the assumptions (a) that the fish eat all prey encountered, and (b) that there is a constant efficiency of conversion. Then for the forage fishes

$$\text{Total consumption} = \alpha h,$$

where $h$ is the concentration of herbivore zooplankton, and $\alpha$ is a predation coefficient.

$$\text{Total assimilation} = \alpha \beta h,$$

where $\beta$ represents the fraction of consumption that can be assimilated (0.8–0.9 in most fishes according to Ricker, 1946). If $\gamma$ is defined as basal plus work metabolism,

$$\text{Growth} = \delta(\alpha \beta h - \gamma),$$

where $\delta$ is the fraction of the excess over maintenance that results in an increase in weight (generally 0.3 or more in small fishes).

The rate of change of the forage animals can now be given as

$$\frac{dF}{dt} = F[\delta(\alpha \beta h - \gamma) - cT], \tag{26}$$

where $F$ is forage animals, and $c$ is the coefficient of capture by potential tuna,

$T$. If the population exists in a steady state in a particular locality, equation (26) reduces to

$$F[\delta(\alpha\beta h - \gamma) - cT] = 0,$$

$$cT = \delta(\alpha\beta h - \gamma),$$

and

$$T = \frac{\alpha\beta\,\delta}{c}\left(h - \frac{\delta\gamma}{c}\right). \tag{27}$$

We must recognize that although zooplankton can be stated in terms of biomass, the available tuna data represent catch per unit of effort, which is presumed to be a function of biomass but cannot be translated easily into absolute terms. Thus equation (27) may be written

$$f(T) = f\left(\frac{\alpha\beta\,\delta}{c}\cdot h - \frac{\delta\gamma}{c}\right), \tag{28}$$

where $f(T)$ is the tuna catch per 100 hooks, and $f$ is a conversion factor relating tuna biomass to catch.

Statistical analysis of zooplankton volumes and tuna catches can be used to derive empirical constants for (28). Pairs of values were read from Sette (1955, fig. 7). The correlation for observations north of the equator was 0.80 (0.01 > $P$ > 0.001). Inclusion of data from south latitudes lowered the correlation to a relatively insignificant level. The regression equation for north latitudes is

$$f(T) = 0.282h - 3.2, \tag{29}$$

so that the statistically computed constants are

$$f(\alpha\beta\delta/c) = 0.282 \tag{30}$$

$$f(\delta\gamma/c) = 3.2. \tag{31}$$

These will be applied to a further consideration of north equatorial waters. They are obviously unsuitable for regional extrapolation.

Although the constants in these equations are a cumbersome conglomeration of terms, they have been kept intact rather than substituting a simpler notation to emphasize the fact that they have biological meaning and are not merely empiricisms. It would be desirable sooner or later to evaluate them in quantitative terms. $\beta$, $\gamma$ and $\delta$ have been measured in experiments with various fishes and presumably could be determined for some of the forage species. $\alpha$ and $c$ are more difficult in that predatory habits are so much a part of the natural environment in which they occur that one would hardly expect to get realistic results under controlled observation. However, equations (27), (29) and (30) provide an indirect method of dealing with these constants provided we know the other physiological coefficients. Let us assume for purposes of discussion that $\beta = 0.8$, $\gamma = 0.02$ (on a daily basis) and $\delta = 0.3$. These are reasonable values for some of the fishes that have been examined and presumably are at least of the right order of magnitude for present purposes.

Equation (31) can be written in the form

$$\delta\gamma/c = 3.2T/f(T), \tag{32}$$

where $T$ is the biomass of potential tuna expressed in any convenient terms of weight per unit volume of water. Substituting known and assumed values and rearranging,

$$c = \frac{0.3 \times 0.02}{3.2T/f(T)} = \frac{0.0019f(T)}{T}. \tag{33}$$

Similarly equation (30) can be written

$$\alpha\beta\ \delta/c = 0.282T/f(T). \tag{34}$$

If $c$ is derived from equation (32) and substituted into (34),

$$\alpha = 0.088\gamma/\beta = 0.0022. \tag{35}$$

Calculated in this way, $\alpha$ and $c$ each have the sense of the fraction of prey consumed each day by a unit biomass of predator.

The value for $\alpha$ seems small, and it remains to be seen whether it will prove to be realistic. There is no frame of reference except temperate waters, which might be quite different. Harvey (1950) found that the biomass of pelagic fish in the English Channel was approximately equal to that of zooplankton. The zooplankton was estimated to have a daily productivity of 10% of the biomass. In the present case the biomass of forage animals would have to equal that of zooplankton in order to utilize some 5–10% of the crop per day at the calculated value of $\alpha$. However, such data as are available (King, Austin and Doty, 1957) suggest that the biomass of forage animals averages little more than a tenth that of the zooplankton.

It seems unlikely that the physiological coefficients used in deriving $\alpha$ in equation (35) are in error by an order of magnitude, but there is a possibility of a mathematical artifact due to oversimplification of the food chain. Suppose that forage animals consist of two trophic levels instead of one and that $F_1$ feeds upon herbivores and $F_2$ upon $F_1$. The origin of the constant in (35) goes back to the correlation between tuna catch and zooplankton, but this is now an even more indirect relationship because $F_1$ intervenes between $h$ and $F_2$. If the biomass of $F_1$ is less than that of $h$, a predation coefficient for $F_2$ that is derived from the constant in the regression equation must be correspondingly larger. Equation (29) may be rewritten

$$f(T) = (0.282h/F_1)F_1 - 3.2,$$

and (35) then becomes

$$\alpha_2 = 0.088\gamma_2 h/\beta_2 F_1, \tag{36}$$

where the subscripts 2 denote that these are coefficients for the $F_2$ group.

A realistic treatment of the food chain would require thorough investigation of such problems. It suffices for the moment to point out that they exist. The remainder of the analysis will continue as before, with the assumption of a single group of forage animals.

There is no indication in the equations that have been presented of an easy way to evaluate the biomass of potential tuna and its predation coefficient. However, several earlier equations have demonstrated that, if either one can be evaluated, the other follows. It is doubtful whether the biomass can be directly determined with any degree of confidence. Alternative methods would require feeding experiments on tuna, which are not likely to be as simple as in the case of small fish. Nevertheless, it seems desirable to examine the matter theoretically to determine what kind of information is needed.

We can write an equation for the rate of change of potential tuna, which has the same general form as (26). It is

$$\frac{dT}{dt} = T[\delta(c\beta'F - \gamma') - M], \tag{37}$$

where $\beta'$, $\gamma'$ and $\delta'$ have the same connotation as in the analysis of forage fish but refer specifically to physiological properties of tuna, and $M$ is a coefficient of total mortality. In the steady state,

$$cT\beta' \, \delta'F = T(\delta'\gamma' + M). \tag{38}$$

$T$ has been retained because expressions for $cT$ can be obtained from equations (27) and (33). These lead to

$$\alpha\beta \, \delta\beta' \, \delta'Fh = T(\delta'\gamma' + M) \tag{39}$$

and $$0.0019f(T)\beta' \, \delta'F = T(\delta'\gamma' + M). \tag{40}$$

If $M$ can be evaluated by the usual fisheries technique of analyzing the age structure of the stock, solutions of $T$ and $c$ will follow without difficulty. It is also apparent that $M$, which is stated in general terms, can be elaborated into a treatment of both natural and fishing mortality.

It will be noted in equation (39) that potential tuna vary in proportion to the product of zooplankton and forage animals. This will lead to a non-linear relation between zooplankton and tuna, if $M$ is constant and if a positive correlation exists between zooplankton and forage animals. The latter is almost certainly the case, judging from data published by King, Austin and Doty (1957).

Steele (*in litt.*) has pointed out a possible variant of this relationship. If there is a significant amount of fishing mortality in certain areas, it may be supposed that $M$ will be proportional to $T$ because the fishermen will tend to congregate in areas of greatest tuna abundance. It then follows from equation (39) that variations in $T$ will more nearly approximate the square root of $Fh$. Thus the more intensive the fishing becomes, the more nearly will linearity between tuna and zooplankton be realized.

The application of equations (38) to (40) is not expected to be entirely straightforward. Some of the physiological coefficients probably are not constants but vary with age and food supply. However, present information indicates that all of the factors tend to vary together in a general slowing down with increasing age and size, and the relation of the coefficients in the equations is such that the variations tend in part to cancel. Thus they can be altered considerably without having a profound effect on the final result. This is readily demonstrated by postulating some more or less reasonable values and applying them to the equations that have been developed.

TABLE I

Hypothetical Tuna Physiological Coefficients

|  | Skipjack or other small tunas | Yellowfin or other similarly large fish |
| --- | --- | --- |
| $\beta'$ | 0.8 | 0.8 |
| $\gamma'$ | 0.01 | 0.005 |
| $\delta'$ | 0.2 | 0.1 |
| $M$ | 0.0033 | 0.001 |

It will be noted in Table I that $\gamma'$ and $\delta'$ are smaller in the small tunas than in forage animals, and there is a further reduction in the large tunas. This is qualitatively typical of cases that have been examined, although the quantities proposed are open to question. $M$ is equal to the observed growth rate on a daily basis of skipjack in the vicinity of Hawaii (Murphy *et al.*, 1958), and the use of this figure for $M$ assumes a steady state in which growth is balanced against total mortality. It seems reasonable to postulate a lower value for the large tunas.

As an average sort of equatorial situation, let us suppose that the zooplankton population has a displacement volume of 25 ml/1000 m³ and the tuna catch is 6/100 hooks. The forage fish population is assumed to be 3.5 ml/1000 m³, which appears to be in accord with the trawl data that are available.

If all of the tunas were of the smaller variety, and if they had no competitors for the available food supply, their potential biomass according to equation (40) would be 1.2 g wet weight/1000 m³. Putting the result into more familiar terms, this would equal 1.6 lb/acre in a stratum 150 m thick. The estimated annual production would be 1.9 lb/acre. Corresponding figures for large tuna would be 2.8 lb/acre for biomass and 1.0 for annual production. This is more or less in accord with the usual concept of the large fishes having a larger total biomass and a slower rate of growth.

The figures given are small compared with estimated crops on some of the temperate-water fishing banks. However, the annual production is within the same range as commercial catches from some pelagic associations that have been described, such as the cisco and lake trout fisheries in Canadian lakes, as well as the Great Lakes fisheries (Rawson, 1952). Thus the theoretical results are at least credible, although not much confidence can be placed in them until physiological coefficients and descriptive aspects of the food chain have been investigated more fully.

Although absolute values have little meaning, it may be of some interest to examine relative changes in potential tuna biomass with different levels of food availability. The quantities of herbivores and forage animals that were used in the example above were first doubled and then halved, and the tuna biomass was computed according to equation (39). Table II shows that a fourfold change in herbivores and forage animals leads to an approximately fifteen-fold variation in potential tuna.

## TABLE II

Potential Tuna (Skipjack) Biomass in g/1000 m³ Computed for Three Levels of Accompanying Populations of Herbivores and Forage Animals

| $h$ | 50 | 25 | 12.5 |
|-----|------|------|------|
| $F$ | 7 | 3.5 | 1.75 |
| $T$ | 4.7 | 1.2 | 0.3 |

## References

Bartlett, M. S., 1957. On theoretical models for competitive and predatory biological systems. *Biometrika*, **44**, 27–42.

Clarke, G. L., 1938. Seasonal changes in the intensity of submarine illumination off Woods Hole. *Ecology*, **19**, 89–106.

Clarke, G. L. and D. J. Zinn, 1937. Seasonal production of zooplankton off Woods Hole with special reference to *Calanus finmarchicus*. *Biol. Bull.*, **73**, 464–487.

Cushing, D. H., 1958. The effect of grazing in reducing the primary production: a review. *Rapp. Cons. Explor. Mer*, **144**, 149–154.

Cushing, D. H., 1959. On the nature of production in the sea. *Min. Agric. Fish. Food, U.K., Fish. Invest.*, **22** (6), 1–40.

Dawes, B., 1930. Growth and maintenance in the plaice (*Pleuronectes platessa*). Part I. *J. Mar. Biol. Assoc. U.K.*, **17**, 103–174.

Dawes, B., 1931. Growth and maintenance in the plaice (*P. platessa* L). Part II. *J. Mar. Biol. Assoc. U.K.*, **17**, 877–947.

Fleming, R. H., 1939. The control of diatom populations by grazing. *J. Cons. Explor. Mer*, **14**, 210–227.

Harvey, H. W., 1950. On the production of living matter in the sea off Plymouth. *J. Mar. Biol. Assoc. U.K.*, **29**, 97–137.

Harvey, H. W., L. H. N. Cooper, M. V. LeBour and F. S. Russell, 1935. Plankton production and its control. *J. Mar. Biol. Assoc. U.K.*, **20**, 407–441.

Holmes, R. W., M. B. Schaefer and B. M. Shimada, 1957. Primary production, chlorophyll, and zooplankton volumes in the tropical eastern Pacific Ocean. *Inter-Amer. Trop. Tuna Comm. Bull.*, **2**, 127–169.

King, J. E., T. S. Austin and M. S. Doty, 1957. Preliminary report on expedition EAST-TROPIC. *U.S. Fish Wild. Serv. Spec. Sci. Rep. Fish.*, **201**, 1–155.

Kokubo, Seiji, 1940. Quantitative studies of the neritic littoral microplankton of Japan collected at sixteen stations ranging from Saghalien to Formosa 1931–1933. *Proc. 6th Pacific Sci. Cong. Calif.*, **3**, 541–564.

Lillick, L. C., 1937. Seasonal studies of the phytoplankton off Woods Hole, Massachusetts. *Biol. Bull.*, **73**, 488–503.

Murphy, G. I., and staff, 1958.

Progress in 1957, Privately printed, Hawaii.

Rawson, D. S., 1952. Mean depth and fish production of large lakes. *Ecology*, **33**, 513–521.

Ricker, W. E., 1946. Production and utilization of fish populations. *Ecol. Monog.*, **16**, 373–391.

Riley, G. A., 1941. Plankton studies. IV. Georges Bank. *Bull. Bingham Oceanog. Coll.*, **7** (4), 1–73.

Riley, G. A., 1946. Factors controlling phytoplankton populations on Georges Bank. *J. Mar. Res.*, **6**, 54–73.

Riley, G. A., 1947. Seasonal fluctuations of the phytoplankton population in New England coastal waters. *J. Mar. Res.*, **6**, 114–125.

Riley, G. A., 1947a. A theoretical analysis of the zooplankton population of Georges Bank. *J. Mar. Res.*, **6**, 104–113.

Riley, G. A., 1956. Oceanography of Long Island Sound, 1952–1954. II. Physical oceanography. *Bull. Bingham Oceanog. Coll.*, **15**, 15–46.

Riley, G. A., 1956a. Review of the oceanography of Long Island Sound. *Deep-Sea Res.*, **3** (Suppl.), 224–238.

Riley, G. A., H. Stommel and D. F. Bumpus, 1949. Quantitative ecology of the plankton of the western North Atlantic. *Bull. Bingham Oceanog. Coll.*, **12** (3), 1–169.

Riley, G. A. and Ruth Von Arx, 1949. Theoretical analysis of seasonal changes in the phytoplankton of Husan Harbor, Korea. *J. Mar. Res.*, **8**, 60–72.

Ryther, J. H., 1956. The measurement of primary production. *Limnol. Oceanog.*, **1**, 72–84.

Ryther, J. H. and C. S. Yentsch, 1957. The estimation of phytoplankton production in the ocean from chlorophyll and light data. *Limnol. Oceanog.*, **2**, 281–286.

Ryther, J. H. and C. S. Yentsch, 1958. Primary production of continental shelf waters off New York. *Limnol. Oceanog.*, **3**, 327–335.

Sette, O. E., 1955. Consideration of midocean fish production as related to oceanic circulatory systems. *J. Mar. Res.*, **14**, 398–414.

Smith, E. V. and H. S. Swingle, 1939. The relationship between plankton production and fish production in ponds. *Trans. Amer. Fish. Soc.*, **68**, 309–315.

Smith, F. E., 1952. Experimental methods in population dynamics: a critique. *Ecology*, **33**, 441–450.

Southwell, R. V., 1946. *Relaxation methods in theoretical physics.* Oxford Univ. Press.

Steele, J. H., 1956. Plant production on the Fladen ground. *J. Mar. Biol. Assoc., U.K.*, **35**, 1–33.

Steele, J. H., 1958. Plant production in the northern North Sea. *Scot. Home Dep. Mar. Res.*, 1958 (7), 1–36.

Steele, O. H., 1961. Primary production. In: *Oceanogrophy*, A.A.A.S. Pub. 67, 519–538.

Steele, J. H. and C. S. Yentsch, 1960. The vertical distribution of chlorophyll. *J. Mar. Biol. Assoc. U.K.*, **39**, 217–226.

Volterra, V, 1928. Variations and fluctuations of the number of individuals in animal species living together. (Translated into English by M. E. Wells.) *J. Cons. Explor. Mer*, **3**, 1–51.

Wangersky, P. J. and W. J. Cunningham, 1957. Time lag in prey–predator population models. *Ecology*, **38**, 136–139.

# 21. FISHERY DYNAMICS—THEIR ANALYSIS AND INTERPRETATION

M. B. SCHAEFER and R. J. H. BEVERTON

## 1. Introduction

Most of the research on the dynamics of exploited fish populations is concerned, directly or indirectly, with investigating and measuring the effects of fishing on the population, with the objective of establishing, eventually, the relation between the fishing activity on the one hand and the resulting sustainable catch on the other. Thus, the fishermen and the fish are treated as a predator–prey system, with the centre of interest to fishery researchers lying in the benefits to the predator.

The basic dynamics of a fishery, that is of the predator–prey system, have been simply outlined by Russell (1931) and may be illustrated, following Ricker (1958), using the diagram in Fig. 1. The total weight, or biomass, of fish

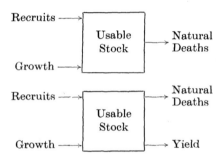

Fig. 1. Diagram of the dynamics of a fish stock (fish of usable sizes), when there is no fishing (*above*) and when there is a fishery (*below*). (After Ricker, 1958.)

in the exploitable phase consists of all fish above the average minimum size at which they are capable of being retained in the fishery, i.e. the size above which they are selected by the fishing gear, or by the selective efforts of the fishermen's operations such as choice of fishing grounds or schools of fish of certain sizes. This fished population, of biomass $P$, loses members by natural deaths as well as by fishing; it is replenished by recruitment of young fish as they grow large enough to be caught and by the growth of the fish already in it.

In the virgin situation, with no fishery, all increments to the stock by recruitment and growth are balanced by losses due to natural mortality. With the imposition of a fishery there is an additional loss from the stock to the predator, the fisherman. This added mortality results in a decrease in the abundance of the fished part of the stock and a shift in its age composition towards the younger fish. Eventually, when the amount of fishing has increased as much as is permitted by the economic conditions of the fishery, a new equilibrium is established in which the catch is balanced by the compensatory changes which have occurred in one or more of the vital rates of natural mortality, growth and recruitment.

[*MS received September, 1960*]      464

A very general model describing these dynamics may be formulated by expressing the relative rate of change in biomass of the fished stock in terms of the component factors influencing it;

$$\frac{1}{P}\frac{dP}{dt} = r(P)+g(P)-M(P)-F(X)+\eta, \tag{1}$$

where $P$ = biomass of the fishable part of the population, $r$, $g$ and $M$ = rates of recruitment, growth and natural mortality respectively; all of these are, in general, functions of the biomass $P$ of the population and its age composition. $F$ = rate of loss due to fishing, which is a function of the fishing effort, $X$. $\eta$ = a (variable) rate of change of stock biomass, independent of both $P$ and $X$, due to variations in external environmental factors. In practice, these effects are eliminated as far as possible by averaging over periods of time so that the model can be regarded as describing events for average environmental conditions.

In the steady state, with the population in equilibrium under average environmental conditions, $dP/dt = 0$ and $\eta = 0$, so that

$$F(X) = r(P)+g(P)-M(P). \tag{2}$$

In these circumstances, the average steady catch, $Y = F(X)\,P$, is equal to the additions due to recruitment and growth, less the losses by natural death, i.e.

$$Y = F(X)\,P = [r(P)+g(P)-M(P)]\,P. \tag{3}$$

Ideally, this general model could be applied by measuring the various elemental rates and establishing from observation their dependence on the biomass and age-composition of the population. In practice, however, a complete description in such terms of the dynamics of exploited fish populations has not yet proved possible, and it is necessary to proceed by making certain kinds of simplifications to the general model, depending on the kind and amount of data available and the particular questions concerning the population and its associated fishery which it is desired to answer.

Two general types of approach have been developed. That most widely used has been to retain the identity of the various elemental rates, to estimate their magnitude and relations as far as is possible from the data available, and to combine them in an appropriate special form of the general model (3), making such simplifying assumptions as are required for practical application. This analytical approach, with which are associated such investigations as those of Baranov (1918), Thompson and Bell (1934), Ricker (1944) and others, has been developed in detail by Beverton and Holt (1957); it is for convenience referred to here as the "Beverton–Holt" approach.

The alternative method does not attempt to distinguish between the elemental rates of recruitment, growth and natural mortality, but considers only their resultant effect as a single function of population size. This approach stems from the concept of the logistic law of population growth (e.g. Pearl, 1930; Lotka, 1925; Gause, 1934) and was applied to the dynamics of a fishery

16—s. II

by Hjort, Jahn and Ottestad (1933), Graham (1935, 1939) and others. It has since been further developed by Schaefer (1954, 1957), and is denoted here as the "Schaefer" approach.

Historically, the development of these two methods has been determined very much by the type of fishery with which the investigators have been concerned and by the amount and kind of data available. In what follows, an account is given of each of them separately, but it is important to note at this stage that the difference between them lies essentially in the kind of assumptions and procedures that are adopted to overcome the lack of certain kinds of information. The objective of both is the same, namely to provide the best means of utilizing the available data to understand and predict the effect of fishing on a stock. Some further comments on the relation between the two approaches, both in concept and application, will be found in the concluding remarks.

## 2. The Beverton–Holt Approach

The pioneer work along this line was that of Baranov (1918), although this was not subsequently taken up in Russia, and did not come to the attention of western scientists until the 1930's. Baranov constructed what amounts to a particular case of the general model represented by (1). He assumed that a constant number of recruits, $R$, enters the fished stock each year [which implies $r$ proportional to $1/P$ in (1)], and that the natural mortality coefficient, $M$, is constant. He incorporated the age-specificity of growth by assuming that the length of a fish increased in proportion to its age and that its weight varied as the cube of length. With these assumptions Baranov derived an equation expressing the length-frequency of the stock corresponding to any given values of the above parameters and also to any value of the fishing mortality coefficient, $F$; and from this equation it is possible to compute the equilibrium catch for various values of $F$. By making similar assumptions about recruitment and mortality, Thompson and Bell (1934) calculated arithmetically the stable-age composition of halibut corresponding to various rates of fishing, and, by introducing the observed relation between age and weight for that species, they computed the equilibrium relation between catch and fishing rate.

The methods developed by Beverton and Holt combine, in effect, the formal mathematical treatment of Baranov with the age-composition treatment of Thompson and Bell. For what they have called their "simple" model (Beverton, 1953), Beverton and Holt retained Baranov's assumption of constant recruitment and a constant natural mortality rate but used a different growth function, because the assumption that the length of a fish increases in proportion to its age is, in fact, valid over only a small part of the growth span, and this assumption causes serious discrepancies when Baranov's model is used to predict the catch at low fishing rates. They also distinguished between the number of young fish comprising each year-class when they first enter the area where fishing is in progress, and so become liable to encounters with fishing gear, and the number which survive to an age at which they have grown large enough to be retained by the particular kind of fishing gear in use. In this way it is possible to develop a model which can be used to predict the effect on the equilibrium catch of a change in the selectivity of the gear as well as in the amount of fishing, a requirement which is of special importance in the trawl fisheries of the North Sea and adjacent waters with which the authors were primarily concerned.

Beverton and Holt's simple model can be derived most conveniently by considering the catch obtained from one year-class throughout its life, since in the steady state, with the same number of recruits entering the exploited area each year, this catch is the same as the average steady catch obtained each year from the whole stock. Suppose $R$ fish are recruited to the exploited area at age $t_r$, but are not retained by the gear until they have reached some greater age $t_c$. During this early "pre-exploited" period of their life their numbers will be reduced by natural mortality only, and with a constant coefficient $M$, the rate of decrease of numbers during this period is

$$dN/dt = -MN. \tag{4}$$

The solution of this equation gives the number surviving to enter the catch at age $t_c$ as

$$Re^{-M(t_c-t_r)}. \tag{5}$$

Thereafter, their numbers will be reduced by the combined operation of both fishing and natural mortality, according to the equation

$$dN/dt = -(F+M)N, \tag{6}$$

and from this and (5) the number surviving to any age $t$ greater than $t_c$ will be

$$N_t = [Re^{-M(t_c-t_r)}]\, e^{-(F+M)(t-t_c)}. \tag{7}$$

If $w_t$ is the average weight of fish at age $t$, the total weight of the year-class at age $t$ will be $N_t \cdot w_t$, and the rate of catch in weight as a function of age becomes

$$dY/dt = FN_t \cdot w_t, \tag{8}$$

where $F$ is the fishing mortality coefficient. In the first instance a linear increase of weight with age was used for the function $w(t)$ (Hulme, Beverton and Holt, 1947), and the same relation has also been used by Doi (1951). This gives a reasonable representation of part of the growth curve but not the whole of it, and does not lead to a model which is particularly convenient for computation. As giving the best simple representation of the general growth pattern of fish, Beverton and Holt later adopted the equation developed by von Bertalanffy (1934, 1938), namely

$$w_t = W_\infty(1-e^{-K(t-t_0)})^3, \tag{9}$$

in which $W_\infty$ is the asymptotic weight to which the fish tends with increasing age, $K$ is a constant which determines the rate (for given units of age) at which the asymptotic weight $W_\infty$ is approached (i.e. the "curvature" of the growth curve), and $t_0$ is an arbitrary constant which is, in effect, the theoretical origin of the age-scale, i.e. the age at which the weight would be zero, if growth followed the von Bertalanffy equation at all ages. Actually, this equation does not always hold for very young fish (well below the exploited ages), so that $t_0$ does not correspond to the real origin in time. Substituting in (8) this expression for $w_t$ and that for $N_t$ from (7), and integrating over the whole life of the fish after they first enter the fished stock at age $t_c$, gives the following equation for the equilibrium catch:

$$Y = FRW_\infty \, e^{-M(t_c-t_r)} \sum_{n=0}^{3} \Omega_n e^{-nK(t_c-t_0)}/(F+M+nK). \qquad (10)$$

In this equation the summation term is a shorthand way of writing the cubic term of the growth equation (9), with

$$\Omega_0 = +1, \quad \Omega_1 = -3, \quad \Omega_2 = +3, \quad \Omega_3 = -1.$$

Beverton and Holt (1957) applied this equation to the plaice and haddock fisheries of the North Sea, obtaining estimates of the parameters it contains from the extensive data available for these two stocks. The methods of estimation cannot be described in detail here, although some of the principles involved may be outlined. The most favourable situation is one in which reliable age-composition data of the catch are available over a period of years, together with statistics of fishing effort in units which are proportional to the fishing mortality rate generated. In a trawl fishery, for example, the total hours fished per year multiplied by the average tonnage of the vessels is a suitable measure of total fishing effort. It is then possible to express the age-composition for each year in units of catch per unit effort, which can be regarded as indices of true abundance, and hence to estimate the total mortality coefficient $(F+M)$ from the rate of decrease in the abundance of year-classes. In the absence of data of fishing effort, only percentage age-compositions can be obtained, but, if these are available over a sufficient period, reasonably good estimates of $(F+M)$ can usually be derived by combining all the data, a procedure which tends to average out the effect of differences in year-class abundance. More difficult is to separate the total mortality into its components, $F$ and $M$. If statistics of fishing effort are available, and if these cover periods during which the amount of fishing has changed considerably, it may be possible to establish the relation between total mortality coefficient $(F+M)$ and fishing effort, which, if effort is measured in proper units, should be a linear one. Extrapolation of the regression of $(F+M)$ on effort to the origin where $F=$ zero therefore provides an estimate of the natural mortality coefficient, $M$. Alternatively, the fishing mortality coefficient $F$ may be estimated independently by tagging experiments, although to obtain good estimates by this method it is necessary that the population of tagged fish should be reasonably representative of that of the stock as a whole, and also that the operation of catching fish for tagging and of attaching a tag to them should not adversely affect their subsequent survival. For a further discussion of the estimation of mortality rates by these and other techniques which can be used in similar circumstances, the reader is referred to the original work of the authors (Beverton and Holt, 1956, 1957).

Estimation of the parameters $K$ and $W_\infty$ of the von Bertalanffy growth equation seldom presents much difficulty if size and age data are available. It could scarcely be expected that any simple mathematical function could give a precise fit to the observed growth of fish under all conditions; unequal availability of food to fish of different sizes, for example, might produce distortions in the growth curve which would not be followed closely by any except a highly complex function. Nevertheless, the simple von Bertalanffy equation has been found to give a satisfactory fit to the growth data of a wide range of fish species, and serious exceptions are rare.

It is obvious from (10) that the catch obtained from a year-class throughout its fished life is proportional to its initial numbers when recruited; and, indeed, it can be shown that the average catch over a period is similarly proportional to the average recruitment during that time. For most purposes it is not, therefore, necessary to attempt to estimate the absolute number of recruits, but instead it is sufficient to use the model to compute the "catch per recruit", since this latter quantity is influenced by changes in the amount of fishing or in the selectivity of the gear in the same proportion as would be the actual average yield over a period.

Having obtained estimates of the parameters in (10), the equation may be used to compute either the equilibrium relation between catch and fishing effort by varying the value of $F$, or the effect of changing the selectivity of the gear by altering the value of the age $t_c$ at which fish enter the catch. Beverton and Holt (1957) give a number of curves of equilibrium catch of this kind for the plaice and haddock fisheries, and from a consideration of the effect of changing both $F$ and $t_c$ together have developed the concept of *eumetric fishing*, in which the gear selectivity is so adjusted that it allows the greatest steady catch to be obtained that is possible with the particular amount of fishing effort which is being generated.

In practice, it is seldom possible to obtain accurate estimates of all the parameters, especially of the natural mortality coefficient $M$, although it may well be feasible to establish a range of values within which the true value is likely to lie. For example, Beverton and Holt (1959) have shown that within certain groups of fish species, notably the clupeoids and the gadiformes, there is a tendency towards a characteristic relation between natural mortality and growth pattern, so that even if no direct estimate of $M$ is available it may nevertheless be possible to establish its likely range within certain limits simply from a knowledge of the growth pattern. The analytical model may then be used to test the effect of this range of uncertainty of the parameter in question on the conclusion which is to be established from the properties of the model. Sometimes it may happen that the answer to the question—for example, whether any increase in the amount of fishing beyond the contemporary level will result in an increased steady catch—is relatively insensitive to the degree of accuracy with which certain parameters of the model can be estimated but is critically affected by that of others. In such cases the properties of the analytical model, as well as establishing in quantitative terms the reliability of the answer, may also serve as a valuable guide to the direction in which future research on the fishery should be conducted.

Ricker (1944), in reviewing the analytical type of population model, pointed out that the assumption that the numbers of recruits, the coefficient of natural mortality and the rate of growth are all independent of population density, as in the above model, cannot be valid over a large range of population size, nor hence over a wide range of fishing effort or gear selectivity. In some circumstances, for example when dealing with the regulation of an established fishery, this may not be a serious drawback, because for economic reasons regulation has to be introduced in stages, each involving only a relatively small change in the fishing activity and hence in population size. Thus the effect of the first steps in regulation itself provides the opportunity of establishing whether the population parameters are being influenced by population density, and if they are, the prediction of the effect of the next step towards the final objective of regulation can be made that much more accurate in the light of the secondary effects which regulation may be having on the dynamic characteristics of the stock. Nevertheless, although it may be sufficient for certain purposes to use an analytical model with density-independent parameters, it is clearly important for a more general understanding and interpretation of the dynamics of exploited fish populations that these secondary effects should be measured and incorporated into the theoretical formulation of the fishery.

Beverton and Holt (1957) have considered at some length the problems involved in the analysis of density-dependent effects. Generally speaking, the effect of population density on growth would usually be expected to be the easiest to detect, partly because changes in growth can be measured with some accuracy where routine age-determination is possible, and partly because growth is directly related to the average amount of food available per fish. The authors have, in fact, detected a relation between growth and density in both the plaice and haddock populations of the North Sea, and have shown in the latter species that it can be adequately represented by a linear relation between the asymptotic length $L_\infty$ [corresponding to the asymptotic weight $W_\infty$ of (10)] and population numbers. There is also some justification from the physiological basis of the von Bertalanffy growth equation for regarding the asymptotic size as being more likely to be influenced by density than the coefficient $K$, although this is a question which has yet to be properly resolved. Now, an expression for population numbers can readily be derived in the same way as before, except that reference to size of fish is omitted, and is

$$N = R\left(\frac{1 - e^{-M(t_c - t_r)}}{M} + \frac{e^{-M(t_c - t_r)}}{F + M}\right); \tag{11}$$

so that if a relation between growth and population size can be established from data, for example, in the simplest case, the linear relation

$$L_\infty = a - bN, \tag{12}$$

where $a$ and $b$ are constants, then the two equations (11) and (12) can be treated as simultaneous equations and together provide a solution for the value of $L_\infty$ corresponding to any particular value of $F$ or $t_c$. Since the value of $W_\infty$ corresponding to $L_\infty$ can readily be determined from the relation between weight and length for the species in question, it is now possible to compute the relation between equilibrium catch and $F$ or $t_c$ from equation (10), with the growth rate varying according to the population density.

Compared with the behaviour of the simple model, introducing the compensatory effect of a density-dependence of growth causes the population biomass, and hence the weight of the catch, to be less sensitive to changes in either $F$ or $t_c$. Sometimes this may lead to quantitatively different conclusions being drawn. In a fairly heavily fished stock, for example, the simple model with density-independent growth might predict that some decrease in fishing effort would result in an increased equilibrium catch, and vice versa; the introduction of a sufficiently marked dependence of growth on density could reverse this conclusion in certain circumstances, because of the retardation of growth that would follow from a reduction in fishing effort, or enhancement of growth from an increase in effort. Examples of the influence of the density-dependence of growth on the theoretical relation between catch and fishing effort, and between catch and gear selectivity, will be found in Beverton and Holt (1957).

The theoretical introduction of a relation between the natural mortality coefficient $M$ and population density in the simple population model presents no special difficulties, and can be treated by methods analogous to those described above for growth. In practice, however, virtually nothing is known about the variation, if any, of the natural mortality rate with density in any major fish population, primarily because of the difficulty of measuring the natural mortality rate with the necessary accuracy. Unlike the density-dependence of growth, in which it can be safely accepted that growth will decrease fairly steadily as density increases, and vice versa, even this assumption may not necessarily be true for the natural mortality rate. It is, perhaps, something of a reassurance to find that in many of the major fish populations of temperate waters in which the natural mortality rate has been measured, it is fairly small, usually less than 30% per year; and some of this is almost certainly due to causes of death such as senility and adverse environmental conditions which are unlikely to be much influenced by population density, if at all. Natural mortality rates of important species of fishes and invertebrates of tropical regions (e.g. tuna, anchovies and shrimp) have, however, been found to be a good deal higher than this. It has to be admitted that knowledge of all aspects of the mortality in fish populations due to causes other than fishing is still fragmentary and is a serious gap in our present understanding of their dynamics.

There remains for consideration the density-dependence of recruitment. This is of quite a different nature to that of growth or natural mortality, because recruitment is a property not of the contemporary population but of an earlier mature population of which the recruits are the progeny. A study of the density-dependence of recruitment therefore concerns the relations, in effect, between successive generations and involves the analysis, both from data and in the theoretical model, of the long-term dynamics and self-regenerating properties of the population.

In the simple model formulated by (10), it is assumed that a constant absolute number of fish is recruited annually, although, as has been mentioned earlier, this assumption can be broadened to allow for a fluctuating recruitment without loss of generality, provided the fluctuations have themselves a constant mean value. It could hardly be expected, however, that this assumption could remain valid over very wide ranges of size of parent population. A population of one mature male and one mature female, even with the enormous fecundity which characterizes many species of fish, could scarcely be expected to produce the same number of recruits as a population consisting of hundreds of millions of mature fish of each sex, which many do. It may therefore seem rather remarkable that for no major fish population has a clear-cut relation between parent stock and subsequent recruitment yet been established. The reasons for this are not, however, far to seek. The main one is that recruitment usually fluctuates very widely from one year to the next, primarily through the influence of varying environmental conditions on the survival of the young fish, and this range of fluctuation is usually much greater than that of the size of the parent stock which has been encompassed by the available data. In this connection it has to be remembered that the fecundity of most marine teleosts, including the pleuronectids, gadoids, clupeoids and scombroids on which most of the world's largest fisheries are based, is very high; a mature female of any of these groups might spawn something between a hundred thousand and several million eggs annually. Yet for the adult population to be maintained in a steady state only a very few of these, perhaps only two or three, are needed to survive to maturity—an event which in many cases roughly coincides with recruitment to the commercial fishery.

With such an enormous mortality intervening between successive adult generations it is hardly surprising to find that recruitment usually fluctuates widely. There are, nevertheless, marked differences in the degree of recruitment fluctuations in various stocks. In the North Sea haddock, for example, the abundance at recruitment of the 30 year-classes which have been sampled since 1920 has varied over a 500-fold range; in contrast, the variation of recruitment to the North Sea plaice population over 26 year-classes has been only six-fold, with an average variation of about 35% above or below the mean (Beverton, 1962). Since the range of adult population size of the two species over the periods in question has been of the same order, between about six- and ten-fold in weight, these observations suggest that a pronounced degree of compensation occurs in the early life of plaice which is absent in haddock. In fact, the fluctuation of recruitment in plaice is sufficiently small for it to be possible to say that recruitment is effectively independent of the parent population over the range of size of the latter which has so far been observed. In the haddock, on the other hand, although there is a slight tendency for the larger recruitments to come from the larger parent stocks, the relation is not significant and the scatter of the recruit values is so great that, from the data available so far, no firm statement can be made about the form of the relation between stock and recruitment or even whether the two are related at all.

The theoretical consequences of introducing a relation between parent stock and subsequent recruitment lead to some important conclusions concerning the requirements for a population to be able to stabilize itself under the influence of fishing. Moran (1950) was the first to investigate this problem, various aspects of which have since been discussed further by Ricker (1954) and Beverton and Holt (1957). It has been shown that postulating a proportional relation between stock and recruitment leads to a highly unstable population; the slightest disturbance from an initial steady state, such as a very small change in the amount of fishing, causes successive generations either to decrease in numbers until the population is extinct, or to increase without limit. In practice, as Ricker (1954) has shown, the superimposition of a large random component of fluctuation in recruitment could cause this progression to be delayed for many generations. Nevertheless, the impression gained from the history of even the very long established fisheries, including those for the Arcto-Norwegian cod and the Atlanto-Scandian herring, of which there are records dating back many hundreds of years, is that although there may have been cycles of abundance, the major fish stocks are broadly stable and are not tending either to extinction or to an ever increasing size.

The essential requirement for stability is that the relation between stock and recruitment should curve away from proportionality, so that the rate of increase of recruitment with stock declines throughout as stock increases. Beverton and Holt (1957) have shown that postulating a linear relation between the natural mortality coefficient $M$ at one or more stages in the pre-recruit life of the fish and their density leads to the simplest form of relation between stock and recruitment which gives stability, namely a curve which bends over continuously and tends to an asymptotic limit of recruitment with increasing stock. Denoting the total egg production of the population by $E$, which to a first approximation can be regarded as proportional to the biomass of mature fish, the equation is

$$R = \frac{1}{\alpha + \beta/E},\tag{13}$$

in which $\alpha$ and $\beta$ are coefficients defining the effective mortality rate and its dependence on density during the pre-recruit phase of the life-history. The limiting recruitment, as $E$ tends to infinity, is seen to be $R = 1/\alpha$; this can be interpreted as the maximum number of fish which the food supply and other characteristics of the environment can allow to survive to the age of recruitment.

Ricker (1954), on the other hand, has proposed a dome-shaped curve, with recruitment increasing to a maximum at some intermediate level of adult stock and then declining, as providing the best general interpretation of the relation between stock and recruitment. A relation of this kind has been found in certain experimental populations, such as *Drosophila*, maintained under restricted conditions of space, where above a certain abundance of adults the medium becomes fouled and so causes the survival of young to be sharply reduced. A tendency towards reduced recruitment at the higher stock levels has also been detected in certain salmon populations, where again the environmental conditions for spawning and for the early life of the young fish are highly restricted compared with those found in the sea. It is, however, a measure of the difficulty of interpreting the data available so far for marine fish populations that no conclusive evidence of a declining recruitment at higher stock levels has yet been demonstrated, remembering that a given set of rather scattered points can often be fitted rather better by a paraboidal-shaped curve than by a simpler asymptotic one. In this connection the reader will find it instructive to read the discussion by Clark and Marr (1955) on the interpretation of the data of stock and recruitment for the Pacific sardine. In the present state of knowledge, the two hypotheses can perhaps best be reconciled by supposing that if the stock size were to increase far enough it would indeed give rise to conditions which caused something like a catastrophic mortality of young fish to occur, but that before that stage was reached the stock–recruit curve would have become nearly flat over a considerable range of stock size.

A dome-shaped stock–recruit relation gives rise to rather complex stability characteristics. On the left of the maximum the population has essentially the same kind of stability as that implied by the asymptotic relation (13), in which a change in the external restraints, such as the amount of fishing, causes the population to increase or decrease to a new steady state. On the descending right-hand limb of the curve, however, self-induced oscillations are set up, the amplitude of which may either gradually decrease with time or remain constant or increase, depending on the particular shape of the curve. Theoretical models of a fishery in which the relation between stock and recruitment is dome-shaped have not yet been developed, but Beverton and Holt (1957) have examined the properties of models in which the relation is of the kind formulated by (13). If the other rates are treated as density-independent, they found that certain important characteristics of the relations between equilibrium catch and $F$ or $t_c$ predicted by the simple model with constant recruitment are scarcely changed, notably the amount of fishing or the gear selectivity required to produce the greatest steady catch, although the magnitude of that catch was markedly affected. If, in addition, growth was also made to vary with density, more profound departures from the properties of the simple model were found, since introducing the variation of recruitment with stock size accentuates the changes in the latter with $F$ or $t_c$, and hence in the growth rate, which in turn affects the spawning potential of the adult stock.

The problems involved in the treatment of recruitment as a factor in the analytical population model have been discussed at some length because, generally speaking, the characteristics of the recruitment to marine fish populations—its degree of fluctuation, its relation to stock size and the influence on it of changing environmental conditions—are the key to the interpretation and prediction of the long-term dynamics of a fishery, whether by the Beverton–Holt or by the Schaefer approach, which will now be described.

## 3. The Schaefer Approach

If, in equations (1) and (2), we combine the rates of recruitment, growth and natural mortality into a single function of $P$, $f(P)$, called "the coefficient of natural increase",[1] we obtain *for average environmental conditions*:

$$\frac{1}{P}\frac{dP}{dt} = f(P) - F(X) \qquad (14)$$

and, for steady-state equilibrium,

$$F(X) = f(P), \qquad (15)$$

$$Y = P\,F(X) = P\,f(P), \qquad (16)$$

where $Y$ is the total catch. $f(P)$ is taken to be a single-valued function of $P$. At the environmental-limited upper value of the average population magnitude, $L$, in the absence of any fishery, it must be equal to zero, since, in this circumstance, $dP/dt = 0$ and $F = 0$. It must increase with decreasing $P$. This is a "negative feedback" term which is necessary to describe the self-regulating property by which the rate of natural increase is regulated appropriately to bring the population into balance again with the imposition of a fishery.

The utility of this approach in fishery dynamics, and in other branches of population dynamics as well, depends on being able to apply a fairly simple approximation to $f(P)$. $f(P)$ may be expressed as a power series, $f(P) = k(L - P - bP^2 - cP^3 \cdots)$. Now, the simplest approximation that we can make is to assume that it is linear with $P$ (Moran, 1954), that is, to take only the first two terms of the series, so that

$$f(P) = k(L - P). \qquad (17)$$

This, of course, corresponds to the Verhulst–Pearl "logistic" or "sigmoid" law of population growth, which provides a useful approximation for a wide variety of organisms, and seems to be also a useful first approximation to the growth in biomass of at least some fish populations.

---

[1] Schaefer (1954) has called the equivalent of $P\,f(P)$ the "rate of natural increase", which was designated $f(P)$ in that paper.

This logistic function, however, gives a relationship which is symmetrical, with a maximum at exactly half the maximum average population size, i.e. at $L/2$. There is reason to believe that the maximum will actually occur, for many if not all populations, at a level of $P$ lower than $L/2$. It will, therefore, be desirable to take more terms than the first two in $f(P)$ if the data of the fishery are sufficient to estimate the parameters involved. On the other hand, the most simple approximation may be adequate for the data available (see, for example, page 478).

We have, then,

$$\frac{1}{P}\frac{dP}{dt} = k(L-P) - F(X) \tag{18}$$

and, if we further assume that fishing mortality rate is proportional to fishing effort, $F = cX$,

$$\frac{1}{P}\frac{dP}{dt} = k(L-P) - cX \tag{19}$$

and, at equilibrium, the catch, $Y_e$, is

$$Y_e = cXP = kP(L-P). \tag{20}$$

Obviously, if we have data on the abundance of the population (catch-per-unit-of-effort, which is proportional to $P$) and the effort or catch at two different steady states, we can from (20) compute the parameters in this equation, and so arrive at an estimate of the relationship between $Y_e$ and $P$. Graham (1935) applied this method, essentially, to the data of the demersal fisheries of the North Sea.

The number of instances in which two or more clearly defined steady-state periods, corresponding to substantially different levels of fishing intensity, are encompassed by the available data are, however, rare. To make the method of wider application it is necessary to adapt it to the analysis of data of fisheries which are not in a steady state. Schaefer (1954) has developed a technique of this kind which enables an approximation to the equilibrium catch to be obtained on a year-to-year basis, and has applied it to the data of the Pacific halibut and the California sardine. The actual catch, $Y$, in each year is divided by its fishing rate, $F$, to estimate the mean population during the year. The level of population at the end of each year is taken as approximately the average of the mean population of that year and the following year, and the difference between the populations at the beginning and end of the year is taken as the increase (or decrease) for that year. That is, the increase in year $i$ is estimated as

$$\frac{\Delta P_i}{\Delta t} = \frac{Y_{i+1}/F_{i+1} - Y_{i-1}/F_{i-1}}{2}$$

and the equilibrium catch at the level of population in year $i$ is estimated as

$$Y_{e.\,i} = Y_i + \frac{\Delta P_i}{\Delta t}.$$

The value of $c$ required for estimating the values of $F$ from $X$ was obtained, in each case, from the data of tagging experiments. The equilibrium catches, $Y_{e.\,i}$, thus estimated can be plotted against population size, $P_i$, regardless of any hypothesis as to the form of the relationship between them, and if a long enough series of data, over a wide enough range of values of $P_i$, is available, an empirical average relationship can be drawn. If, however, the model of (20) is assumed, a parabola can be fitted to estimate the relationship.

Likewise, designating $Y_i/X_i$ (catch-per-unit-of-effort) by $U_i$, and noting that $U_i = cP_i$, we have from (19) for year $i$:

$$cX_i = k\left(L - \frac{U_i}{c}\right) - \frac{1}{U_i}\frac{\Delta U_i}{\Delta t} + \epsilon, \tag{21}$$

where

$$\frac{\Delta U_i}{\Delta t} = \frac{Y_{i+1}/X_{i+1} - Y_{i-1}/X_{i-1}}{2},$$

$(1/U_i)\,\Delta U_i/\Delta t$ being an estimate of $(1/P_i)\,\Delta P_i/\Delta t$, and where $\epsilon$ is a population-independent random variable, to take account of variable environmental conditions.

From a sufficiently long series of data on catch and catch-per-unit-of-effort, the parameters, including $c$, may be estimated by statistical procedures, without reference to tagging data. Schaefer (1957) applied this method to the data of the yellowfin tuna fishery of the Eastern Pacific, and thus estimated for the model of (19) and (20) the average relationships between fishing effort, catch-per-unit-of-effort, and total catch for steady states. He rearranged the terms in (21), accumulating some constants for simplicity of notation, and obtained

$$\frac{1}{c}\frac{1}{U_i}\frac{\Delta U_i}{\Delta t} = a(M - U_i) - X_i + \frac{\epsilon}{c}. \tag{22}$$

Summing over $n$ years,

$$\frac{1}{c}\sum_{i=1}^{n}\frac{1}{U_i}\frac{\Delta U_i}{\Delta t} = \sum_{i=1}^{n} a(M - U_i) - \sum_{i=1}^{n} X_i, \tag{23}$$

since the expectation of $\sum\epsilon$ is zero.

Since positive values of $\Delta U$ are, on the average, associated with positive values of the right side of (22), and conversely, he also obtained, summing over $m$ years,

$$\frac{1}{c}\sum_{i=1}^{m}\left|\frac{1}{U_i}\frac{\Delta U_i}{\Delta t}\right| = \sum_i [a(M - U_i) - X_i] - \sum_i [a(M - U_i) - X_i]. \tag{24}$$

$$\Delta U_i > 0 \qquad\qquad \Delta U_i < 0$$

He had data to estimate $\Delta U_i/\Delta t$, $U_i$ and $X_i$ for each of 20 years. Taking the sums in (23) for the first ten years and the second ten years separately, and the sums in (24) for the whole series, he then had three simultaneous linear equations to solve for the three parameters, $1/c$, $a$ and $M$. The solution was made by a method of iteration, but we may note that the same result may be more easily obtained by a straightforward simultaneous solution of the three equations.

It was found that the estimate of $c$ by this means, for the tuna fishery, had a very low precision but, on the other hand, that there was little difference in the relationships among the variables for a rather wide range of values of $c$, so that even an approximate estimate of this constant gives as good an estimate of the steady-state relationships among $X$, $U$ and $Y$ as the data can provide.

Equation (20) predicts that the equilibrium relationship between effort and catch-per-unit-effort should be linear, and in the same paper the author tested this on the yellowfin tuna data from 1934 to 1955. No significant deviation from linearity could be detected, so that over the observed range of population size, which included sizes near to the maximum (unfished) level down to about half this magnitude, the simple linear interpretation of the function $f(P)$ was a sufficient approximation. This is a valuable confirmation of the Schaefer approach in at least one fishery within the limits of the data, but it is, of course, possible that a more complex function of $f(P)$ might be required to fit a wider range of data.

In this connection it has to be remembered that the simple law relating rate of population increase to population size formulated by (14) implies certain assumptions about the dynamics of fish populations which are, to a greater or lesser degree, unrealistic, and which, if the degree of departure from reality be large, can limit its usefulness. For some fisheries these will not constitute serious sources of error, but for others they may; some aspects of these matters have been discussed by Watt (1956), Beverton and Holt (1957) and Ricker (1958).

There are, in fact, two assumptions about the fish population and its application to data implied in (14):

(1) That the rate of natural increase responds immediately to changes in population density. That is, delayed effects of changes in population density on rate of natural increase, such as the effects of the time-lag between spawning and recruitment of resulting progeny, are ignored.

(2) That the rate of natural increase at a given weight of population is independent of any deviation of the age-composition of the population from the steady-state age-composition at that population weight.

Neither of these conditions is exactly fulfilled by populations of fishes or other multicellular organisms. The effects of intraspecies competition on individuals between the egg and the age of entry into the catchable stock involve some time lag. Some of the effects of time lags on models similar to those used here have recently been examined by Wangersky and Cunningham (1957).

With respect to the second assumption, although certain factors, such as fecundity and perhaps mortality in many cases, are approximately proportional to total biomass, growth is highly age-specific; thus the age-structure of the population, as well as its total biomass, must enter to some degree into its potential for increase.

It is to be noted that if the model is applied to data for steady states only, these assumptions are not necessary. In this case the age-structure of the population is uniquely determined by its biomass. Also, since in the steady state the population during any year is, on the average, of the same size and age-structure as during any subsequent year, the effect of time lag on recruitment, etc. is eliminated. Actual fisheries are, however, seldom in steady states, so we must deal with transient states, in which these effects appear.

It is important in this connection to note that the changes in fishing effort in many commercial fisheries are often gradual, so that the displacement from steady state is not large. Furthermore, some kinds of fishes, and other organisms, especially those of the tropics, such as peneid shrimp and yellowfin tuna, enter the catchable stock very young, mature very early, and have a very short life-span. For such species, the time lag between spawning and recruitment is small, and effects of changes in size composition due to changes in fishing effort are less important than for organisms of greater life-span. At the other extreme there are fisheries based exclusively on the mature phase of populations in which maturity does not occur until a relatively high age is reached. A case in point is the Norwegian Lofoten fishery for mature arctic cod, which are not recruited to the fishery until they are eight or nine years of age, on the average.

The rate of natural increase is determined not only by the magnitude and age-structure of the population, but also by many environmental factors. Variation in these will cause departures from the rate of increase that would occur under average environmental conditions. In applying the model to the data from the fishery, we assume that the effects of variable environmental factors on recruitment, survival, and growth are random, or at least that they are not correlated with population changes due to changes in fishing effort, so that they may be averaged out. In other words, we assume that the mathematical expectation of $\epsilon$ in equation (21) is zero.

Variations in the fishes' environment may also affect their behaviour so as to make them more or less accessible to capture. It is, again, implicitly assumed in the application that these effects are random, or at least uncorrelated with changes in fishing effort, so that the mathematical expectation of the constant $c$ remains the same for different levels of fishing effort (and population size).

## 4. Concluding Remarks

As was mentioned at the beginning of this chapter, the fact that the two approaches to the analysis and prediction of fishery dynamics have evolved hitherto independently of each other is largely because of the different situations, both as regards the type of fishery and the kind of data available, with which the investigators were confronted. The reader who has stayed with us to this point will probably see now why this should be. Thus Beverton and Holt were concerned primarily with the development of techniques which could be used for the assessment of the need for, and effects of, regulation of the North Sea demersal fisheries; assessment of the effect of mesh regulation was one of the main requirements, although this could not be undertaken without reference to the effect of changes in fishing effort as well. For some of these fisheries, such as those for plaice and haddock, good commercial statistics of catch and effort were available for a number of years, but the fisheries had long since passed their developmental stage and, except for the transitory effects of the two wars, the data covered only small changes in population size which could be ascribed to changes in fishing effort. On the other hand, age-determination of these species is relatively easy and extensive age-composition and other research data were available for both of them. In contrast, Schaefer has been primarily concerned with establishing the relation between catch and fishing effort in the Pacific tuna fishery. Good statistics of catch and effort for this fishery are available almost from its inception, and therefore cover that highly critical range of population size near to the virgin state which for most of the North Sea stocks is lost in antiquity. Age-determination, however, is difficult in tropical species, at least by conventional methods on a routine basis, so that most of the data required for the Beverton–Holt approach were not easily available. It will be appreciated that the two approaches would not have been interchangeable between the two situations.

Of course, if the development of a fishery from its beginning to a grossly overfished state were fully and reliably documented, and the influence of external causes of trend and variability in the data were either small or could be allowed for from other information, the relation between steady catch and fishing effort could be established directly without recourse to theory of any kind. It would still be necessary, however, to use some form of analytical approach to predict the effect of changes in gear selectivity if these had not already been experienced, and perhaps also to detect as quickly as possible the influence of possible changes in the dynamics of the stock due to environmental changes. In practice, however, this is an ideal which no actual fishery approaches; therefore some kind of theoretical treatment and assumptions are needed to bridge the gap between past experience and future prediction. In the Schaefer approach these assumptions are made directly about the dynamic properties of the stock and fishery as a whole, at the primary level of information, as it were; in the Beverton–Holt approach the assumptions are introduced at the secondary level of information and concern the factors responsible for the dynamics rather than the dynamics themselves. Certain assumptions are, however, common to both—notably that of a proportional relationship between fishing effort and the fishing mortality rate it causes. Indeed, for any proper understanding of the dynamics of a fishery it is essential that the characteristics of the fishing effort—the fishing power of the vessels and gear and their distribution relative to the fish population—should be thoroughly investigated so as to be able to measure fishing effort in units which are as closely proportional as possible to fishing mortality rate.

Certain practical difficulties arising in the treatment and interpretation of data are also present in both approaches, although their consequences may differ. The most important of these, in general, concerns the effect of fluctuations in recruitment which are uncorrelated with size of parent population. It has been mentioned earlier that these are often so great that they completely mask the underlying relation between stock and recruitment—whatever that may be. In the Beverton–Holt approach this means that the only possible way of treating recruitment in the theoretical model is to regard it as fluctuating about a mean which is independent of population size; this may not be the truth, but in such a case no more information can be extracted from the data available. The analytical model can still be used, nevertheless, for prediction on a "catch per recruit" basis, and for many purposes this is a useful procedure until such time as sufficient data have accumulated to enable the definitive relation between stock and recruitment to be established. Fluctuations in recruitment present similar difficulties in the Schaefer method; they cause a scatter to appear in a plot of catch against population size and this may make it difficult to fit any theoretical relation between them with any precision. Again, the conclusions which can be drawn are correspondingly uncertain. It follows, in fact, that if no relation between stock and recruitment can be detected for incorporation in a Beverton–Holt model, then an analysis of catch and effort data for the same fishery over the same period by the Schaefer method would be equally inconclusive as far as this particular relationship is concerned.

It would obviously be valuable to be able to compare the results of applying both the Beverton–Holt method and the Schaefer method to the same fishery, but at the present time there are few, if any, in which the data required for both methods are equally reliable. Such a comparison is now being made for the yellowfin tuna of the Eastern Tropical Pacific, for which the growth and natural mortality parameters have only recently been estimated, and will be reported upon in the near future. A comparison can also be made theoretically, however, and leads to certain important conclusions. It will be remembered that in the Schaefer method the coefficient of natural increase is taken as being a linear function of population biomass. This results in a paraboidal curve relating equilibrium catch to population size, with the maximum catch occurring at exactly half the maximum population size in the virgin state. Now a relationship between catch and population, in which different values of the latter are generated by changes in the amount of fishing, is also implied by the Beverton–Holt type of model. A formal mathematical treatment of this problem is not practicable, owing to the complex nature of the equations involved, but graphical solutions may readily be obtained from which certain general conclusions can be established. Thus it is found that the simple model with constant parameters comparable in magnitude to those estimated for plaice and haddock gives catch/population curves in which the maximum catch occurs at about one-third of the virgin population size. This shift of the maximum to the left compared with the parabola is the result partly of maintaining the recruitment constant despite the reduced size of the population and partly to the enhanced relative growth rate of the smaller and younger fish which predominate in the stock under these conditions compared with those in the virgin stock. These two factors combine to cause the rate of natural increase of the stock to be rather greater at low population levels compared with high ones than is implied by (17), so the maximum catch is obtained by fishing the population rather more heavily and down to a smaller fraction of its virgin size. If, however, the relation between stock and recruitment defined by (13) is incorporated into the model, combined with a density-dependent growth, the maxima of the curves are shifted to the right, and with a certain degree of dependence of recruitment on stock a very close replica of a parabola is obtained. With the most marked degree of change of recruitment with stock that is compatible with stability, the maximum of the curve occurs even higher, at about 60% of the virgin population size. The studies on which these findings are based are not exhaustive and, indeed, are still in progress, but they are sufficient to show that the simple parabola implied in the Schaefer approach is certainly not inconsistent with the properties of the more complex and, in a sense, more "realistic", of the Beverton–Holt models. Possibly, in some fisheries, recruitment may be sufficiently independent of stock for the Schaefer approach to tend to underestimate the extent to which the virgin stock should be reduced to obtain the maximum catch—the North Sea plaice may be a case in point—but that is a matter for future research.

# References

Baranov, F. I., 1918. On the question of the biological basis of fisheries. (In Russian.) *Izvest. nauch.-issl. Ikthiol. Inst.*, **1**, 81–128.

Bertalanffy, L. von, 1934. Untersuchungen über die Gesetzlichkeit des Wachstums. 1 Teil. Allgemeine Grundlage der Theorie; mathematische und physiologische Gesetzlichkeit des Wachstums bei Wassertieren. *Arch. Entw. Mech. Org.*, **131**, 613–52.

Bertalanffy, L. von, 1938. A quantitative theory of organic growth. (Inquiries on Growth Laws. II). *Hum. Biol.*, **10**, 181–213.

Beverton, R. J. H., 1953. Some observations on the principles of fishery regulation. *J. Cons. Explor. Mer*, **19**, 56–68.

Beverton, R. J. H., 1962. The long-term dynamics of certain North Sea fish populations. In *The Exploitation of Natural Animal Populations*. British Ecological Society, Symposium Number 2 (Edited by Le Cren, E.D. and Holdgate, M. W.) Oxford, 242–259.

Beverton, R. J. H. and S. J. Holt, 1956. A review of methods for estimating mortality rates in exploited fish populations, with special reference to sources of bias in catch sampling. *Rapp. Cons. Explor. Mer*, **140**, Pt. I, 67–83.

Beverton, R. J. H. and S. J. Holt, 1957. On the dynamics of exploited fish populations. *Fish. Invest., Lond.*, Ser. 2, **19**.

Beverton, R. J. H. and S. J. Holt, 1959. A review of the lifespan and mortality rates of fish in nature, and their relation to growth and other physiological characteristics. In Ciba Foundation, Colloquia on Ageing. *The Lifespan of Animals*, **5**, (edited by Wolstenholme, G. E. W. and O'Connor, M.) London, 142–177.

Clark, F. N. and J. C. Marr, 1955. Population dynamics of the Pacific sardine. *Prog. Rep. Calif. Coop. Oceanic Fish. Invest.*, 1953–55, 11–48.

Doi, T., 1951. A mathematical consideration on the analysis of annual yield of fish and its application to "Buri" (*Seriola quinqueradiata*). *Central Fish. Stat., Japan, Contrib.* 1948–49, No. 117.

Gause, G. F., 1934. *The Struggle for Existence*. Williams & Wilkins, Baltimore.

Graham, M., 1935. Modern theory of exploiting a fishery, and application to North Sea trawling. *J. Cons. Explor. Mer*, **10**, 263–274.

Graham, M., 1939. The sigmoid curve and the overfishing problem. *Rapp. Cons. Explor. Mer*, **110**, 15–20.

Hjort, J., G. Jahn and P. Ottestad, 1933. The optimum catch. Essays on population. *Hvalrad. Skr.*, No. 7, 92–127.

Hulme, H. R., R. J. H. Beverton and S. J. Holt, 1947. Population studies in fisheries biology. *Nature*, **159**, 714–715.

Lotka, A. J., 1925. *Elements of Physical Biology*. Williams & Wilkins, Baltimore.

Moran, P. A. P., 1950. Some remarks on animal population dynamics. *Biometrics*, **6**, 250–258.

Moran, P. A. P., 1954. The logic of the mathematical theory of animal populations. *J. Wildlife Mgmt.*, **18**, 60–66.

Pearl, R., 1930. *The Biology of Population Growth*. Knopf, New York.

Ricker, W. E., 1944. Further notes on fishing mortality and effort. *Copeia*, 1944, No. 1, 23–44.

Ricker, W. E., 1954. Stock and recruitment. *J. Fish. Res. Bd. Canada*, **11**, 559–623.

Ricker, W. E., 1958. Handbook of computations for biological statistics of fish populations. *Bull. Fish. Res. Bd. Canada*, No. 119.

Russell, E. S., 1931. Some theoretical considerations on the "Overfishing" Problem. *J. Cons. Explor. Mer*, **6**, 3–27.

Schaefer, M. B., 1954. Some aspects of the dynamics of populations important to the management of the commercial marine fisheries. *Bull. Inter-Amer. Trop. Tuna Comm.*, **1**, 26–56.

Schaefer, M. B., 1957. A study of the dynamics of the fishery for yellowfin tuna in the Eastern Tropical Pacific Ocean. *Bull. Inter-Amer. Trop. Tuna Comm.*, **2**, 245–285.

Thompson, W. F. and F. H. Bell, 1934. Biological statistics of the Pacific halibut fishery. (2) Effect of changes in intensity upon total yield and yield per unit of gear. *Rep. Intern. Fish. Comm.*, No. 8.

Wangersky, P. J. and W. J. Cunningham, 1957. Time lag in population models. *Cold Spring Harbor Symposia on Quantitative Biology*, **24**, 329–338.

Watt, K. E. F., 1956. The choice and solution of mathematical models for predicting and maximizing the yield of a fishery. *J. Fish. Res. Bd. Canada*, **13**, 613–645.

# V. OCEANOGRAPHICAL MISCELLANEA

## 22. SEASONAL CHANGES IN SEA-LEVEL

### June G. Pattullo

### 1. Introduction

The study of variations in sea-level has a surprisingly long history; papers on this subject were appearing more than 200 years ago. However, the examination of the *seasonal* variation requires at least a year's record of reliable observations, all referred to the same fixed point on land. Data for this purpose became adequate earliest in northern Europe but only towards the end of the 19th century. Soon a lively controversy was under way as to the causes of the variations observed, and in 1926 Gallé summarized the work of various European writers who, in the first decades of this century, had analyzed local data and had advanced hypotheses relating them to meteorological or oceanographic factors. Japanese students were early in the field as well, and were certainly among the first to give their principal attention to the seasonal term (Nagaoka, 1908; Omari, 1908). From 1900 to 1940 papers appeared at the rate of one or two a year. At this point the first world-wide collection of monthly mean-sea-level data appeared in print (I.A.P.O., 1940). As a result both new emphasis and new scope were given to the problem.

This chapter will summarize principally the results of the work during the years since 1940, with full recognition of the fact that earlier workers had already suggested all of the interrelationships between sea-level and other physical phenomena that we will mention. A few of the earlier papers are included in the references and these publications will lead the interested reader to other sources.

### 2. The Observed Seasonal Variations

The International Association of Physical Oceanography (I.A.P.O.) of the International Union of Geodesy and Geophysics had recognized the need for more uniformity and availability of sea-level data, and, in 1933, it appointed a Committee on Mean Sea-Level and its Variations. This committee undertook to compile and publish monthly and annual values of mean sea-level at gauges throughout the world. The first publication appeared in 1940 (I.A.P.O., 1940) and additional compilations have been published at intervals since then (I.A.P.O., 1950, 1953, 1958, 1959). The presentation of the data in such useful form made it possible to examine, for the first time, ocean-wide variations in sea-level. Two papers appeared almost simultaneously summarizing the

[*MS received July, 1960*]                485

large-scale features that could be detected (Lisitzin, 1955; Pattullo, Munk, Revelle and Strong, 1955).

Lisitzin and the American group agreed in their conclusions that long sections of the coastlines undergo similar seasonal fluctuations in sea-level, i.e. the variations are "coherent" over rather long distances. Furthermore, at the same latitudes on opposite sides of the oceans (and, as far as could be determined, on islands in midocean) the seasonal variations are similar. Lisitzin found evidence of four principal *zones* in both the Atlantic and Pacific Oceans; the Americans agreed as to three areas but did not identify the fourth. Within each zone all gauges showed similar variations, but there was a marked difference from zone to zone. From high northern latitudes to about 45°N, the maximum elevations were observed in winter (December); from 45°N to the equator maxima occurred earlier, usually in September; from the equator to about 45°S maxima occurred in March, and (Lisitzin only) south of 45°S the maximum was usually in June.

These results were convincing evidence that at least some of the factors causing the changes must be of large areal extent. Lisitzin offered the suggestion that changes in oceanic circulation could effect the observed results. Pattullo *et al.* proposed no explanation, but pointed out that (1) astronomical tides are apparently not large enough to be the principal cause, (2) atmospheric pressure effects at the gauges do not seem to be large enough either, (3) between 40°N and 40°S the observed changes bear considerable similarity to changes in level that can be deduced if only the volume, not the mass, of sea-water changes seasonally at any given point, and (4) outside the latitudinal band 40°N to 40°S changes in water volume are *not* large enough to account for observed changes (see also *steric levels*, Volume I, Chapter 16).

In an attempt to clarify the situation further, several dozen additional tide gauges were installed during the International Geophysical Year (I.G.Y.), 1957 and 1958, and operation of some of them was continued through the International Geophysical Cooperation (I.G.C.), 1959. The sites were chosen, whenever possible, at oceanic islands, with the explicit purpose of providing data of oceanic character to compare with the more numerous records already taken along coastlines. At the same time, temperature and salinity measurements were made, to a depth of a few hundred meters, in deep water offshore from each gauge. It was hoped that these collections would make possible a better comparison between changes in water volume and recorded changes in elevation of the sea surface.

The results of the tide-gauge measurements are summarized in Figs. 1–4. To keep the description as nearly synoptic as possible, I.G.Y. (and some I.G.C.) data have been used where available. Where adequate I.G.Y. data either were not collected or have not been published in time for inclusion in this volume, data collected in earlier years have been added, in order to complete the picture. These non-synoptic data are principally in the Southern Hemisphere; they include the Antarctic data, gauges along the west coast of South America and the east coast of Africa, and all gauges in Indonesia and New Zealand.

Both the distribution of gauges and the large variations in level with time and with space make it questionable whether contours of the sort given here are worthwhile. Our reasons for presenting the results in this way are (1) to make the large-scale, general features easy to comprehend, and (2) to indicate clearly the areas where deviations are particularly large or irregular.

The principal features shown on Fig. 1, the chart for March, are that sea-level is lower than its annual mean elevation almost everywhere in the Northern Hemisphere, but higher than average almost everywhere south of the equator. Appreciable exceptions in the Northern Hemisphere occur in the Arabian Sea, in the Gulf of Siam and between 40° and 60°N in both the Pacific and the Atlantic Oceans. In the Southern Hemisphere the only negative values of any

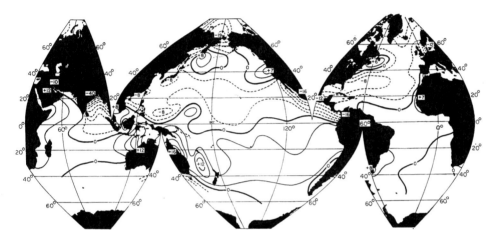

Fig. 1. Deviations from mean annual sea-level, averaged for the months February, March and April. Heavy lines, zero deviation; light solid lines, positive deviations (sea-level higher than the annual mean); dashed lines, negative deviations. Contour interval 3 cm.

magnitude occur along the south coast of Australia. The largest deviations observed anywhere in the world occur in the Bay of Bengal; deviations from the annual mean at some gauges there are − 40 cm. Gauges along the coast of Mexico and Central America also show large negative values, − 16 cm, as does the one on the northeast tip of Siberia, − 19 cm. The largest positive values observed anywhere occur around Australia, particularly on the northeast coast, + 16 cm. An interesting detail small in area but having large deviations is found in the Indian Ocean. The variations east of Sumatra are of different sign from those on the west. Deviations in the Bay of Bengal are negative while those in the Gulf of Siam are positive. This sharp reversal in this short distance can be found in all four seasons.

If we now compare this chart with Fig. 2, the chart for September, we find almost identical general features, except that, where deviations were positive in March, they are negative in September, and vice versa. Also, the locations of the maximum deviations are virtually the same—the Bay of Bengal has

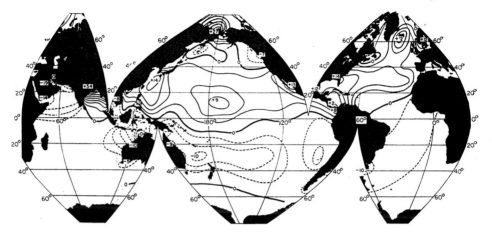

Fig. 2. August, September and October deviations. (See Fig. 1.)

deviations up to $+54$ cm; the west coast of Mexico, $+13$ cm; and north-eastern Siberia, $+27$ cm. Parts of the Australian coast and one offshore island have the relatively large negative deviation of $-10$ cm. Principal exceptions to this almost "mirror image" relationship between March and September are: (1) large positive values occur in September along the southeastern United States and at Iceland that are not equalled in magnitude by their negative counterparts in March, and (2) deviations in southern Australia are negative in both seasons.

The patterns for the other two quarters, however, are neither quite so simple nor quite so similar to each other.

In June (Fig. 3) most but not all of the gauges in the central parts of the oceans, both north and south of the equator, show negative deviations. Large positive deviations occur in three principal areas: along all of the northern

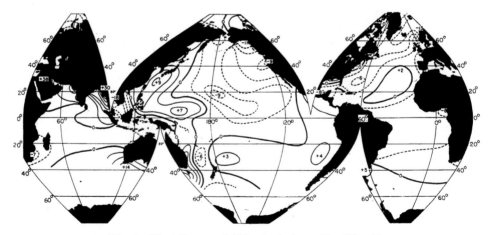

Fig. 3. May, June and July deviations. (See Fig. 1.)

Indian Ocean except the Gulf of Siam, in the western third of the Pacific both north and south of the equator, and around the southern half of Australia. Small positive deviations (between 0 and +5 cm) occur at gauges in the central South Pacific, along part of the eastern seaboard of the United States, and at several locations in the tropical and subtropical Atlantic as indicated by the zero contour there. The deviations of largest magnitude are: −18 cm in the Gulf of Siam, −13 cm along northern Norway, +30 cm in the Bay of Bengal and +14 cm along southern Australia.

To a first approximation, the December chart (Fig. 4) shows positive deviations where negative ones appear in June, and vice versa, but one would hardly like to call these charts mirror images of each other. Largest positive values in December are +20 cm in the Gulf of Siam and +16 cm along Norway; negative deviations reach −26 cm in the northern part of the Bay of Bengal and −10 cm

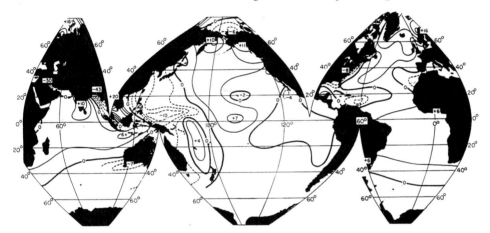

Fig. 4. November, December and January deviations. (See Fig. 1.)

in southern Australia. Noticeable discrepancies from any mirror-image relationship occur around India, on the Pacific coast of Central America, and, perhaps, in parts of the Central South Pacific. All three of these areas have positive deviations both in December and in June; the deviations are small, and therefore somewhat indeterminate, except around India.

During the I.G.Y. the Arctic Ocean was ringed for the first time with tide gauges, although the distribution was sparse and, of course, no gauges could be installed except attached to the land-masses around the edge of this body of water. Gauges located north of the Arctic Circle include ten in the northern part of Norway, plus the following:

| Murmansk | 68° 58′N | 33° 03′E |
|---|---|---|
| Russkaya Gavan | 76° 14′N | 62° 32′E |
| Cape Chelyuskin | 77° 43′N | 104° 17′E |
| Cape Schmidt | 68° 55′N | 179° 28′E |
| Point Barrow | 71° 20′N | 156° 46′W |
| Resolute | 74° 41′N | 94° 53′W |

We cannot, with such a distribution of data, be certain as to the behavior of the central portion of this ocean. Fig. 5a does illustrate, however, that all of these gauges show negative deviations in March, with the extreme value recorded at Cape Schmidt ($-19$ cm) as noted in the discussion of Fig. 1. Fig. 5b (September) contains no negative values within the Arctic proper, and

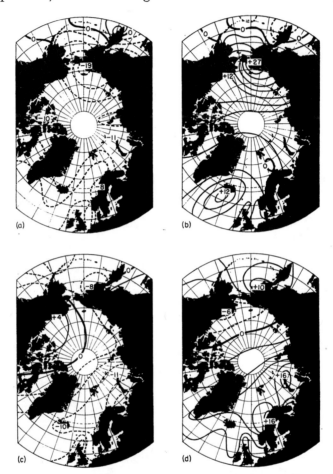

Fig. 5. Legend as for Fig. 1. (a) February, March, April; (b) August, September, October; (c) May, June, July; (d) November, December, January.

again the largest deviation, $+27$ cm, occurs at Cape Schmidt. Figs. 5c and 5d, for June and December respectively, are nearly but not quite mirror images of each other: the Siberian coast north of the Aleutians has small negative deviations in both seasons while a small portion of the Canadian archipelago apparently has positive values at both times. The largest deviations on these two charts are, for June, $-12$ cm along Norway and $-11$ cm at Russkaya Gavan; for December, $+16$ cm at these same locations.

## 3. Factors Influencing Variations in Sea-Level

Many explanations of variations in sea-level have been proposed; Rossiter has summarized them for all long-period changes in Volume I, Chapter 16.

There has been, unfortunately, some confusion in discussions of cause and effect in this field, due principally to the fact that meteorological and oceanographic phenomena are so inextricably interdependent. It appears more suitable at this point, instead of attempting to search out all "initial causes", to consider one by one the other physical phenomena to which changes in sea-level must somehow be related. This approach offers the advantages that these accompanying phenomena can in most cases be measured, although few have yet been studied in any detail, and one can hope that eventually the causes of each will be thoroughly explored. The following table lists the various phenomena and the maximum seasonal rise in sea-level that is known or estimated to accompany each.

### TABLE I

Seasonal Deviation in Sea-Level (cm) and Associated
Phenomena

|  |  | 0 | 10 | 20 |
|---|---|---|---|---|
| (a) | Fall in local atmospheric pressure (total air-mass over oceans constant) |  |  | x |
| (b) | Increase in heat content of water (mass of sea-water constant) |  |  | x |
| (c) | Decrease in salinity of water (mass of sea-water constant) |  |  | x |
| (d) | Increase in speed of onshore component of wind |  |  | ? |
| (e) | Increase in speed of longshore component of current (sign of effect depends on direction of flow) |  | ? |  |
| (f) | Approach of annual or semiannual high of astronomic tide | x |  |  |
| (g) | Increase in total mass of water in oceans | x |  |  |
| (h) | Decrease in mixing of water | ? |  |  |

The amplitudes given are estimated for the open ocean or exposed coastline; in restricted bodies of water, local conditions must be more completely considered. Where "x" appears in the column the estimates have been derived from the following sources: (a) Lisitzin and Pattullo (1961): (b), (c) and (f), Pattullo et al. (1955); and (g) van Hylckama (1956). Question marks represent the author's estimates as to the probable maximum amplitudes associated with the phenomena indicated.

If these estimates have any reality at all, it appears that the larger of the variations in sea-level with season are related to the first five phenomena listed. Terms (a), (b) and (c) considered together constitute what has been called the *isostatic* part of sea-level variations, because they are not accompanied by changes in the pressure field on the floor of the deep sea. Changes in air pressure are followed rather quickly by compensating changes in sea-surface elevation, so that the total mass of air plus water resting on the sea floor remains a constant (Groves, 1957). The effects of phenomena (b) and (c) may be computed either separately or as a combined effect by using measurements of temperature and salinity at various depths. The results are usually added together and called *steric levels* since they represent seasonal changes in level related to changes in the *volume* of the water only. Therefore, (a), (b) and (c), taken together, comprise all those changes and only those changes in sea-level that can occur with no net change of pressure anywhere on the sea floor, as stated above. Lisitzin and Pattullo (1961) have examined the average magnitude of the isostatic term in the Pacific Ocean, and have compared it with the observed variation in sea-level in the same ocean during the I.G.Y. Their results show that, over most of the open ocean, the recorded variations in sea-level and the isostatic variations agree well. Exceptions to this generality occur at certain places near coastlines and in the tropical western Pacific. The largest differences between recorded and isostatic levels occur along the mainland coast of Mexico, northeastern Australia, and, occasionally, the coast of North America near the border between Canada and the United States. The pressure and steric terms attain the same amplitudes, but *not* at the same localities. Pressure effects are a maximum between 40° and 60°N; steric levels show largest deviations between 40°N and 40°S.

The implications of this result are that the variation in total mass over any part of the deep-sea floor must be small (a finding of some interest to geophysicists), and that seasonal changes in the mass of water at a given location can be estimated with some reliability from atmospheric pressure variations alone.

This pressure effect is a rather clear-cut case where one can readily discuss cause and effect, since the changing forces accompanying changes in atmospheric pressure can be computed if the field of pressure is known. "Causes" of the steric deviations are not so easy to identify. For example, Pattullo (1957) has shown that about two-thirds of the seasonal change in heat content in the oceans of the Northern Hemisphere can be accounted for by local heating alone. However, many problems still remain unsolved as to the relationships between wind and thermal structure; one must consider at least the following: set-up (which may add more warm surface waters to the area), currents (permanent and transient), and mixing. That is, one can determine values for (b) quite readily, but its causes are not yet well understood.

The isostatic term is by no means equal to the observed deviations everywhere. The island areas of the tropical Pacific show marked, sometimes erratic, variations in both recorded and steric levels and the agreement between the two cannot be considered satisfactory (the pressure term is relatively unimportant at these latitudes). These are regions of complicated hydrography and topography; it appears that more years of observation and of detailed analyses will be necessary before we will understand precisely what is happening in these areas. The other principal areas of discrepancy in the Pacific Ocean are interesting in that they occur just where recorded deviations in sea-level have been noted to be large—especially along northeastern Australia and Mexico. Both of these are tropical to subtropical coastlines and, at least along sections of the shoreline, have shallow water offshore. The evidence suggests that, in these areas, non-isostatic effects related to phenomena $(d)$ or $(e)$ may be of considerable magnitude.

Oceans other than the Pacific have not yet been adequately analyzed for the isostatic term, although a little work on steric levels in some areas, and considerable work on pressure effects in others, has been done. At least part of the very large deviations observed in the northern Indian Ocean are steric in behavior, since changes in specific volume related to changes in salinity are large (LaFond and Rao, 1954). This is the only relatively open area identified to date where salinity changes are the predominant factor in a large steric term. It seems reasonable to conclude that the real *cause* here is the monsoon, and that the active agents are the precipitation that falls during the period of onshore winds and the monsoon winds themselves. However, a complete quantitative explanation has not yet been given.

A number of studies have been published where other effects in addition to changes in pressure and volume have been considered for a particular area. A complete list would require a very long bibliography; a few typical examples are given here. The large and complicated variations in Indonesian waters have been studied in detail by Wyrtki (in press) and he relates them almost entirely to changes induced by the monsoon. In this area this involves not only onshore and offshore wind stress and changes in precipitation and evaporation, but also all of the complications of varying currents carrying different water-masses through restricted passages into semi-enclosed, shallow basins. Barber (1957) found that recorded sea-levels at Prince Rupert, British Columbia, Canada, corresponded well to near-shore variations in isostatic levels, and he was, moreover, able to attribute the changes in water volume at that locality to variations in *water-mass* induced by the seasonally varying winds. Here the term *water-mass* is used in the customary way to indicate the types of water in the column, after the definition of Helland-Hansen (1916). Balay (1958) has studied the seasonal variations in level along Argentina. He notes that the recorded variations are quite similar for long distances along this coast, although all of the associated phenomena vary considerably with latitude. He has found that the third harmonic of the annual variation is consistently present, with an amplitude of 2.5 cm, and attributes this to the influence of the continental shelf on the combination of the annual and semi-annual wave. Chase (1952) has continued the work along the southeastern coast of the United States. He finds a high correlation between seasonal variation in sea slope between Charleston and Bermuda (and hence in sea-level at the gauges) and the strength of the winds in the North Atlantic gyre. A major part of the work on particular areas, however, continues to emanate from the countries of northern Europe. Lisitzin, in particular, has made a number of studies, principally in sections of the

Baltic Sea, and has pointed out that in this area water supply, wind stress, currents and sea-ice must all be considered (Lisitzin, 1946, 1957).

## 4. Summary

We have in this discussion considered the changes in sea-level from season to season; let us now summarize them by area.

### A. Subtropics

In subtropical latitudes, extreme deviations from the annual mean occur near the times of the equinoxes. Levels are high north of the equator in September and south of the equator in March and vice versa; ranges of 10 to 15 cm are typical. This variation is due principally to seasonal variations in heat storage in the oceans, induced, to a large extent, by the seasonal variations in local heating.

### B. Tropics

In the tropics, deviations are largest at the times of the solstices, but whether the level is lower or higher than the annual mean depends on the longitude. The central parts of the oceans experience high sea-levels in December, while closer to the coastlines levels are low. In June the deviations are similarly distributed but of opposite sign. As in the subtropics, the deviations at some of the island locations seem related to variations in heat content. However, even in those localities where this relationship is observed, it has not yet been determined whether local heating or advective effects are the predominant influence.

### C. Subarctic

North of 40°N in the Atlantic and Pacific oceans, subarctic localities have highest levels most frequently in December and January; the magnitude of the annual range is, again, about 15 cm. These deviations are induced almost completely (except for some local coastal effects) by the seasonal variation in atmospheric pressure. Corresponding variations of opposite sign in other latitudes, which must obviously take place, are not as readily detected because the effect is not concentrated in an equally small area.

### D. Arctic Ocean

Gauges in the Arctic Ocean all show high levels in September and low levels in March. During December levels are high, except along Alaska. The same pattern, with signs reversed, is observed in June. No observations are available except for the coastal stations, and no explanation of the variations has been given.

## E. Regional Effects

Regional effects are superimposed on all of these general variations, and, indeed, at times are the largest and most interesting features observed. Monsoon effects are large and complicated around all of the coastlines of southeast Asia from Arabia to Japan. In these waters, precipitation, wind and advection of different water-masses throughout the year are all important. Local effects of similar types must be considerable around the tropical coasts of Mexico and Australia, and have been shown to be of great influence on the shores of northern Europe.

## F. Isostasy

Finally, for most of the deep-water areas of the open ocean, the seasonal variation in sea-level is part of an isostatic variation in the distribution of mass in the hydrosphere and troposphere. That is, the total mass of air plus water above a point on the sea floor is very nearly constant throughout the year. However, the fraction of the mass that is water (as opposed to air) and also the specific volume of the water change with season, and both effects are reflected in changes in elevation of the air–sea interface. Along most coastal areas, non-isostatic changes of relatively local origin must be appreciable and obviously cannot be summarized here. The climatology of each such region must be examined in detail before a complete discussion can be given.

## 5. Conclusions

Although many able workers have devoted themselves to the study of seasonal variations in sea-level, only a few facts can be considered definitely established. The principal features of the large-scale variations are now evident, and about a half dozen local areas have been thoroughly studied. Much work has been done in correlating changes in sea-level to changes in meteorological phenomena, in particular to variations in atmospheric pressure and in local winds. A little is known about oceanographic phenomena such as the variation in specific volume of the water throughout the year. On the other hand, relatively little has been accomplished as yet in the study of seasonal variations in currents, or in distinguishing between the barotropic and baroclinic portions of these. Only a few workers have been able to make any quantitative estimate as to the contribution of advection to the observed changes in local sea-water characteristics which result in what we have called *steric levels*. Observations of set-up in response to winds have been, perhaps necessarily, confined to enclosed basins. It is to be hoped that the exchange of data and ideas initiated in 1933 by the International Association of Physical Oceanography, and the added impetus provided by the opportunities of the International Geophysical Year, will lead to fruitful work on some of these problems.

# References

Balay, M., 1958. Variations saisonnières du niveau moyen de la Mer Argentine. *Rev. Hydrog. Intern.*, 32 pp.

Barber, F., 1957. The effect of the prevailing winds on the inshore water masses of the Hecate Strait region, B.C. *J. Fish. Res. Bd. Canada*, 14, 945–952.

Chase, J., 1952. A comparison of certain wind, tide gauge and current data. *Tech. Rep. Woods Hole Oceanog. Inst.*, Ref. No. 52–79.

Gallé, P., 1926. On the relation between departures from normal in the strength of the trade-winds of the Atlantic Ocean and those in water level and temperature in the northern European seas. *Koninkl. Ned. Akad. Wetenschap., Proc. Sec. Sci.*, 17, 1147–1158.

Groves, G., 1957. Day to day variation of sea level. *Meteor. Monog.*, 2 (10), 32–45.

Helland-Hansen, B., 1916. Nogen hydrografiske metoder. *Skand. Nat. möte*, Kristiania, Oslo.

Hylckama, T. van, 1956. The water balance of the earth. Drexel Institute of Technology, Laboratory of Climatology, Pub. in *Climatology IX* (2), 117 pp Centerton, New Jersey.

I.A.P.O., 1940. Monthly and annual mean heights of sea-level, up to and including the year 1936. Pub. Sci. No. 5.

I.A.P.O., 1950. Monthly and annual mean heights of sea-level, 1937 to 1946, and unpublished data for earlier years. Pub. Sci. No. 10.

I.A.P.O., 1953. Monthly and annual mean heights of sea-level, 1947 to 1951, and unpublished data for earlier years. Pub. Sci. No. 12.

I.A.P.O., 1958. Monthly and annual mean heights of sea-level, 1952 to 1956, and unpublished data for earlier years. Pub. Sci. No. 19.

I.A.P.O., 1959. Monthly and annual mean heights of sea-level for the period of the International Geophysical Year (1957 to 1958), and unpublished data for earlier years. Pub. Sci. No. 20.

LaFond, E., 1939. Variations of sea level on the Pacific coast of the United States. *J. Mar. Res.*, 2, 17–29.

LaFond, E. and R. Prasado Rao, 1954. Changes in sea level at Visakhapatnam on the east coast of India. *Mem. Andhra Univ., Oceanog.*, 1. Waltair, India.

Lisitzin, E., 1946. The relations between wind, current and water level in the Gulf of Bothnia. *Soc. Sci. Fennica, Comment. Phys-Math.*, 13 (6), 37 pp.

Lisitzin, E., 1955. Les variations annuelles du niveau des océans. Bulletin d'Information, Comité d'Océanographie et d'Etudes des Côtes, No. 6.

Lisitzin, E., 1957. On the reducing influence of sea ice on the piling-up of water due to wind stress. *Soc. Sci. Fennica, Comment. Phys-Math.*, 20 (7), 11 pp.

Lisitzin, E. and J. Pattullo, 1961. The principal factors influencing the seasonal oscillation of sea level. *J. Geophys. Res.*, 66, 845–852.

Nagaoka, H., 1908. Apparent seasonal variation of sea-level. *Proc. Tokyo Math. Phys. Soc.*, 4, 382–385.

Omari, F., 1908. On the annual variation of the height of sea-level along Japanese coasts, 2nd Paper. *Bull. Earthquake Invest. Comm.*, 2, 35–50. Tokyo.

Pattullo, J., 1957. The seasonal heat budget of the oceans. Doctoral dissertation, University of California, Los Angeles.

Pattullo, J., W. Munk, R. Revelle and E. Strong, 1955. The seasonal oscillation in sea level. *J. Mar. Res.*, 14, 88–155.

Wyrtki, K., in press. Physical oceanography of southeast Asian waters.

# 23. BATHYSCAPHS AND OTHER DEEP SUBMERSIBLES FOR OCEANOGRAPHIC RESEARCH

R. S. Dietz

## 1. Introduction

There are many types of devices and submersibles for penetrating the sea's third dimension. In order to circumscribe the subject for the purpose of this chapter, the writer proposes to consider here only those capable of descending to considerable depths and designed primarily for oceanographic research. This, at the outset, eliminates from mention military submarines, for example, although considerable oceanography has been accomplished by some of these. The arctic under-ice voyages of the S.S.N. *Nautilus* and S.S.N. *Skate* provide remarkable examples.

The relative word "deep" can, of course, be construed in many ways as applied to submersibles. In its shallowest sense, it could refer to the working depths attained by unprotected divers. But to study deep mid-water life *in situ* one must descend below the "twilight zone"—or roughly below 800 m—where animals show strong photic response. To study the sea floor deeper than the fringing continental shelves (7% of the sea bed), a plunge of only 120 m is required; but only another 15% of the sea floor lies between the shelves and 3000 m. The average depth of the sea is 3800 m. For generally exploring the abyssal ocean floor (outside of the hadal trench depths which comprise only 1% of the sea floor between 6000 m and 11,000 m) one needs a craft with a 6000-m capability.

Marine scientists recognize that the classical methods of data collection with instruments lowered on ropes and wires fall far short of the needs of modern oceanography. In the past decade, geophysical methods using acoustics, magnetics, gravity, etc. have revolutionized concepts of the ocean. In all likelihood, the next break-through will be ushered in by manned invasion of the sea with deep-research vehicles. Prototypes have already been successful and the development of sophisticated crafts is gaining impetus.

The oceanographer must use his senses in combination with understanding and intuition to unravel the nature of the deep ocean. As well as to measure, he needs to touch, smell and hear; and most of all *to see*. This he can do only bluntly through any remote extension of his senses into the abyss by instruments. Devices and tools lowered on long wires often have discrimination and precision but they cannot integrate the overall environment. Close-at-hand seeing is especially vital in the largely observational sciences of biology and geology; less so in marine chemistry and physics. From a deep ship, the ecologist can survey the animals of mid-water space and study their behavioral patterns. The biologist's use of nets from ships has been aptly compared to a blind man making a butterfly collection. In a classic experiment, Helland-Hansen exposed a photographic plate at the end of a 1000-m wire for 80 min; he found blackening. This finding was duly reported and textbooks for several

[*MS received September, 1960*]

decades recorded that appreciable sunlight apparently penetrated to that depth. But had this experiment been made under direct observation from a bathyscaph, it would have been realized that it was doubtless vitiated by the flashes and scintillations of luminescent zooplankton.

One can list several categories in which any deep-research craft may be especially useful. First and most important, it allows direct visual inspection of the mid-water environment and the sea floor. Secondly, such a craft potentially enables selective sampling under visual control instead of remotely, blindly and randomly. The oceanographer's concept of the general environment in which he is sampling and of how his sampler is operating may be far from reality. A third use is as a deep, quiet and stable platform for the scientist and his instruments. The chief value here lies in experiments needing elaborate electronic devices which cannot be readily self-contained, so that long electric cables must be used for deep measurements. Lowering a 3-mile-long multi-conductor cable from a surface ship is such a formidable task that few such experiments have ever been conducted. Fourthly, many geophysical measurements such as magnetic surveys, sub-bottom acoustic penetration, etc. can best be conducted in close proximity to the sea bed.

## 2. Diving Chambers

Historically, diving chambers have provided oceanographers with their first means for penetrating the depths; hence brief mention of them is made here. Although such observation chambers are commonly employed for salvage, etc., they have never enjoyed wide oceanographic use—or at least few scientific publications have resulted. The Barton–Beebe bathysphere (Fig. 1), the French

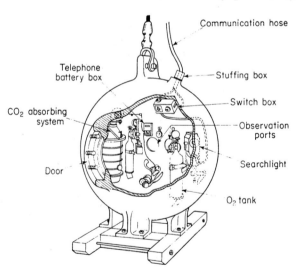

Fig. 1. A cutaway drawing of the Barton–Beebe bathysphere used for diving off Bermuda to 3000 ft in the early thirties. (Redrawn from Beebe, 1935.)

Galeazzi diving chamber and the Japanese *Kuroshio* are exceptions. Beebe predicted the day would soon arrive when there would be scores of bathyspheres. But, to date, the only direct lineal descendent has been Barton's benthoscope. In it he briefly dived off California in 1948 attaining a maximum depth of 4500 ft.

The bathysphere used in the early thirties was a 54-in. i.d. steel sphere weighing 2.5 tons in air. Roughly 30 dives were made off Bermuda to a maximum depth of 3028 ft. Details of the scientific results and of the bathysphere's construction are recorded in Beebe's (1935) book, *Half-Mile Down*. Beebe reported a plethora of fishes and other sea life far in excess of that seen by any subsequent observers; unfortunately no photographic confirmation of his observations was obtained. Access to the sphere for two observers was made through a heavy steel door; a view of the sea through two ports of fused quartz. The sphere was lowered by a 1-in. wire rope and a power cable supplied lighting and telephone communication. Watertight integrity proved troublesome; leaks through the door, the cable lead-through and the windows occurred from time to time but without serious results.

More recently, the Japanese constructed the diving chamber *Kuroshio* to explore the waters around Japan for oceanographic and fishery research (Fig. 2). Using the *Oshoro Maru* as the mother-ship, the *Kuroshio* completed 380 dives to a maximum depth of 205 m between 1951 and 1957. Inoue (1959) mentions 21 papers which indicate the wide variety of studies made. These researches emphasize especially fish behavior, plankton ecology, bioluminescence and the operation of fish nets, trawls and various types of traps. Studies of light penetration, sea-floor geology and suspended matter were also accomplished.

As demonstrated to the writer in a descent to 120 m off Ito not far from Tokyo in 1953, the standard technique is to lower the chamber to within about 3 m off the bottom. The chamber is then permitted to drift along with the surface ship; constant height over the bottom is maintained through telephonic communications. The sea floor is well illuminated with several floodlights; minute animals of the mid-water, even as small as copepods, could be observed by looking directly down a strong beam of light.

In 1959, plans were under consideration for reconstructing the *Kuroshio* into a tethered submarine craft which would cruise within the limits of an "umbilical cord" supplying propulsion power from a small mother ship. Returning to the surface, the craft would be towed rather than hoisted aboard. Some of the basic specifications proposed were: height 3 m, length 7 m, diameter of cylindrical chamber 1.48 m, length of manned cylinder 5 m, gross weight 9 tons, diving depth 200 m, complement 4 persons, endurance 10 hours, cable diameter 30 mm, underwater velocity 2.5 knots, 10 windows, 3 peepholes and 7 floodlights. There would also be various communications equipment for sounding devices, scientific apparatus, etc. If ever realized, this craft will be a novel and interesting addition to the world fleet of underwater research vehicles.

For many years, French marine scientists (e.g. Trégouboff, 1955, 1956) have made use of the Galeazzi diving chamber aboard the research vessel *Élié Monnier* descending as deep as 300 m. The chamber is, of itself, buoyant and sinks because of an attached ballasting weight which can be jettisoned if the cable breaks. There is constant telephonic communication with the surface ship. The quiet clear waters of the Mediterranean make this region ideal for visual undersea observations.

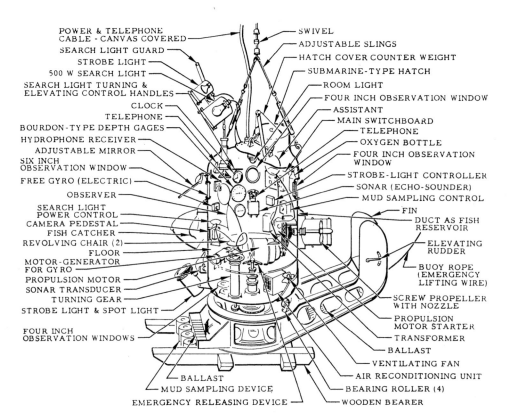

Fig. 2. Outside view of the Japanese oceanographic research diving chamber, *Kuroshio*, completed in 1952 and subsequently extensively used in Japanese shelf waters to 200 m. (By courtesy of Richard Terry, Ph.D.)

## 3. Bathyscaphs

Since the completion of the prototype bathyscaph FNRS-2 in 1948, bathyscaphs have remained the only operational untethered deep ships, so that somewhat fuller treatment is accorded them here. The FNRS-2 of A. Piccard successfully completed an unmanned test dive to 1380 m off Dakar in 1948, but waves caused damage to her float preventing further operations. Acquired by the French Navy, the cabin of the FNRS-2 was subsequently incorporated into the new bathyscaph, the FNRS-3. The submersible has subsequently operated in the French Mediterranean, off North Africa, off Portugal and off Japan, captained by Houot. In 1954 the FNRS-3 attained its maximum depth of 4050 m in the Atlantic off Dakar. In principle, the FNRS-3 is similar to the *Trieste* which is discussed at some length below. For fuller details regarding the FNRS-3, the reader is referred to Houot and Willm (1955). Numerous papers, chiefly biological and especially by Pérès, have appeared (Bernard, 1955; Pérès and Picard, 1956; Pérès, Picard and Ruivo, 1957; Pérès, 1958, 1959; Trégouboff, 1956, 1958).

A new French bathyscaph is reported to be under active construction by the French Navy at Toulon which will be capable of dives to 11,000 m. It will be identical in principle to the FNRS-3. The new sphere will measure 2.1 m

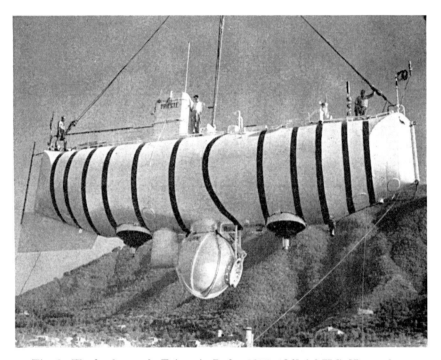

Fig. 3. The bathyscaph *Trieste* in Italy, 1957. (Official U.S. Navy photo.)

inside diameter and the shell will be 15 cm thick as contrasted with 2.0 m and 9 cm for the FNRS-3. It will be forged of a special nickel–chromium–molybdenum steel; the FNRS-3 was cast. The sphere will have a safety factor of 2 at 11,000 m; that is, when under a static pressure of 1155 kg/cm$^2$, it will be at only half of its elastic limit. The reinforcement that exists in the FNRS-3 sphere in the vicinity of the port and the door will be avoided since strains in these zones are not easily calculated; the sphere of the new bathyscaph will have a constant thickness. The ports will be of very small dimensions so that observations will have to be made through them with the assistance of a special optical system, permitting a reproduction of the exterior visual field. The displacement will be at least double that of the FNRS-3. The storage batteries will be four to eight times as powerful so as to permit considerable horizontal propulsion. It is expected that the new bathyscaph will be launched in early 1961 followed by test dives in the vicinity of Toulon (see p. 514). Subsequently, it will be converted into an efficient oceanographic research craft for scientific diving, including descents to maximum oceanic depths.

The *Trieste* (Figs. 3–5), built by A. Piccard and operated by J. Piccard, was completed in Italy in 1953. Twenty-two dives were completed between that date and 1956 in the Tyrhennian Sea to a maximum depth of 3700 m. Twenty-six more dives were made off Italy under the aegis of the U.S. Navy's Office of Naval Research in 1957. Subsequently the craft was transferred to the U.S.

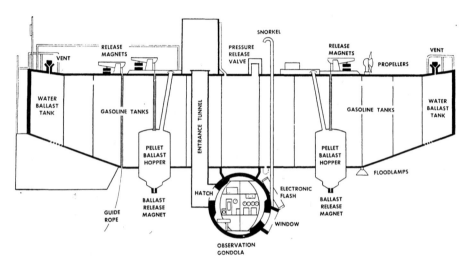

Fig. 4. Longitudinal section of the bathyscaph *Trieste*. (Official U.S. Navy photo.)

Navy operation in the Pacific off San Diego, California, and then to Guam where ultra-deep dives in the hadal depths of the Mariana Trench were undertaken. These abyssal plunges attained depths of 5530 m, 5852 m, 7025 m and to the nadir of the earth with J. Piccard and D. Walsh at 10,910 m (35,800 ft) in the Challenger Deep on 23 January, 1960. As a non-conventional and highly experimental craft the scientific results have been modest (e.g. Piccard and Dietz, 1957; Dietz, Lewis and Rechnitzer, 1958; Botteron, 1958; Dietz, 1959; Rechnitzer, 1959; Mackenzie, 1960; Piccard and Dietz, 1961). In contrast to the largely biological program of the FNRS-3, research from the *Trieste* has encompassed gravity, acoustics, marine geology, as well as biology.

Since the completion of the FNRS-2, bathyscaphs remain the only deep ships which have been fully realized so that a brief description of principle and construction of the *Trieste* is given here. For a detailed description of the *Trieste*, the reader is referred to A. Piccard (1956).

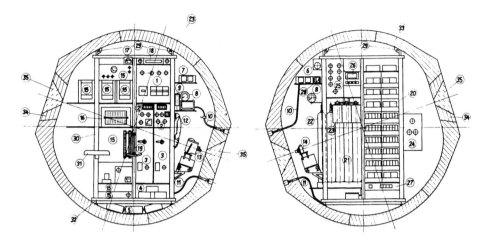

Fig. 5. Diagram of the interior arrangement of the bathyscaph *Trieste*'s Krupp sphere showing starboard and port views. (Drawing by Ateliers de Constructions Mécanique de Vevey, Switz.)

1. Control panel for lights.
2. Control panel for the ballast and miscellaneous equipment.
3. Control panel for the propulsion motors.
4. Diverse equipments.
5. Reserve alkali for absorption of carbon dioxide.
6. Chronometer.
7. Instrument for recording the amount of ballast expended.
8. Pressure gauges.
9. Instrument to measure the carbon dioxide level inside the sphere.
10. High-pressure tube for connecting the pressure gauges with the outside water.
11. Leadthrough for electric cables entering the cabin from the outside.
12. Telephone.
13. Motion picture camera.
14. Still camera.
15. Various oceanographic instruments.
16. Multi-channel recorder.
17. Water-current meter.
18. Electric thermometer.
19. Pendulum to indicate any departure of the cabin from a level position.
20. Battery rack and silver–zinc batteries.
21. Bottles of compressed air for blowing the entrance chamber.
22. Oxygen for breathing.
23. Alkali for carbon-dioxide absorption.
24. Electric control panel.
25. Acoustic telephone (wireless).
26. Ultrasonic echo-sounder.
27. Various electronic oceanographic equipment.
28. Tachometer for measuring vertical speed of ascent or descent.
29. Indirect interior lighting.
30. Supports of the cabin's instrument rack.
31. Seat.
32. Flooring of the cabin.
33. Shell of the sphere.
34. Entrance door.
35. Windows.

### A. *Trieste's Cabin*

The crew of the *Trieste* is encapsulated in a spherical watertight chamber. This cabin or gondola is forged of steel alloyed with nickel, chromium and molybdenum. Two spheres have been constructed for the *Trieste*, both with an exterior diameter of 2.18 m. The first was forged in two hemispherical sections by Terni in Italy in 1952. The second was forged especially for "Project Nekton" by Krupp at Essen, Germany, in 1958. The Terni sphere weighs 10 metric tons in air and 5 tons in water; it is 9 cm in thickness, reinforced to 15 cm around the portholes. The Krupp sphere weighs 13 tons in air and 8 tons in water; it is 12 cm thick, reinforced to 18 cm around the portholes.

Watertightness against the great deep-sea pressure is effected simply by the careful machining of the bearing surfaces so that they fit precisely, giving a metal-to-metal seal. The low-pressure seal for the Terni sphere is effected by an external band and the two hemispheres are gripped together by clamps. The three pieces of the Krupp sphere were originally bonded together with epoxy resin but this glue subsequently failed on surfacing from a dive to 18,100 ft (Dive 61). The sphere was then held together by six metal bands gripping two metal rings.

For viewing the ocean, there are portholes fore and aft. The axis of these windows, and the sphere as well, is inclined 18° to enable an easy view of the sea floor through the forward window. The windows are made of "Plexiglas" which has the advantage of yielding without brittle fracturing as does glass or fused quartz. Strength and optical considerations dictated that the plastic blanks should have the form of a frustrum of a cone with an approximately 90° apical angle. For the Terni sphere, the window dimensions were as follows: 15 cm thick, 40 cm wide on the outside and 10 cm wide on the interior. The window is set directly against a metal seat machined into the sphere; a little lubricant is smeared along the interface. These simple windows have never leaked a drop of water nor otherwise failed.

The aft porthole is seated in the middle of a heavy vault-like door which has the shape also of the frustrum of a cone. The inside diameter of the door is 43 cm for the Terni sphere, permitting a man to slide easily into the sphere from an antechamber. The heavy door is carefully balanced by springs to allow easy opening and closing. A single small screw is sufficient to hold it in place when the external pressure is slight. This door has never leaked or jammed.

Twelve holes are drilled in the wall of the sphere around the forward porthole. Eight of these are for passage of electrical cables; two are used for snorkels for emergency ventilation on the surface in the event of the passengers being unable to escape from the cabin after a dive. Two other lead-throughs are for high-pressure tubing for pressure gauges and compressed air to blow out the entrance chamber. The stouter Krupp sphere is very similar to the original one. The following differences are noteworthy; an inside diameter of 1.94 m; a weight of 13 tons; inside window diameter of 6 cm; inside door diameter of 40 cm and outside diameter of 57.3 cm.

### B. Gasoline-Filled Float

Since the steel sphere is non-buoyant by several tons, a float filled with gasoline is needed for buoyancy. This float is of thin sheet steel only thick enough to withstand wave buffeting and the rigors of towing. Sea-water is permitted to enter through a hole on the bottom of the envelope so that the gasoline is in free communication with the surrounding sea. The pressure is thus always equalized inside and out. Gasoline, being lighter than and immiscible with water, floats on the top.

Originally the float was divided into fourteen compartments. The two end compartments were air tanks which were flooded for diving. The remaining twelve were gasoline compartments, containing 28,000 gallons (106 m³). For "Project Nekton" the total volume of the float was increased by 6300 gallons. At the same time, the two inside partitions were removed leaving 10 gasoline compartments. This provided an increase in buoyancy to offset the greater weight of the new sphere and the additional ballast needed for ultra-deep diving. Originally 50 ft long, this modification increased the float length to 58 ft. The weight of the modified float when empty is 16 metric tons; the diameter 11.5 ft.

### C. Ballast for Buoyancy Control

The bathyscaph *Trieste* is an up-and-down craft with very limited horizontal maneuverability. Vertical traverses are effected by buoyancy control with jettisonable ballast. First of all, to provide the necessary negative buoyancy for diving, water ballast (fixed liquid ballast) is flooded into the two end compartments and into the entrance shaft or *sas*. Then additional negative buoyancy, which can be finely adjusted by the pilot, is effected by releasing gasoline (jettisonable liquid ballast) from the central maneuvering compartment. This compartment has a capacity of 4 m³ or about 1150 gallons with a total lifting ability of 1.4 tons.

Positive buoyancy is obtained by the release of iron pellets (jettisonable solid ballast) like BB shot—the type used industrially for sand blasting. The shot is held in two large tubs which have a funnel-shaped construction at the bottom. Once the mechanical keeper is removed for diving, the shot is held in place by the magnetic field induced by an electromagnet. When the magnet is energized, the shot is frozen into a solid plug. When the current is cut, it falls freely through the "electrochute" like sand through an hourglass. The current in normal operation is manually cut by the pilot. In the event of power failure, the ballast is jettisoned automatically, returning the *Trieste* to the surface. Since the ballast must be a magnetic material, iron ballast is used in place of lead. As originally constructed, the ballast tubs held 9 tons of ballast weight in air. The capacity was increased to 16 tons for the ultra-deep "Project Nekton" dives.

The guide rope and the ballast tubs themselves (fixed solid ballast) are attached to the bathyscaph by means of electromagnets, so that they can be jettisoned at will or in emergency. If, for example, the guide rope becomes fouled on the bottom, it can be cast off; the ballast tubs or silos themselves can be jettisoned if the shot funnel should become plugged for any reason.

Inherently the bathyscaph is dynamically unstable since gasoline is more compressible than water. For example, if the surface temperature is 30°C and at a depth of 35,800 ft the temperature is 0°C, then the water is compressed 5.3% and normal aviation gasoline is compressed 13%. The differential compressibility is then roughly 8%. Therefore, once descending, a bathyscaph will fall increasingly faster; the ascent is correspondingly accelerated. To offset this increase in the weight of the bathyscaph, about one ton of ballast must be jettisoned for each thousand meters. However, the temperature of the gasoline is also important, especially as this undergoes compressional heating during descent and expansional cooling during ascent. Hence the rate of descent is an important factor for establishing the amount of deballasting needed. If the gasoline temperature stays above the ambient sea-water temperature, less ballast is expended.

## D. Exterior Devices and Arrangements

A conning tower in the middle of the float protects the airlock hatch and the crew while topside. Stabilizing keels are present to prevent excessive rolling and are placed along the interior underlines of the gasoline compartments. This is in contrast to the usual exterior position of bilge keels. These internal keels are very effective since they act against the gasoline rather than against the rolling seas. Searchlights present a special problem since a pressure-proof metal case for a normal incandescent light would weigh several hundred pounds. This lighting problem is solved by utilizing pencil-sized 500-V mercury-vapor lights. The strong fused-quartz envelope of this light is directly exposed to the ambient sea pressure. Although originally designed to be cooled by a water stream and for operating under zero pressure, these lights provide the *Trieste* with excellent floodlights. In addition, there are two small normal incandescent lights protected in steel cases—one outside the forward window and the second inside the antechamber.

Propellers, driven by electric motors of 2 h.p. each, enable the craft to move sluggishly forward, backward and to turn horizontally. These motors are immersed in a bath of oil which is exposed to the ambient pressure of the sea. A maximum speed of about one knot can be attained for a short distance.

On the conning tower are several important instruments for navigational and scientific use. These include a tachometer, an externally sensitive vertical and horizontal current-meter, and a transducer for the underwater telephone. To accommodate the needs of oceanographic research, the *Trieste* was designed with sufficient buoyancy to accommodate 2.5 tons of scientific equipment.

## E. Interior Arrangement of the Sphere

The interior diameter of the original sphere was 6.7 ft; this is reduced to 6.4 ft for the Krupp sphere. About half of the space is occupied by equipment so that the habitable space is severely restricted permitting a crew of only two. Dive 65 to the Challenger Deep lasted 9 h but it apparently placed no severe strain on the divers. The ultimate length of stay is posed only by the ability of the divers to stay awake—and for safety reasons the need to complete a dive during the daylight portions of a single day. If circumstances dictated, it would be possible to remain submerged overnight.

The forward and aft portions of the sphere are kept clear for the door and the viewing ports. The rest of the interior is surrounded by tiers of racks for operational and scientific equipment. Located on the starboard side is an underwater telephone, the echo-sounder, the Draegger canisters for carbon dioxide absorption, two oxygen bottles, and the compressed air for emptying the entrance tube. The air regeneration is accomplished in the following manner: the air in the cabin is circulated through a filter with soda lime which extracts carbon dioxide, preventing it from building up to poisonous levels; new oxygen is bled into the cabin from cylinders of compressed oxygen; excess moisture is absorbed by silica gel. Gas-masks are carried along in the event of noxious gases being generated by electrical shorts, etc.

Near the door on the starboard side is a rack of silver–zinc batteries for supplying all the current, there being no external batteries as yet, although the installation of these is contemplated. Each important circuit has a separate set of batteries. There are two sets for the 12-V ballast circuits; there is one set of 24-V for miscellaneous small equipments. Circuits of 250 V operate each propulsion motor. These two sets can be placed in series to provide a 500-V circuit for the mercury-vapor lights. There is also a 28-V circuit for special scientific equipment. The total capacity of these batteries is 20 kWh.

On the port side are located various controls such as ballast control rheostats, motor switches, light switches, magnet controls, etc. Ciné and still cameras are installed which can be moved in position in front of the window along the guide bar. There are also timers for controlling deballasting, etc.

The foregoing description of the *Trieste* applies as of mid 1960; she is presently undergoing extensive modifications.


## 4. Diving Saucer

The Cousteau "diving saucer" (Cousteau, 1960) is a new submersible craft with great promise for future oceanographic research (Fig. 6). It was built by Jean Mollard under the direction of J.-Y. Cousteau at Marseilles, for use from the small oceanographic ship, *Calypso*. In early 1960, it made its debut—a 4-h dive to 300 m, its maximum operational depth. This craft was built to fill the need for a small and agile sea craft approaching the maneuverability and dexterity of the scuba diver but not limited to his 60-m working depth. Another specification was that it be small enough to be lifted aboard the *Calypso* by a hydraulic crane and stowed below deck.

The pressure hull is a flattened or oblate sphere of 2-cm thick steel, spun in two halves and then welded together. The manned capsule accomodates two persons lying prone, the pilot and a scientific observer. A second fiberglass fairing, open to the sea, surrounds the equator of the cabin forming an inner and outer hull for the power assemblies, etc. The craft is 2 m in diameter and 1.5 m thick; the total weight in air is 3500 kg.

Power is provided by externally mounted lead–acid batteries operating hydrojets forced at high speeds through nozzles instead of propellers. The jets are said to allow great maneuverability and speeds up to 1.5 knots. The craft is ballasted with 250 kg of water and iron blocks. One of the two iron weights is dropped off routinely; the other in an emergency. Trim is maintained by mercury which can be pumped to various reservoirs. The weight of the diving payload must be carefully controlled; diving is commenced with a negative buoyancy of 20 kg. Both still and movie cameras are installed; mercury-vapor lamps provide external lighting. A hydraulic three-fingered prehensile arm

Fig. 6. Cousteau's "diving saucer" for plunges to 300 m. (*Nat. Geog. Mag.* photo.)

permits the collection of specimens. Emergency escape is possible by donning scuba (self-contained underwater breathing apparatus) and pressurizing the hull to the ambient sea pressure so that the top hatch can be opened.

The present "diving saucer" is limited to 300 m with a tested safety factor of three. It is thus confined to the continental shelf and the upper portions of the continental slope. Construction is underway of a new model with a spherical cabin of the same volume as the prototype. This will extend the depth range to 1000 m. More efficient hydrojets will be used pumping a greater volume of water at a lower pressure than at present.

## 5. Buoyant-Hulled Deep Ships or Bathynauts

Although the existing bathyscaphs have made a remarkable contribution toward direct exploration of the deep sea, it is evident that they are somewhat

obsolescent in terms of the capabilities of modern technology. Their deep draft
and blimp shape make them difficult to tow, moor or lift aboard a mother ship.
They primarily have the capability of sinking and rising vertically with very
limited horizontal maneuverability. These disadvantages can, of course, be
partially overcome by a more sophisticated design that would add more
propulsion power and retract the cabin into the float to form a single spindle-
shaped body. But it seems even more likely that the future trend in deep ships
will be to construct them with *buoyant* and *cylindrical* hulls obviating the need
for any flotation substances such as gasoline, concentrated liquid ammonia or
lithium. A sphere, as used in the bathyscaph, is the ideal geometric form for
resisting high pressures. A cylindrical hull, however, can be far better organized
for habitability, housekeeping needs, and for the operational and scientific
equipment. For proper streamlining, a submersible should have a length roughly
eight times as great as the diameter. Such a cigar-shape is readily achieved with
a cylinder but to achieve it with a sphere one would need several connected in
tandem.

A new term is needed to describe such new submersibles, as these will be an
important departure from the bathyscaph principle; neither can the conventional submarine design be extrapolated to resist the abyssal pressures. As
the writer has suggested (1959), the term *bathynaut* (from Greek roots for
"deep ship") as contrasted with *bathyscaph* (from Greek roots meaning "deep
boat") would seem appropriate.

At present (mid-1960) no bathynauts exist. But the *Aluminaut* (Fig. 7) of

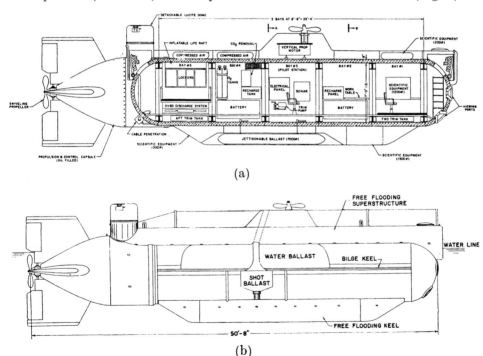

(a)

(b)

Fig. 7. Inboard (a) and outboard (b) longitudinal profiles of the proposed 15,000-ft deep
submersible, the *Aluminaut*.

the Reynolds Aluminum and Metals Co. is scheduled for completion in 1961. As described by Wenk *et al.* (1960): "it appears feasible to build an oceanographic research submarine of aluminum alloys that can operate at 15,000 ft depth with safety, mobility, range and an instrumentation payload to satisfy even the most demanding research mission.... Such a vehicle ... would set the stage for an entirely new generation of deep-diving submarines with unlimited implications for exploration and utilization of the sea and its resources".

Resembling in gross outline a miniature submarine, the *Aluminaut* will have a 50-ft overall length and a 10-ft beam. The manned pressure hull will be a 6-in. thick aluminum-alloy-stiffened cylinder, 33 ft long and 7 ft in diameter. Since the selected aluminum alloy cannot be welded, the ring sections of the hull will be bolted and bonded together. The craft will have a displacement of 75 tons and an excess of buoyancy of 23 tons over primary hull weight. Silver–zinc cells have been tentatively selected to provide the entire power requirements. A maximum speed of 4.8 knots with a total range of 100 miles is contemplated.

To ensure adequate freeboard for surface handling, controlled positive or negative buoyancy for diving, and metacentric stability, a combination of three different ballast systems will be used. These systems will employ water and steel shot for normal operations and a jettisonable lead keel for emergency ascent. The hull will be less compressible than sea-water, allowing a stable descent. Since the boat will actually become lighter with depth, the *Aluminaut* will have the almost legendary capability (in folk-lore wrecks sink only part way to the sea bed) of remaining suspended in equilibrium at some mid-depth.

For tight maneuverability, a swiveling propeller and rudder will be installed. Constant depth will be achieved with stern planes, ballast and a vertical propeller. To avoid penetrating the pressure hull with a rotating shaft, the electric propulsion motors will be externally mounted and bathed in a nonconducting fluid. In fact, an entire stern capsule will be filled with silicone oil which will be pressure-equalized to the sea.

Although capable of cruising directly out of port for moderate distances, it seems more likely that the *Aluminaut* will be transported to the dive site aboard a mother ship or towed at speeds up to 10 knots. Once on station, the crew of three will dive the craft under power, reaching maximum depth in 90 min. The boat will then cruise along the bottom making various oceanographic measurements with its 2-ton payload of scientific instruments. Visual studies will also be possible through two truncated conical plastic windows mounted in the bow.

## 6. Deep-Research Vessel (D.R.V.)

Another deep oceanographic craft under design study by Firth Pierce and colleagues at the U.S. Naval Ordnance Test Station, China Lake, California, is the D.R.V. or Deep-Research Vessel. Fig. 8 explains fairly well its tentative design as currently conceived. An ammonia solution (55%) is used in place of gasoline because of its low compressibility and non-flammability. The single manned compartment, placed forward, is a 9-ft sphere of die steel; the hull is of titanium. The main buoyancy tanks are sealed off from water but are sufficiently compressible to equalize the pressure of the ammonia solution with the ambient sea-water pressure. A 40-h.p. d.c. motor will make possible a top speed of 6 knots. Silver–cadmium batteries will permit 100-mile cruises at 3 knots. The plan is to conduct cruises as long as 48 h duration. Capable of descending to 36,000 ft, the D.R.V. will be without depth limitation in the oceans of the world. With its large hydrophone, storage bins, laboratory pit,

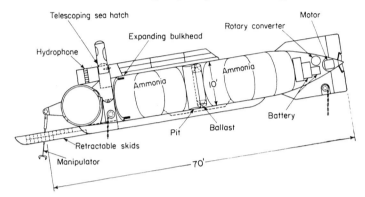

Fig. 8. Tentative design as currently conceived of the D.R.V. (Deep-Research Vessel).

manipulator, etc., the D.R.V. would have scientific as well as cruising range that the present bathyscaphs lack.

## 7. Remote Underwater Manipulator (RUM)

Thus far we have considered only manned craft; unmanned vehicles may serve a vital function in future deep-sea oceanography. The RUM (Fig. 9) is a robot undersea crawler conceived by V. Anderson of Scripps Institution of Oceanography and constructed under the sponsorship of the Office of Naval Research. The prototype, completed in 1960, has a tank chassis modified for underwater mobility. It is equipped with a prehensile arm, four underwater television cameras, and a special sonar for guidance. It is contemplated that this robot will be capable of performing a wide variety of oceanographic missions such as the installation and maintenance of heavy bottom-mounted instruments at depths of up to 20,000 ft.

The robot tank draws its power through a lightweight coaxial cable which is connected with a mobile generator on shore. This cable also carries the television signals, thus enabling the remote observer to exercise control over the craft during operation. A large reel is mounted atop the vehicle permitting 5 miles of scope; using 15 h.p., RUM is capable of maintaining a speed of 3 mph on firm level bottom; it can climb a 60% grade and surmount obstacles as high as 2.5 ft. RUM has a payload carrying capability of 1000 lb in water without exceeding a ground pressure of 1.20 lb/in² over the track area; the total weight is 12 tons in air and 4 tons in water. All components of the vehicle are filled with oil, except for the electronic equipment (lights, television cameras, etc.) which is housed in pressure chambers, thus eliminating the need for a pressure resistant hull as used in manned operations.

All maneuvers are controlled by the remote observer in the mobile van through the controls at his disposal. He can even apply brakes; and, if RUM runs into solid obstructions, the brakes set automatically. His eyes are the RUM's four television cameras housed in cylindrical steel tubes only 3 in. in diameter and 14 in. long. The sensing element is a Vidicon tube, and mercury-vapor lamps provide the light. Two cameras view the sea bed so the operator can see the terrain to steer the robot; the other two cameras monitor the

Fig. 9. A view of RUM (Remote Underwater Manipulator) on the beach at La Jolla, California. This robot crawler is designed for working at great oceanic depths. (Official U.S. Navy photo.)

mechanical arm. The closed-circuit TV system can transmit either a two-dimensional or a three-dimensional image of either high or low resolution.

RUM and the "diving saucer" are the only deep oceanographic vehicles which have an external prehensile capability. RUM's arm is similar to those used in atomic laboratories for handling the transuranium elements; it is capable of picking up anything from a starfish to a large rock. The zinc-coated stainless steel arm is, like the main body, a sealed, oil-filled unit with a special reservoir for compensating the difference in compressibility between oil and water. All the actions of the human arm are duplicated, if not improved upon

—a hand that opens and closes, a wrist that rotates in any direction, an elbow that pivots, and a shoulder that both pivots and rotates. The arm is at the end of a hydraulically actuated boom, mounted at the aft center of the tank. It can reach out to 15 ft and perform jobs requiring skill and dexterity. The boom provides three additional motions—boom rotation, boom pivot and boom flex. The boom also has a hook for lifting objects weighing up to 5000 lb, depending, of course, on the angle.

In early 1960, the first shallow tests with RUM were conducted in San Diego Bay and also through the breaker zone at La Jolla, California. As expected, the preliminary tests with this prototype were not completely successful but much useful information was gained. The construction of a lighter model built of aluminum alloys is presently contemplated; this will reduce RUM's tendency to bog down in a soft bottom. Also, a helicopter type of rotor is planned which will give RUM a swimming as well as a crawling capability. This will also permit its operation directly from an oceanographic mother ship as well as from shore.

Robot vehicles, such as RUM, will doubtlessly play an important role in future oceanography. As a supplement to manned vehicles, they prove invaluable, since they might conceivably perform any heavy or unusually dangerous deep-sea work. Whether or not robots will seriously compete with or replace manned bathynauts is a moot question. With the advent of more sophisticated automated techniques, this is a problem of the not-too-distant future.

## 8. Concluding Remarks

To be useful to the oceanographer, deep-diving craft, both in mid-water and especially along the bottom, will have to remain small and maneuverable. The ability to move slowly and to hover is more important than speed; even in the crystal-clear bottom water, visibility cannot be extended for more than 150 ft. To obtain information on obstructions farther out, scanner sonars are needed. To carry out their oceanographic missions, the crafts will have to be equipped with portholes, closed-circuit TV, prehensile arms, lifting hooks, a wet winch, sonar sensors and numerous other devices. Animals of the deep are mostly minute; seeing them properly will necessitate careful attention to lighting and the external optical systems.

With modern metallurgy it is apparently feasible to construct buoyant stiffened or reinforced cylindrical hulls capable of operating at abyssal depths, even though the spherical cabin may remain the best solution for penetrating to hadal trench depths. A reinforced steel cylinder, still light enough to remain buoyant, can be built to operate at 3000 m with the usual submarine safety factor of 1.5. The *Aluminaut* studies have apparently shown that the light aluminum alloys are even more favorable; they permit the construction of a stiffened cylindrical hull for operating at 4600 m. The strong light metal titanium offers attractive possibilities. A titanium-hulled craft could operate at the normal maximum depth of the sea of 5500 m. The metallurgical problems posed by using titanium are difficult. Fabrication of a hull with the light metal beryllium would open up even the greatest depths of the seas to bathynauts. Beryllium, however, is so expensive and its metallurgy so exotic (it is even poisonous to man) that its use seems unlikely for some time to come.

Nuclear power is, of course, the ultimate answer to the propulsion problem as such power permits the construction of a submersible with unlimited underwater cruising ability. The stay in the abyss would be limited only by the stamina and human frailty of the crew. As yet, however, there is no deep craft even in the planning stage that contemplates using this new power source.

Deep-ocean diving seems, in a sense, a forbidding prospect under the crushing pressures of 600 bars or more. Yet the rigors of the deep are small as compared with those of outer space, and, with the backing of unlimited funds, giant strides have already been taken in its invasion by man. The oceans are a mere film covering two-thirds of the earth, many thousands of times as wide as they are deep. Through their direct approach, adventurous scuba-diving scientists have provided new insight concerning the shallow fringes. That oceanographers will soon be invading the abyss in deep ships, there can be no doubt at all.

## Note Added in Proof

Two years have elapsed since the submission of this manuscript (June 1960) and the present time (June 1962). It is desirable therefore to bring some of the developments mentioned up to date.

The new French bathyscaph, christened *Archimède*, was launched in July, 1961, at the Toulon Naval Base. Since that date she has undergone tests in the Mediterranean Sea and has been outfitted with scientific instrumentation. In April, 1962, the *Archimède* was sent to Japan for a series of deep dives into the fauna-rich Kurile Trench. Houot, Willm, De Lauze, Pérès and Japanese scientists will participate. A dive into the Vityaz Deep, 10,542 m and the third deepest point in the world, will be made. According to a news report, the first dive was completed off Onagawa, Japan, on May 22 to 4800 m. A 1963 diving campaign is planned for the Puerto Rico Deep.

The completion of the *Aluminaut* has been deferred beyond original expectations so that its realization lies still in the indefinite future. Considerable difficulty has been encountered in welding or glueing aluminum hull sections together; also a small model failed under a pressure test, thus requiring considerable designing.

A plethora of new designs of a large variety of deep craft have been made, particularly by various aircraft companies in the U.S.A., but as far as the writer knows, no prototypes have been built.

The RUM vehicle has continued to undergo tests and modifications close to shore and in shallow water on an experimental rather than an operational basis. The helicopter modification of a RUM-type vehicle has undergone design studies by Hughes Aircraft Corp., but to the writer's knowledge no prototype has been built.

(On July 15, 1962, Houot and Willm made a dive in the bathyscaph *Archimède* to 31,350 ft in the Kurile Trench 120 miles south of Urup Island. The dive lasted from 0850 to 1810 hours, or 9 hours and 20 minutes, with three hours being spent on the bottom).

(The opinions and assertions contained herein are those of the author only, and are not to be construed as representing the Navy Department or the naval service at large.)

## References

Beebe, W., 1935. *Half-Mile Down*. John Lane, London, 206 pp.

Bernard, F., 1955. Densité du plancton vu au large de Toulon depuis le bathyscaphe FNRS-3. *Bull. Inst. Océanog. Monaco*, **1063**, 1–16.

Botteron, G., 1958. Étude de sediment revoltes au cours de plongées avec le bathyscaphe *Trieste* au large de Capri. *Bull. Univ. Lausanne*, **124**, 19 pp.

Cousteau, J.-Y., 1960. Diving saucer takes to the deep. *Nat. Geog. Mag.*, **117**, no. 4, 571–586.

Dietz, R. S., 1959. 1100-meter dive in the bathyscaphe *Trieste*. *Limnol. Oceanog.*, **4**, 94–101.

Dietz, R. S., R. V. Lewis and A. B. Rechnitzer, 1958. The bathyscaph. *Sci. Amer.*, **198**, 27–33.

Houot, G. and P. Willm, 1955. *2,000 Fathoms Down*. E. P. Dutton and Co., New York, 192 pp.

Inoue, N., 1959. *The undersea observation vessel Kuroshio: Its history and plan of reconstruction*. Japan.

Mackenzie, K. V., 1960. Formulas for the computation of sound speed in sea water. *J. Acoust. Soc. Amer.*, **32**, 100–104.

Pérès, J. M. 1958. Trois plongées dans le canyon du Cap Sicie, effectuées avec le bathyscaphe FNRS-3 de la Marine Nationale. *Bull Inst. Océanog. Monaco*, **1115**, 21 pp.

Pérès, J. M., 1959. Deux plongées au large du Japon avec le bathyscaphe français FNRS-3. *Bull. Inst. Océanog. Monaco*, **1134**, 28 pp.

Pérès, J. M. and J. Picard, 1956. Nouvelles observations biologique effectuées avec le bathyscaphe FNRS 3 et considérations sur le système aphotique de la Méditerranée. *Bull. Inst. Océanog. Monaco*, **1075**, 1–16.

Pérès, J. M., J. Picard and M. Ruivo, 1957. Résultats de la campagne de recherches du bathyscaphe FNRS-3. *Bull. Inst. Océanog. Monaco*, **1092**, 29 pp. (Recounts dives off Portugal.)

Piccard, A., 1956. *Earth, sky and sea*. Oxford Univ. Press, New York, 192 pp.

Piccard, J. and R. S. Dietz, 1957. Oceanographic observation by the bathyscaph *Trieste* (1953–56). *Deep-Sea Res.*, **4**, 221–229.

Piccard, J. and R. S. Dietz, 1960. *Seven miles down*. G. P. Putnam & Sons, New York, 249.

Rechnitzer, A. B., 1959. The 1957 diving program of the bathyscaph *Trieste*. *U.S. Navy Electronics Lab. Rep.* 941, 21 pp.

Trégouboff, G., 1955. Sur l'emploi de la tourelle submersible Galeazzi pour des observations biologique sous-marine à faibles profondeurs. *Bull. Inst. Océanog. Monaco*, **1070**, 5 pp.

Trégouboff, G., 1956. Prospection biologique sous-marine dans la région de Villefranche-sur-Mer en Juin 1956. *Bull. Inst. Océanog. Monaco*, **1085**, 24 pp.

Trégouboff, G., 1958. Prospection biologique sous-marine dans la région de Villefranche-sur-Mer au cours de l'année 1957. Plongées en bathyscaphe. *Bull. Inst. Océanog. Monaco*, **1117**, 37 pp.

Wenk, E., R. C. DeHart, P. Mandel and R. Kissinger, 1960. An oceanographic research submarine of aluminium for operation to 15,000 ft. *Trans. Roy. Inst. Naval Architects* **102**, 555–578.

# 24. DEEP-SEA ANCHORING AND MOORING

## John D. Isaacs

## 1. Introduction

The following relates to mooring light devices in the deep ocean at depths from 100 to 3000 fathoms, and is intended as an introduction to those planning research approaches. For those who plan to design and install such deep-sea moorings, a more elaborate treatment is available (Isaacs et al., 1962).

This discussion is based on about eight years of deep-mooring experience at the Scripps Institution of Oceanography from which, at times, as many as thirty stations have been moored and maintained in large deep-sea areas.

## 2. Historical Background

The history of deep moorings dates back at least as far as 1888–1889 when the U.S. Coast and Geodetic Survey Steamer *Blake*, under the command of J. E. Pillsbury, carried out studies of the Gulf Stream. During these investigations the *Blake* was anchored at several locations in depths to 4000 m. Subsequently, other vessels were anchored in great depths in the North Atlantic and elsewhere. Some vessels, including the *Meteor*, have been anchored in depths greater than 4500 m for periods up to two weeks, and the *Meteor* was anchored in 5500 m for two days. Although most of these efforts employed anchors weighing 400 to 500 lb, the *E. W. Scripps* was anchored successfully in depths of about 1600 m using a Danforth anchor of only 40 lb.

Scopes (length of mooring rope : depth of sea) of 1.1 to 1.6 have been used in the open sea and scopes of 2 to 3 have been used in anchoring in the strong currents of the Gulf Stream.

In recent years, a somewhat newer type of mooring has been introduced. This is a "taut-mooring" developed at Scripps and described by Bascom (1953).

Fig. 1 shows a more recent mooring of this type, designed by J. D. Isaacs, R. P. Huffer and L. W. Kidd, that is adapted for use in any depth of water.

Referring to Fig. 1, the station consists of (1) a dead weight on the bottom; (2) a mooring wire of high strength steel; (3) a special float that provides high tension in the line and which is submerged 15 to 100 fm below the surface; and (4) a buoyant pennant connecting to the surface buoy, a skiff, which contains the recorders.

An equally satisfactory type of mooring with rather different applications has been developed and used by Allyn Vine of Woods Hole Oceanographic Institution (*in litt.*). This is shown in Fig. 2, which is self-explanatory. The essential feature of this mooring system is the buoyant fiber rope of which the lower half of the mooring cable is composed. This eliminates the need for a submerged buoy. Where the instrumentation is principally for observations on the surface, and where close station-keeping is not important, this type of mooring has the advantages of simplicity and ease of installation.

A third type of mooring is the free-moored instrument without a surface float.

[*MS received October, 1960*]　　　　516

This type of instrument platform or vehicle is the most variable in design, for the separation of the instrument from the bottom may be either only a few inches or many thousands of meters.

Such instrumentation is usually operated in a series installation spanning a

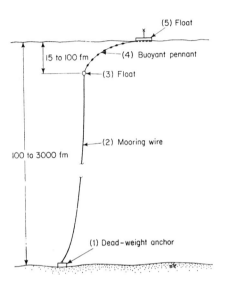

Fig. 1. Taut mooring.

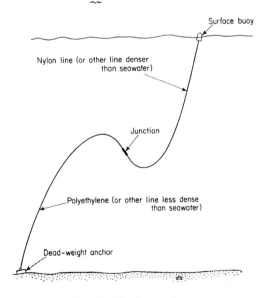

Fig. 2. Slack mooring.

recording period of from a few hours to a few days. The instruments rise to the surface at a pre-set time and are recovered. One such mooring is described by Isaacs and Schick (1960) and its essentials are shown in Fig. 3. The ordinary

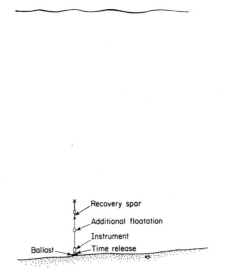

Fig. 3. Free instrument vehicle and mooring.

pay-load is about 25 lb, but additional loads in increments of 75 to 100 lb are obtainable by adding further buoyancy in the form of rubber drums of gasoline or diesel oil.

### 3. Problems of Taut-Mooring

The remainder of this discussion is concerned almost entirely with the taut-mooring, for the reason that such moorings provide great potential for deep-sea measurements and yet require the greatest discretion in their design and use.

The primary intent of deep-sea mooring is to maintain a surface float in some geographical position with the simplest effective system.

The forces of relatively steady magnitude that tend to displace the system are primarily the horizontal drag forces imposed on the components by wind and current. These are principally applied near the surface.

Intermittent impulsive or oscillating forces, both horizontal and vertical, are imposed by the waves, and these forces dictate many of the design characteristics of the system. The initial force available to the designer for resisting the displacement of the system is almost solely shear force on the bottom material (although other possibilities exist).

The problem thus is to conduct the horizontal drag forces, principally near the surface, some miles vertically so that they are resisted by the bottom shears, and to do this in a manner that is compatible with the other forces and design restraints.

Insofar as the mechanical components of the mooring are concerned, the conditions that impose force or other destructive influences on the mooring are:

(a) steady or quasi-steady currents,
(b) waves—particularly combers,
(c) electrolysis,
(d) sharks, and
(e) miscellaneous: fouling organisms, fishermen and collisions.

### A. Currents

The strong currents of the deep ocean are apparently restricted to the upper or Eckman layer. This may be as much as 400 m thick, and currents of the order of 2 to 5 knots may be common. Much below the level of the mixed layer, currents of as much as 0.5 knot are rare.

The current profile of an intended site imposes basic design requirements on a mooring, and it is important to have some information on this.

A rule of thumb in the design of moorings that has been used at Scripps is to accept the normal measured surface current as extending through the mixed layer undiminished. A current of 0.5 knot is assigned to the inflection point in the thermocline, and a 0.2-knot current at the bottom. These currents are then interpolated linearly and used for design.

Obviously, a measured current profile is preferable, but this ordinarily is not available until the first moorings have been used.

This procedure is conservative to the degree that currents are not in the same direction. Indeed submerged countercurrents cancel some of the stress in the mooring.

The calculation is non-conservative to the degree that transient currents exceed the design current in the region. This is accommodated by factors of safety in the design.

### B. Waves

Waves, and particularly deep-sea-breaking waves (combers), result in the most demanding requirements of moorings with surface floats.

It is clear that if particular mooring requirements are met by subsurface components alone, with no surface float, many of the problems of deep-sea mooring can be eliminated or ameliorated. Such wholly submerged floats can even be useful for station keeping by use of sonar.

Undoubtedly the greatest forces that naturally affect a mooring in the deep sea are those resulting from deep-sea combers striking the surface float. Although no defensible deep-sea measurements have been made, it is apparent that the float is struck by a cascade of water moving essentially at the wave (phase) velocity. In the case of a wave of an 8-sec period, this velocity is about 41 ft/sec.

This cascade persists for about one-third of a wavelength, or 100 ft in the example.

Obviously a design adequate to withstand this motion will withstand the extent of motion of the common swell, but not necessarily its ceaseless repetition.

The comber, then, imposes the stress and excursion requirements on the mooring, and the common waves and swell impose the requirements for preventing chafing.

It is essential that the surface float has sufficient restraint so that it is pulled slowly through the comber and is not carried to its terminus; otherwise it is likely to be struck by the subsequent comber, and a succession of these events can transmit very high stresses to the mooring system in general.

In addition, a very lightly restrained surface float can "surf-board" and broach-to on a comber, and be carried even farther than the terminus of the comber. Hence, the next comber may strike it in the most vulnerable situation, that is, when it is drifting back into position and possibly broadside.

In storms in the deep sea, the dominant waves and winds are most likely to be in the same direction. Consequently, the wind stress can be considered as imposing an initial load and restoring force on the surface float that is carried out of position by a breaking wave. This condition does not obtain near fast-moving storm centers of course, but a mooring has survived the close passage of the eye of one of the fast-moving "chubasco" of Mexican water, where wind velocities were estimated at about 80 knots.

### C. Other Environmental Factors

Other environmental factors that influence the design of a mooring are: the electrical conductivity of sea-water, the low oxygen tension, sharks, seaweed rafts, ice, seals and sea-lions, rough-bottom debris and boat traffic.

The electrical conductivity makes it mandatory that no dissimilar metals be in electrical contact in sea-water.

The oxygen tension decreases with depth to a minimum in the deep ocean. The oxygen tension rarely falls below 0.4 ml $O_2$/l. in the open ocean, but it may drop to 0.15 ml $O_2$/l. or lower in trenches and basins. Some metals, including some stainless steels, are inadequately protected at such tensions.

Sharks are voracious creatures of the deep sea that will strike or bite at any object that appears attractive. A small float on a line or a tag-end of line or tape on an instrument cable frequently is bitten off or damaged by sharks. It is essential that every component appear continuous with the system to minimize shark bites. That is, tags, pigtails and small floats of different color from the line should be avoided.

Seaweed rafts appear close in to coasts, usually at oceanic fronts. A surface pennant will become fouled by such rafts, and if the raft is large enough it will carry the mooring away. Nipa rafts of immense dimensions are present in some parts of the South Pacific. Where this problem is likely to be severe, and the mooring must survive for long periods of time, the mooring may have to be designed with no surface float.

Where sea-ice is present it is not possible to employ moorings with surface floats. Ice build-up on the float may occur in sleet or black-ice storms at high latitudes. If these factors are hazards, the float should be designed to survive capsizement.

Near rocky coasts, seals and sea-lions are sometimes troublesome in that they leap aboard and bask on any large enough surface float. They may break antennae or even capsize the float. They can be discouraged by designing the float without flat horizontal surfaces.

On the bottom in the deep ocean, there are a number of sharp objects, debris that has been jettisoned, wrecks, rock outcroppings and nodules. In general, therefore, any part of the mooring cable that is allowed to touch and drag on the bottom must be high quality flexible wire rope several times stronger than the main mooring cable.

Perhaps the most difficult problem stems from human intervention. If the surface floats on a mooring are conspicuous, visually or by radar, passing craft will pull off course to investigate. If there are any floating lines about the float they are very likely to be picked up on the hull or in the screws. In addition, fishermen will tie up to the moorings, attempt to hoist them, or cut them free. In many cases this recovery is intended to be helpful. Regardless of the warnings or information painted on the float, seamen cannot believe that a float in 2000 or 3000 fm is moored there.

The importance of these environmental factors can perhaps be indicated by the fact that of those fifteen or so moorings that have been recovered off stations by Scripps Institution, only about two appeared to have failed from direct natural stress. All others showed evidence of having been fouled by ships, cut free by fishermen, or in several cases bitten off by sharks. A broken shark tooth was recovered embedded in one parted pennant.

### 4. Remarks on Components of Moorings

The components will next be discussed followed by general design problems.

#### A. Surface Floats

The translocation of horizontal forces on and near an oscillating surface to the bottom cannot be carried out simply, and many restrictions to the manner in which it is carried out are determined by the type of surface float.

Two general types of surface float are feasible: (1) a surface-following or responding float; that is, a float such as a skiff or nun-buoy that tends to move with the waves, and (2) a vertically stable float, such as a spar or "hydrometer-shaped" float that tends to remain in a vertical plane even in the presence of waves.

Each of these types of surface float has its specific advantages and disadvantages. For the purpose of this discussion, however, the difficulty with the properly designed spar is the problem of handling it and coming alongside, as its motion in a seaway is incompatible with that of any conventional craft.

The surface pennant to a responding float (skiff) must ordinarily lead almost horizontally away from the float, or, if nearly vertical, it must be *highly elastic*. This is because when the float is subject to white caps or combers its vertical downward acceleration approaches that of gravity. Hence, any vertical string depending *only* upon gravity forces for tension will slack and jerk in a way that will greatly decrease its life, if, indeed, it survives at all.

The pennant to a non-responding float (spar) must tend away almost vertically because, if horizontal, it is subjected to severe flexing or abrasion at the point of attachment from the vertical motion of water by the relatively non-moving spar.

Resonance of a spar to the waves is a condition to be avoided.

Another requirement of the surface float is that it possess sufficient displacement to support its instrument and power load. We have always used floats that would also support the weight of the mooring wire and a swamped submerged float to prevent the entire mooring being lost if the submerged float were to leak.

The characteristics for a responding float are best met by a skiff or boat shape, and at the Scripps Institution we have used decked glass-reinforced plastic skiffs about 16 ft long for the surface float.

### B. Pennants and Instrument Wires

The pennant is the member of the mooring assembly that connects the surface float to the submerged float or the main mooring component.

The pennant to a non-responding spar necessarily tends almost vertically and is difficult to relate to a submerged float because slack during calm weather will foul the submerged float.

The pennant to a responding-type float must either tend horizontally or, if vertical, be highly elastic, as previously discussed.

Both types of pennants are in use. A common pennant is a nylon float line that tends upward in a floating catenary. In slack weather this floats forward off the surface float and greatly increases the possibility of the mooring being run down by a passing vessel. This situation is intensified by the fact that in slack weather the visual and radar detection range of vessels is increased and the curious passing navigator is also more likely to investigate the contact.

The instrument line is then lowered independently over the stern of the float using a harp of shock cord to provide elasticity. This system is limited to instrument lines that are much shallower than the submerged float except in weather situations that are constant, such as the trades, otherwise the pennant and instrument line will foul.

Experimental vertically tending elastic pennants are now being tested. The pennant is made of nylon-covered shock cord with the instrument line married to it at intervals in S-shaped loops.

Advantages of the latter system are that only one line is used for both mooring and recording, and the terminations of the instrument line remain at an approximately constant depth as the scope increases. On the basis of recent tests, this system appears to be successful.

### C. Subsurface Floats

The subsurface floats have usually consisted of spherical steel tanks (commercial butane tanks) of approximately 3-ft diameter and $\frac{1}{8}$-in., or greater, wall thickness. Some cylindrical steel tanks have been used. In use, the subsurface floats are normally pressurized to about the ambient pressure at their proposed depth. In a recent test (February, 1958) an unpressurized spherical steel tank, 34 in. in diameter and having a wall thickness of $\frac{7}{32}$ in., survived submersion to a depth of 500 m.

The drag of the submerged float is small in comparison with the other drags of the system and streamlining has not been contemplated. Streamlining

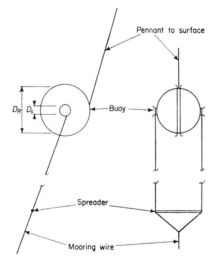

Fig. 4. Yo-yo buoy—two views.

involves an asymmetry and orientation in the system that is disadvantageous where any wave action occurs because the float will continuously rotate.

An interesting modification of the spherical buoy is now being tested that will pay out an additional wire as soon as the maximum working stress of the pennant is reached, thereby allowing added scope in extreme weather and also permitting the recovery of the pennant and instrument wire by merely pulling on it. This float functions as shown in Fig. 4 and has been called the Yo-yo buoy. The stored cable is wound around the large diameter, $D_R$, of the drum and the bridle is wound around two smaller spools, $D_s$, on opposite sides. When the tension in the pennant exceeds $D_s/D_R$ times the buoyancy, $B$, the float winds *down* the bridle releasing a length of wire $D_R/D_s$ times the distance travelled until the stress in the pennant falls below $(D_s/D_R) \times B$. Models and experiments of this system appear successful, and it is being installed in deep-sea systems.

## 5. Mooring Cables

One of the problems of anchoring or of installing a mooring in deep water is the relationship between the ultimate tensile strength for a steel wire of uniform cross-section and its allowable length in the sea when bearing an additional load equal to its own weight.

Using this allowable length criterion, wire of an ultimate tensile strength of 100,000 lb/in² can be safely used to a depth of about 1700 fm, or in about 30% of the oceans. A wire of an ultimate tensile strength of 180,000 lb/in², however, can be employed to 3000 fm, or used in 99% of the sea area. The wire used in recent work had an ultimate tensile strength of about 260,000 lb/in².

The above relates solely to the quasi-static stresses produced in lowering the mooring wire. Other dynamic stresses become important as soon as the anchor reaches bottom. Also, an allowance must be made for weakening of the wire by handling and by corrosion.

The area presented by such a wire to the horizontal drag forces can be quite large. For example, 15,000 ft of $\frac{1}{8}$-in. diameter wire presents a projected area of about 156 ft² of form drag area, or about that of a large barge. Fortunately, water velocities at great depths are low, and hence the large area presented is not a great problem.

Such horizontal forces must be resisted at the anchor. Thus, as horizontal forces increase, the anchor weight and the lowering stresses must be increased. Also, the excursions of the moorings are functions of the tensile stresses and the horizontal drags. Hence, the strength/drag ratio of the wire is important. This can be expressed as:

$$R = \frac{s\pi t^2}{4C_d t},\tag{1}$$

or, $R \sim st$ for constant range of $C_d$ and wire construction, where $R =$ strength/drag ratio, $C_d =$ coefficient of drag, $t =$ thickness of wire, and $s =$ ultimate strength of wire, lb/in².

The strength/drag ratio can be considered as a rough measure comparing the *minimum* obtainable inclinations from the vertical of systems utilizing different size mooring cables of the same construction, and hence this ratio is an inverse comparative measure of horizontal displacement.

This can be seen from the following force diagram:

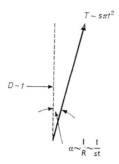

where $D$ =drag force, and $T$ =tension.

A somewhat more discriminating comparison of mooring cables considers their weight in water. For taut-wire moorings of small inclinations, the weight (or buoyancy) can be considered to act in the same direction as the tension. Thus,

$$T \sim s\pi t^2 - (\rho - 1)(Lt^2 \pi 62.4)/144 \tag{2}$$

and $D \sim t$, where $D$ =drag.

The tension/drag ratio is

$$R' \sim t[s - 0.4(\rho - 1)L], \tag{3}$$

where $L$ =length in feet, and $\rho$ =specific gravity of material.

For the purposes of comparison of the tension/drag ratio consider Table I showing materials with the same tension/drag ratio.

TABLE I

| Material | Strength, lb/in$^2$ | Specific gravity | Mooring depth, ft | Thickness, in. | $R'$ |
|---|---|---|---|---|---|
| Piano wire | 280,000 | $\sim 8$ | 20,000 | 0.12 | 28,000 |
| Wire rope | 125,000 | $\sim 8$ | 20,000 | 0.39 | 28,000 |
| Nylon | 30,000 | $\sim 1$ | 20,000 | 0.93 | 28,000 |

Thus, to attain the same tension/drag ratio under these circumstances as for a $\frac{1}{8}$-in. piano wire, the common cable will weigh ten times as much, and the nylon eight times as much, in air.[1]

For this most critical component of a deep-ocean taut-moored system, the two to three miles of mooring wire, two types of wire have been used, a solid

[1] Taut nylon moorings with no submerged float are now undergoing tests at Scripps. They appear to be very satisfactory in regions of low current.

Bethanized steel wire and a stranded Bethanized steel wire. The solid wire size has varied from 0.082-in. diameter to 0.120-in. diameter. The stranded wire size has varied from 0.123-in. to 0.200-in. diameter.

Experimental stations using a special (1 ×7, long lay) stranded wire with a tensile strength of 260,000 lb/in² have been installed. Experience indicates that such stations have a life expectancy at least equal to, if not greater than, stations using a solid wire. A few well-spaced butt welds can be used in fabricating the individual strands, for the loss in strength is reduced by the number of strands.

There are several advantages in using stranded wire. The stranded wire will take a smaller radius bend than the solid wire, it is simpler to unspool because of a reduced tendency to "cut in" to the underlying layers, it is possible to make the stranded wire in extremely long continuous lengths without special processing, and it has good tension/drag and strength/weight ratios.

Special machined wire clamps are used to hold the mooring wire without (or with minimum) loss of breaking strength. These clamps can be secured to the standing part of the wire while it is under tension.

## 6. The Anchors

In anchors a material having a high density is usually desirable both for its lowering characteristics and for its compactness. Various kinds of anchors have been used. Some have been solid steel cubes with pad eyes welded on to the top and bottom and some have been ordinary railroad-car wheels. As moorings increase in size and complexity, a more sophisticated design of anchor will be required if the size of the mooring wire is to be kept at a minimum. For our installations, gravity anchors have been designed so that the net vertical reaction against the bottom (i.e. weight of anchor in water and weight of wire in water minus the buoyancy of floats) is at least equal to 1.4 times the sum of the expected horizontal forces. This has led to satisfactory performance on a relatively flat bottom. For a sloping bottom the net reaction must be increased as follows:

$$(W - F_v) = F_h / (0.7 - \tan \star H)$$

With these assumptions a simple gravity anchor becomes impractical on bottoms of slopes exceeding about twenty degrees, and anchors with hooks, flukes or grapnels must be used.

The anchor must be designed so that stability to overturning is acceptable. This requires a flat anchor with a low point of attachment.

## 7. Fittings

For moorings designed to last as long as possible, special shackles, insulated by micarta sleeves to minimize electrolytic action in the mooring components, have been used. The relative efficiency of such measures for this purpose is not known exactly but it is apparent that, as a minimum, it prolongs the life of connecting shackles by preventing abrasive damage to the protective zinc coating.

## 8. Performance

A well-designed mooring of this nature will have a life of many months, barring accident, in the conditions of the Central Pacific. Surface floats and instrument strings can be exchanged without disturbing the basic mooring.

By proper design, excursion (i.e. horizontal motion) ordinarily can be maintained at less than 10% of the depth, or about 1% if critically important, under extremes of weather. Dip, that is vertical motion, can easily be maintained below 1% of the depth.

Cost of such mooring, exclusive of instrumentation and ship time, is about $700 for mooring components and an equal amount for the surface float.

## References

Bascom, W. N., 1953. A deep-sea instrument station. *Scripps Inst. Oceanog. Rep.*, 53–38.

Isaacs, J. D., James L. Faughn, George B. Schick and Marston C. Sargent, 1962. Deep-sea mooring. *Bull. Scripps Inst. Oceanog. Univ. Calif.* In press.

Isaacs, J. D. and G. B. Schick, 1960. Deep-sea free-instrument vehicle. *Deep-Sea Res.*, 7, 61–67.

# AUTHOR INDEX

Authors' names of Chapters in this book and the page numbers at which these Chapters begin are printed in **heavy type**, page numbers of citations in the text are in ordinary type and those of bibliographical references listed at the ends of the chapters (including joint authors) are in *italics*.

Where the same author is mentioned in more than one chapter it may be convenient to differentiate by subject between the successive ranges of page numbers by consulting the list of contents at the beginning of the book.

Names associated with known apparatus, equations, laws, principles, etc. are not entered here but in the Subject Index.

# SUBJECT INDEX

Titles of Parts and Chapters and the page numbers at which they begin are printed in **heavy type**. Names of ships and of genera and species are printed in *italics*.

Where a single index entry (such as a geographical name) is followed by a string of page numbers the reader should consult the list of contents at the beginning of the book to differentiate them by subject.

The letters ff after a page number denote either the beginning of a chapter subdivision on the subject in question or that the subject is mentioned again on one or both of the next two following pages.

This book is a facsimile edition authorized by
Harvard University Press.  It is produced
on acid-free archival paper which meets
the requirements of ANSI/NISO Z39.48-1992.

Facsimile created by Acme Bookbinding,

Charlestown, Massachusetts